Dark Matter in Astrophysics and Particle Physics

Dark 2009

Proceedings of the
7th International
Heidelberg Conference

Dark Matter in Astrophysics and Particle Physics

Dark 2009

Christchurch, New Zealand
18–24 January 2009

Editors

Hans Volker Klapdor-Kleingrothaus
Heidelberg, Germany

Irina V. Krivosheina
Heidelberg, Germany, and Nishnij Novgorod, Russia

NEW JERSEY • LONDON • SINGAPORE • BEIJING • SHANGHAI • HONG KONG • TAIPEI • CHENNAI

Published by

World Scientific Publishing Co. Pte. Ltd.
5 Toh Tuck Link, Singapore 596224
USA office: 27 Warren Street, Suite 401-402, Hackensack, NJ 07601
UK office: 57 Shelton Street, Covent Garden, London WC2H 9HE

British Library Cataloguing-in-Publication Data
A catalogue record for this book is available from the British Library.

DARK MATTER IN ASTROPHYSICS AND PARTICLE PHYSICS
Proceedings of the Seventh Heidelberg International Conference on Dark 2009

ISBN-13 978-981-4293-78-5
ISBN-10 981-4293-78-4

Printed by Fulsland Offset Printing (S) Pte Ltd, Singapore

PREFACE

The Seventh HEIDELBERG International Conference on Dark Matter in Astrophysics and Particle Physics, DARK2009, took place at Christchurch, New Zealand, January 18-24, 2009. It was, after Capetown 2002, Texas 2004 and Sydney 2007, the fourth conference of this series held outside Germany, and the third held in the southern hemisphere. The first three conferences were held at Heidelberg, Germany, in 1996, 1998 and 2000 (see Fig. 1).

Fig. 1. Geography of Heidelberg DARK Conferences, from 1996 till 2009.

Dark Matter is still one of the most exciting and central fields of astrophysics, particle physics and cosmology. The conference covered, as usual for this series, a wide range of topics, theoretical and experimental. They range from the expectations for dark matter and TeV scale physics of the new accelerator LHC, to direct dark matter search underground, to cosmology research by the Wiggle Z Dark Energy Survey, the Cosmic Ray Positron Excess found the the PAMELA, ATIC and EGRET missions, the mass of dark matter and dark stars at the center of the Galaxy , the potential for indirect dark matter search by the terrestrial detectors ICECUBE and ANTARES, and with imaging atmospheric Cherenkov telescopes, to search for dark matter annihilitation in X-rays, and many others. After the presentation of the evidence for cold dark matter by the DAMA/LIBRA experiment, and the many underground and space experimental efforts to search for cold dark matter, one of the new highlights was certainly the presentation about a possible connection between neutrino mass and dark energy. In a cloud of massive fermions interacting by exchange of a light scalar field, the effective mass and the total energy density eventually increases with decreasing density. In this regime, the pressure-density relation can approximate that required for dark energy. Applying this phenomenon to the expansion of the Universe with a very light scalar field leads to the result that Majorana neutrinos of a mass of about 0.3 eV (as observed by neutrinoless nuclear double beta decay) may be consistent with current observations of dark energy. Theoretically, among many others, also presentations about a fifth family as origin of dark matter, extra dimensions and dark matter, and non-standard Wigner classes and dark matter found large interest.

An overview of the topics of the conference has been recently published in CERN Courier 2009. We add this report at the end of this preface. All presentations given at the conference can be viewed at *www.klapdor-k.de/Conferences/Program09.htm.* We are confident that the Proceedings of this conference will provide a useful overview of this exciting field of research - its current status and the future prospects - and of ist fundamental connections to various frontier disciplines of particle physics and cosmology. We hope that this book can also serve as a kind of handbook for students.

The organizers express their thanks to all colleagues from many countries, who contributed so actively to the success of the meeting. Our thanks go the University of Canterbury, New Zealand, and in particular to the Head of the Department of Physics, Prof. Roger Reeves, fort his great help and the generous financial support. We thank to all people who contributed in one way or another to the organisation of the conference, and in creating a pleasant and inspiring atmosphere during the meeting. We thank in particular Ms. Rosalie Reilly and Ms. Rhondda Sullivan from the Department of Physics of Canterbury University for the untiring help in the organisation. We are highly indebted to Matthias Danninger for his decisive help in the electronic presentations at the conference. Our thanks go also to Dr. Dharam Vir Ahluwalia and his students Dimitri Schritt and Cheng-Yang Lee for their support in the preparation of the conference. HVKK thanks to the Scientific Secretary of this conference, Dr. Irina Krivosheina, who was responsible for much of the exciting program and for the preparation of these Proceedings.

Hans V. Klapdor-Kleingrothaus
Chairman of HEIDELBERG
Dark Matter Conferences
and of DARK 2009
Heidelberg, Germany

Irina Vladimirovna Krivosheina
Scientific Secretary of DARK 2009
Heidelberg - Nishnij Novgorod
Germany - Russia

New Zealand meeting looks at dark matter

Participants from around the world gathered in Christchurch, New Zealand, for the Dark 2009 conference in January. **Hans Volker Klapdor-Kleingrothaus** reports.

The 7th Heidelberg International Conference on Dark Matter in Astrophysics and Particle Physics – Dark 2009 – was held at Canterbury University in Christchurch on 18–24 January. The event saw 56 invited talks and contributions, which provided an exciting and up-to-date view of the development of research in the field. The participants represented well the distribution of dark-matter activities around the world: 25 from Europe, 11 from the US, 5 from Japan and Korea, 14 from Australia and New Zealand, and 1 from Iran. The programme covered the traditionally wide range of topics, so this report looks at the main highlights.

The conference started with an overview of searches for supersymmetry at the LHC and dark matter by Elisabetta Barberio of the University of Melbourne. To date, the only evidence for cold dark matter from underground detectors is from the DAMA/LIBRA experiment in the Gran Sasso National Laboratory, as Pierluigi Belli from the collaboration explained. This experiment, which looks for an expected seasonal modulation of the signal for weakly interacting massive particles (WIMPs), now has a significance of 8.4 σ. Unfortunately, all other direct searches for dark matter do not currently have the statistics to look for this signal. Nevertheless, Jason Kumar from Hawaii described how testing the DAMA/LIBRA result at the Super-Kamiokande detector might prove interesting.

Later sessions covered other searches for dark matter. Tarek Saab from Florida gave an overview of ongoing direct searches in underground laboratories, including recent results from the Cryogenic Dark Matter Search experiment in the Soudan mine, and Nigel Smith of the UK's Rutherford Appleton Laboratory presented results from the ZEPLIN III experiment in the Boulby mine. Irina Krivosheina of Heidelberg and Nishnij Novgorod discussed the potential offered by using bare germanium detectors in liquid nitrogen or argon for dark-matter searches, on the basis of the results from the GENIUS-Test-Facility in the Gran Sasso National Laboratory. Chung-Lin Shan of Seoul National University reported on how precisely WIMPs can be identified in experimental searches in a model-independent way.

Searching for signals from dark-matter annihilation in X-rays and weighing supermassive black holes with X-ray emitting gas were subjects for Tesla Jeltema of the University of California Observatories/Lick Observatory and David Buote of the University of California, Irvine. Stefano Profumo of the University of California, Santa Cruz, provided an overview of fundamental physics with giga-electron-volt gamma rays. Iris Gebauer of Karlsruhe addressed the excess of cosmic positrons indicated by the Energetic Gamma Ray Experiment Telescope, which are still under discussion, as well as the new anomalies observed by the Payload for Antimatter Matter Exploration and Light-Nuclei Astrophysics (PAMELA, p12) satellite experiment and the Advanced Thin Ionization Calorimeter (ATIC) balloon experiment. These results and the limits that they set on some annihilating dark matter (neutralino or gravitino) models were also discussed by Kazunori Nakayama of Tokyo and Koji Ishiwata of Tohoku.

Participants of DARK 2009 at the monument for Robert Falcon Scott, who started his Antarctic expedition from Christchurch in 1910.

Other presentations outlined results and prospects for the AMANDA, IceCube and ANTARES experiments, which study cosmic neutrinos – though there is still a long way to go before they have conclusive results. Emmanuel Moulin of the Commissariat à l'Énergie Atomique/Saclay presented results from imaging atmospheric Cherenkov telescopes, in particular the recent measurements from HESS, which exploited the fact that dwarf spheroidal ▷

galaxies, such as Canis Major, are highly enriched in dark matter and are therefore good candidates for its detection. Unfortunately, the results do not yet have the sensitivity of the Wilkinson Microwave Anisotropy Probe in restricting either the minimal supersymmetric Standard Model or Kaluza–Klein scenarios.

Leszek Roszkowski of Sheffield gave an overview of supersymmetric particles (neutralinos) as cold dark matter, while scenarios of gravitino dark matter and their cosmological and particle-physics implications were presented by Gilbert Moultaka of the University of Montpellier and Yudi Santoso of the Institute for Particle Physics Phenomenology, Durham. Dharam Vir Ahluwalia of the University of Canterbury put the case for the existence of a local fermionic dark-matter candidate with mass-dimension one, on the basis of non-standard Wigner classes. However, as the proposed fields, as outlined in detail by Ben Martin of Canterbury, do not fit into Steven Weinberg's formalism of quantum-field theory, this suggestion led to dispute between other experts. An interesting candidate for dark matter was presented by Norma Susanna Mankoc-Borstnik of the University of Ljubljana, who proposed a fifth family as candidates for forming dark matter.

Dark energy and the cosmos

Dark energy was a major topic at the conference. Chris Blake of Swinburn University of Technology in Melbourne presented the prospects for the WiggleZ survey at the Anglo-Australian Telescope, the most sensitive experiment of this kind, and Matt Visser of Victoria University in Wellington gave a cosmographic analysis of dark energy. On the theoretical side there are diverging approaches to dark energy, including attempts to explain it in a "radically conservative way without dark energy", as David Wiltshire of Canterbury University, Christchurch, explained.

A particular highlight was the presentation by Terry Goldman of Los Alamos, which discussed a possible connection between sterile fermion mass and dark energy. His conclusion was that a neutrino with mass of 0.3 eV could solve the problem of dark energy. This possibility was qualitatively supported by results of non-extensive statistics in astroparticle physics that Manfred Leubner of the University of Innsbruck presented, in the sense that dark energy is expected to behave like an ordinary gas. Goldman's suggestion is also of interest with respect to the final result of the Heidelberg–Moscow double-beta-decay experiment, reported by Hans Klapdor-Kleingrothaus, which predicts a Majorana neutrino mass of 0.2–0.3 eV.

Danny Marfatia of the University of Kansas discussed mass-varying neutrinos in his presentation about phase transition in the fine structure constant. He proposed that the coupling of neutrinos to a light scalar field might explain why $\Omega_{\text{dark energy}}$ is of the same order as Ω_{matter}. Possible connections between dark matter and dark energy with models of warped extra dimensions and the hierarchy problem were outlined by Ishwaree Neupane of the University of Canterbury and Yong Min Cho of Seoul National University.

Dark mass and the centre of the galaxy was the topic of a special session in which Andreas Eckart of the University of Cologne presented recent results on the luminous accretion onto the dark mass at the centre of the Milky Way. Patrick Scott of Stockholm University discussed dark stars at the galactic centre, while Benoit Famaey of the Université Libre de Bruxelles and Felix Stoehr of the Space Telescope European Coordinating Facility/ESO in Garching discussed the distribution of dark and baryonic matter in galaxies. Primordial molecules and the first structures in the universe were the topics addressed by Denis Puy of the Univesité Montpellier II. Youssef Sobouti of the Institute of Advanced Studies on Basic Science in Zanjan, Iran, presented a theorem on a "natural" connection between baryonic dark matter and its dark companion, while Matthias Buckley of the California Institute of Technology put forward ideas about dark matter and "dark radiation".

Gravity also came under scrutiny. David Rapetti of SLAC explored the potential of constraining gravity with the growth of structure in X-ray galaxy clusters, while Agnieszka Jacholkowska of IN2P3/Centre National de la Recherche Scientifique gave an experimental view of probing quantum-gravity effects with astrophysical sources. In a special session on general relativity, Roy Patrick Kerr of Canterbury University gave an interesting historical lecture entitled "Cracking the Einstein Code".

To conclude, the lively and highly stimulating atmosphere of Dark 2009 reflected a splendid future for research in the field of dark matter in the universe and for particle physics beyond the Standard Model. The proceedings will be published by World Scientific.

Further reading

For the presentations at Dark 2009, see www.klapdor-k.de/Conferences/Program09.htm.

Résumé

La Nouvelle-Zélande se penche sur la matière noire

Des spécialistes du monde entier se sont réunis en janvier 2009 à Christchurch (Nouvelle-Zélande) pour la 7e Conférence internationale d'astrophysique et de physique des particules sur la matière noire. Quelque 56 communications et contributions passionnantes ont permis de faire un tour d'horizon des derniers développements de la recherche dans le domaine de la matière noire. Cette conférence, généralement biennale, rassemble des chercheurs travaillant dans les domaines de la cosmologie, l'astrophysique, la physique des particules et la physique nucléaire. Cette année, les thèmes comprenaient la recherche (directe ou indirecte) de la matière noire, divers aspects de l'énergie sombre, la structure à grande échelle et la gravité quantique.

Hans Volker Klapdor-Kleingrothaus, *Heidelberg*.

Fig. 2. 1. David Buote 2. Tamara Davis 3. David Wiltshire 4. Matthais Danninger 5. Iris Gebauer 6. Gianfranco Gentile 7. Tarek Saab 8. Michel H.G Tytgat 9. Irina Krivosheina 10. Stefano Profumo 11. Norma Susana Mankoc Borstnik 12. Ben Martin 13. Dimitri Schritt 14. Gilbert Moultaka 15. Yudi Santoso 16. Andrew Beckwith 17. Seon-Hee Seo 18. Alfio Rizzo 19. Adam Gillard 20. Elisabetta Berberio 21. Patrick Scott 22. Javier Grande 23. ??? 24. Nigel Smith 25. Benoit Famaey 26. Dharam Vir Ahluwalia 27. Csaba Balazs 28. Felix Stoehr 29. Matthew Buckley 30. Matt Visser 31. Paul Scovell 32. Tesla Jeltema 33. Hans Volker Klapdor-Kleingrothaus 34. Holger Motz 35. David Rapetti 36. Manfred Leubner 37. Chung-Lin Shan 38. Jason Kumar 39. Lydia Philpott 40. Cheng Yang Lee 41. Kazunori Nakayama 42. Peter Smale 43. Agnieszka Jacholkowska 44. Emmanuel Moulin 45. Koji Ishiwata

CONTENTS

Preface v

Direct Search for Cold Dark Matter I

Signals from the Dark Universe: Where We Are, Where We Are Going
R. Bernabei, P. Belli, F. Montecchia, F. Nozzoli, F. Cappella, A. d'Angelo, A. Incicchitti, D. Prosperi, R. Cerulli, C. J. Dai, H. L. He, H. H. Kuang, X. H. Ma, X. D. Sheng and Z. P. Ye 3

From DAMA/LIBRA to Super-Kamiokande
J. Kumar 18

A Light Scalar WIMP, The Higgs Portal and DAMA
M. H. G. Tytgat 28

SUSY/SUGRA Phenomenology in Dark Matter, LHC and Perspectives, Extra Dimensions and Dark Matter/Dark Energy

SUSY Searches at LHC and Dark Matter
E. Barberio (for the ATLAS Collaboration) 39

Dark Matter and Supersymmetry
L. Roszkowski 51

Scenarios of Gravitino Dark Matter and their Cosmological and Particle Physics Implications
G. Moultaka 66

The Phenomenology of Gravitino Dark Matter Scenarios in Supergravity Models
Y. Santoso 77

Likelihood Analysis of the Next-to-Minimal Supergravity Motivated Model
C. Balázs and D. Carter 87

Hierarchy Problem and Dilatonic Dark Matter
Y. M. Cho 97

Dark Energy and Dark Matter in Models with Warped Extra Dimensions
I. P. Neupane 116

Hot Dark Matter, Neutrino Mass and Dark Energy

Nuclear Double Beta Decay, Fundamental Particle Physics, Hot Dark Matter, and Dark Energy
H. V. Klapdor-Kleingrothaus and I. V. Krivosheina 137

Neutrino Mass, Dark Matter and Baryon Asymmetry via TeV-Scale Physics
M. Aoki, S. Kanemura and O. Seto 170

A Possible Connection between Massive Fermions and Dark Energy
T. Goldman, G. J. Stephenson, Jr., P. M. Alsing and B. H. J. Mckellar 180

Nonextensive Statistics in Astro-Particle Physics: Status and Impact for Dark Matter/Dark Energy Theory
M. P. Leubner 194

Cosmic Ray Positron Excess

PAMELA and ATIC Anomalies in Decaying Gravitino Dark Matter Scenario
K. Ishiwata, S. Matsumoto and T. Moroi 209

Indirect Dark Matter Searches versus Cosmic Ray Transport Model Uncertainties
I. Gebauer 218

Signatures of Dark Matter Annihilation in the Light of PAMELA/ATIC Anomaly
K. Nakayama 233

X-Rays and Dark Matter, GeV Gamma Rays

Fundamental Physics with GeV Gamma Rays
S. Profumo 243

Searching for Dark Matter Annihilation in X-Rays and Gamma-Rays
T. E. Jeltema and S. Profumo 255

Weighing Super-Massive Black Holes with X-Ray–Emitting Gas
D. A. Buote, P. J. Humphrey, F. Brighenti, K. Gebhardt and W. G. Mathews 264

Dark Energy, Dark Matter and Dark Radiation

Dark Matter and Dark Radiation
L. Ackerman, M. R. Buckley, S. M. Carroll and M. Kamionkowski 277

Cosmographic Analysis of Dark Energy
M. Visser and C. Cattoën 287

Dark Mass at the Center of the Galaxy

Luminous Accretion onto the Dark Mass at the Centre of the Milky Way
A. Eckart, M. García-Marín, S. König, D. Kunneriath, K. Mužić, C. Straubmeier, G. Witzel and M. Zamaninasab 303

The Darkstars Code: A Publicly Available Dark Stellar Evolution Package
P. Scott, J. Edsjö and M. Fairbairn 320

Relativistic Signatures at the Galactic Centre
E. M. Howard 328

Dark and Baryonic Matter in Galaxies

Non-Standard Baryon-Dark Matter Interactions
B. Famaey and J.-P. Bruneton 335

Annihilation in the Milky Way: Simulated Distribution of Dark Matter on Small Scales
F. S. Stoehr 344

Primordial Molecules and First Structures
D. Puy 350

Dark Companion of Baryonic Matter in Spiral Galaxies
Y. Sobouti 356

Fifth Family, Dark Energy Perturbations and Dark Matter

Offering a Mechanism for Generating Families: The Approach Unifying Spins and Charges Predicts a New Stable Family Forming Dark Matter Clusters
N. S. Mankoč Borštnik 365

Dark Energy Perturbations and a Possible Solution to the Coincidence Problem
J. Grande, A. Pelinson and J. Solà 380

Gravitation, Dark Energy, Dark Matter

Dark Energy without Dark Energy: Average Observational Quantities
D. L. Wiltshire 397

Search for Quantum Gravity Signature with Photons from Astrophysical Sources
A. Jacholkowska and J. Bolmont 413

Constraining Dark Energy and Gravity with X-Ray Galaxy Clusters
D. Rapetti, S. W. Allen, A. Mantz, R. G. Morris, H. Ebeling, R. Schmidt and A. C. Fabian 426

Dark Matter or Modified Dynamics? Hints from Galaxy Kinematics
G. Gentile 440

Dark Matter and Non-Standard Wigner Classes

Quantum Fields, Dark Matter and Non-Standard Wigner Classes
A. B. Gillard and B. M. S. Martin 451

Indirect Search for Dark Matter

Dark Matter Searches with Imaging Atmospheric Cherenkov Telescopes
E. Moulin 459

Constraints on Dark Matter WIMPS Models with H.E.S.S. Observations of the Canis Major Overdensity
M. Vivier (for the H.E.S.S. Collaboration) 471

Recent Results and Status of IceCube
S. H. Seo (for the IceCube Collaboration) 482

Search for Dark Matter with AMANDA and Ice Cube Detectors
A. Rizzo (for the IceCube Collaboration) 494

Indirect Search for Dark Matter with the ANTARES Neutrino Telescope
H. Motz (for the ANTARES Collaboration) 504

Direct Search for Cold Dark Matter II

How Precisely Could We Identify WIMPs Model-Independently with Direct Dark Matter Detection Experiments
M. Drees and C.-L. Shan 521

Hot on the Tail of the Elusive WIMP: Direct Detection Dark Matter Searches Enter the 21st Century
T. Saab (for the CDMS Collaboration) 535

The ZEPLIN-III Veto Detector
P. R. Scovell (for the ZEPLIN-III Collaboration) 546

Exotics

Hypothetical Dark Matter/Axion Rockets: Dark Matter in Terms of Space Physics Propulsion
A. Beckwith 557

List of Participants 563

Authors Index 575

Canterbury Rooms 2

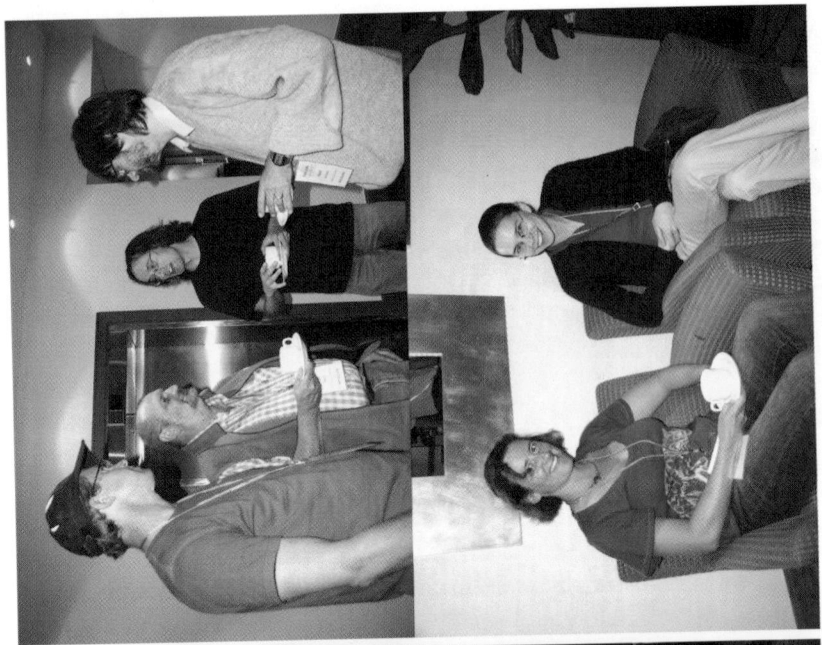

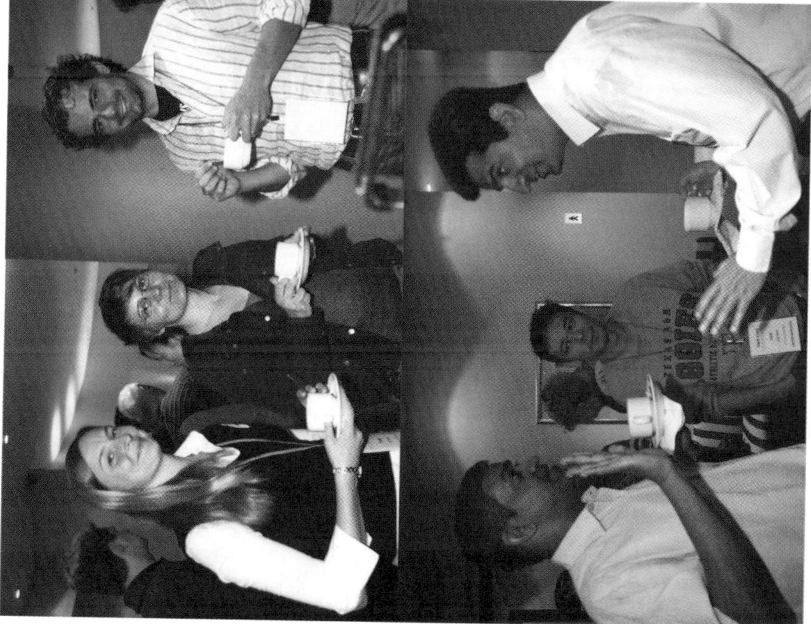

CANTERBURY 3 ROOM

PART I

Direct Search for Cold Dark Matter I

SIGNALS FROM THE DARK UNIVERSE: WHERE WE ARE, WHERE WE ARE GOING

R. BERNABEI, P. BELLI, F. MONTECCHIA and F. NOZZOLI

Dip. di Fisica, Università di Roma "Tor Vergata" and INFN, sez. Roma "Tor Vergata", I-00133 Rome, Italy

F. CAPPELLA, A. d'ANGELO, A. INCICCHITTI and D. PROSPERI

Dip. di Fisica, Università di Roma "La Sapienza" and INFN, sez. Roma, I-00185 Rome, Italy

R. CERULLI

Laboratori Nazionali del Gran Sasso, I.N.F.N., Assergi, Italy

C.J. DAI, H.L. HE, H.H. KUANG, X.H. MA, X.D. SHENG and Z.P. YE

IHEP, Chinese Academy, P.O. Box 918/3, Beijing 100039, China

Arguments on the investigation of the Dark Matter particles in the galactic halo are addressed. Recent results obtained by the DAMA/LIBRA set-up – exploiting the model independent annual modulation signature for Dark Matter (DM) particles – are shortly summarized. In fact, the DAMA project is an observatory for rare processes and it is operative deep underground at the Gran Sasso National Laboratory of the I.N.F.N. Its main apparatus is at present the DAMA/LIBRA set-up, consisting of $\simeq$ 250 kg highly radiopure NaI(Tl) detectors. Its first results confirm those obtained by the former DAMA/NaI, supporting the evidence for Dark Matter presence in the galactic halo at 8.2 σ C.L.; in addition, the cumulative data satisfy all the many peculiarities of the DM annual modulation signature. Future perspectives are also addressed.

Keywords: Scintillation detectors; Dark Matter; Underground Physics.

1. Introduction

With the present technology the only reliable signature able to point out, in a model independent way, the presence of Dark Matter (DM) particles in the galactic halo and sensitive to wide ranges both of DM candidates and of interaction types, is the DM annual modulation signature. This signature – originally suggested in the middle of '80 in ref. [1] – exploits the effect of the Earth revolution around the Sun on the number of events induced by the Dark Matter particles in a suitable low-background set-up placed deep underground. In fact, as a consequence of its annual revolution, the Earth should be crossed by a larger flux of Dark Matter particles around $\sim$ 2 June (when its rotational velocity is summed to the one of the solar system with respect to the Galaxy) and by a smaller one around $\sim$ 2 December (when the two velocities are subtracted). This offers an efficient model independent signature, able to test a large interval of cross sections and of halo densities.

The expected differential rate as a function of the energy, dR/dE (see also ref. [2–8] for some discussions), depends on the DM particle velocity distribution and on the Earth's velocity in the galactic frame, $\vec{v}_e(t)$. Projecting $\vec{v}_e(t)$ on the galactic plane, one can write: $v_e(t) = v_\odot + v_\oplus cos\gamma cos\omega(t-t_0)$. Here $v_\odot$ is the Sun's velocity with the respect to the galactic halo ($v_\odot \simeq v_0 + 12$ km/s and v_0 is the local velocity whose value is in the range 170-270 km/s [9, 10]); $v_\oplus = 30$ km/s is the Earth's orbital velocity around the Sun on a plane with inclination $\gamma = 60^o$ with the respect to the galactic plane. Furthermore, $\omega= 2\pi/\mathrm{T}$ with T=1 year and roughly $\mathrm{t}_0 \simeq 2^{nd}$ June (when the Earth's speed is at maximum). The Earth's velocity can be conveniently expressed in unit of v_0: $\eta(t) = v_e(t)/v_0 = \eta_0 + \Delta\eta cos\omega(t-t_0)$, where – depending on the assumed value of the local velocity – η_0=1.04-1.07 is the yearly average of η and $\Delta\eta$ = 0.05-0.09. Since $\Delta\eta \ll \eta_0$, the expected counting rate can be expressed by the first order Taylor approximation:

$$\frac{dR}{dE}[\eta(t)] = \frac{dR}{dE}[\eta_0] + \frac{\partial}{\partial\eta}\left(\frac{dR}{dE}\right)_{\eta=\eta_0} \Delta\eta \cos\omega(t-t_0). \quad (1)$$

Averaging this expression in a k-th energy interval one obtains:

$$S_k[\eta(t)] = S_k[\eta_0] + [\frac{\partial S_k}{\partial\eta}]_{\eta_0} \Delta\eta cos\omega(t-t_0) = S_{0,k} + S_{m,k} cos\omega(t-t_0); \quad (2)$$

the contribution from the highest order terms less than 0.1%. The DM annual modulation signature is very distinctive since the corresponding signal must simultaneously satisfy all the following requirements: the rate must contain a component modulated according to a cosine function (1) with one year period (2) and a phase that peaks roughly around $\simeq 2^{nd}$ June (3); this modulation must only be found in a well-defined low energy range, where DM particle induced events can be present (4); it must apply only to those events in which just one detector of many actually "fires" (*single-hit events*), since the DM particle multi-interaction probability is negligible (5); the modulation amplitude in the region of maximal sensitivity must be $\lesssim$7% for usually adopted halo distributions (6), but it can be larger in case of some possible scenarios such as e.g. those in refs. [11, 12]. Only systematic effects or side reactions able to fulfil these 6 requirements and to account for the whole observed modulation amplitude could mimic this signature [a]; thus, no other effect investigated so far in the field of rare processes offers a so stringent and unambiguous signature.

It is worth noting that the corollary questions related to the exact nature of the DM particle(s) (detected by means of the DM annual modulation signature) and to the astrophysical, nuclear and particle Physics scenarios require instead subsequent model dependent corollary analyses, as those performed e.g. in refs. [2–8, 13]. On the other hand, one should stress that it does not exist any approach in direct and indirect DM searches which can offer information on the nature of the candidate independently on assumed astrophysical, nuclear and particle Physics scenarios.

In the following, we will just briefly summarize the first results on the Dark Matter particles obtained by DAMA/LIBRA, exploiting over four annual cycles the model independent DM annual modulation signature (exposure of 0.53 ton×yr). The result has also been combined together with the previous data collected over 7 annual cycles by DAMA/NaI (0.29 ton×yr). The whole available data correspond to 11 annual cycles for a total exposure of 0.82 ton×yr.

2. The DAMA Project

DAMA is an observatory for rare processes and it is operative deep underground at the Gran Sasso National Laboratory of the I.N.F.N. The DAMA project is mainly based on the development and use of low background scintillators [2–8, 13–21]; the main aim is the direct detection of DM particles in the galactic halo by investigating the model independent annual modulation signature. Profiting of the low background features of these set-ups, many rare processes are also investigated obtaining very competitive results. The main experimental set-ups are: i) DAMA/NaI ($\simeq$ 100 kg of highly radiopure NaI(Tl)) which completed its data taking on July 2002 [14, 17]; ii) DAMA/LXe ($\simeq$ 6.5 kg liquid Kr-free Xenon enriched either

[a]It is worth noting that the DM annual modulation is not – as often naively said – a "seasonal" variation and it is not a "winter-summer" effect. In fact, the DM annual modulation is not related to the relative Sun position, but it is related to the Earth velocity in the galactic frame. Moreover, the phase of the DM annual modulation (roughly 2nd June) is well different than those of physical quantities (such as temperature of atmosphere, pressure, other meteorological parameters, cosmic rays flux, ...) instead correlated with seasons.

in ^{129}Xe or in ^{136}Xe) [22, 23]; iii) DAMA/R&D, devoted to tests on prototypes and to small scale experiments [24]; iv) the new second generation DAMA/LIBRA set-up ($\simeq$ 250 kg highly radiopure NaI(Tl)) in operation since March 2003 [20, 21]. Moreover, in the framework of devoted R&D for radiopure detectors and photomultipliers, sample measurements are carried out by means of the low background DAMA/Ge detector (installed deep underground since more than 10 years); the detector has also been used for some small scale experiments [25].

3. The DAMA/LIBRA Results

Highly radiopure NaI(Tl) scintillators have offered and offer many competitive features to effectively investigate the DM annual modulation signature, such as e.g.: i) high duty cycle; ii) well known technology; iii) large masses; iv) no safety problems; v) the lowest cost with the respect to every other considered technique; vi) necessity of a relatively small underground space; vii) reachable high radiopurity by material selections and protocols, by chemical/physical purifications, etc.; viii) feasibility of well controlled operational conditions and monitoring, ix) feasibility of routine calibrations down to few keV in the same conditions as the production runs; x) high light response (that is keV threshold reachable); xi) absence of the necessity of re-purification or cooling down/warming up procedures (implying high reproducibility, high stability, etc.); xii) absence of microphonic noise and an effective noise rejection at threshold (time decay of NaI(Tl) pulses is hundreds ns, while that of noise pulses is tens ns); xiii) wide sensitivity to both high and low mass DM candidates and to many different interaction types and astrophysical, nuclear and particle Physics scenario; xiv) possibility to effectively investigate the DM annual modulation signature in all the needed aspects; xv) possibility to achieve significant results on several other rare processes; xvi) etc.

These arguments motivated the development and use of highly radiopure NaI(Tl) scintillators for the DAMA/NaI and DAMA/LIBRA target-detectors. The competitivity of these set-ups is based on the reached intrinsic radiopurity (obtained after very long and accurate work for the selection of all low radioactive materials, for the definition of suitable protocols, etc.), on the large sensitivity to many of the DM candidates, of the interaction types and of astrophysical, nuclear and particle Physics scenarios, to the granularity of the set-ups, to the data taking up to the MeV scale (even though the optimization is made for the lowest energy region), to the full control of the running conditions, etc.

The DAMA/NaI set up and its performances are described in ref. [2, 3, 15, 16], while DAMA/LIBRA set-up and its performances in ref. [20]. Here we just summarized that: i) the detectors' responses range from 5.5 to 7.5 photoelectrons/keV; ii) the hardware threshold of each PMT is at single photoelectron (each detector is equipped with two low background photomultipliers working in coincidence); iii) energy calibration with X-rays/γ sources are regularly carried out down to few keV; iv) the software energy threshold of the experiment is 2 keV. The DAMA/NaI experiment collected an exposure of 0.29 ton$\times$yr over 7 annual cycles [2, 3, 16], while DAMA/LIBRA has released so far an exposure of 0.53 ton$\times$yr collected over 4 annual cycles [21]; thus, the total exposure of the two experiments is 0.82 ton$\times$yr, which is orders of magnitude larger than the exposure typically collected in the field.

Several analyses on the model-independent investigation of the DM annual modulation signature have been performed (see ref. [21] and references therein); here just few arguments are reminded. In particular, Fig. 1 shows the time behaviour of the experimental residual rates [b] for *single-hit* events collected by DAMA/NaI and by DAMA/LIBRA in the (2–4), (2–5) and (2–6) keV energy intervals. The superimposed curves represent the cosinusoidal functions behaviors $A \cos \omega(t - t_0)$ with a period $T = \frac{2\pi}{\omega} = 1$ yr and with a phase $t_0 = 152.5$ day (June 2nd), while the modulation amplitudes, A, have been obtained by best fit over the DAMA/NaI and DAMA/LIBRA data. When the period and the phase parameters are released in the fit, values well compatible with those expected for a DM particle induced effect are obtained [21]. Summarizing, the cumulative analysis of the *single-hit* residual rate favours the presence of a modulated cosine-like behaviour with proper features at 8.2 σ C.L. [21].

The same data of Fig.1 have also been investigated by a Fourier analysis, obtaining a clear peak corresponding to a period of 1 year [21]. The same analysis in other energy region shows instead only aliasing peaks. Similar result is obtained when comparing the *single-hit* residuals in the (2–6) keV with those in other energy interval; in fact, a clear modulation is present in

---

[b] These residual rates are calculated from the measured rate of the *single-hit* events (obviously corrections for the overall efficiency and for the acquisition dead time are applied) after subtracting the constant part: $< r_{ijk} - flat_{jk} >_{jk}$. Here r_{ijk} is the rate in the considered i-th time interval for the j-th detector in the k-th energy bin, while $flat_{jk}$ is the rate of the j-th detector in the k-th energy bin averaged over the cycles. The average is made on all the detectors (j index) and on all the energy bins (k index) which constitute the considered energy interval. The weighted mean of the residuals must obviously be zero over one cycle.

the lowest energy interval, while it is absent just above [21]. In particular, in order to verify absence of annual modulation in other energy regions and, thus, to also verify the absence of any significant background modulation,

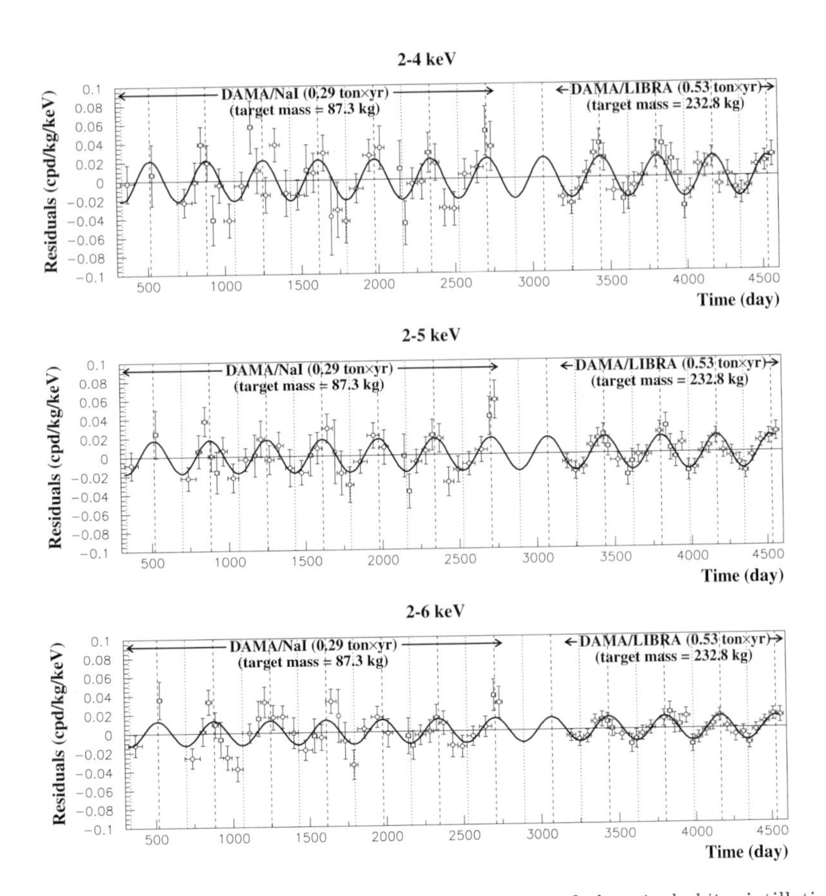

Fig. 1. Experimental model-independent residual rate of the *single-hit* scintillation events, measured by DAMA/NaI and DAMA/LIBRA in the (2 – 4), (2 – 5) and (2 – 6) keV energy intervals as a function of the time. The zero of the time scale is January 1st of the first year of data taking of DAMA/NaI. The experimental points present the errors as vertical bars and the associated time bin width as horizontal bars. The superimposed curves are the cosinusoidal functions behaviors $A\cos\omega(t-t_0)$ with a period $T=\frac{2\pi}{\omega}=1$ yr, with a phase $t_0=152.5$ day (June 2nd) and with modulation amplitudes, A, equal to the central values obtained by best fit over the whole data, that is: (0.0215 ± 0.0026) cpd/kg/keV, (0.0176 ± 0.0020) cpd/kg/keV and (0.0129 ± 0.0016) cpd/kg/keV for the (2 – 4) keV, for the (2 – 5) keV and for the (2 – 6) keV energy intervals, respectively. The dashed vertical lines correspond to the maximum of the signal (June 2nd), while the dotted vertical lines correspond to the minimum. The total exposure is 0.82 ton × yr. For details see [21].

the energy distribution measured during the data taking periods in energy regions not of interest for DM detection has also been investigated. In fact, the background in the lowest energy region is essentially due to "Compton" electrons, X-rays and/or Auger electrons, muon induced events, etc., which are strictly correlated with the events in the higher energy part of the spectrum. Thus, if a modulation detected in the lowest energy region would be due to a modulation of the background (rather than to a signal), an equal or larger modulation in the higher energy regions should be present. The data analyses have allowed to exclude the presence of a background modulation in the whole energy spectrum at a level much lower than the effect found in the lowest energy region for the *single-hit* events [21].

A further relevant investigation has been done by applying the same hardware and software procedures, used to acquire and to analyse the *single-hit* residual rate, to the *multiple-hits* one. In fact, since the probability that a DM particle interacts in more than one detector is negligible, a DM signal can be present just in the *single-hit* residual rate. Thus, this allows the test of the background behaviour in the same energy interval of the observed positive effect. In particular, Fig. 2 shows the residual rates of the *single-hit* events measured over the four DAMA/LIBRA annual cycles, as collected in a single annual cycle, together with the residual rates of the *multiple-hits* events, in the considered energy intervals. A clear modulation is present in the *single-hit* events, while the fitted modulation amplitudes for the *multiple-hits* residual rate are well compatible with zero: $-(0.0004 \pm 0.0008)$ cpd/kg/keV, $-(0.0005 \pm 0.0007)$ cpd/kg/keV, and $-(0.0004 \pm 0.0006)$ cpd/kg/keV in the energy regions $(2-4)$, $(2-5)$ and $(2-6)$ keV, respectively. Similar results were previously obtained also for the DAMA/NaI case [3]. Thus, again evidence of annual modulation with proper features, as required by the DM annual modulation signature, is present in the *single-hit* residuals (events class to which the DM particle induced events belong), while it is absent in the *multiple-hits* residual rate (event class to which only background events belong). Since the same identical hardware and the same identical software procedures have been used to analyse the two classes of events, the obtained result offers an additional strong support for the presence of a DM particle component in the galactic halo further excluding any side effect either from hardware or from software procedures or from background.

The annual modulation present at low energy has also been shown by depicting the differential modulation amplitudes, $S_{m,k}$, as a function of the energy; the $S_{m,k}$ is the modulation amplitude of the modulated part of

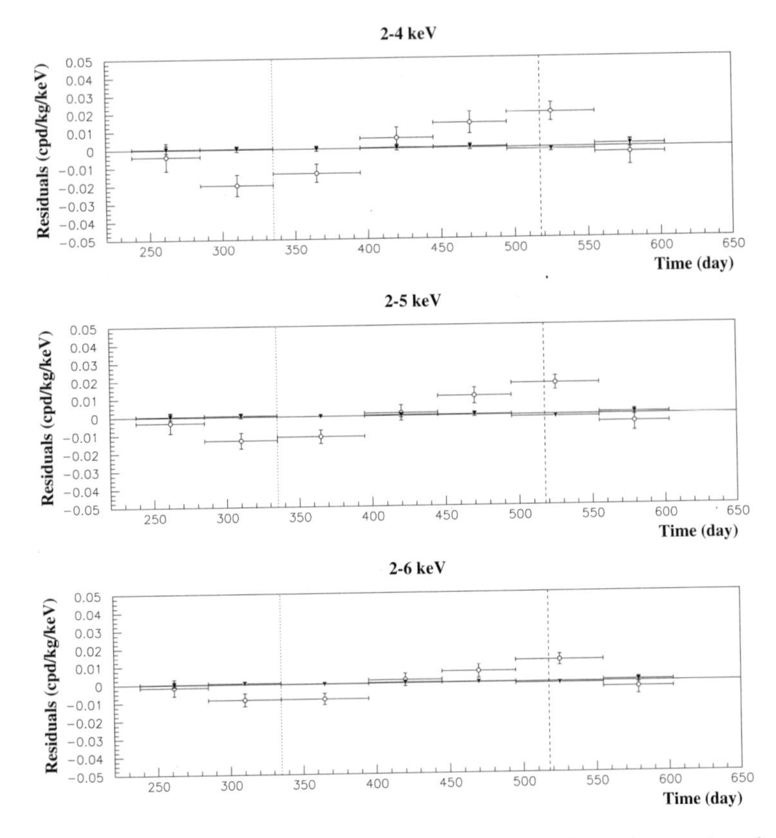

Fig. 2. Experimental residual rates over the four DAMA/LIBRA annual cycles for *single-hit* events (open circles) (class of events to which DM events belong) and for *multiple-hits* events (filled triangles) (class of events to which DM events do not belong), in the energy intervals (2 – 4), (2 – 5) and (2 – 6) keV, respectively. They have been obtained by considering for each class of events the data as collected in a single annual cycle and by using in both cases the same identical hardware and the same identical software procedures. The initial time of the scale is taken on August 7th. The experimental points present the errors as vertical bars and the associated time bin width as horizontal bars. See ref. [21]. Analogous results were obtained for the DAMA/NaI data [3].

the signal (see above) obtained by maximum likelihood method over the data, considering T =1 yr and t_0 = 152.5 day. In Fig. 3 the measured $S_{m,k}$ for the total exposure (0.82 ton×yr, DAMA/NaI and DAMA/LIBRA) are reported as function of the energy. It can be inferred that positive signal is present in the (2–6) keV energy interval, while $S_{m,k}$ values compatible with zero are present just above. In fact, the $S_{m,k}$ values in the (6–20) keV energy interval have random fluctuations around zero with χ^2 equal to 24.4 for 28 degrees of freedom.

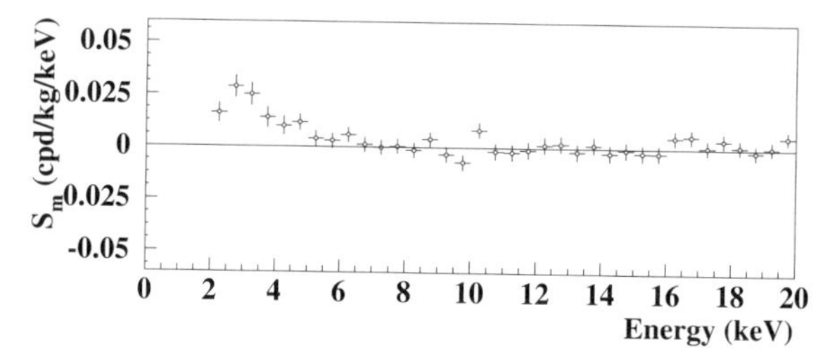

Fig. 3. Energy distribution of the $S_{m,k}$ variable for the total exposure of DAMA/NaI and DAMA/LIBRA: 0.82 ton×yr. A clear modulation is present in the lowest energy region, while $S_{m,k}$ values compatible with zero are present just above. In fact, the $S_{m,k}$ values in the (6–20) keV energy interval have random fluctuations around zero with $\chi^2/d.o.f.$ equal to 24.4/28. See ref. [21].

It has also been verified that the measured modulation amplitudes are statistically well distributed in all the crystals, in all the annual cycles and in the energy bins; these and other discussions can be found in ref. [21].

It is also interesting the results of the analysis performed by releasing the assumption of a phase $t_0 = 152.5$ day in the procedure to evaluate the modulation amplitudes from the data of the seven annual cycles of DAMA/NaI and the four annual cycles of DAMA/LIBRA. In this case alternatively the signal has been written as: $S_{0,k} + S_{m,k} \cos\omega(t - t_0) + Z_{m,k} \sin\omega(t - t_0) = S_{0,k} + Y_{m,k} \cos\omega(t - t^*)$. Obviously, for signals induced by DM particles one would expect: i) $Z_{m,k} \sim 0$ (because of the orthogonality between the cosine and the sine functions); ii) $S_{m,k} \simeq Y_{m,k}$; iii) $t^* \simeq t_0 = 152.5$ day. In fact, these conditions hold for most of the dark halo models; however, it is worth noting that slight differences can be expected in case of possible contributions from non-thermalized DM components, such as e.g. the SagDEG stream [13] and the caustics [26].

Fig. 4–*left* shows the 2σ contours in the plane (S_m, Z_m) for the (2–6) keV and (6–14) keV energy intervals and Fig. 4–*right* shows, instead, those in the plane (Y_m, t^*). The best fit values for the (2–6) keV energy interval are (1σ errors): $S_m = (0.0122 \pm 0.0016)$ cpd/kg/keV; $Z_m = -(0.0019 \pm 0.0017)$ cpd/kg/keV; $Y_m = (0.0123 \pm 0.0016)$ cpd/kg/keV; $t^* = (144.0 \pm 7.5)$ day; while for the (6–14) keV energy interval are: $S_m = (0.0005 \pm 0.0010)$ cpd/kg/keV; $Z_m = (0.0011 \pm 0.0012)$ cpd/kg/keV; $Y_m = (0.0012 \pm 0.0011)$ cpd/kg/keV and t^* obviously not determined (see Fig. 4). These results confirm those achieved by other kinds of analyses. In particular, a modulation amplitude is present in the lower energy intervals and the period and the phase agree with those expected for DM induced signals. For more discussions see ref. [21]

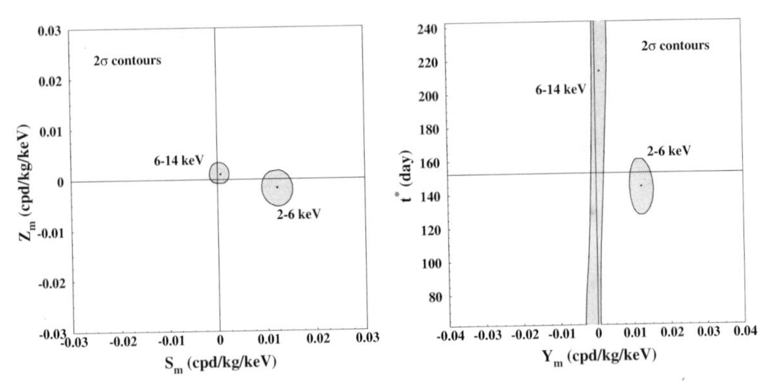

Fig. 4. 2σ contours in the plane (S_m, Z_m) (*left*) and in the plane (Y_m, t^*) (*right*) for the (2–6) keV and (6–14) keV energy intervals. The contours have been obtained by the maximum likelihood method, considering the seven annual cycles of DAMA/NaI and the four annual cycles of DAMA/LIBRA all together. A modulation amplitude is present in the lower energy intervals and the period and the phase agree with those expected for DM induced signals. See ref. [21].

Both the data of DAMA/LIBRA and of DAMA/NaI fulfil all the requirements of the DM annual modulation signature.

As previously done for DAMA/NaI [2, 3], careful investigations on absence of any significant systematics or side reaction effect in DAMA/LIBRA have been quantitatively carried out and reported in details in ref. [21].

In order to continuously monitor the running conditions, several pieces of information are acquired with the production data and quantitatively analyzed. No modulation has been found in any possible source of systematics or side reactions for DAMA/LIBRA as well; thus, cautious upper limits (90% C.L.) on the possible contributions to the DAMA/LIBRA measured modulation amplitude have been estimated and are summarized in Table 1. It is important to stress that - in addition - none able to mimic the signature has been found or suggested by anyone over more than a decade. In fact, they cannot account for the measured modulation amplitude and contemporaneously satisfy all the requirements of the signature. For detailed quantitative discussions on all the related topics and for results see ref. [21] and refs. therein. Just as an example we remind here the case of muons, whose flux has been reported by the MACRO experiment to have a 2% modulation with phase around mid–July. In particular, it has been shown that not only this effect would give rise in the DAMA set-ups to a quantitatively negligible contribution (see in the 2008 publication list and refs. therein), but some of the six requirements necessary to mimic the annual modulation signature – namely e.g. the conditions of presence of modulation just in the

Table 1. Summary of the results obtained by investigating all possible sources of systematics and side reactions in the data of the DAMA/LIBRA four annual cycles. None able to give a modulation amplitude different from zero has been found; thus cautious upper limits (90% C.L.) on the possible contributions to the measured modulation amplitude have been calculated and are shown here. It is worth noting that none of them is able to mimic the DM annual modulation signature, that is none is able to account for the whole observed modulation amplitude and to contemporaneously satisfy all the requirements of the signature. For details see ref. [21]. Analogous results were obtained for DAMA/NaI [2, 3].

Source	Main comment (also see ref. [20])	Cautious upper limit (90%C.L.)
Radon	Sealed Cu Box in HP Nitrogen atmosphere, 3-level of sealing	$< 2.5 \times 10^{-6}$ cpd/kg/keV
Temperature	Air conditioning + huge heat capacity	$< 10^{-4}$ cpd/kg/keV
Noise	Efficient rejection	$< 10^{-4}$ cpd/kg/keV
Energy scale	Routine + intrinsic calibrations	$< 1 - 2 \times 10^{-4}$ cpd/kg/keV
Efficiencies	Regularly measured	$< 10^{-4}$ cpd/kg/keV
Background	No modulation above 6 keV; no modulation in the (2 – 6) keV *multiple-hit* events; this limit includes all possible sources of background	$< 10^{-4}$ cpd/kg/keV
Side reactions	From muon flux variation measured by MACRO	$< 3 \times 10^{-5}$ cpd/kg/keV
In addition: no effect can mimic the signature		

single-hit event rate and of the phase value – would also fail. Moreover, even the pessimistic assumption of whatever (even exotic) possible cosmogenic product – whose decay or de-excitation or whatever else (with mean-life τ) might produce: i) only events at low energy; ii) only *single-hit* events; iii) no sizeable effect in the *multiple-hits* counting rate – cannot give rise to any side process able to mimic the investigated DM signature. In fact, not only this latter hypothetical process would be quantitatively negligible, but in addition its phase – as it can be easily derived – would be (much) larger than July 15th, and therefore well different from the one measured by the DAMA experiments and expected by the DM annual modulation signature ($\simeq$ June 2nd). Thus, any possible effect from muons can be safely excluded.

Summarizing, DAMA/LIBRA has confirmed the presence of an annual modulation satisfying all the requirements of the DM annual modulation signature, as previously pointed out by DAMA/NaI; in particular, the evidence for the presence of DM particles in the galactic halo is cumulatively supported at 8.2 σ C.L.

4. On Corollary Quests and on Comparisons

As regards the corollary investigation on the nature of the DM candidate particle(s) and related astrophysical, nuclear and particle Physics scenarios, it has been shown – on the basis of the DAMA/NaI result – that the obtained model independent evidence can be compatible with a wide set of possibilities (see e.g. ref. [2–8] and in literature, for example see [27]); many others are also open. Obviously, this is also the case when the DAMA/NaI and DAMA/LIBRA data are considered all together (see e.g. [21]); an updating is foreseen.

It is worth noting that no other experiment exists, whose result can be directly compared in a model-independent way with those by DAMA/NaI and DAMA/LIBRA. In particular, let us also point out that results obtained with different target materials and/or different approaches cannot intrinsically be directly compared among them even when considering the same kind of candidate and of coupling, although apparently all the presentations generally refer to cross section on nucleon.

In particular, claims for contradictions made by experiments insensitive to the DM annual modulation signature, using different target materials and approaches, exploiting marginal exposures, having well different sensitivities to various DM candidate and interactions, etc. have by the fact no impact even in the single arbitrary scenario they consider without accounting for experimental and theoretical uncertainties, using often crude approximation in the calculation, etc. Moreover, as pointed out (see for example [28]), some critical points exist on relevant experimental aspects (energy threshold, energy scale, multiple selection procedures, stabilities, etc.). A relevant argument is also the methodological robustness [29].

It is worth noting that, whenever an experiment using the same identical target material and methodological approach would be available in future, as usual in whatever field of Physics a serious comparison would require – in every case – e.g. a deep investigation of the radiopurity of all the part of the different set-ups, of their specific performances in all the aspects, of the detailed procedures used by each one, of the used exposures, etc.

Finally, as regards the indirect detection searches, let us note that also no direct model-independent comparison can be performed between the results obtained in direct and indirect activities, since it does not exist a biunivocal correspondence between the observables in the two kinds of experiments. Anyhow, if possible excesses in the positron to electron flux ratio and in the γ rays flux with respect to some simulations of the hypothesized contribution, which is expected from standard sources, might be interpreted in terms of Dark Matter (but e.g. available more complete handling of some aspects of the simulations [30] and the pulsars contribution [31] should be included), this would also be not in conflict with the effect observed by DAMA experiments.

5. Already Performed and Planned Upgradings

During September 2008 the first upgrading of the DAMA/LIBRA set-up has been realized and the shield has been opened in HP Nitrogen atmosphere. This has allowed the increase of the exposed mass since one detector has been recovered by replacing a broken PMT. Moreover, a new optimization of some PMTs and HVs has been done. Finally, a total replacement of the used transient digitizers with new ones, having better performances, has been realized and a new DAQ with optical fibers has been installed and put in operation. The data taking has been restarted on October 2008.

In particular, the model independent results achieved by the DAMA/LIBRA set-up has pointed out the relevance to lower the software energy threshold used by the experiment. Thus, the replacement of all the PMTs with other ones with higher quantum efficiency has been planned; this will also improve – as evident – other significant experimental aspects.

A larger exposure collected by DAMA/LIBRA (or possibly by DAMA/1ton; see later) and the lowering of the 2 keV energy threshold in the data analysis will improve the corollary information on the nature of the DM candidate particle(s) and on the various related astrophysical, nuclear and particle Physics scenarios. Moreover, it will also allow the investigation – with high sensitivity – of other DM features, of second order effects and of several rare processes other than DM. In particular, some of the many topics – not yet well known at present and which can affect whatever model dependent result and comparison – are: i) the velocity and spatial distribution of the Dark Matter particles in the galactic halo; ii) the effects induced on the Dark Matter particles distribution in the galactic halo by contributions from satellite galaxies tidal streams; iii) the effects induced on the Dark Matter particles distribution in the galactic halo by the existence of caustics; iv) the detection of possible "solar wakes", the gravitational focusing effect of the Sun on the Dark Matter particle of a stream (two kinds of enhancements in the flow are expected: "spike" and "skirt"); v) the investigation of possible diurnal effects; vi) the study of possible structures as clumpiness with small scale size; vii) the coupling(s) of the Dark Matter particle with the ^{23}Na and ^{127}I and its nature; viii) the scaling laws and cross sections.

In addition, it is worth noting that ultra low background NaI(Tl) scintillators can also offer the possibility to achieve significant results on several other rare processes as already done e.g. by the former DAMA/NaI apparatus [17–19].

Finally, we mention that a third generation R&D effort towards a possible NaI(Tl) ton set-up has been funded by I.N.F.N.

6. Conclusions

The highly radiopure NaI(Tl) DAMA set-ups have investigated the presence of DM particles in the galactic halo by exploiting the DM annual modulation signature; they have achieved a model independent evidence at 8.2 σ C.L.

The collection of larger exposures (with the upgraded DAMA/LIBRA or possibly with DAMA/1ton) can allow to significantly investigate several open aspects on the nature of the candidate particle(s) and on the various related astrophysical, nuclear and particle Physics as well as other DM features and second order effects.

References

1. K.A. Drukier et al., Phys. Rev. D 33 (1986) 3495; K. Freese et al., Phys. Rev. D 37 (1988) 3388.
2. R. Bernabei el al., La Rivista del Nuovo Cimento 26 n.1 (2003) 1-73.
3. R. Bernabei et al., Int. J. Mod. Phys. D 13 (2004) 2127.
4. R. Bernabei et al., Int. J. Mod. Phys. A 21 (2006) 1445.
5. R. Bernabei et al., Int. J. Mod. Phys. A 22 (2007) 3155.
6. R. Bernabei et al., Eur. Phys. J. C 53 (2008) 205.
7. R. Bernabei et al., Phys. Rev. D 77 (2008) 023506.
8. R. Bernabei et al., Mod. Phys. Lett. A 23 (2008) 2125.
9. P. Belli et al., Phys. Rev. D 61 (2000) 023512.
10. P.J.T. Leonard and S. Tremaine, Astrophys. J. 353 (1990) 486; C.S. Kochanek, Astrophys. J. 457 (1996) 228; K.M. Cudworth, Astron. J. 99 (1990) 590
11. D. Smith and N. Weiner, Phys. Rev. D 64 (2001) 043502; D. Tucker-Smith and N. Weiner, Phys. Rev. D 72 (2005) 063509.
12. K.Freese et al. astro-ph/0309279; Phys. Rev. Lett. 92 (2004) 11301.
13. R. Bernabei et al., Eur. Phys. J. C 47 (2006) 263.
14. R. Bernabei et al., Phys. Lett. B 389 (1996) 757; R. Bernabei et al., Phys. Lett. B 424 (1998) 195; R. Bernabei et al., Phys. Lett. B 450 (1999) 448; P. Belli et al., Phys. Rev. D 61 (2000) 023512; R. Bernabei et al., Phys. Lett. B 480 (2000) 23; R. Bernabei et al., Phys. Lett. B 509 (2001) 197; R. Bernabei et al., Eur. Phys. J. C 23 (2002) 61; P. Belli et al., Phys. Rev. D 66 (2002) 043503.

15. R. Bernabei et al., Il Nuovo Cim. A 112 (1999) 545.
16. R. Bernabei et al., Eur. Phys. J. C 18 (2000) 283.
17. R. Bernabei et al., Phys. Lett. B408 (1997) 439; P. Belli et al., Phys. Lett. B460 (1999) 236; R. Bernabei et al., Phys. Rev. Lett. 83 (1999) 4918; P. Belli et al., Phys. Rev. C60 (1999) 065501; R. Bernabei et al., Il Nuovo Cimento A112 (1999) 1541; R. Bernabei et al., Phys. Lett. B 515 (2001) 6; F. Cappella et al., Eur. Phys. J.-direct C14 (2002) 1; R. Bernabei et al., Eur. Phys. J. A 23 (2005) 7.
18. R. Bernabei et al., Eur. Phys. J. A 24 (2005) 51.
19. R. Bernabei et al., Astrop. Phys. 4 (1995) 45; R. Bernabei, in the volume *The identification of Dark Matter*, World Sc. Pub. (1997) 574.
20. R. Bernabei et al., Nucl. Instr. & Meth. A 592 (2008) 297.
21. R. Bernabei et al., Eur. Phys. J. C 56 (2008) 333.
22. P. Belli et al., Astropart. Phys. 5 (1996) 217; P. Belli et al., Nuovo Cim. C 19 (1996) 537; P. Belli et al., Phys. Lett. B 387 (1996) 222; Phys. Lett. B 389 (1996) 783 (err.); P. Belli et al., Phys. Lett. B 465 (1999) 315; P. Belli et al., Phys. Rev. D 61 (2000) 117301; R. Bernabei et al., New J. of Phys. 2 (2000) 15.1; R. Bernabei et al., Phys. Lett. B 493 (2000) 12; R. Bernabei et al., Nucl. Instr. & Meth A 482 (2002) 728; R. Bernabei et al., Eur. Phys. J. direct C 11 (2001) 1; R. Bernabei et al., Phys. Lett. B 527 (2002) 182; R. Bernabei et al., Phys. Lett. B 546 (2002) 23. R. Bernabei et al., in the volume *Beyond the Desert 2003*, Springer, Berlin (2003) 365; R. Bernabei et al., Eur. Phys. J. A 27, s01 (2006) 35.
23. R. Bernabei et al., Phys. Lett. B 436 (1998) 379.
24. R. Bernabei et al., Astropart. Phys. 7 (1997) 73; R. Bernabei et al., Nuovo Cim. A 110 (1997) 189; P. Belli et al., Astropart. Phys. 10 (1999) 115; P. Belli et al., Nucl. Phys. B 563 (1999) 97; R. Bernabei et al., Nucl. Phys. A 705 (2002) 29; P. Belli et al., Nucl. Instr. & Meth A 498 (2003) 352; R. Cerulli et al., Nucl. Instr. & Meth A 525 (2004) 535; R. Bernabei et al., Nucl. Instr. & Meth A 555 (2005) 270; R. Bernabei et al., Ukr. J. Phys. 51 (2006) 1037; P. Belli et al., Nucl. Phys. A 789 (2007) 15; P. Belli et al., Phys. Rev. C 76 (2007) 064603; P. Belli et al., Phys. Lett. B 658 (2008) 193; P. Belli et al., Eur. Phys. J. A 36 (2008) 167;
25. P. Belli et al., Nucl. Instr. & Meth. A 572 (2007) 734; Nucl. Phys. A 806 (2008) 388; to appear on the Proceed. of NPAE 2008, INR-Kiev.
26. F.S. Ling, P. Sikivie and S. Wick, Phys. Rev. D 70 (2004) 123503.
27. A. Bottino, N. Fornengo, and S. Scopel, Phys. Rev. D 67 (2003) 063519; A. Bottino, F. Donato, N. Fornengo, and S. Scopel, Phys. Rev. D 69 (2003) 037302; Phys. Rev. D 78 (2008) 083520.
28. R. Bernabei et al., arXiv:0806.0011[astro-ph].
29. R. Hudson, Found. Phys. 39 (2009) 174-193.
30. F. Donato et al., arXiv:0810.5292; T. Delahaye et al., arXiv:0809.5268.
31. S. Profumo, arXiv:0812.4457.

FROM DAMA/LIBRA TO SUPER-KAMIOKANDE

JASON KUMAR

*Department of Physics and Astronomy, University of Hawai'i,
Honolulu, HI 96822, USA*

We consider the prospects for probing low-mass dark matter with the Super-Kamiokande experiment. We show that upcoming analyses including fully-contained events with sensitivity to dark matter masses from 5 to 10 GeV can test the dark matter interpretation of the DAMA/LIBRA signal. We consider prospects of this analysis for two light dark matter candidates: neutralinos and WIMPless dark matter.

1. Introduction

The DAMA/LIBRA experiment has seen, with 8.2σ significance,[1] an annual modulation[2] in the rate of scattering events, which could be consistent with dark matter-nucleon scattering. Much of the region of dark matter parameter space that is favored by DAMA is excluded by null results from other direct detection experiments, including CRESST,[3] CDMS,[4] XENON10,[5] TEXONO,[6,7] and CoGeNT.[8] On the other hand, astrophysical uncertainties[9,10] and detector effects[11] may open up regions that can simultaneously accommodate the results from DAMA and these other experiments.

1.1. *A New Window for Light Dark Matter*

The sensitivity of direct detection experiments is largely determined by the kinematics of non-relativistic elastic nuclear scattering. If the nucleus recoil energy is measured, the momentum transfer can then determined, which in turn determines the dark matter mass (for a standard dark matter velocity distribution). This connection allows a direct detection experiment to convert an observed event rate into a detection of dark matter with a particular mass, or alternatively, convert a lack of a signal into an exclusion of a region of $(m_{DM}, \sigma_{\rm SI})$ parameter space.

But the procedure described above implicitly contains the uncertainties which could potentially lead to consistency between the dark matter interpretation of the DAMA/LIBRA signal and the lack of signal at other detection experiments. All direct detection experiments have a recoil energy threshold; nuclei recoiling with energies below this threshold cannot be detected. Low nuclei recoil energies imply a low dark matter mass. Thus, the sensitivity of direct detection experiments tends to suffer at low mass, as seen if Fig. 1. DAMA has a recoil energy threshold which is lower than many other direct detection experiments, thus raising the possibility that its signal is a result of elastic scatter with a light dark matter candidate.

However, there are some direct detection experiments, such as TEXONO[6,7] and CoGeNT,[8] which also have very low energy thresholds, and whose negative results would seem to exclude the light dark matter region which would be preferred by the DAMA signal. But there are more uncertainties at issue. One example is the measurement of the recoil energy. DAMA uses a NaI crystalline scintillator which converts the recoil energy of the nuclei into light, which is measured. But, a fraction of the recoil energy is transferred to phonons in the lattice and lost. So the measured energy must be scaled by a quenching factor to determine the actual nucleus recoil energy. But the DAMA experiment has recently noted the existence of the "channeling effect," a property of crystalline scintillators wherein a ion moving through certain channels in the lattice will lose no energy to phonons. In this case, scaling by the old quenching factor would overestimate the recoil energy, and thus the dark matter mass. The DAMA collaboration has adjusted its analysis to account for this effect, and their resulting preferred parameter region is not excluded by any other experiment.[11,12]

On the other hand, any uncertainty in the dark matter velocity distribution would alter the dark matter mass which should be inferred from any particular measured recoil energy distribution. If local streams of dark matter alter the velocity distribution from that expected in our local neighborhood, then this could also potentially shift the mass region preferred by the DAMA recoil energy signal.[9,10]

These effects are detailed in Fig. 1. The application of both these effects to the DAMA signal is somewhat controversial. It is not at all clear that a dark matter stream with sufficient velocity to shift the DAMA preferred region appreciably is reasonable from the point of view of astrophysics. And the effect of channeling on a scintillator like DAMA for the $1 - 10$ GeV energy range is currently being studied and cross-checked by various groups.

As if this controversy were not enough, three groups[13–15] have analyzed the spectrum of modulations within the recoil energy bins which DAMA reports. Naturally, they all reach different conclusions, ranging from ruling out the dark matter interpretation of DAMA to declaring it completely consistent with the spectrum of the DAMA signal.

It is far from clear what DAMA is actually seeing. What is clear, however, is that if DAMA is seeing dark matter, one preferred region of parameter space has dark matter mass in the range $m_X \sim 1 - 10$ GeV and spin-independent proton scattering cross section $\sigma_{\rm SI} \sim 10^{-5} - 10^{-2}$ pb. This is a mass region where several different experimental effects can push in different directions, and potentially create a window where dark matter could be observed at DAMA while not being ruled out by other experiments.

Moreover, there are a variety of theoretical models which attempt to explore this region of parameter space. Although neutralinos have been proposed as an explanation,[16] such low masses and high cross sections are not typical of weakly-interacting massive particles (WIMPs), and alternative candidates have been suggested to explain the DAMA signal.[17–24]

1.2. *Cross-checking DAMA*

The current state of affairs also makes it abundantly clear that complementary experiments are likely required to sort out the true nature of this result. Other direct detection experiments may play this role. In this paper,[25] we note that corroborating evidence may come from a very different source, namely, from the indirect detection of dark matter at Super-Kamiokande (Super-K). In contrast to direct detection experiments, which rapidly lose sensitivity at low masses, Super-K's limits remain strong for low masses. But in contrast to other indirect detection experiments, which can only be compared to DAMA after making astrophysical assumptions which are highly uncertain (such as the cuspiness of the dark matter density profile near the galactic center), Super-K offers a way of testing the DAMA result which is largely model independent. Super-K is therefore poised as one of the most promising experiments to either corroborate or exclude many dark matter interpretations of the DAMA/LIBRA data.

In Sec. 2, we show the relation between the DAMA and Super-K event rates. In Sec. 3, we show that there is significant potential for Super-K to extend its reach to dark matter masses from 5 to 20 GeV and provide sensitivity that is competitive with, or possibly much better than, direct detection experiments. In Sec. 4, we apply our analysis to two specific dark matter candidates that have been proposed to explain DAMA: neutralinos[16] and WIMPless dark matter.[18–20] We present our conclusions in Sec. 5.

2. Bounding σ_{SI} with Super-Kamiokande

Super-K can probe dark matter in the Sun or Earth's core annihilating to standard model (SM) particles, which subsequently emit neutrinos. Muon neutrinos then interact weakly at or near the detector to produce muons, which are detected at Super-K. The observed rate of upward-going muon events places an upper bound on the annihilation rate of dark matter in the Sun or the Earth's core. For low-mass dark matter, the dominant contribution to neutrino production via dark matter annihilation is from the Sun,[26] on which we focus.

The total annihilation rate is

$$\Gamma = \frac{1}{2} C \tanh^2[(aC)^{\frac{1}{2}} \tau] \,, \tag{1}$$

where C is the capture rate, $\tau \simeq 4.5$ Gyr is the age of the solar system, and $a = \langle \sigma_{ann.} v \rangle / (4\sqrt{2} V)$, with $\sigma_{ann.}$ the total dark matter annihilation cross section and V the effective volume of WIMPs in the Sun ($V = 5.7 \times 10^{30}\,\text{cm}^3 (1\,\text{GeV}/m_X)^{3/2}$).[26–28] If $\langle \sigma_{ann.} v \rangle \sim 10^{-26}$ cm^3 s^{-1} (to get the observed dark matter relic density), then for the range of parameters considered here, the Sun is in equilibrium[26,29,30] and $\Gamma \approx \frac{1}{2} C$. WIMP evaporation is not relevant if $m_X \gtrsim 4\,\text{GeV}$.[27–29]

The dark matter capture rate is[26]

$$C = \left[\left(\frac{8}{3\pi} \right)^{\frac{1}{2}} \sigma \frac{\rho_X}{m_X} \bar{v} \frac{M_B}{m} \right] \left[\frac{3}{2} \frac{\langle v^2 \rangle}{\bar{v}^2} \right] f_2 f_3 \,. \tag{2}$$

The first bracketed factor counts the rate of dark matter-nucleus interactions: σ is the dark matter-nucleus scattering cross section, ρ_X / m_X is the local dark matter number density, m is the mass of the nucleus, and M_B is the mass of the capturing object. The velocity dispersion of the dark matter is $\bar{v}$, and $\langle v^2 \rangle$ is the squared escape velocity averaged throughout the Sun. The second bracketed expression is the "focusing" factor that accounts for the likelihood that a scattering event will cause the dark matter particle to be captured. The parameters f_2 and f_3 are computable $\mathcal{O}(1)$ suppression factors that account for the motion of the Sun and the mismatch between X and nucleus masses, respectively. $f_3 \sim 1$ for solar capture.[26] The capture rate is thus a completely computable function of σ / m_X. Assuming $\rho_X = 0.3$ GeV cm^{-3}, $\bar{v} \sim 300 \frac{\text{km}}{\text{s}}$, $\frac{3}{2} \frac{\langle v^2 \rangle}{\bar{v}^2} \sim 20$,[26] and taking $f_2 \sim f_3 \sim 1$, one finds $C \sim 10^{29}$ (σ / m_X) GeV pb^{-1} s^{-1}.

The major remaining particle physics uncertainty is the neutrino spectrum that arises from dark matter annihilation. Assuming the dark matter annihilates only to SM particles, a conservative estimate for neutrino production may be obtained by assuming that the annihilation to SM particles is dominated by $b\bar{b}$ production for $m_b < m_X < M_W$, by $\tau\bar{\tau}$ production for $m_W < m_X < m_t$, and by W, Z production for $m_X > m_t$.[31]

Super-K bounds the ν_μ-flux from dark matter annihilation in the Sun. Since the total annihilation rate is equal to the capture rate, this permits Super-K to bound the dark matter-nucleon scattering cross section using Eq. (2). Fig. 1 shows the published bounds from Super-K, limits from other dark matter direct detection experiments and the regions of $(m_X, \sigma_{\rm SI})$ parameter space favored by the DAMA signal given astrophysical and detector uncertainties. As evident from Fig. 1, the published Super-K bounds (solid line) do not test the DAMA-favored regions. But we will see that consideration of the full Super-K event sample provides significant improvement and extends Super-K's sensitivity to low masses and the DAMA-favored regions.

3. Projection of Super-K Sensitivity

As shown in Fig. 1, Super-K currently reports dark matter bounds only down to $m_X = 18$ GeV. For heavier dark matter, it is estimated[32] that more than 90% of the upward-going muons will be through-going (i.e., will be created outside the inner detector and will pass all the way through it). However, one can study dark matter at lower masses by using stopping, partially contained, or fully contained muons, that is, upward-going muons that stop within the detector, begin within the detector, or both[a].

Our strategy for projecting dark matter bounds from these event topologies is as follows: we begin by conservatively assuming that the measured neutrino spectrum at low energies matches the predicted atmospheric background. In any given bin with N neutrino events, the 2σ bound on the number of neutrinos from dark matter annihilation is then $2\sqrt{N}$. This bounds the dark matter annihilation rate to neutrinos, which, for a conservative choice of the neutrino spectrum, implies a bound on the capture rate, and thus on σ/m_X. To include experimental acceptances and efficiencies, we scale our results to the Super-K published bound at $m_X = 18$ GeV, assuming these effects do not vary greatly in extrapolating down to the $5-10$ GeV range of interest.

The annihilation of dark matter particles X typically produces neutrinos with $E_\nu/m_X \sim \frac{1}{3} - \frac{1}{2}$. The muons produced by weak interactions of ν_μ lie in

[a]Testing the dark matter interpretation of DAMA has with Super-K through-going muon data has also been considered.[28]

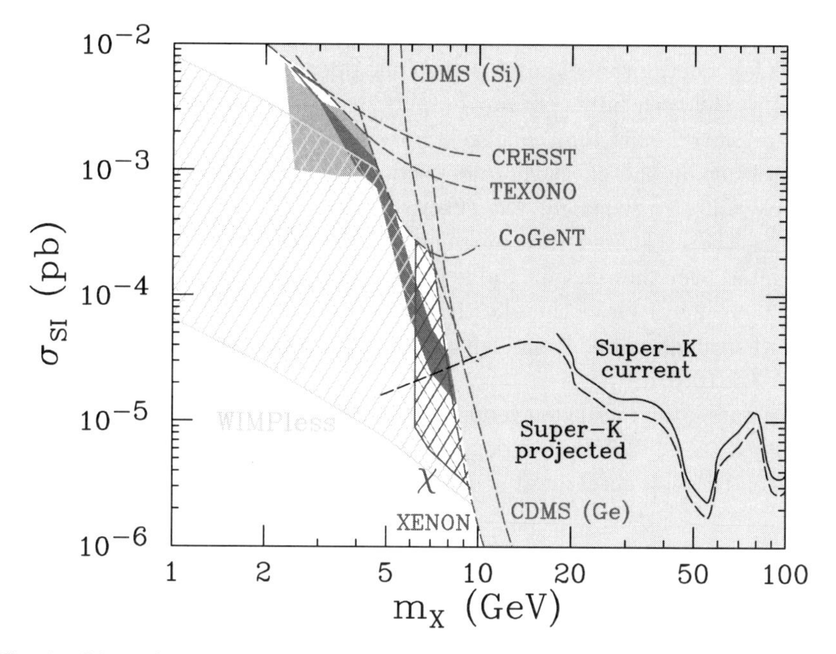

Fig. 1. Direct detection cross sections for spin-independent X-nucleon scattering as a function of dark matter mass m_X. The black solid line is the published Super-K exclusion limit,[32] and the black dashed line is our projection of future Super-K sensitivity. The magenta shaded region is DAMA-favored given channeling and no streams,[12] and the medium green shaded region is DAMA-favored at 3σ given streams but no channeling.[10] The light yellow shaded region is excluded by the direct detection experiments indicated. The dark blue cross-hatched region is the prediction for the neutralino models[16] and the light blue cross-hatched region is the parameter space of WIMPless models with connector quark mass $m_Y = 400$ GeV and $0.3 < \lambda_b < 1.0$. Other limits come from the Baksan and MACRO experiments,[32–34] though they are not as sensitive as Super-K.

a cone around the direction to the Sun with rms half-angle of approximately $\theta = 20^\circ \sqrt{10 \text{ GeV}/E_\nu}$.[35] Bounds on dark matter with $m_X = 18$ GeV were set using neutrinos with energies $E_\nu \sim 6 - 9$ GeV.[32] The event sample used consisted of 81 upward through-going muons within a 22° angle of the Sun collected from 1679 live days.

For masses $m_X \sim 5 - 10$ GeV, the ν_μ are typically produced with energies between $2 - 4$ GeV. At these energies the detected events are dominantly fully-contained events,[36] so we use this event topology with only events within the required cone around the Sun. The number of events is

$$N_{solar} = N \frac{1 - \cos\theta}{2} , \tag{3}$$

where N is the total number of fully-contained muon events expected in the $2-4\,\text{GeV}$ energy range and θ is the cone opening angle. Super-K expects $N_{solar} = 168$ such fully contained events per 1000 live days.[36]

We convert this limit on event rate to a limit on the neutrino flux by dividing by the effective cross section for the Super-K experiment in the relevant energy range. The effective cross-section can be estimated by dividing the estimated rate of events by the predicted atmospheric flux, integrated over the relevant range of energies.[36] For fully contained events with $E_\nu \sim 2-4$ GeV, the effective cross section is $\sim 2.1 \times 10^{-8}$ m^2. For upward through-going events with $E_\nu \sim 8$ GeV, the effective cross-section is $\sim 1.7 \times 10^{-8}$ m^2.

Assuming the neutrino events are detected primarily in either the fully-contained ($2-4$ GeV) or through-going sample (~ 8 GeV), one can set 2σ limits on the time-integrated neutrino flux due to dark matter annihilation:

$$\begin{aligned} \Phi_{\rm FC}^{\rm max} &= \frac{2\sqrt{N_{\rm FC}}}{2.1\times 10^{-8}\ {\rm m}^2} \sim 1.6\times 10^9\ {\rm m}^{-2}\sqrt{\frac{N_{\rm days}}{1679}} \\ \Phi_{\rm TG}^{\rm max} &= \frac{2\sqrt{N_{\rm TG}}}{1.7\times 10^{-8}\ {\rm m}^2} \sim 1.0\times 10^9\ {\rm m}^{-2}\sqrt{\frac{N_{\rm days}}{1679}}\,, \end{aligned} \tag{4}$$

where $N_{\rm FC} = 168\,(N_{\rm days}/1000)$ and $N_{\rm TG} = 81\,(N_{\rm days}/1679)$ are the number of fully-contained and through-going events within the angle and energy ranges, respectively, scaled to $N_{\rm days}$ live days.

The ratio of these flux limits obtained from the fully-contained and through-going samples are then equal to the ratio of σ/m_X in the $5-10$ GeV regime to the same quantity at 18 GeV. We find

$$\frac{1.6\times 10^9\ {\rm m}^{-2}}{1.0\times 10^9\ {\rm m}^{-2}} \sim \left(\frac{\sigma_{5-10}}{m_{5-10}}\right)\left(\frac{\sigma_{18}}{18\ {\rm GeV}}\right)^{-1}\,, \tag{5}$$

where σ_{5-10} is the Super-K bound on the dark matter nucleon cross-section for a dark matter particle with mass in the range $5-10$ GeV, and σ_{18} is the bound for a dark matter particle with mass 18 GeV. In Fig. 1 this projected Super-K bound is plotted, assuming 3000 live days of the SK I-III run. This bound gets better at lower energies and may beat other direct detection experiments.

4. Prospects for Various Dark Matter Candidates

We now consider specific examples of theoretical models that have been proposed to explain the DAMA result. We first consider neutralino dark matter. Although neutralinos typically have larger masses and lower cross sections than required to explain the DAMA signal, special choices of supersymmetry parameters may yield values in the DAMA-favored region.[16]

The region of the $(m_X, \sigma_{\rm SI})$ plane spanned by these models[16] that do not violate known constraints is given in Fig. 1. We see that if Super-K's limits can be extended to lower mass, it could find evidence for models in this class. Note, however, that many of these models have $\rho < 0.3$ GeV cm^{-3}, and for these models Super-K's bound on the cross section will be less sensitive.

WIMPless dark matter provides an alternative explanation of the DAMA/LIBRA signal.[19] These candidates are hidden sector particles that naturally have the correct relic density.[18] In these models, the dark matter particle X couples to SM quarks via exchange of a particle Y that is similar to a 4th generation quark. The Lagrangian for this interaction is

$$\mathcal{L} = \lambda_f X \bar{Y}_L f_L + \lambda_f X \bar{Y}_R f_R \ . \tag{6}$$

The Yukawa couplings λ_f are model-dependent, and it is assumed that only the coupling to 3rd generation quarks is significant, while the others are Cabbibo-suppressed.[b] One finds that the dominant nuclear coupling of WIMPless dark matter is to gluons via a loop of b-quarks (t-quark loops are suppressed by m_t). The X-nucleus cross section is given by[19]

$$\sigma_{\rm SI} = \frac{1}{4\pi} \frac{m_N^2}{(m_N + m_X)^2} \left[\sum_q \frac{\lambda_b^2}{m_Y - m_X} \left[Z B_b^p + (A - Z) B_b^n \right] \right]^2 , \tag{7}$$

where Z and A are the atomic number and mass of the target nucleus N, and $B_b^{p,n} = (2/27) m_p f_g^{p,n} / m_b$, where $f_g^{p,n} \simeq 0.8$.[37,38]

In Fig. 1, we plot the parameter space for WIMPless models with $m_Y = 400$ GeV and $0.3 < \lambda_b < 1.0$. These models span a large range in the $(m_X, \sigma_{\rm SI})$ plane, and overlap much of the DAMA-favored region. We see that Super-K's projected sensitivity may be sufficient to discover a signal that corroborates DAMA's. But WIMPless models illustrate an important caveat to the analysis above; if there are hidden decay channels, then the annihilation rate to SM particles is only a fraction of Γ_{tot}, and Super-K's sensitivity is reduced accordingly.

---

[b] This is a reasonable assumption and is consistent with small observed flavor-changing neutral currents.

5. Summary

The DAMA/LIBRA signal has focussed attention on the possibility of light dark matter, and alternative methods for corroborating or excluding a dark matter interpretation are desired. We have shown that Super-K, through its search for dark matter annihilation to neutrinos, has promising prospects for testing DAMA at low mass.

Using fully contained muon events, we expect that current super-K bounds may be extended down to $M_{DM} \sim 5 - 10$ GeV, and can test light dark models (such as neutralino models[16] and WIMPless models[19]). We have the intriguing prospect that the DAMA/LIBRA signal could be sharply tested by an indirect detection experiment in the near future.

Acknowledgments

We are grateful to Hank Sobel, Huitzu Tu and Hai-Bo Yu for discussions, and especially to Jonathan Feng, John Learned and Louis Strigari. This work was supported by NSF grants PHY–0239817,0314712,0551164 and 0653656, DOE grant DE-FG02-04ER41291, and the Alfred P. Sloan Foundation. JK is grateful to CERN and the organizers of Strings '08, where part of this work was done, and to the organizers of Dark 09 for their hospitality.

References

1. R. Bernabei *et al.*, [DAMA Collaboration], arXiv:0804.2741 [astro-ph]; Riv. Nuovo Cim. **26N1**, 1 (2003); Int. J. Mod. Phys. D **13**, 2127 (2004).
2. A. K. Drukier, K. Freese and D. N. Spergel, Phys. Rev. D **33**, 3495 (1986); K. Freese, J. A. Frieman and A. Gould, Phys. Rev. D **37**, 3388 (1988).
3. G. Angloher *et al.*, Astropart. Phys. **18**, 43 (2002).
4. D. S. Akerib *et al.*, [CDMS Collaboration], Phys. Rev. Lett. **96**, 011302 (2006); Z. Ahmed *et al.*, [CDMS Collaboration], arXiv:0802.3530 [astro-ph].
5. J. Angle *et al.*, [XENON Collaboration], Phys. Rev. Lett. **100**,021303(2008).
6. S. T. Lin *et al.*, [TEXONO Collaboration], arXiv:0712.1645 [hep-ex].
7. F. T. Avignone, P. S. Barbeau and J. I. Collar, arXiv:0806.1341 [hep-ex].
8. C. E. Aalseth *et al.*, arXiv:0807.0879 [astro-ph].
9. See, *e.g.*, M. Brhlik and L. Roszkowski, Phys. Lett. B **464**, 303 (1999); P. Belli, R. Bernabei, A. Bottino, F. Donato, N. Fornengo, D. Prosperi and S. Scopel, Phys. Rev. D **61**, 023512 (2000).
10. P. Gondolo and G. Gelmini, Phys. Rev. D **71**, 123520 (2005).
11. R. Bernabei *et al.*, Eur. Phys. J. C **53**, 205 (2008).
12. F. Petriello and K. M. Zurek, arXiv:0806.3989 [hep-ph].
13. S. Chang, A. Pierce and N. Weiner, arXiv:0808.0196 [hep-ph].
14. M. Fairbairn and T. Schwetz, arXiv:0808.0704 [hep-ph].
15. C. Savage, G. Gelmini, P. Gondolo and K. Freese, arXiv:0808.3607 [astro-ph].

16. A. Bottino, F. Donato, N. Fornengo and S. Scopel, Phys. Rev. D **68**, 043506 (2003); Phys. Rev. D **77**, 015002 (2008); arXiv:0806.4099 [hep-ph].
17. D. Tucker-Smith and N. Weiner, Phys. Rev. D **64**, 043502 (2001); Phys. Rev. D **72**, 063509 (2005).
18. J. L. Feng and J. Kumar, Phys. Rev. Lett. **101**, 231301 (2008).
19. J. L. Feng, J. Kumar and L. E. Strigari, Phys. Lett. B **670**, 37 (2008).
20. J. L. Feng, H. Tu and H. B. Yu, JCAP **0810**, 043 (2008).
21. R. Foot, arXiv:0804.4518 [hep-ph].
22. M. Y. Khlopov and C. Kouvaris, arXiv:0806.1191 [astro-ph].
23. S. Andreas, T. Hambye and M. H. G. Tytgat, arXiv:0808.0255 [hep-ph].
24. E. Dudas, S. Lavignac and J. Parmentier, arXiv:0808.0562 [hep-ph].
25. J. L. Feng, J. Kumar, J. Learned and L. E. Strigari, arXiv:0808.4151 [hep-ph].
26. A. Gould, Astrophys. J. **321**, 571 (1987).
27. A. Gould, Astrophys. J. **321**, 560 (1987).
28. D. Hooper, F. Petriello, K. M. Zurek and M. Kamionkowski, arXiv:0808.2464 [hep-ph].
29. K. Griest and D. Seckel, Nucl. Phys. B **283**, 681 (1987) [Erratum-ibid. B **296**, 1034 (1988)].
30. M. Kamionkowski, K. Griest, G. Jungman and B. Sadoulet, Phys. Rev. Lett. **74**, 5174 (1995).
31. G. Jungman and M. Kamionkowski, Phys. Rev. D **51**, 328 (1995).
32. S. Desai *et al.* [Super-Kamiokande Collaboration], Phys. Rev. D **70**, 083523 (2004) [Erratum-ibid. D **70**, 109901 (2004)].
33. T. Montaruli [MACRO Collaboration], arXiv:hep-ex/9905020.
34. C. de los Heros, et al., arXiv:astro-ph/0701333.
35. G. Jungman, M. Kamionkowski and K. Griest, Phys. Rept. **267**,195(1996).
36. Y. Ashie *et al.* [Super-Kamiokande Collaboration], Phys. Rev. D **71**, 112005 (2005).
37. H. Y. Cheng, Phys. Lett. B **219**, 347 (1989).
38. J. R. Ellis, J. L. Feng, A. Ferstl, K. T. Matchev and K. A. Olive, Eur. Phys. J. C **24**, 311 (2002).

A LIGHT SCALAR WIMP, THE HIGGS PORTAL AND DAMA

MICHEL H.G. TYTGAT

Service de Physique Théorique
Université Libre de Bruxelles
Boulevard du Triomphe, CP225, Brussels 1050, Belgium

In these proceedings, we report on the possible signatures of a light scalar WIMP, a dark matter candidate with $M_{DM} \sim$ few GeV and which is supposed to interact with the Standard Model particles through the Higgs. Its existence may be related to the annual modulation observed by DAMA/LIBRA.

1. Introduction

There are many models of dark matter but the most acclaimed is the neutralino. However the most economical extension of the Standard Model with dark matter consists simply in adding a real singlet scalar field S,

$$\mathcal{L} \ni \frac{1}{2}\partial^\mu S \partial_\mu S - \frac{1}{2}\mu_S^2 S^2 - \frac{\lambda_S}{4} S^4 - \lambda_L H^\dagger H S^2 \tag{1}$$

where $H = (h^+ \,(h+iG_0)/\sqrt{2})^T$ is the Higgs doublet. This Lagrangian has a discrete Z_2 symmetry, $S \to -S$ and, if this symmetry is not spontaneously broken, the S particle is a dark matter candidate,[1–3] with mass

$$m_S^2 = \mu_S^2 + \lambda_L \mathrm{v}^2, \tag{2}$$

where $\mathrm{v} = 246$ GeV. This is also one of the simplest instance of dark matter through the Higgs portal, a general scheme according to which a hidden sector interact with ordinary matter through the Higgs sector of the Standard Model.[4] Also the singlet scalar extension effectively encompasses many other models with extra fields, for instance the so-called Inert Doublet Model (IDM).[5–8] The IDM is a model with two Higgs field, one odd under a discrete symmetry which is introduced to prevent FCNC. The extra Higgs has also couplings to electroweak gauge bosons but decoupling of the extra scalars states (one neutral and one charged in the present case) one has just to make sure not to violate LEPI bounds on isospin breaking. This is turns out to be natural in the IDM, thanks to a hidden custodial $SU(2)$ symmetry.[9]

The phenomenology of the theory (1) has been much discussed in the litterature, but the focus has somewhat been on heavy or moderately heavy dark matter candidates, say in the 50 GeV to a few TeV range. In the present proceeding we consider a lighter candidate, with $m_S \lesssim 10$ GeV. This possibility has received little attention, but is nevertheless both viable and phenomenologically very interesting.

One of our motivation is the DAMA/NaI and DAMA/Libra experiments, which have observed an annual modulation in the rate of single scattering events, with 8.2 σ significance (combined).[10] This signature is supposed to be one of the landmark of dark matter-nucleon interactions, the modulation being due of the combined motion of the Sun and of the Earth with respect to the halo of dark matter of the Galaxy (generically considered to be non-rotating).[11] The DAMA/NaI detector, and its successor, DAMA/Libra, consist of sodium iodide (NaI) crystals and use scintillation to measure the nuclei recoil energy. While the all the other dark matter experiments work hard on eliminating the possible background, the strategy of DAMA is essentially to exploit the possible annual modulation. Most of the background is supposedly eliminated by focusing on single hit events, but of course contamination by mundane radioactivity is still expected to exist. The DAMA data are impressive and no explanation but dark matter really exists. Nevertheless the interpretation in terms of elastic collisions of nuclei in the detector with dark matter from the halo is challenged by the null results of various other direct detection experiments, at least those that are probing similar dark matter mass and cross section ranges.

Elastic scattering has been addressed in various works, possibly taking into account the possible uncertainties regarding the properties of the dark matter halo, or of the interaction of dark matter with nuclei.[12–15] All these works assert that the interpretation of the DAMA results in term of the elastic scattering of dark matter is inconsistent with the null results of other direct detection experiments (CDMS,[16,17] XENON10,[18] TEXONO,[19] CRESST,[20] COUPP[21] and CoGenT[22]), at least if all events the dama events are 'quenched'. Quenching relates to the fact that, after a collision with dark matter, a recoiling nuclei may loose energy either through electrons (which are responsible for scintillation) or through collisions with other nuclei (heat, phonons). The energy measured (which is expressed in electron equivalent keV, or keVee) is thus typically smaller than the true nuclei recoil energy, $E_{ee} = QE_{\rm recoil}$, where Q is the so-called quenching factor, $Q \lesssim 1$. In a crystal however, like in NaI, some of the recoiling nuclei events may occur along the axis of the crystal, in which case collisions with nuclei are ineffective and $Q \approx 1$. This effect is called channelling.[23] If channelling is taken into account, it may be possible to explain the DAMA results (at 3σ) with a light $m_{DM} \lesssim 7-8$ GeV candidate, with a cross section (normalized to a nucleon) $\sigma_n \approx 10^{-5}$ pb.[12,15] One should emphasize that the fit to data is not very good since the chi-square is minimum for values which are excluded by the other experiments. Furthermore this interpretation implies that the background is small in the relevant region of recoil energies. This is embarrassing but, yet, the possibility that many models, including supersymmetric ones, may actually explain the DAMA data is not excluded.[24–30] Although we do not address the possibility here, we should also mention the very interesting possibility of explaining the DAMA data (with a substantially improved fit) in terms of the inelastic scattering of dark matter particles.[31–34] This requires some finely tuned models, but beautifully fits the data.

In the sequel, we focus on elastic scattering and report on the phenomenology of a singlet scalar (including the IDM).[28] For the sake of reference, we refer to the range quoted in the work of Pietrello and Zurek,[12] corresponding to

$$3 \times 10^{-5}\text{pb} \lesssim \sigma_n^{SI} \lesssim 5 \times 10^{-3}\text{pb},$$

and

$$3\text{GeV} \lesssim m_{DM} \lesssim 8\text{GeV}.$$

This range, which is based on the two bins version of the DAMA data for the spectrum of modulated events,[12] is clearly larger than the one based on the full set of data.[15] If anything, like if channelling turns out to be inoperant, this range is representative of a class of models which are not yet excluded by any experiments.

2. Direct Detection and Relic Abudance

For the model (1), the only processes relevant for direct detection and to fix the relic abundance (we consider the standard thermal freeze-out) are those of Fig.1. This is also true for the IDM provided $m_{DM} \ll m_{Higgs}, m_{Z,W}$. Since these processes have the same dependence in the coupling λ_L and the Higgs mass, both processes are closely related.[2,28] Concretely, if we fix the cross section assuming a cosmic abundance given by cosmological observations (and thermal freeze-out), then the direct detection cross section

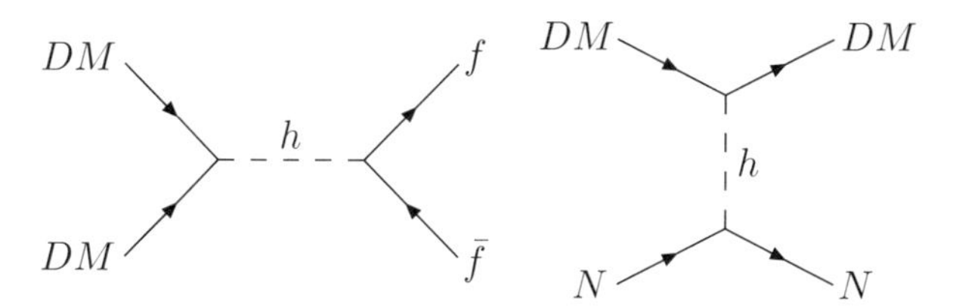

Fig. 1. Higgs exchange diagrams for the DM annihilation (a) and scattering with a nucleon (b).

is fixed, modulo the uncertainty in the coupling of the Higgs to nucleons, parameterised by f, with $fm_N \equiv \langle N|\sum_q m_q \bar{q}q|N\rangle = g_{hNN}$v. Here we take $f = 0.30$ as central value, and vary it within the rather wide range $0.14 < f < 0.66$.[28] As Fig.2 shows, compatibility between DAMA (regions in red) and WMAP (black region) is possible. Conversely, one may say that the red regions are not excluded by existing dark matter detection experiments. Notice that the coupling $|\lambda_L|$ tened to be large, but is still perturbative. Incidentally, from (2) we see that $\mu_2 \sim v$ so that a light m_{DM} potentially poses a (small) hierarchy problem. This may perhaps explain why this simple model is systematically overlooked in the current litterature even though it has many interesting consequences.

3. Indirect Detection

The dark matter candidates considered here have both large cross sections and a large abundance compared to more mundane, heavier dark matter candidates and their annihilations in the Galaxy may lead to some interesting signals.

In Fig. 3 we show the flux from the annihilation of the dark matter candidate into photons at the centre of the Galaxy. The mass of the DM candidate puts it in the energy range of EGRET data and of the Fermi/Glast satellite. Fig.3 we show the predicted flux of gamma rays from the galactic centre for a sample of scalar DM with parameters which are consistent both with DAMA and WMAP and we compare to EGRET data.[28,35] It is interesting that the predicted flux is of the order of magnitude of the observed flux at the lowest energies that have been probed by EGRET. A tentative conclusion is that observations by Fermi/GLAST might constrain the model, modulo the usual uncertainties regarding the profile of the dark matter at the centre of the Galaxy. Similar predictions have been reached in the framework of so-called WIMPless models (see also the talk by Jason Kumar at this conference).[27,30]

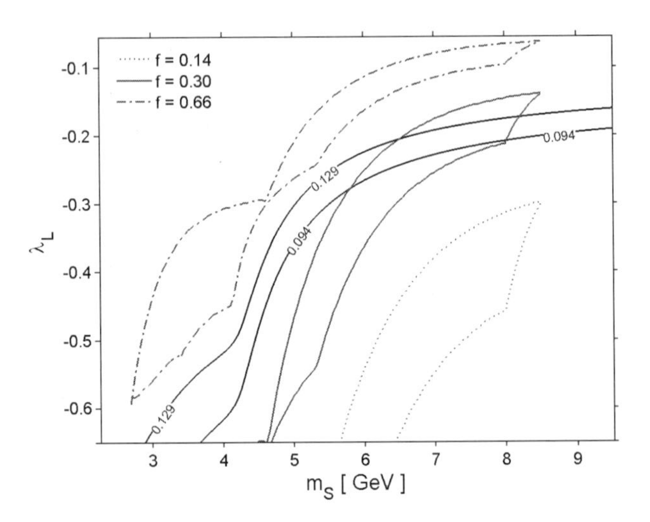

Fig. 2. For $m_h = 120$ GeV, values of m_S and λ_L consistent with WMAP $0.094 < \Omega_{DM}h^2 < 0.129$ (solid black lines), and which match the direct detection constraints (two bins fit to DAMA of Petriello and Zurek[12]).

Dark matter may also be captured in the core of the Sun, where its annihilation may be observed by neutrino detectors. In Fig. 4 we show the limit which may be expected from Super-Kamiokande.[13,36,37] The dominant source of neutrinos is annihilation, through the Higgs, into tau-antitau pairs (see also the talk by Kumar at this conference)

Finally we may consider the production and propagation of antiparticles coming from the annihilation of dark matter in the Galaxy, a possibility which may be constrained using the new data on the positron and antiproton fluxes in cosmic rays. The flux of positrons and antiprotons is quite large for the candidates considered here (for the same reasons given at the beginning of this section the flux of positrons (and for that matter, other cosmic ray components, including antiprotons). However the fluxes fall in an energy range where solar modulation severely limits the possibilities to constrain the model. In Fig.4 we show the predictions of the model for the positron fraction, and Fig.5 shows the $\bar{p}/p$ ratio.[38] The signal in positrons is rather weak, unless there is a boost factor (say BF=10). The signature into antiproton is however quite significant and, for instance, a boost factor of order 10 is clearly excluded, a conclusion which is probably robust, even without knowing precisely the impact of solar modulation. A better understanding of solar modulation is nevertheless desirable and

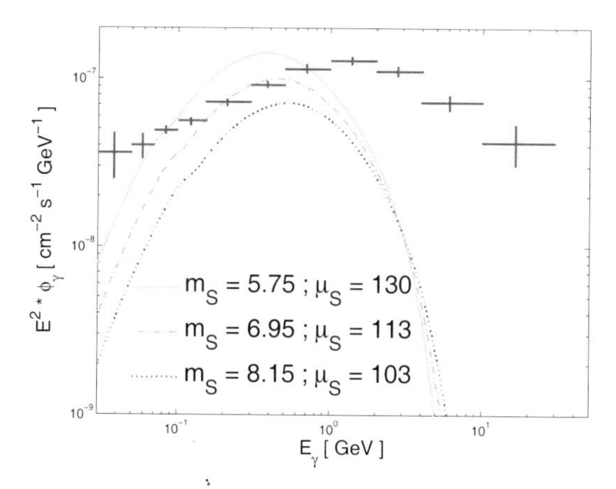

Fig. 3. Flux of gamma rays from the galactic center from the annihilation of a scalar DM consistent with DAMA, compared with EGRET data ($m_h = 120$ GeV and using a NFW profile).

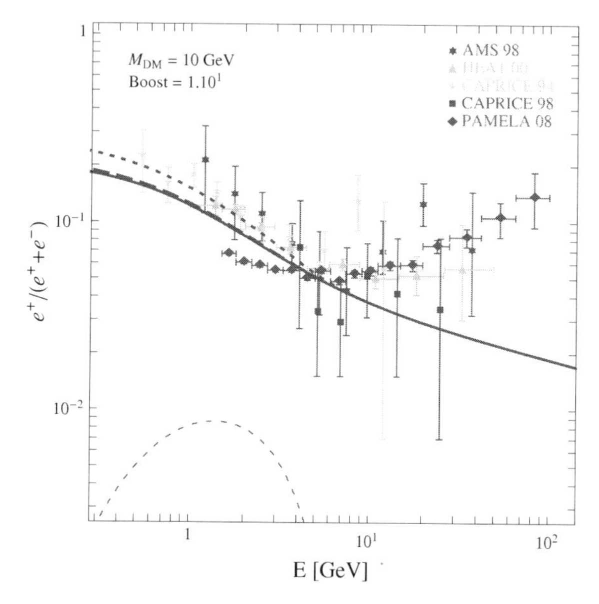

Fig. 4. Positron fraction compared to Pamela and other relevant data. The signal is weak and within the region affected by solar modulation. We boost the signal by a factor of 10 (short dash blue line).

perhaps more severe constraints could be obtained in this way. Interestingly the model also predicts a substantial production of antideuteron.[38] The

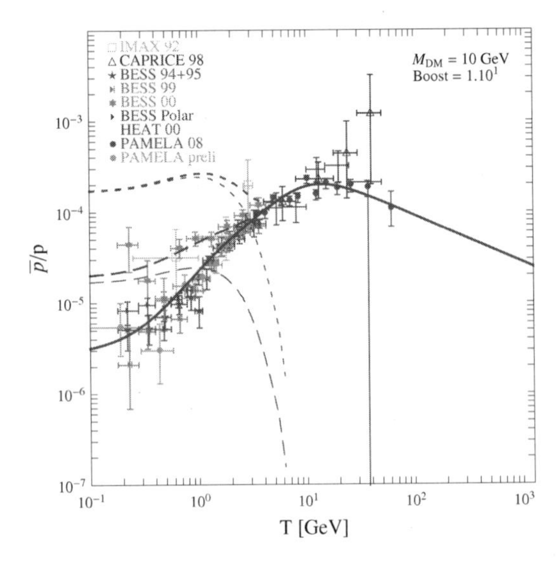

Fig. 5. Flux of antiprotons to protons for a $m_{DM} = 10$ GeV scalar singlet. We show the contribution for no boost factor (long dashed) and for a boost factor of 10 (short dashed). The latter is clearly excluded.

production of anti-deuteron by spallation in cosmic rays is typically small and is predicted to fall for kinetic energies below 1 GeV per nucleon. Given its large abundance, a light WIMP may give a substantial contribution to the flux of anti-deuteron at low energies.[25,39] For the IDM candidate with MDM = 10 GeV, we obtain, using the DarkSUSY routines, an anti-deuteron flux at $T_{\bar{D}} = 0.25$ GeV/n of $9 \cdot 10^{-7}$ (GeV/n s sr m^2)) (for BF = 1), which is below the upper limit of $1.9 \cdot 10^{-4}$ (GeV/n s sr m^2) set by the BESS experiment, but above the expected acceptance of the future AMS-02 and GAPS experiments, which are $4.5 \cdot 10^7$ (GeV/n s sr m^2) and $1.5 \cdot 10^7$ (GeV/n s sr m^2) respectively. Thus anti-deuteron data might turn out to give the strongest constraint on the light WIMP dark matter candidate considered here.

4. A Light Scalar At The LHC

In the present model, the coupling between the Higgs and the dark matter particle is large. This leads to a large Higgs boson decay rate to a scalar DM pairs at the LHC.[2,28] For example, for $m_S = 7$ GeV and $\lambda_L = -0.2$ and for a Higgs of mass 120 GeV we get the branching ratio $BR(h \rightarrow SS) = 99.5\%$, while for $m_h = 200$ GeV and $\lambda_L = -0.55$ we get $BR(h \to SS) = 70\%$. This reduces the visible branching ratio accordingly, rendering the Higgs boson basically invisible at LHC for $m_h = 120$ GeV, except possibly for many years of high luminosity data taking. Such a dominance of the invisible DM channel is a clear prediction of the framework, although it poses a challenge to experimentalists.[40]

5. Conclusion

A light, $m \sim$ few GeV, singlet scalar dark matter candidate interacting through the Higgs is perhaps not very motivated theoretically speaking but it may have a very interesting phenomenology. If its relic abundance is fixed by thermal freeze-out, all its cross sections are also essentially fixed. Its elastic scattering with nucleons is in a range consistent with the modulation observed by DAMA and its annihilations into various by-products are within reach of current gamma ray, neutrinos and cosmic ray detectors. A rather clear cut prediction is the production of a rather large flux of antideuteron in cosmic rays. Also, it predicts that a light Higgs is essentially invisible at the LHC.

Acknowledgments

The results reported here are based on work done in collaboration with Sarah Andreas, Thomas Hambye, Emmanuel Nezri, Quentin Swillens and Gilles Vertongen. The work of the author is supported by the FNRS and the Belgian Federal Science Policy (IAP VI/11).

References

1. J. McDonald, *Phys. Rev.* **D50**, 3637 (1994).
2. C. P. Burgess, M. Pospelov and T. ter Veldhuis, *Nucl. Phys.* **B619**, 709 (2001).
3. V. Barger, P. Langacker, M. McCaskey, M. J. Ramsey-Musolf and G. Shaughnessy, *Phys. Rev.* **D77**, p. 035005 (2008).
4. B. Patt and F. Wilczek (2006).
5. E. Ma, *Phys. Rev.* **D73**, p. 077301 (2006).
6. R. Barbieri, L. J. Hall and V. S. Rychkov, *Phys. Rev.* **D74**, p. 015007 (2006).
7. L. Lopez Honorez, E. Nezri, J. F. Oliver and M. H. G. Tytgat, *JCAP* **0702**, p. 028 (2007).
8. W. H. Press and D. N. Spergel, *Astrophys. J.* **296**, 679 (1985).
9. T. Hambye and M. H. G. Tytgat, *Phys. Lett.* **B659**, 651 (2008).
10. R. Bernabei *et al.*, *Eur. Phys. J.* **C56**, 333 (2008).

11. K. Freese, J. A. Frieman and A. Gould, *Phys. Rev.* **D37**, p. 3388 (1988).
12. F. Petriello and K. M. Zurek, *JHEP* **09**, p. 047 (2008).
13. C. Savage, G. Gelmini, P. Gondolo and K. Freese (2008).
14. M. Fairbairn and T. Schwetz, *JCAP* **0901**, p. 037 (2009).
15. C. Savage, K. Freese, P. Gondolo and D. Spolyar (2009).
16. D. S. Akerib *et al.*, *Phys. Rev. Lett.* **96**, p. 011302 (2006).
17. Z. Ahmed *et al.*, *Phys. Rev. Lett.* **102**, p. 011301 (2009).
18. J. Angle *et al.*, *Phys. Rev. Lett.* **100**, p. 021303 (2008).
19. S. T. Lin *et al.*, *Phys. Rev.* **D79**, p. 061101 (2009).
20. G. Angloher *et al.*, *Astropart. Phys.* **18**, 43 (2002).
21. E. Behnke *et al.*, *Science* **319**, 933 (2008).
22. C. E. Aalseth *et al.*, *Phys. Rev. Lett.* **101**, p. 251301 (2008).
23. R. Bernabei *et al.*, *Eur. Phys. J.* **C53**, 205 (2008).
24. R. Foot, *Phys. Rev.* **D78**, p. 043529 (2008).
25. A. Bottino, F. Donato, N. Fornengo and S. Scopel, *Phys. Rev.* **D78**, p. 083520 (2008).
26. E. Dudas, S. Lavignac and J. Parmentier, *Nucl. Phys.* **B808**, 237 (2009).
27. J. L. Feng, J. Kumar and L. E. Strigari, *Phys. Lett.* **B670**, 37 (2008).
28. S. Andreas, T. Hambye and M. H. G. Tytgat, *JCAP* **0810**, p. 034 (2008).
29. Y. G. Kim and S. Shin (2009).
30. J. Kumar (2009).
31. S. Chang, G. D. Kribs, D. Tucker-Smith and N. Weiner, *Phys. Rev.* **D79**, p. 043513 (2009).
32. D. Tucker-Smith and N. Weiner, *Phys. Rev.* **D64**, p. 043502 (2001).
33. Y. Cui, D. E. Morrissey, D. Poland and L. Randall, *JHEP* **05**, p. 076 (2009).
34. J. March-Russell, C. McCabe and M. McCullough, *JHEP* **05**, p. 071 (2009).
35. S. D. Hunter *et al.*, *Astrophys. J.* **481**, 205 (1997).
36. S. Andreas, M. H. G. Tytgat and Q. Swillens, *JCAP* **0904**, p. 004 (2009).
37. J. L. Feng, J. Kumar, J. Learned and L. E. Strigari (2008).
38. E. Nezri, M. H. G. Tytgat and G. Vertongen, *JCAP* **0904**, p. 014 (2009).
39. F. Donato, N. Fornengo and P. Salati, *Phys. Rev.* **D62**, p. 043003 (2000).
40. O. J. P. Eboli and D. Zeppenfeld, *Phys. Lett.* **B495**, 147 (2000).

PART II

SUSY/SUGRA Phenomenology in Dark Matter, LHC and Perspectives, Extra Dimensions and Dark Matter/Dark Energy

SUSY SEARCHES AT LHC AND DARK MATTER

E. BARBERIO, for the ATLAS collaboration
The University of Melbourne,
Parkville, Victoria, 3010, Australia
E-mail: barberio@unimelb.edu.au

Supersymmetric models with R-parity conservation provide an excellent candidate for Dark Matter, the Lightest Supersymmetric Particle, which will be searched for with the ATLAS detector at the Large Hadron Collider (LHC). Based on recent simulation studies, we present the discovery potential for Supersymmetry (SUSY) with the first few fb^{-1} of ATLAS data, as well as studies of the techniques used to reconstruct decays of SUSY particles at the LHC. We further discuss how such measurements can be used to constrain the underlying Supersymmetric model and hence to extract information about the nature of Dark Matter.

1. Introduction

Several astronomical observations have hinted at the existence of non-baryonic matter in the Universe, the so-called *dark matter*. These observations tell us that dark matter constitutes about 90% of the matter density of the universe, where baryonic matter contributes only around 10 %. Phenomena implying the presence of dark matter include the rotational speeds of galaxies and gravitational lensing of background objects by galaxy clusters and the behaviour of the Bullet cluster. The latter provides strong evidence that dark matter must be a Weakly Interacting Massive Particle (WIMP). Furthermore, precision measurements of the power spectrum fluctuations in the cosmic microwave background from WMAP,[1] strongly disfavour warm dark matter.

Despite these advancements, there are a lot of questions we still need to answer. Do fundamental particles comprise the bulk of the dark matter? If so, is there a symmetry from which these particles originate? How and when were these particles produced? There are many experiments that are attempting to answer these questions.

Astrophysical experiments attempt to detect dark matter by searching for dark matter annihilation processes in the galaxy using land-based gamma ray telescopes or space-based satellites. Particle physics experiments aim to create and study dark matter in the laboratory. One of the latter is the Large Hadron Collider (LHC) at CERN, in Geneva, which will start taking data towards the end of 2009. The purpose of this proceeding is to describe how the LHC can help to understand the nature of dark matter.

The Standard Model (SM) of particle physics, while being a very successful description of the observed particles and their interactions, does not contain any particle that can explain dark matter. However, there are many extensions of the SM that can.

One popular extension of the SM is Supersymmetry (SUSY), in which a new symmetry between bosons and fermions is introduced. This symmetry leads to new particles, not yet observed. All existing particles of the SM would have partners called *sparticles*: each boson would have a fermonic partner and each fermion a bosonic one.

In SUSY there exists a new quantum number, R-parity, under which SM particles are even whilst SUSY particles are odd. In the following we consider the phenomenology of SUSY models where R-parity is conserved. This has two important phenomenological consequences. Firstly, sparticles can only be produced in pairs, and secondly, the lightest SUSY particle (LSP) is stable and escapes detection in high-energy physics detectors. Thus, the LSP is a natural WIMP candidate. Within SUSY there are several possibilities for WIMP candidates, depending on the SUSY model.

The most obvious feature of SUSY is that none of the superpartners have been discovered yet and hence SUSY must be a broken symmetry, with the masses of the superpartners being much larger than their SM counterparts. It is assumed that the spontaneous breaking of SUSY occurs in what is called the hidden sector. How this soft breaking is done defines the SUSY model and its phenomenology. We will focus on the assumption that supersymmetry exists in nature as the Minimal Supersymmetric Standard Model (MSSM). In this model, SUSY breaking is implemented by including explicit soft mass terms for the SUSY particles in the MSSM multiplets. These terms contain a vast number of free parameters that spoil the predictive power of this model. To reduce these, some specific assumptions for the SUSY breaking are adopted, giving models defined by a small number of parameters at the SUSY breaking scale. We will discuss here in detail the mSUGRA model, where SUSY breaking is mediated by gravitational interaction. In this model the LSP is the lightest neutralino.

Assuming the equality of various soft parameters at the pre-SUSY breaking energy scale (universality), the MSSM phase space can be described by five soft SUSY breaking free parameters at the unification scale:[2] the Higgs field mixing, μ, the universal scalar mass m_0, the universal gaugino mass $m_{1/2}$, the universal trilinear coupling A_0 and $\tan\beta = v_1/v_2$ the ratio between the vacuum expectation values of the two Higgs doublets. From these parameters, the mass spectrum of the superpartners, the cross-sections of their production as well as the branching ratios of their decays can be calculated. In these proceedings, we will use the MSSM as an example to show how measurements at the LHC can be used to calculate the relic density. Even within this reduced set of parameters, different regions of the MSSM parameter space have different LSP's, with possible options including the gluino, sneutrino, gravitino and the lightest neutralino. The last particle in this list is an admixture of the superpartners of the neutral SM gauge bosons, and is the subject of the majority of studies.

Supersymmetric extensions of the Standard Model, with scales around 1 TeV, give a rich spectrum of SUSY particles in the mass range to be explored by the LHC. At LHC energies mostly pairs of squarks or gluinos will be produced, which then subsequently decay via cascades into the LSP. Typical event topologies at the LHC are multi jet events with zero or more leptons and missing transverse energy due to the LSPs. This SUSY signal is relatively easy to find, with relatively small SM backgrounds. The main problem is to disentangle the underlying model using the observations.

2. Sparticle Production at the LHC

At the LHC, squarks and gluinos are produced via strong processes, hence their production will have a large cross section. For example, with 100 pb^{-1} of data, we expect about 100 events with squarks of 1 TeV mass. Direct production of charginos, neutralinos and sleptons occur via electroweak processes, hence the production cross sections are much smaller. They are produced much more abundantly in squark and gluino decays.

The strongly interacting sparticles (squarks, gluinos) which dominate the production are much heavier than the weakly interacting and SM particles, giving long decay chains to the LSP and large mass differences between SUSY states. Consequently, searches for Supersymmetry at the LHC concentrates on cascade decays, which will produce spectacular events with many jets, leptons and a lot of missing transverse energy, making it relatively easy to extract a SUSY signal from the SM background.

If the LHC discovers signatures that look like R-parity conserving SUSY signatures, the procedure to study their relation to Dark Matter is the following:

- 1st Step: Look for deviations from the Standard Model, for example, in the multi-jet plus E_T^{miss} signature.
- 2st Step: Is it SUSY? If so, establish the SUSY mass scale using inclusive variables, e.g. effective mass distributions defined as:[3] $M_{eff} = \sum_{i=1}^{n} p_T^{jet,i} + \sum_{i=1}^{m} p_T^{lep,i} + E_T^{miss}$ where n and m are the number of jets and lepton in the events.

  Relevance to Dark Matter:

  - Inclusive studies: Verify if the discovered signal provides a possible Dark Matter candidate.
  - Exclusive studies: Model-independent calculation of LSP mass and compare with direct searches.
- 3rd Step: Which SUSY flavour is it? Determine model parameters, selecting particular decay chains and use kinematics to determine mass combinations.

  Relevance to Dark Matter:

  - Model-dependent calculation of relic density.

In ATLAS the detailed SUSY analysis a set of benchmark points in the mSUGRA framework were chosen, with the aim of exploring sensitivity to a wide class of of final-state signatures. In particular, the predicted cosmological relic density of neutralinos was chosen to be consistent with WMAP measurements.[1] The points chosen are described in Reference 3.

2.1. *Inclusive Searches*

The chosen benchmark points provide a wide range of possible decay topologies. However, they share some common features, for example for all these points the gluino mass is less than 1 TeV and decay products of strongly interacting SUSY particles will contain two LSP's and a number of SM particles, in particular highly energetic quarks and gluons. Typical SUSY signatures are therefore based on large missing transverse energy and hard jets. Detailed studies have been carried out for a wide range of channels, including 0-, 1-, multi-lepton modes, as well as signatures involving τ-and b-jets. Selection criteria are discussed in Reference 3. After applying the selection criteria, SUSY production is evident in the distribution of the effective mass, defined as the scalar sum of E^T_{miss} and the transverse momentum of each of the requested particles. Figure 1 shows the effective mass distribution for a set of benchmark scenarios and for the SM background in the 0-lepton and 1-jet channels. Channels with leptons will have smaller signal, but better signal to background conditions, providing a more robust

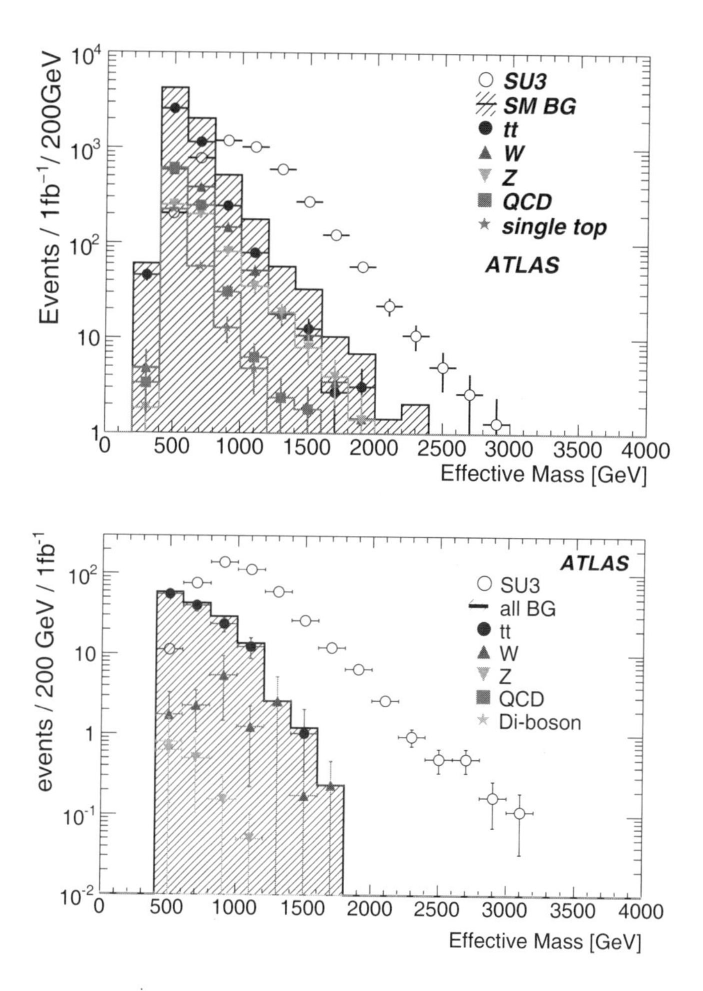

Fig. 1. Effective mass distributions for the background processes and for an example SUSY benchmark point, SU3, in the 0-lepton mode (top) and 1-lepton mode (bottom) for an integrated luminosity of 1 fb^{-1}. The black circles show the SUSY signal. The black histogram shows the sum of all Standard Model backgrounds; also shown in different colours are the various components of the background.[3]

discovery potential, specially in early data when the uncertainty on the backgrounds are large. In most cases, a noticeable excess of events is observable at high effective mass values with 1fb^{-1} of integrated luminosity, an amount of data that is expected to be collected within the first one or two years of LHC running.

The plot of Figure 2 shows the discovery contours for the same luminosity in the $(m_0, m_{1/2})$ parameter space of the Minimal Supergravity model for the 0-lepton and 1-lepton mode.[3] The remaining parameters are fixed to $\tan\beta = 10$, $A_0 = 0\ GeV$ and positive μ. Squark and gluino masses of the order of up to 1 TeV can be reached. The current limit from Tevatron is of the order of 400 GeV.

If the SUSY mass scale is in the sub-TeV range, early LHC data will likely be sufficient to claim a discovery of new physics, although new physics does not strictly mean SUSY as other new physics scenarios may have similar features and properties. To distinguish different scenarios and to determine the full set of model parameters within one scenario, as many measurements of the new observed phenomena as possible are needed. This

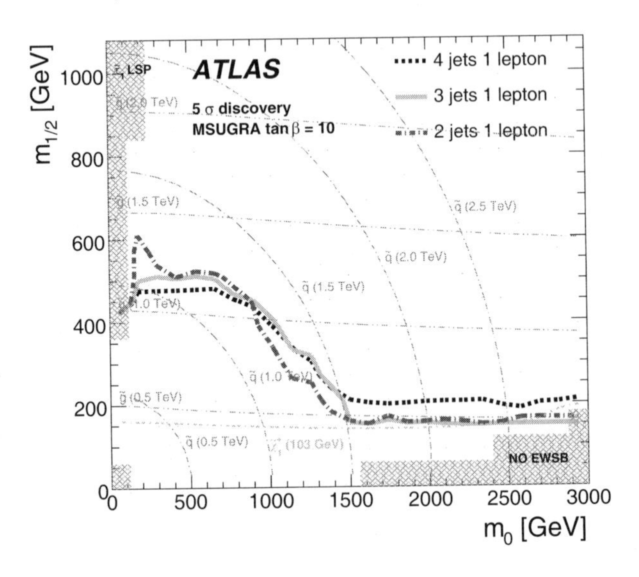

Fig. 2. The 1 fb^{-1} 5σ reach contours for the1-lepton analyses with various jet requirements as a function of m_0 and $m_{1/2}$ for the $\tan\beta = 10$ mSUGRA scan. The horizontal and curved grey lines indicate the gluino and squark masses respectively in steps of 500 GeV.[3]

includes the precise measurement of masses, spins and CP properties of the newly observed particles.

3. Mass Measurement and Parameter Determination

After the discovery of new physics beyond the SM, many measurements of the production process and particle properties are needed to pin-down the exact model of new physics. For example, the masses of the new particles can be used to distinguish between different SUSY models.

Due to the escaping LSPs in every SUSY event, no mass peaks can be reconstructed and masses must be measured by other means. The LHC strategy is to look for kinematic endpoints in the invariant mass distributions of visible decay products. For example, in mSUGRA models, the main source of mass information can be provided by $\tilde{\chi}_2^0$ decays. When kinematically accessible the $\tilde{\chi}_2^0$ can undergo sequential two-body decays to a $\tilde{\chi}_1^0$ via a right-handed slepton, $\tilde{\chi}_2^0 \rightarrow \tilde{\ell}^+\ell^- \rightarrow \tilde{\chi}_1^0\ell^+\ell^-$, for example in the SU3 benchmark point [a]. Due to the scalar nature of the slepton, the invariant mass of the two leptons $m_{\ell\ell}$ from this decay chain exhibits a triangular shape with a sharp drop-off at a maximal value $m_{\ell\ell}^{max}$, Fig 3(top). The position of this endpoint depends on the masses of the sparticles involved:

$$m_{\ell\ell}^{max} = m_{\chi_2^0}\sqrt{1 - \left(\frac{m_{\ell_R}}{m_{\chi_2^0}}\right)^2}\sqrt{1 - \left(\frac{m_{\chi_1^0}}{m_{\ell_R}}\right)^2}$$

The endpoint is measured from the di-lepton mass distribution.[3] Combinatorial background from SM and other SUSY processes can be estimated from data and subtracted using the flavour-subtraction method. The flavour subtraction method is based on the fact that the signal contains two opposite-sign same-flavour leptons, while the background leptons can be of the same flavour or of different flavour with the same probability.

In some SUSY parameter space region (co-annihilation region), the $\tilde{\chi}_2^0$ can decay to both left and right sleptons, for example in the SU1 benchmark point[b], giving a double dilepton invariant mass edge structure, Fig. 3(bottom), with the two edges about 50 GeV apart. The expected sensitivity for electrons and muons is listed in Tab. 1.

---

[a]SU3: m_0 = 100 GeV, $m_{1/2}$ = 300 GeV, A_0 = -300, $\tan\beta = 6$, $\mu > 0$. Bulk region, the LSP annihilation happens through the exchange of light sleptons.

[b]SU1: m_0 = 70 GeV, $m_{1/2}$ = 350 GeV, A_0 = 0, $\tan\beta = 10$, $\mu > 0$. Coannihilation region where $\tilde{\chi}_1^0$ annihilate with near-degenerate $\tilde{\ell}$.

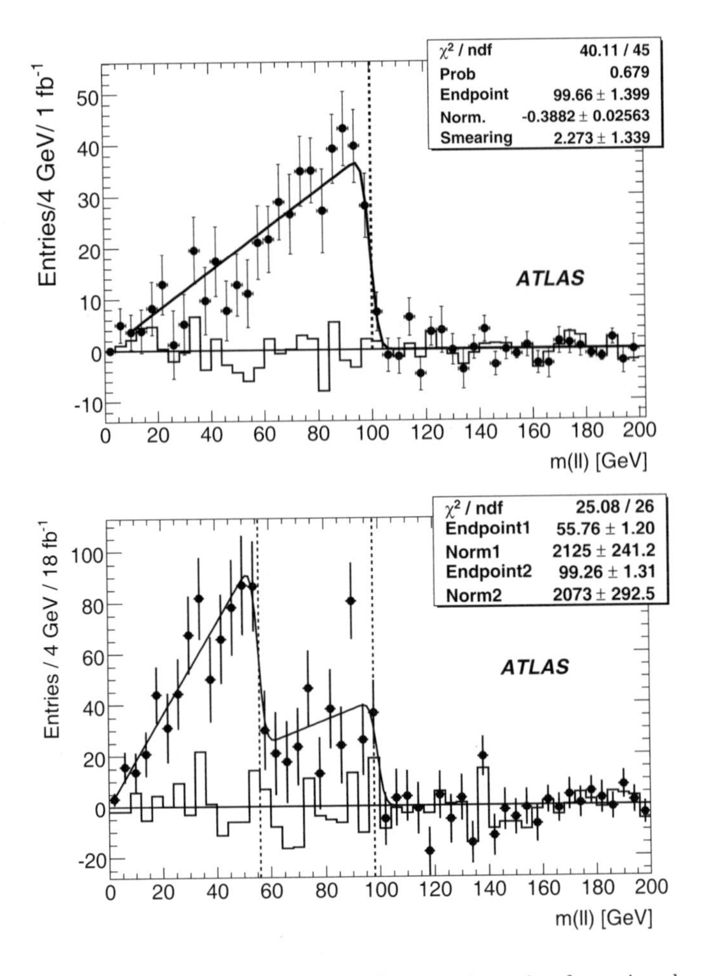

Fig. 3. Distribution of invariant mass after flavour subtraction for various benchmark points with an integrated luminosity of 1 fb^{-1}. The line histogram is the Standard Model contribution while the points are the sum of Standard Model and SUSY contributions. The fitting function is superimposed and the expected position of the endpoint is indicated by a dashed line. Top: Two-body decay with $\tilde{\chi}_2^0$ decaying to right-sleptons for 1 fb^{-1} (SU3), bottom two-body decay with $\tilde{\chi}_2^0$ decaying to both left- and right-sleptons for 18 fb^{-1} (SU1).[3]

A similar analysis can be performed by replacing the electrons and muons with taus. Due to the additional neutrinos from the tau decay, the visible di-tau mass distribution is no longer triangular. This complicates measurements of the spectrum endpoint. A solution to this problem is to fit a suitable function to the trailing edge of the visible di-tau mass spec-

Table 1. Reconstructed endpoint. The first error is statistical and the second is mainly due to the lepton energy scale and the ratio of the electron and muon reconstruction efficiency.

benchmark point	true $m_{\ell\ell}$ mass [GeV]	expected $m_{\ell\ell}$mass [GeV]	luminosity [fb^{-1}]
SU1	56.1	$55.8 \pm 1.2 \pm 0.2$	18
SU1	97.9	$99.3 \pm 1.3 \pm 0.3$	18
SU3	100.2	$99.7 \pm 1.4 \pm 0.3$	1

trum and use the inflection point as an endpoint sensitive observable, which can be related to the true endpoint using a simple MC based calibration procedure. Note that the third error is due to the SUSY-model dependent polarization of the two taus. On the other hand, the influence of the tau polarization on the di-tau mass distribution can be used to measure the tau polarization from the mass distribution and distinguish different SUSY models from each other.

In all cases the $m_{\ell\ell}$ endpoint can be measured without a bias although the required integrated luminosity is quite different. Furthermore, the fit function to extract the endpoint(s) needs to be adjusted to the underlying mass spectrum. The expected sensitivity is summarized in Tab. 1 including the assumed luminosity.

By including the jet produced in association with the $\tilde{\chi}_2^0$ in the $\tilde{q}_L$ decay, several other endpoints of measurable mass combinations are possible. All these measured mass combinations can be used to extract the underlying high mass model parameters.

From the end point measurements we can derive the SUSY mass spectra and parameters. From the SUSY mass spectrum we can obtain the mSUGRA parameters and the unification scale. The procedure is described in Reference3, where is discussed as an example, the mSUGRA benchmark points SU3 for a luminosity of 1 fb^{-1}. Here we will give only the results of a fit of the mSUGRA parameters for the SU3 point, which are listed in Tab. 2. With 1 fb^{-1} the reconstruction of part of the SUSY mass spectrum will only be possible for favourable SUSY scenarios and with some assumptions about the decay chains involved.

4. Spin Measurement

Measurements of the number of new particles and their masses will give us enough information to extract model parameters for one of the SUSY models. However, the mass information alone will not be enough to distinguish

Table 2. Results of a fit of the mSUGRA mass spectra for the the SU3 point. This fit is for $sig(\mu) = +1$. The uncertainty due to assumption son theoretical uncertainties is also shown.

Parameter	true value	fitted value	uncertanty
$\tan\beta$	6	7.4	4.6
m_0	100 GeV	98.5 GeV	$\pm$9.3 GeV
$m_{1/2}$	300 GeV	317.7 GeV	$\pm$6.9 GeV
A_0	-300 GeV	445 GeV	$\pm$408GeV

different new physics scenarios. For example Universal Extra Dimensions with Kaluza-Klein parity can have a mass spectrum very similar to the one of certain SUSY models. However, the spin of the new particles is different and can be used to discriminate between models.

One possibility is to use two-body slepton decay chain described above. The charge asymmetry of lq pairs can be used to measure the spin of $\tilde{\chi}_2^0$, while the shape of dilepton invariant mass spectrum measures slepton spin.[4] The first lepton in the decay chain is called the *near* lepton while the other is called the *far* lepton.

The the invariant masses $m_{ql^{near}}$ charge asymmetry A is defined as:

$$A = \frac{s^+ - s^-}{s^+ + s^+}$$

where $s^{\pm} = d\sigma/dm_{ql^{near(\pm)}}$.

In general is not possible to distinguish between the ear and the far lepton, but only $m_{\tilde{q}l^{near}}$ can be measured, diluting A. The expected asymmetry for SU3 is shown in Fig. 4 for a luminosity of 30 fb^{-1}.

5. Determining Relic Density from LHC Data

For neutralinos to account for the observed dark matter, their density at a certain time in the expansion of the early Universe would have had to become low enough to cease annihilation, leaving relic cold dark matter. Inflationary models of the universe along with astronomical data from experiments like WMAP and SDSS[5] can be used to put limits on the rates of neutralino production and annihilation. There are four main mechanisms that can occur to cease anihilation:

(1) Slepton exchange, which is suppressed unless the slepton masses are lighter than approximately 200 GeV.

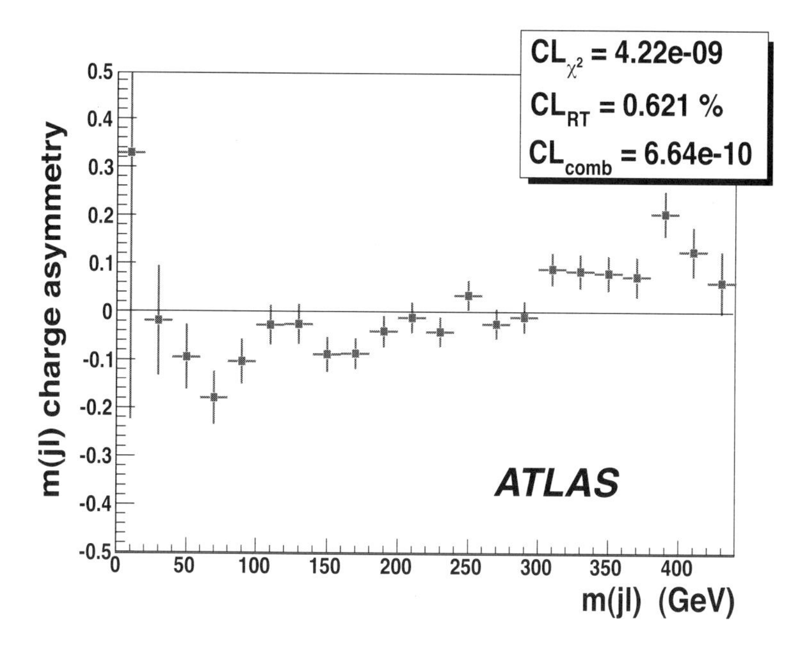

Fig. 4. Expected charge asymmetry A for SU3 and 30 fb^{-1}.

(2) Annihilation to vector bosons, that occurs when the neutralino LSP acquires a significant wino or higgsino component.
(3) Co-annihilation with light sleptons, which happens when there are suitable mass degeneracies in the sparticle spectrum.
(4) Annihilation to third-generation fermions that is enhanced when the heavy Higgs boson A is almost twice as massive as the LSP.

To reproduce the observed relic density, the model parameters must ensure efficient annihilation of the neutralinos in the early universe. In the mSUGRA scenario this is possible in restricted regions of the parameter space where annihilation is enhanced either by a significant higgsino components in the lightest neutralino or through mass relationships.

There are various strategies for determining the relic density; we will discuss here the one presented in Reference 6, where they target the weak scale parameters relevant to the relic density calculation. Endpoints are used to constrain sparticle masses, which are then used to constrain the neutralino mixing matrix, obtaining $\tan\beta$ dependent values of the mixing parameters. They then constrain the slepton sector using a ratio of branching fractions that is sensitive to the stau mixing parameters:

$BR(\tilde{\chi}_2^0 \to \tilde{l}_R l)/BR(\tilde{\chi}_2^0 \to \tilde{\tau}_1 \tau)$. Finally, they consider constraints on the Higgs sector, even if their benchmark point is in a region in which the LHC is expected to produce only the lightest (SM-like) Higgs boson. They obtain a relic density distribution as a function of m_A, but show ways of improving their measurement by placing a lower limit of 300 GeV on m_A due to its non-observation in cascade decays. This assumption provides an improvement in their control over the relic density, and they obtain a final value of:

$$\Omega_\chi h^2 = 0.108 \pm 0.01(stat + sys)^{+0.00}_{-0.002}(m(A))^{+0.001}_{-0.011}(\tan\beta)^{+0.002}_{-0.005}(m(\tilde{\tau}_2))$$

In these proceedings we discussed some examples of a dark matter search at the LHC, however other measurements are possible.[3] If the mass differences in cascade decays will be small, the measurement of the sparticle mass will be challenging. In such a case, one would hope to be able to constrain the SUSY Lagrangian from other measurements, but the LHC may prove insufficient to accomplish this.

6. Summary

We have reviewed recent work that explains how to use the LHC to learn about dark matter, using supersymmetry as an example. The LHC is an excellent discovery machine, with a wide search reach for observing SUSY WIMP candidates in inclusive channels. It has been shown that the LHC may be capable of determining the dark matter relic density with a precision of approximately 10% but that this is highly dependent on the underlying SUSY model.

However, there are questions that a collider can never address, for example how much of the observed astrophysical dark matter is comprised of WIMPS. Furthermore, we would know we have produced a WIMP candidate if we know its lifetime. For these reasons, direct and indirect experiments are complementary to the collider program.

References

1. D. N. Spergel at. al., *Astrophys. J. Suppl.* **148** (2003) 175.
2. S. Dawson (1996) hep-ph/9612229.
3. ATLAS Collaboration, Expected Performance of the ATLAS Experiment, Detector, Trigger and Physics *CERN-OPEN-2008-020*, Geneva (2008).
4. A. J. Barr *Phys. Lett.* **B596** (2004) 205.
5. M. Tegmark et al. *Physical Review* **D69** (2004) 10350.
6. M. M. Nojiri, G. Polesello and D. R. Tovey *J. High Energy Phys.* **063**(2006) 063 and hep-ph/0512204.

DARK MATTER AND SUPERSYMMETRY

LESZEK ROSZKOWSKI

Department of Physics and Astronomy, University of Sheffield Sheffield, S3 7RH, UK, and The Andrzej Soltan Institute for Nuclear Research, Warsaw, Poland.

Dark matter in the Universe is likely to be made up of some new, hypothetical particle which would be a part of an extension of the Standard Model of particle physics. After a general introduction, in this talk I present in the framework of two popular unified supersymmetric models Bayesian statistics results for neutralino dark matter in the context of direct detection in underground experiments, and in indirect detection modes of relevance to Fermi and Pamela. While prospects for direct detection look excellent, those for Fermi strongly depend on poorly known cuspiness of the galactic halo in the central region of the Milky Way. Positron flux in those models appears to be significantly below the Pamela result, which on the other hand, may be explained in terms of nearby pulsars.

Keywords: Supersymmetry, dark matter.

1. A Few Basic Questions

A wide variety of observations point towards the existence of large amounts of invisible, or dark, matter (DM) in the Universe. It is presumably cold (i.e., non-relativistic at the time of matter dominance) and is distributed in space in the form of both extended halos around visible (baryonic) parts of large scale structures (LSS), most likely as a fairly smooth component. This picture is supported by numerical simulations of LSS and from interpretations of observational evidence in terms of galactic halos, gravitational lensing by galaxy clusters, etc. Assuming a LCDM concordance model, studies of CMB anisotropies by WMAP and other experiments also currently provide the most precise determinations of the CDM relic abundance $\Omega_{\rm CDM}h^2 = 0.1099 \pm 0.0062$.[1]

Before plunging into a more detailed discussion of DM candidates and their properties, it is worth posing a few basic questions.

- Is the evidence for DM convincing? – I think yes but it comes only through its gravitational effects.
- Is DM made up of particles? – This is suggested by clustering but otherwise otherwise is an assumption.
- Is DM made up of only/predominantly one species? – This is an economical assumption but is not rooted in any evidence.
- Is DM cold? – The cold DM (CDM) framework seems to be remarkably successful in reproducing in numerical simulations the observed distribution of matter. Some alleged problems of CDM that have recently been claimed are probably not insurmountable, maybe even not very serious.
- Has DM been detected yet? – There have been various hints and anomalies claimed in a number of astrophysical observations but to me their DM explanation is not convincing, nor often even necessary.

What can the WIMP be? In Fig. 1, adopted from Ref. 2, I have included some of the most popular, and also most robust, candidates. The main suspect remains the neutralino χ. Experimental bounds on its mass are actually not too strong, nor are they robust: they depend on a number of assumptions. In minimal SUSY (the so-called MSSM) 'in most cases' $m_\chi \gtrsim 70 - 100$ GeV, but the bound can be also much lower. Theoretically, because of the fine tuning argument, one expects its mass to lie in the range of several tens or hundreds of GeV. More generally, $m_\chi \gtrsim$ few GeV from $\Omega_\chi h^2 \lesssim 1$ (the so–called Lee-Weinberg bound[4]) and $m_\chi \lesssim 300$ TeV from unitarity.[5] Neutralino interaction rates are generally suppressed relative to σ_{EW} by various mixing angles in the neutralino couplings. In the MSSM they are typically between $\sim 10^{-7}\sigma_{EW}$ (from current experimental upper bounds from direct detection searches) and $\sim 10^{-10}\sigma_{EW}$, although could be even lower in more complicated models where the LSP would be dominated for example by a singlino (fermionic partner of an additional Higgs singlet under the SM gauge group). This uncertainty of the precise nature of the neutralino is reflected in Fig. 1 by showing both the smaller (dark blue) region of minimal SUSY and an extended one (light blue) with potentially suppressed interaction strengths in non–minimal SUSY models and other "new physics" models (see below).

The neutralino is a particularly well motivated case of a WIMP. On the other hand, one should remember that in the box marked with the generic name "WIMP" (light blue) one can accommodate not just the neutralino but also several other stable states appearing in various extensions of the

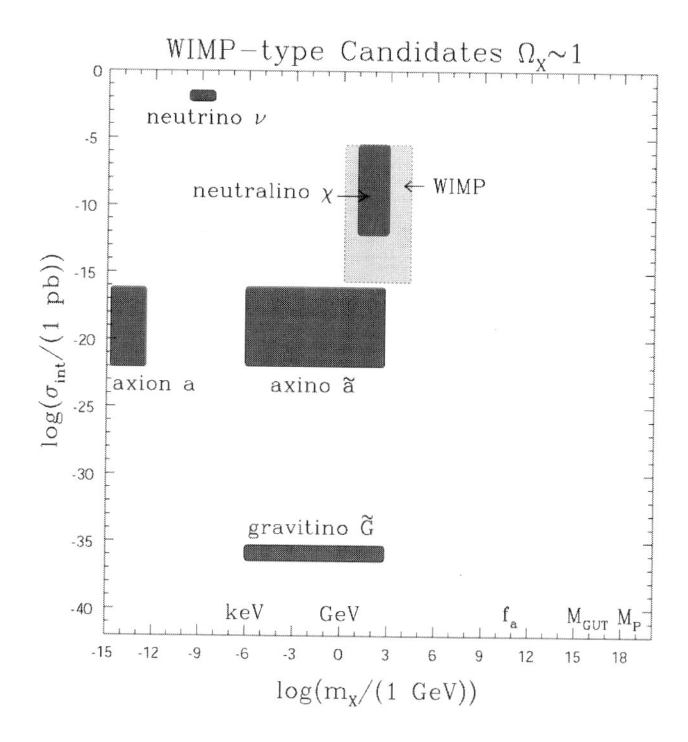

Fig. 1. A schematic representation of some well–motivated WIMP–type particles for which a priori one can have $\Omega \sim 1$. $\sigma_{\rm int}$ represents a typical order of magnitude of interaction strength with ordinary matter. The neutrino provides hot DM which is disfavored. The box marked "WIMP' stands for several possible candidates, e.g., Kaluza–Klein particles. This figure originally appeared in Ref. 2.

SM, e.g., lightest Kałuża-Klein (KK) state[6] from warped/universal extra dimensions, and several other quite interesting candidates: multiple (UPT) DM, little Higgs DM, mirror DM, shadow DM, sequestered DM, secluded DM, flaxino DM, Higgs portal DM, inflation and DM, etc etc. They are no nonsense but not superior to the neutralino either. What is certainly clear from Fig. 1 that a particle physics explanation of DM requires one to go beyond the SM of particle interactions.

There are some other potentially cosmologically relevant relics out there whose interactions would be much weaker than electroweak. One well–known example is the axion – a light neutral pseudoscalar particle which is a by–product of the Peccei–Quinn solution to the strong CP problem. Its interaction with ordinary matter is suppressed by the PQ scale $\sim (m_W/f_a)^2\sigma_{EW} \sim 10^{-18}\sigma_{EW} \sim 10^{-20}\,{\rm pb}$ ($f_a \sim 10^{11}$ GeV), hence extremely tiny, while its mass $m_a \sim \Lambda_{QCD}^2/f_a \sim \left(10^{-6} - 10^{-4}\right)$ eV if $\Omega_a \sim 1$. The axion, despite being so light, is of CDM–type because it is produced by the non–thermal process of misalignment in the early Universe.

In the supersymmetric world, the axion has its fermionic superpartner, called axino. Its mass is strongly model–dependent but, in contrast to the neutralino, often not directly determined by the SUSY breaking scale $\sim$ 1 TeV. Hence the axino could be light and could naturally be the LSP, thus stable. An earlier study concluded that axinos could be *warm* DM with mass less than 2 keV.[7] More recently it has been pointed out more massive axinos quite naturally can be also *cold* DM as well, as marked in Fig. 1. Relic axinos can be produced either through thermal scatterings and decays involving gluinos and/or squarks in the plasma, or in out-of-equilibrium decays of the next–to-LSP, e.g. the neutralino.[8] The first mechanism is more efficient at larger reheat temperatures $T_{\rm R} \gtrsim 10^4$ GeV, the second at lower ones.

Lastly, there is a gravitino – the fermionic superpartner of the graviton – which arises by coupling SUSY to gravity. The gravitino relic abundance can be of order one[9] but one has to also worry about the so–called gravitino problem: heavier particles, like the NLSP, will decay to gravitinos very late, around 10^7 sec after the Big Bang, and the associated energetic photons may cause havoc to BBN products. The problem is not insurmountable but constraints from BBN are quite severe and basically rule out the possibility of the neutralino being the NLSP,[10] unless it is rather light, below 1 GeV.[11] In Fig. 1 the gravitino is marked in the mass range of keV to GeV and gravitational interactions only, although keV–gravitinos have actually strongly enhanced couplings via their goldstino component.

It is clear that WIMP's interactions can be very strongly suppressed relative to the weak strength. EWIMPs (or superWIMPs) such as axions, axinos or gravitinos are perfectly well motivated and acceptable as CDM. The reason is that in the early Universe (E)WIMPs can be produced in more than one way. In particular, cosmologically relevant axions would be produced via the mis-alignment mechanism and axinos and gravitinos in either the next-to-lightest superpartner (NLSP) decays after its freeze-out, or else in scatterings in the hot plasma at high reheating temperatures $T_{\rm R}$.

The number of *well-motivated* WIMP candidates for CDM is in the end not so large. Interestingly, they seem linked to some popular scales of new physics: the SUSY breaking scale of 1 TeV, the PQ scale and the GUT/Planck scale.

The way I view it is that it is clearly fairly easy to invent a WIMP candidate. It is much (!) harder to construct a self-consistent and not ad-hoc looking extension of the SM which would accommodate a WIMP DM in a natural way. It is even much harder to produce a "lasting" model of "new physics". In this respect SUSY seems to have few competitors, if any.

2. Neutralino – The Prime Suspect

The DM candidate that has attracted perhaps the most wide-spread attention of both the theoretical and experimental communities is the neutralino. If it is the lightest SUSY particle, it is massive and stable, if one assumes R-parity. A perfect candidate for a WIMP!

The are some good reasons why the neutralino WIMP χ is so popular:

- it is a part of a well-defined and well-motivated framework of SUSY
- its observable properties are calculable;
- its relic density $\Omega_\chi h^2$ computed from freeze-out can agree with the correct value, ~ 0.1. Although one has to remember that typically the spread is more like $10^{-4} - 10^3$. In this sense it is a bit of an exaggeration to claim that the neutralino WIMP relic density "naturally" reproduces the right value.
- the neutralino is stable due to some discrete symmetry (normally, R-parity);
- it is testable with today's experiments, both dark matter and collider;
- no obviously superior competitor (both to SUSY and to χ) has been accepted;

Much literature has been devoted to the neutralino as DM, including a number of comprehensive and topical reviews (see, e.g., Refs. 2,12). Here I will only summarize the main results and comment on some recent developments and updates.

The neutralino χ is a neutral Majorana particle, the lightest of the mass eigenstates of the fermionic partners of the gauge and Higgs bosons: the bino, wino and the neutral higgsinos. Neutralino properties as DM and ensuing implications for SUSY spectra are quite model dependent but certain general conclusions can be drawn. First, its cosmological properties are very different depending on a neutralino type. The relic abundance $\Omega_\chi h^2$ of gaugino-like (mostly bino-like) neutralinos can be brought into the "WMAP range" of around 0.1 in three ways: in the "bulk" region $\Omega_\chi h^2$ is primarily determined by the (lightest) sfermion exchange in the LSP annihilation into $f\bar{f}$: $\Omega h^2 \propto m_{\tilde{f}}^4/m_\chi^2$. In the "funnel" region $\Omega_\chi h^2$ is reduced via a broad resonance in neutralino pair-annihilation due to pseudoscalar

Higgs A exchange. Finally, in some regions the mass difference between the neutralino LSP and the NLSP (typically the stau, and sometimes the stop) is rather small, allowing for efficient coannihilation between the two species. Remarkably, just such a cosmologically preferred gaugino type of neutralino typically *emerges* in a grand-unified scenario with additional assumptions that the mass parameters of all spin-zero particles are equal at the unification scale $\sim 10^{16}$ GeV. What one finds there is that the lightest bino-like neutralino emerges as essentially the only choice for a neutral LSP.[3]

On the other hand, higgsino-like neutralino is in comparison much less attractive, because its relic abundance is very small both because of very efficient pair-annihilation into ZZ, WW and $t\bar{t}$ and because of coannihilation with $\chi_1^{\pm}$ and χ_2^0. However, the higgsino with larger mass ($m_\chi \gtrsim 500$ GeV) remains cosmologically acceptable although this mass range can be considered as less natural.[13]

One also arrives at the same conclusions in the case of 'mixed', or well-tempered,[14] neutralinos composed of comparable fractions of gauginos and higgsinos. This is because, in order to produce sufficient relic abundance, it has to be rather heavy.[13]

In the general MSSM, basically a supersymmetrized version of the SM, there are over 100 free parameters, leading to poor predictive power of the model. For example, for the cross sections of spin-independent interactions with detector material one can obtain basically almost any value down to $\sigma_p^{SI} \sim 10^{-11}$ pb, and below. However, successful gauge coupling unification within SUSY has shifted the attention to unified SUSY models where not only the gauge coupling unify but also other parameters of the model.

The are a whole multitude of more or less well motivated effective low-energy SUSY models, e.g., the MSSM, the Constrained MSSM (CMSSM), split SUSY framework, the Next-to-MSSM (NMSSM), the Non-Universal Higgs Model (NUHM), $SO(10)$ GUT based models, string inspired models, etc, etc, each having some interesting features of one sort or another. At this point it is not clear which one, if any, has been chosen by Nature. The good news is that many of them possess, at least in principle, at least partially different phenomenological and DM properties. The bad news is that in many cases they may be indistinguishable in a real experimental environment. In order to learn this lesson, there seems to be no other way, however, than to explore the models one by one and compare them.

In this talk I will present dark matter properties of some of those models, namely the CMSSM and the NUHM. The presentation will be based on an ongoing program[15,16,19,26] of exploring those models in terms of Bayesian statistics which I will describe below. For a most recent analysis of the Constrained NMSSM (not covered here), see Refs. 17,18.

2.1. *The Constrained MSSM (CMSSM)*

The most popular SUSY framework these days is the the Constrained MSSM (CMSSM)[3] which includes the minimal supergravity model (mSUGRA).[20] In this scheme one defines all SUSY parameters at the unification scale M_{GUT} and next employs the Renormalization Group Equations (RGEs) to evolve them down and compute the couplings and masses in an effective theory valid at the electroweak scale. The small number of parameters makes the CMSSM a popular framework for exploring SUSY phenomenology, especially with the lightest neutralino often being the lightest supersymmetric particle. Assuming it to be the dominant component of CDM in the Universe allows one to apply the DM relic density determination by WMAP and other experiments as a strong constraint on the CMSSM parameters.

In the CMSSM, at the GUT scale the soft masses of all the sleptons, squarks and Higgs bosons have a common scalar mass m_0, all the gauginos unify at the common gaugino mass $m_{1/2}$, and so all the tri-linear terms assume a common tri-linear mass parameter A_0. In addition, at the electroweak scale one selects $\tan\beta$, the ratio of Higgs vacuum expectation values and $\mathrm{sgn}(\mu)$, where μ is the Higgs/higgsino mass parameter whose square is computed from the conditions of radiative electroweak symmetry breaking (EWSB). The independent parameters of the CMSSM are thus four continuous parameters:

$$m_{1/2}, \quad m_0, \quad A_0, \quad \tan\beta, \tag{1}$$

plus the sign of μ. The CMSSM and its extensions has been extensively studied in a large number of papers. Here I would like to present several results obtained by applying Bayesian statistics to studying the CMSSM and similar models. This is a recent development[15,16,21] which has proved particularly powerful and promising. Initially it employed a very efficient Markov Chain Monte Carlo (MCMC) scanning technique of multi-dimensional parameter spaces, but more recently even more efficient approaches have been successfully applied, especially a Nested Sampling (NS) scanning technique.[22] Basically, usual fixed grid scans are restricted to exploring only a subset of parameters, typically $m_{1/2}$ *vs.* m_0 for fixed values of A_0 and $\tan\beta$, and also for fixed values of SM parameters, e.g., the top mass m_t, which

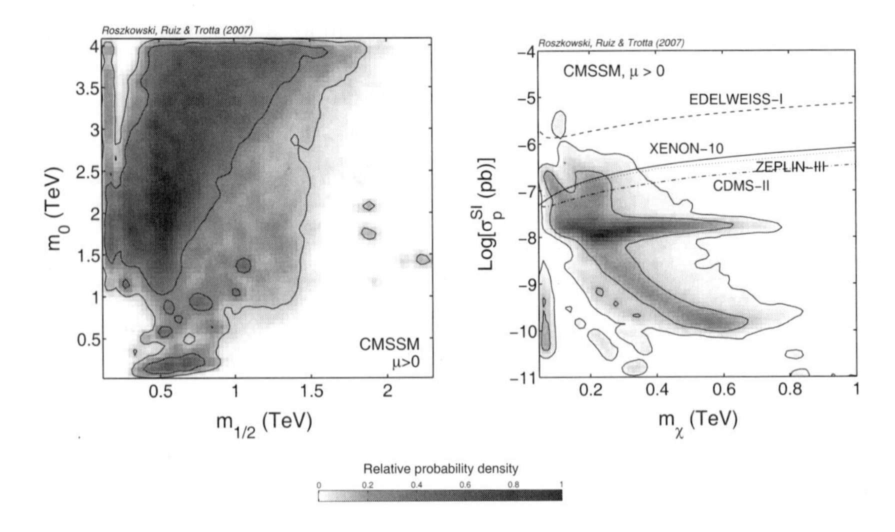

Fig. 2. The 2D posterior relative probability density for the CMSSM parameters $m_{1/2}$ and m_0 (left panel) and the corresponding one for the dark matter spin-independent detection cross section σ_p^{SI} and the neutralino mass m_χ (right panel). In both panels $\mu > 0$ and flat priors have been assumed. The inner (outer) solid contours delimit the regions of 68% and 95% total probability, respectively. Some current experimental upper bounds are also shown. Taken from Ref. 16, for an update and a discussion of prior dependence see Ref. 23.

can have a large impact on results especially at large m_0. Also, the model's predictions for observable quantities, e.g., the relic abundance $\Omega_\chi h^2$ of the neutralino, or Higgs and superpartner masses are compared with experimental data in a simplified way: if the predicted values is within some arbitrary range, typically $1\,\sigma$ or 90% CL then the point is treated as "allowed"; otherwise it is rejected. Theoretical errors are also typically ignored.

Because the MCMC approach is much more efficient than fixed grid scans, one can perform a multi-dimensional exploration of the parameter space. In our case we vary simultaneously the four CMSSM parameters (for a fixed sign of μ) and additionally four SM (nuisance) parameters: the pole top mass, the bottom mass $m_b(m_b)^{\overline{MS}}$, $\alpha_s(M_Z)^{\overline{MS}}$ and $\alpha_{\rm em}(M_Z)^{\overline{MS}}$, the last three evaluated in the MS-bar scheme. Also, because in the likelihood function of Bayesian statistics one can compare the model's predictions with experimental data (usually assuming Gaussian distributions) and obtain a range of weights (rather than a simplistic "allowed/disallowed" pattern), the resulting Bayesian posterior probability maps are much more informative. For more details about our procedure, see Refs. 15,16.

The left panel of Fig. 2 shows 2-dimensional (2D) relative posterior probability maps in the usual plane of $m_{1/2}$ and m_0, assuming $\mu > 0$. The inner (outer) contours delimit the regions of 68% and 95% total probability, respectively, after marginalizing over the additional six dimensions. One can clearly see a rather large high probability region at large m_0 which corresponds to the focus point region.[24,25] In addition, at low $m_0 \ll m_{1/2}$ one can see the elongated stau coannihilation region and patches of the pseudoscalar Higgs A funnel. All the known features of the CMSSM parameter space are therefore reproduced but Bayesian statistics allows us to see the relative "sizes" and "weights" of each region.

The right panel of Fig. 2 shows the corresponding 2D relative posterior probability maps for the dark matter spin-independent detection cross section σ_p^{SI} *vs.* the neutralino mass m_χ. As in the left panel, the inner (outer) contours delimit the regions of 68% and 95% total probability, respectively. The horizontal branch corresponds to the focus point region while the banana-shape region comes from the stau coannihilation and the A funnel. Some recent 90% experimental upper limits are also shown but have not been included in constraining the parameter space. It is encouraging that DM direct detection experiments are already probing the upper regions of the most probable ranges of σ_p^{SI} are already explored by experiment. It is expected that the current generation of detectors will be sensitive down to (and perhaps below) the mark of 10^{-8} pb, which will allow them to explore some of the most favored ranges of σ_p^{SI}. Otherwise future one-tonne detectors will be aiming at reaching down to some 10^{-10} pb and perhaps even below. At the same time, direct LHC searches will be able to explore m_χ up to some $400 - 500$ GeV, or $m_{1/2}$ roughly up to 1 TeV. Clearly, direct DM searches are not only overlapping with and complementary to LHC reach but also reach to much larger WIMP mass ranges which will never be accessible to the LHC but which remain within the range of predictions of the CMSSM and similar models. Roughly similar conclusions have been obtained by several other studies using mostly fixed-grid scans, but without providing vital information about the relative probability of each point.

These results do depend on a choice of priors, or in other words initial ranges of input parameters (1), and also on whether one scans them linearly (flat priors) or some functions of them, e.g., $\log(m_{1/2})$, $\log(m_0)$ (log priors). Prior dependence remains substantial even for the CMSSM, indicating that the model is still under-constrained by the current data. (Needless to say, similar prior dependence is also implicitly present in the usual fixed-grid scans.) What is encouraging is that the favored ranges of σ_p^{SI} are actually

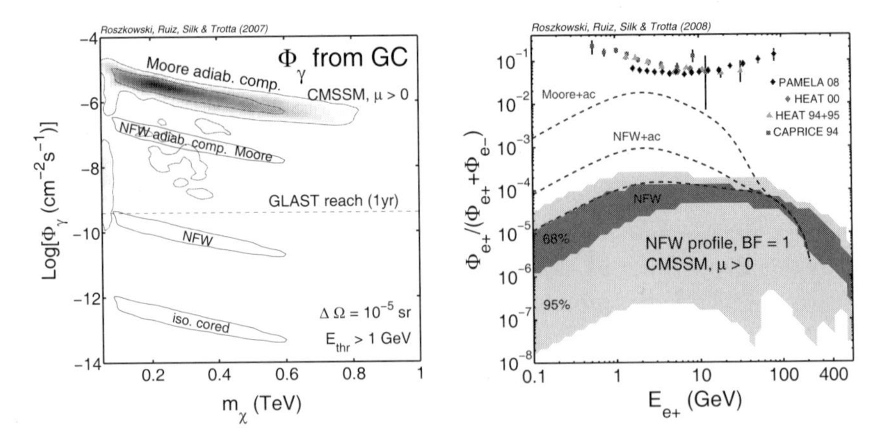

Fig. 3. Left panel: Predicted 2D posterior relative probability density maps in the CMSSM for the total flux of diffuse gamma radiation from the Galactic Center (GC) vs. m_χ, assuming GLAST/Fermi parameters whose 1 year reach has also been indicated. Right panel: Predicted 2D posterior relative probability density maps in the CMSSM for the positron flux fraction NFW profile with a boost factor BF=1 and a specific choice of propagation model. In both panels the inner red (outer yellow) regions delimit the regions of 68% and 95% total probability, respectively. In the right panel we also show for comparison some of the current data, including PAMELA. To illustrate the dependency of the spectral shape at low energies on the halo model, we plot the spectrum for the same choice of CMSSM parameters (with $m_\chi = 229$ GeV) for three different halo models as indicated. In absence of a large boost factor, the signal appears too small to be detected by PAMELA. Updated from Ref. 26.

fairly independent of the above prior choices, although, unsurprisingly, the log prior favors lower mass ranges.[23]

Indirect detection of DM is another promising way of looking for traces of DM annihilation in the Milky Way. Of particular interest are prospects of Fermi (formerly GLAST) and of Pamela both of which have recently been put into orbit.

Fermi's prospects in testing the predictions of the CMSSM are shown in the left panel of Fig. 3 where we plot the total flux of diffuse gamma radiation from the Galactic Center vs. m_χ, using Fermi's angular resolution and sufficiently high photon energy threshold. The inner red (outer yellow) regions delimit the regions of 68% and 95% total probability, respectively, for the halo profile of Moore, et al., with an additional effect of adiabatic compression due to the presence of baryons making the profile at small radius $\propto r^{-1.65}$ very steep. For some other halo models only the contours of the 68% total probability have been marked for comparison. One can see that, for a very steep cusp (or even less cuspy profiles like the one of

Moore, et al., or of NFW with adiabatic compression) in the Galactic Center nearly all the 95% total probability range as predicted by the CMSSM will be explored. On the other hand, with less steep profiles, e.g., the NFW one, the CMSSM will escape Fermi's exploration. In other words, Fermi's ability to test the CMSSM and similar models via diffuse gamma radiation from the Galactic Center will strongly depend on the cuspiness of the DM halo profile at small radii.

Detecting positron flux from WIMP annihilation in the local regions of the halo is one of the main targets of Pamela which has recently announced an unexpected rise in the positron fraction at high energies, above a few GeV.[27] The CMSSM predictions for the quantity are shown in the right panel of Fig. 3 for the NFW halo model with the boost factor of one. As in the left panel, the inner red (outer yellow) regions delimit the regions of 68% and 95% total probability. As one can see, SUSY contributions are a few orders of magnitude too small. Even taking the highest possible values (blue dashed lines) for specific halo models falls short of coming close to the Pamela spectrum, nor is the shape similar, even if one wanted to assume an enormous enhancement in the boost factor (BF).

2.2. *The Non-Universal Higgs Model (NUHM)*

In the CMSSM one may be missing some features of unified models with less restrictive boundary conditions at the unification scale. In particular, the assumption of Higgs (soft) mass unification with those of the sfermions does not seem strongly motivated since the Higgs and matter fields belong to different supermultiplets. One explicit example where this is the case is a minimal $SO(10)$ supersymmetric model ($\mathrm{MSO_{10}SM}$),[28] which is well motivated and opens up a qualitatively new region of parameter space.[29] Models like this provide a good motivation for exploring a wider class of phenomenological models in which the soft masses m_{H_u} and m_{H_d} (as defined at the GUT scale) of the two Higgs doublets are treated as independent parameters and which come under the name of the Non-Universal Higgs Model (NUHM)

In the NUHM, there are therefore six continuous free parameters:

$$m_{1/2}, \quad m_0, \quad A_0, \quad \tan\beta, \quad m_{H_u} \quad \text{and} \quad m_{H_d}, \tag{2}$$

plus again the sign of μ.

The enlarged parameter space leads to a much richer phenomenology. In particular there are more choices for the LSP than in the CMSSM and there are more coannihilation channels as well. This makes a full exploration of

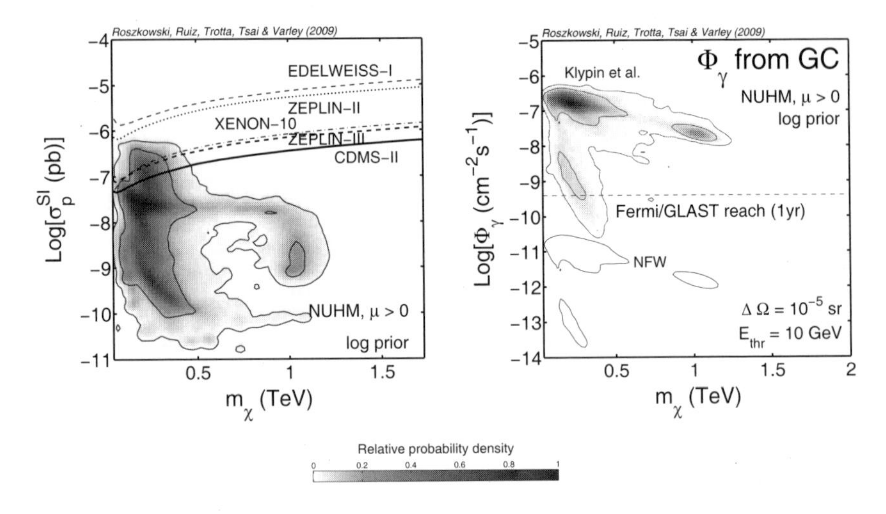

Fig. 4. Left panel: The same as in the right panel of Fig. 2 but for the NUHM and the log prior. Right panel: the same as in the left panel of Fig. 3 but for the NUHM and the log prior.

the NUHM even more challenging than that of the CMSSM. In fact, it is difficult to imagine how one can do that without an efficient algorithm of the MCMC type.

A first Bayesian analysis of NUHM has been performed in Ref. 19, where, along with four SM nuisance parameters, altogether a ten-dimensional parameter space was scanned over. As before in the CMSSM, all relevant constraints from colliders and cosmology were applied.

The results for the cross sections of spin-independent interactions of dark matter are presented in the left panel of Fig. 4. By comparing with the analogous plot for the CMSSM (right panel in Fig 2), we can see some common features and also some differences. At fairly low $m_\chi \lesssim 500$ GeV there is a rapidly dropping, nearly vertical, high-probability (68%) band, which, like in the CMSSM, results from the pseudoscalar A funnel and the stau coannihilation regions. On the other hand, at $m_\chi \sim 1$ TeV one finds an "island" of 68% total probability with $\sigma_p^{SI} \sim 10^{-9}$ pb, where the neutralino WIMP is mostly a higgsino. This is a rather new feature of the NUHM in the assumed range of $m_{1/2}$ and m_0, which distinguishes it from the CMSSM. (In the CMSSM this feature can also be found but at significantly larger values of the above parameters.) The "size" and relative significance of both the low mass and the higgsino 1 TeV higgsino regions rather strongly depends on the prior. (With a flat prior, the latter is much more prominent. This is another reflection of the fact that, with presently available data, the NUHM, like other unified models, remains under-constrained.) Nevertheless, the existence of the new higgsino region is generic. To a large extent it results from a mild focusing mechanism that is operational in the NUHM, see Ref. 19 for more details.

NUHM predictions for the total flux of diffuse gamma radiation from the Galactic Center vs. m_χ are presented in the right panel of Fig. 4, in a way analogous to how this was done in the left panel of Fig. 3 for the CMSSM. Our conclusions are also broadly similar: for cuspy enough halo models (like the Klypin model, see[19] for more details), Fermi will explore the whole 68% total probability region, and a large part of the 95% region, but for less cuspy models, like the NFW one, signal in the NUHM is highly unlikely.

Like in case of the CMSSM, in the NUHM positron flux ratio of relevance to Pamela or ATIC (not shown here, but presented in Ref. 19) falls way below the claimed measurements.

2.3. *Summary*

In summary, we are still waiting for a genuine signal of WIMP dark matter, be it the neutralino or some other WIMP. In the currently most popular scheme of the CMSSM prospects for DM detection in direct searches are very encouraging, although this may take a few more years and a new generation of larger detectors. In another unified SUSY model with less restrictive assumptions at the unification scale, the NUHM, one finds some features that are common with the CMSSM but also a new higgsino-like region at $m_\chi \sim 1$ TeV. In both models, the cross section for spin-independent interactions falls into the range $10^{-10}\,\mathrm{pb} \lesssim \sigma_p^{SI} \lesssim 10^{-7}\,\mathrm{pb}$, the upper end being already probed by currently operating experiments, with the whole range to be covered with upcoming 1 tonne detectors.

Indirect detection signals are much more dependent on (still poorly known) halo properties. Prospects for Fermi in the CMSSM and NUHM strongly depend on how cuspy the halo model is. For strongly cuspy models the flux from the Galactic Center should be rather clearly visible over most of the high-probability regions, but for the NFW and less cuspy models, a signal is highly unlikely.

As regards Pamela, in both SUSY models presented here the predicted positron fraction, assuming the boost factor of one, is orders of magnitude too tiny to account for the claimed range of measured flux. On top of it, at positron energies above a few GeV, it falls down, contrary to the experimental result. Therefore, even assuming a huge boost factor would not be of much help. On the other hand, the Pamela result has been shown to be consistent with purely astrophysical explanations in terms of nearby pulsars,[30] and, even if it is genuine, may not need require the existence of any exotic dark matter properties. At present, while clearly inconsistent with simple unified SUSY models, the alleged Pamela positron fraction result does not, in my opinion, diminish the attractiveness of low-energy SUSY in the slightest.

References

1. J. Dunkley et al. [The WMAP Collaboration], arXiv:0803.0586 [astro-ph].
2. L. Roszkowski, *Pramana* **62** (2004) 389 [hep-ph/0404052].
3. G. L. Kane, C. F. Kolda, L. Roszkowski and J. D. Wells, *Phys. Rev.* **D 49** (1994) 6173 [hep-ph/9312272]. See also R.G. Roberts and L. Roszkowski, *Phys. Lett.* **B 309** (1993) 329.
4. B.W. Lee and S. Weinberg, *Phys. Rev. Lett.* **39** (1977) 165.
5. K. Griest and M. Kamionkowski, *Phys. Rev. Lett.* **64** (1990) 615.
6. G. Servant and T. Tait, *Nucl. Phys.* **B 650** (2003) 391 and *New J. Phys.* **4** (2002) 99; D. Hooper and G.D. Kribs, *Phys. Rev.* **D 67** (2003) 055003.
7. K. Rajagopal, M.S. Turner and F. Wilczek, *Nucl. Phys.* **B 358** (1991) 447.
8. L. Covi, J.E. Kim and L. Roszkowski, *Phys. Rev. Lett.* **82** (1999) 4180 [arXiv:hep-ph/9905212]; L. Covi, H.B. Kim, J.E. Kim and L. Roszkowski, *J. of High Energy Phys.* **0105** (2001) 033 [arXiv:hep-ph/0101009].
9. H. Pagels and J.R. Primack, *Phys. Rev. Lett.* **48** (1982) 223; S. Weinberg, *Phys. Rev. Lett.* **48** (1982) 1303; J. Ellis, A.D. Linde and D.V. Nanopoulos, *Phys. Lett.* **B 118** (1982) 59; *Phys. Lett.* **B 443** (1998) 209; and several more recent papers.
10. J.L. Feng, A. Rajaraman and F. Takayama, *Phys. Rev. Lett.* **91** (2003) 011302 [arXiv:hep-ph/0302215].
11. L. Roszkowski, R. Ruiz de Austri and K.-Y. Choi, *J. of High Energy Phys.* **0508** (2005) 080 [hep-ph/0408227]; D.G. Cerdeño, K.-Y. Choi, K. Jedamzik, L. Roszkowski and R. Ruiz de Austri, *J. Cosmology and Astroparticle Phys.* **0606** (2006) 005 [hep-ph/0509275]; S. Bailly, K.-Y. Choi, K. Jedamzik and L. Roszkowski, JHEP **0905**, 103 (2009) [arXiv:0903.3974].
12. See, e.g., G. Jungman, M. Kamionkowski and K. Griest, *Phys. Rept.* **267** (1996) 195; C. Muñoz, *Int. J. Mod. Phys.* **A 19** (2004) 3093 [hep-ph/0309346]; G. Bertone, D. Hooper and J. Silk, *Phys. Rept.* **405** (2005) 279 [hep-ph/0404175].
13. L. Roszkowski, Phys. Lett. **B 262** (1991) 59.
14. N. Arkani-Hamed, A. Delgado and G. F. Giudice, *Nucl. Phys.* **B 741** (2006) 108.
15. R. Ruiz de Austri, R. Trotta and L. Roszkowski, *J. of High Energy Phys.* **05** (2006) 002 [hep-ph/0602028]; L. Roszkowski, R. Ruiz de Austri and R. Trotta, *J. of High Energy Phys.* **04** (2007) 084 [hep-ph/0611173].

16. L. Roszkowski, R. Ruiz de Austri and R. Trotta, *J. of High Energy Phys.* **07** (2007) 075 [arXiv:0705.2012].
17. Daniel E. Lopez-Fogliani, L. Roszkowski, R. Ruiz de Austri, and T. Varley, arXiv:0906.4911.
18. C. Balazs and D. Carter, arXiv:0906.5012.
19. L. Roszkowski, R. Ruiz de Austri, R. Trotta, S. Tsai and T. Varley, arXiv:0903.1279.
20. A. Chamseddine, R. Arnowitt and P. Nath, *Phys. Rev. Lett.* **49** (1982) 970; R. Barbieri, S. Ferrara and C. Savoy, *Phys. Lett.* **B 119** (1982) 343; L. J. Hall, J. Lykken and S. Weinberg, *Phys. Rev.* **D 27** (1983) 2359; for a review, see P. Nath, hep-ph/0307123.
21. B. C. Allanach and C. G. Lester, *Phys. Rev.* **D 73** (2006) 015013 [hep-ph/0507283]; B. C. Allanach, C. G. Lester and A. M. Weber, *J. of High Energy Phys.* **0612** (2006) 065 [hep-ph/0609295].
22. F. Feroz and M. P. Hobson Mon. Not. Roy. Astron. Soc. **384** 449 (2008); F. Feroz, M. P. Hobson and M. Bridges, arXiv:0809.3437.
23. R. Trotta, et al., *J. of High Energy Phys.* **0812** (2008) 024 [arXiv:0809.3792].
24. K. L. Chan, U. Chattopadhyay and P. Nath, *Phys. Rev.* **D 58** (1998) 096004 [hep-ph/9710473].
25. J. L. Feng, K. T. Matchev and T. Moroi, *Phys. Rev. Lett.* **84** (2000) 2322 [hep-ph/9908309] and *Phys. Rev.* **D 61** (2000) 075005 [hep-ph/9909334].
26. L. Roszkowski, R. Ruiz de Austri, J. Silk and R. Trotta, *Phys. Lett.* **B 671** (2009) 10 [arXiv:0707.0622 [hep-ph]].
27. O. Adriani et al., arXiv:0810.4995.
28. T. Blazek, R. Dermisek and S. Raby, *Phys. Rev. Lett.* **88** (2002) 111804 [hep-ph/0107097]; *Phys. Rev.* **D 65** (2002) 115004 [hep-ph/0201081].
29. R. Dermisek, S. Raby, L. Roszkowski and R. Ruiz De Austri, *J. of High Energy Phys.* **037** (2003) 0304 [hep-ph/0304101]; *J. of High Energy Phys.* **029** (2005) 0509 [hep-ph/0507233].
30. D. Hooper, P. Blasi and P. D. Serpico, JCAP **0901** (2009) 025 [astro-ph/0810.1527]; S. Profumo, astro-ph/0812.4457.

SCENARIOS OF GRAVITINO DARK MATTER AND THEIR COSMOLOGICAL AND PARTICLE PHYSICS IMPLICATIONS

G. MOULTAKA*

Laboratoire de Physique Théorique et Astroparticules
UMR5207–CNRS, Université Montpellier II
Place E. Bataillon, F–34095 Montpellier Cedex 5, France
**E-mail: Gilbert.Moultaka@lpta.univ-montp2.fr*
www.univ-montp2.fr

I report on some scenarios where the gravitino is the dark matter and the supersymmetry breaking mediated by a gauge sector.

Keywords: Supersymmetry; Dark Matter; Gravitino; BBN.

1. Introduction

Providing a viable particle candidate for the non-baryonic dark matter (DM) in the Universe has become one of the main test requirements for model building beyond the standard model (SM) of particle physics. It is well-known (see for instance [1]) that an electrically neutral particle, weakly interacting with the primordial plasma and with a mass of order the electroweak scale ($\lesssim \mathcal{O}(1\,\text{TeV})$) would have today a relic density $\Omega \simeq 1$, provided it is stable or sufficiently long-lived, thus putting Ωh^2 in the ballpark of the WMAP results [2]. It should then not come so much as a surprise that most scenarios beyond the SM can provide potential solutions to the dark matter mystery, even less take it as an indication for their particular physical relevance. Rather, one should keep in mind that *i)* the abovementioned estimate of Ω assumes a simple thermal history of the early Universe *ii)* only a few classes of the proposed scenarios beyond the SM are theoretically framed in what was their motivation in the first place, i.e. solve the shortcomings of the standard model of particle physics.

We take hereafter point *ii)* as our guiding principle and address the question of dark matter from that point of view. We will then see that the assumption of point *i)* does not always apply in typical parameter space regions of the scenarios under consideration.

2. Supersymmetric Extensions

The supersymmetric (SUSY) extensions of the SM are among the most fashionable examples of *ii)*, including (at least) the ingredients of the minimal supersymmetric standard model (MSSM). Yet not all of them provide a DM candidate in the configuration *i)*. However, supersymmetry breaking being the trigger of the electroweak symmetry breaking, it is justified to study on the same footing the physical consequences of different SUSY breaking and mediation scenarios. For instance, in gravity mediated SUSY breaking scenarios [3], [4], [5] it is natural to expect the gravitino mass $m_{\tilde{G}}$ to be of order the electroweak scale, thus leaving room for a massive neutral weakly interacting particle such as a Neutralino to be the lightest SUSY particle. If stable, such a particle would perfectly fit point *i)* and provide a very good DM candidate. This tremendously studied scenario since the work of [6], [7], as natural as it may look, still relies on two crucial assumptions: the lightest susy particle (LSP) is not electrically charged (typically such as the tau slepton ($\tilde{\tau}$)) *and* there is a residual R-parity guaranteeing the stability (or at least a sufficiently long lifetime) of the lightest Neutralino. Theoretically, these two assumptions are not necessarily favored [a] since they can strongly depend on the actual dynamical mechanism underlying SUSY breaking, which is still poorly understood. An alternative option which has attracted much attention in recent years is to take the gravitino as the LSP, another logical possibility within the context of gravity mediated supersymmetry breaking, see for instance [8].

In this presentation we put the focus on a different kind of scenarios where the SUSY breaking and its mediation to the supersymmetric standard model is realized through some gauge interactions [9–17], [18–20]. The models originating from this class of gauge mediated susy breaking (GMSB) scenarios are phenomenologically as compelling as the gravity mediated ones, and have similar theoretical uncertainties. An important difference however is that here the gravitino is very light and *necessarily* the LSP, rather than this being a possibility among others. As usual, one can concoct exceptions. (See for instance [21] for a model where the gravitino is not the LSP, bringing the case back to the Neutralino DM configurations.) We stick however to the generic GMSB, assuming $m_{\tilde{G}} \lesssim \mathcal{O}(1 \text{ GeV})$; in this

[a] Apart from the requirement itself of tayloring a DM candidate!

case the DM issue is somewhat tricky, and in particular does not quite fit point *i)*.

3. The Phenomenological GMSB

We recall hereafter the main phenomenological ingredients of the gauge mediated susy breaking models based on the assumption that the leading contribution to the dynamical susy breaking is originating from some strongly coupled gauge sector (SBGS), screened (often dubbed 'hidden' or 'secluded') from the visible MSSM sector by some intermediate non-gauge interactions [18–20]. The various sectors are schematized as boxes in Fig. 1 and the interactions among them indicated by the arrows. The two messenger sectors are formed of matter chiral superfields $\hat{\Phi}_M, \hat{\overline{\Phi}}_M$ and $\hat{\phi}_m, \ldots$, having rather similar status; they have gauge interactions respectively with the visible MSSM sector and the hidden susy breaking sector, and non gauge interactions, through the superpotential, with an intermediate spurionic chiral superfield $\hat{S}$. $\hat{\Phi}_M$ and $\hat{\overline{\Phi}}_M$ have quark-like or lepton-like charges under $SU(3)_c \times SU(2)_L \times U(1)_Y$, $\Phi \sim (3, 1, -\frac{1}{3})$ or $(1, 2, \frac{1}{2})$, $\overline{\Phi}_M \sim (\bar{3}, 1, \frac{1}{3})$ or $(1, 2, -\frac{1}{2})$. To preserve gauge coupling unification these fields are usually put into larger gauge group multiplets, *e.g.* $\mathbf{5}+\bar{\mathbf{5}}$ or $\mathbf{10}+\overline{\mathbf{10}}$ of $\mathrm{SU}(5)_{\mathrm{GUT}}$, and $\mathbf{16}+\overline{\mathbf{16}}$ of $\mathrm{SO}(10)_{\mathrm{GUT}}$. The other messengers $\hat{\phi}_m$ are charged under some gauge group $\mathcal{G}$ through which they feel the properties of the susy breaking (secluded) sector.[b] Furthermore, the messenger fields on both sides are assumed to interact only indirectly via $\hat{S}$ through the superpotential $W \supset W_S + \Delta W(\hat{S}, \hat{\phi}_i) + W_{\mathrm{MSSM}}$ where, W_{MSSM} is the visible sector superpotential, $W_S = \kappa \hat{S} \hat{\Phi}_M \hat{\overline{\Phi}}_M + \frac{\lambda}{3} \hat{S}^3$ and $\hat{S}$ is neutral under all the gauge groups involved.

In the sequel we will be mainly interested in the three sectors on the right-thand side of Fig. 1. The conditions under which it is justified to ignore the effects of the two other sectors in the early Universe will be touched upon in Sec. 5. From the phenomenological point of view, all we need to assume here about these two sectors is that they cooperate to give non-zero vacuum expectation values (vev), $\langle S \rangle$ and $\langle F_S \rangle$, respectively to the scalar and F-term components of $\hat{S}$. This gives a supersymmetric mass $M_f = \kappa \langle S \rangle \equiv M_X$, to the (Dirac) fermion component as well as SUSY breaking mass spectrum $M_{s_\pm} = M_X (1 \pm \frac{\kappa \langle F_S \rangle}{M_X^2})^{1/2}$ to the mass eigenstates of

[b]We do not enter here the fascinating question of susy breaking through non-perturbative gauge interaction phenomena supposed to occur in the latter sector, which were studied since the early eighties (see [22, 23] for reviews) and rejuvinated recently [24].

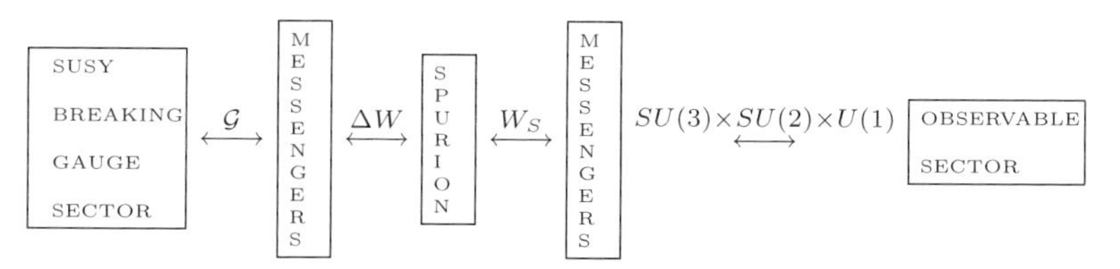

Fig. 1. generic structure of GMSB model sectors.

the scalar components of the $\hat{\Phi}_M, \hat{\overline{\Phi}}_M$ fields. The amount of susy breaking transmitted to the messenger/MSSM sectors, $\langle F_S \rangle$, is in general only a fraction of the total amount of the SUSY breaking in the SBGS which we denote $\langle F_{\rm TOT} \rangle$. The fermionic component ψ_S of $\hat{S}$ will then carry a fraction of the goldstino in the form $\psi_S = \frac{\langle F_S \rangle}{\langle F_{\rm TOT} \rangle} \tilde{G} + \dots$. It will thus contribute to the coupling of the massive gravitino to matter via its spin-$\frac{1}{2}$ component. Last but not least, the SUSY breaking is communicated to the visible sector through the gauge interactions of the messengers, leading to gaugino and scalar soft masses in the MSSM respectively at the 1$-$ and 2$-$loop levels in the form $M_i \sim \left(\frac{\alpha_i}{4\pi}\right) \frac{<F_S>}{M_X}$ and $\tilde{m}_a^2 \sim \left(\frac{\alpha_a}{4\pi}\right)^2 \left(\frac{<F_S>}{M_X}\right)^2$ (where $\alpha_{i,a}$ denote the SM gauge couplings or combinations thereof, and we have dropped for simplicity detailed flavor, messenger number and loop dependent coefficients). Assuming a typical grand unified group, the full MSSM and messenger spectrum and couplings depend uniquely on three continuous and one discrete parameters, namely M_X, $\frac{<F_S>}{M_X} (\equiv \Lambda)$, $\tan\beta$ (the ratio of the two higgs doublet vevs), and $N_{\rm mess}$ the number of quark-like/lepton-like messenger multiplets of some GUT group. [c]

Finally we note that the model defined so far possesses a discrete symmetry implying the conservation of the number of messengers in each physical process. An important consequence is that the lightest messenger particles (LMP) with mass M_{s_-} will be stable due to such a symmetry. As we will see in Sec. 5 such stable particles can have a dramatic cosmological effect. Furthermore, depending on the GUT group multiplets they belong to, the mass degeneracy of these LMPs can be lifted by quantum corrections [25, 26] leading to an LMP with very specific quantum numbers; e.g. $\tilde{\nu}_L$-like or $\tilde{e}_L$-like, if in a $\mathbf{5}+\bar{\mathbf{5}}$ of SU(5), an electrically charged SU$(2)_L$ singlet, if in a $\mathbf{10}+\overline{\mathbf{10}}$ of SU(5), and an MSSM singlet, if in a $\mathbf{16}+\overline{\mathbf{16}}$ of SO(10).

[c] For simplicity we fix the couplings relevant to the interactions with the spurion as $\kappa, \lambda \simeq 1$ in W_S, and choose the MSSM μ-parameter to be positive; note also that the trilinear soft parameter A_0 is vanishing to leading loop order.

4. Coupling to Supergravity

The gravitino being the gauge field of local supersymmetry and the superpartner of the graviton, its proper inclusion in the model of the previous section necessitates the coupling of the latter to supergravity. The ensuing rich structure allows to determine fairly uniquely all the couplings of the gravitino to the other states (such that flat space-time GMSB models are retrieved in the limit of infinite Planck mass) [27]. Coupling to supergravity has some other benefits: –the massless spin-$\frac{1}{2}$ goldstino, originating from the spontaneous SUSY breaking in the SBGS, appears as a mixture of the fermions of all chiral (resp. vector) supermultiplets whose F-terms (resp. D-terms) develop vevs, and ceases to be a physical state since it mixes automatically with the massless spin-$\frac{3}{2}$ gravitino giving a mass to the latter. This is the origin of the relation between ψ_S and $\tilde{G}$ noted in the previous section, where $\langle F_{\rm TOT}\rangle$ includes all F-term and D-term vevs –the requirement of an (almost) vanishing cosmological constant, together with that of SUSY breaking, leads to the general relation $\langle F_{TOT}\rangle \simeq \sqrt{3}\; m_{\tilde{G}}\; m_{\rm Pl}$, where $m_{\rm Pl}$ denotes the reduced Planck mass ($\simeq 2.4\times 10^{18}$GeV). Combined with the qualitative relation $G_F^{-1/2} \sim \frac{\langle F_S\rangle}{M_X}$ between the electroweak scale and the two mass scales at one's disposal in the visible sector, one finds $m_{\tilde{G}} \sim G_F^{-1/2}\frac{M_X}{\sqrt{3}m_{\rm Pl}k}$ (where $k \equiv \langle F_S\rangle/\langle F_{\rm TOT}\rangle$), that is typically a very light gravitino in gauge mediated models since $M_X \ll m_{\rm Pl}$ (as compared to gravity mediation where it becomes of order the electroweak scale with $M_X \sim O(m_{\rm Pl})$) –since GMSB generates soft gaugino masses M_i, the spontaneous SUSY breaking is accompanied by a spontaneous breaking of a continuous R symmetry. The latter leads to an R-axion which is phenomenologically problematic. However, the requirement mentioned previously for the cosmological constant is actually achieved through an additive constant $W_0 = m_{\tilde{G}} m_{\rm Pl}^2$ to the superpotential W of Sec. 3 (if there are no Planck scale vevs), thus breaking explicitly the R symmetry and giving possibly very large masses to the R-axion [28] –including gravity suggests that some discrete symmetries valid in flat space could be violated by non-perturbative quantum gravitational effects, involving for instance black hole physics [29]. In particular, the accidental messenger number conservation responsible for the stabilitly of the LMP can be lost, leading to Planck suppressed decays into MSSM particles through effective (non-)renormalizable messenger number violating (but gauge invariant [30]) operators. Such operators can appear either in the superpotential or in the Kähler potential, the latter being further organized into holomorphic or non-holomorphic in the fields; e.g., taking the SU(5) GUT particle content and messengers in $\mathbf{5}_M + \bar{\mathbf{5}}_M$, one can have $K_{\rm hol} \supset \mathbf{5}_M \bar{\mathbf{5}}_F$, $\frac{1}{m_{Pl}} \times \{\bar{\mathbf{5}}_M \bar{\mathbf{5}}_{F,H} \mathbf{10}_F\,,\, \mathbf{5}_M \mathbf{10}_F \mathbf{10}_F \ldots\}$ and $K_{\rm non-hol} \supset \bar{\mathbf{5}}^\dagger_M \bar{\mathbf{5}}_F$, $\frac{1}{m_{\rm Pl}} \times \{\bar{\mathbf{5}}^\dagger_M \bar{\mathbf{5}}_{H,F} \mathbf{10}_F\,,\, \bar{\mathbf{5}}_M \bar{\mathbf{5}}^\dagger_H \mathbf{10}_F\,,\, \bar{\mathbf{5}}^\dagger_M \mathbf{10}_F \mathbf{10}_F \ldots\}$, [31], or in the $\mathbf{16}_M + \overline{\mathbf{16}}_M$ of SO(10) GUT, one can have operators such as $K_{\rm hol} \supset \overline{\mathbf{16}}_M \mathbf{16}_F$, $\frac{1}{m_{\rm Pl}} \times \{\mathbf{16}_M \mathbf{16}_F \mathbf{10}_H\,,\, \overline{\mathbf{16}}_M \mathbf{16}_F \mathbf{45}_H \ldots\}$ or $K_{\rm non-hol} \supset \frac{1}{m_{\rm Pl}} \times \{\mathbf{16}^\dagger_M \mathbf{16}^\dagger_F \mathbf{10}_H\,,\, \overline{\mathbf{16}}^\dagger_M \mathbf{16}_F \mathbf{10}_H\,,\, \ldots \mathbf{16}^\dagger_M \mathbf{16}_F \mathbf{45}_H \ldots\}$ [32]. We note here that the supergravity features discussed above lead to an important difference between $K_{\rm hol}$ and $K_{\rm non-hol}$ after SUSY breaking: the $K_{\rm hol}$ contributions go effectively in the superpotential with an extra $m_{\tilde{G}}$ suppression, i.e. $W \supset m_{\tilde{G}} \times K_{\rm hol}$. As we will see in the following section, the above operators will play an important role in the cosmological fate of the LMP.

5. The Cosmological Set-Up

As noted at the end of Sec. 3 the LMP is stable within the minimal GMSB scenarios. If such a particle is produced at the end of inflation, i.e. $T_{\rm RH} \gtrsim M_{s_-}$, with $T_{\rm RH}$ the reheat temperature, then it will typically overclose the Universe with a relic density $\Omega_M h^2 \simeq 10^5 \left(\frac{M_{s_-}}{10^3 TeV}\right)^2$ [25, 33], unless its mass is finely adjusted. Of course, one can avoid this 'cosmological messenger problem' either assuming the LMP to be much lighter than $\sim 10^3$TeV or that it is simply too heavy to be produced in the early Universe. However, given our present ignorance of the actual value of $T_{\rm RH}$ that can range from 1MeV up to the GUT scale, and a rough idea about the messenger mass scale $\gtrsim 10^5$GeV [d] the LMP is expected to be generically present in the very early Universe. As we will argue, its presence can even play an important role in making the gravitino a viable DM candidate [33, 34], [31, 35].

In the mass range we consider, $\mathcal{O}(1\text{ keV}) \leq m_{\tilde{G}} \leq \mathcal{O}(1\text{ GeV})$, the gravitino is easily produced through scattering in the thermal bath (see [36–38] and references therein). Due to its gravitationally suppressed coupling, and in particular that of its spin-$\frac{1}{2}$ component which scales as $(m_{\tilde{G}} m_{\rm Pl})^{-1}$, the leading contribution to the thermal component of its relic density reads $\Omega^{\rm th}_{\tilde{G}} h^2 \simeq 0.32 \left(\frac{T_{\rm RH}}{10^8\text{ GeV}}\right) \left(\frac{10\text{ GeV}}{m_{\tilde{G}}}\right) \left(\frac{m_{1/2}}{1\text{ TeV}}\right)^2$. (Here $m_{1/2}$ denotes generically a common value of the gaugino soft masses M_i.) This illustrates one of the various facets of the so-called *gravitino problem*. The dependence on $T_{\rm RH}$, a paramter so far still poorly connected with the particle physics modelling,

---

[d] Indeed, requiring the MSSM soft masses to be $\lesssim 1$TeV implies $\frac{\langle F_S\rangle}{M_X} \lesssim 10^5$GeV. Furthermore, $M^2_{s-} \geq 0$ imposes $\langle F_S\rangle \leq M^2_X$, thus leading to $M_X \gtrsim 10^5$ GeV which gives the mass scale of the LMP, barring fine-tuned values.

is theoretically annoying as it requires a high level of adjustment, with basically no other observational consequences than providing an observationally consistent abundance for the gravitino if it is to play the role of DM. Perhaps more importantly, depending on the values of $m_{\tilde{G}}$ and $m_{1/2}$ (and other parameters of the MSSM), the gravitationally suppressed decay into (or of) the gravitino, depending on whether it is the LSP or the next to LSP (NLSP), can equally strongly affect the success of the standard Big Bang nucleosynthesis (BBN) predictions; we comment further these issues at the end of the section. On top of $\Omega_{\tilde{G}}^{\rm Th}$ the gravitino abundance can have substantial non-thermal contributions from the decay of whatever heavier relic particles, if such decays occur after these particles have dropped out of thermal equilibrium. For instance, if only MSSM particles are present, one gets a non-thermal contribution $\Omega_{\tilde{G}}^{\rm non\text{-}th}h^2 = \Omega_{\rm NLSP}h^2 \frac{m_{\tilde{G}}}{m_{\rm NLSP}}$, where $\Omega_{\rm NLSP}h^2$ is the abundance of the essentially thermally produced NLSP which can be a neutralino or a stau, akin to point *i)* of Sec. 1. $\Omega_{\tilde{G}}^{\rm non\text{-}th}h^2$ is often taken as the main source of gravitino abundance in scenarios of gravitino DM with $m_{\tilde{G}} \gtrsim 150$GeV (motivated by gravity mediation) [8], forgetting altogether the uncertainties from $\Omega_{\tilde{G}}^{\rm th}h^2$. We stress here that in GMSB scenarios one cannot play successfully a similar game since, due to the lightness of the gravitino, the above $\Omega_{\tilde{G}}^{\rm non\text{-}th}h^2$ cannot account alone for the observations as illustred in Fig. 2 where the scan extends up to $m_{\tilde{G}} = 100$GeV and the NLSP is a stau. It is then interesting to note that for a gravitino $\gtrsim 1$GeV (and $m_{\rm stau} \approx 200$GeV, a typical configuration for a not too fine-tuned GMSB) one needs a thermal component with $T_{\rm RH} \gtrsim 5 \times 10^6$ GeV in order to reach a suitable gravitino DM abundance. Such values of $T_{\rm RH}$ become of order the LMP mass suggesting that the LMP (and perhaps other heavier states of the messenger/spurion sectors) will be present in the early Universe. If so, a different thermal history may occur, modifying the usual MSSM based estimates. This brings us to the crux of the scenario: $T_{\rm RH}$ can be anywhere all the way up to very large values. Part or all of the GMSB sectors (Fig. 1) are thus present early on in the thermal bath and contribute to the thermal production of the gravitino which is then typically very large. As stressed at the beginning of this section the LMP decouples from the thermal bath with a very large abundance causing potentially an overcloser problem. However the LMP is likely to decay through Planck suppressed or gravitino suppressed operators as discussed at the end of Sec. 4. Such late decays occur typically after the LMP freeze-out and would substantially dilute the gravitino abundance through entropy release if they occur at a temperature where the LMP dominates the Universe *and* after the gravitino has decoupled from the thermal bath. Thus, the scenario entails the calculation of the LMP thermal relic density yield Y_M and messenger decay width Γ_M, and a comparison among its freeze-out temperature T_M^f, decay temperature $T_{\rm dec} \sim \Gamma_M^{1/2}$, matter domination temperature $T_{MD} \simeq \frac{4}{3} M_{s_-} \times Y_M$ as well as the gravitino freeze-out temperature. One finds a substantial part of the parameter space such that the diluted gravitino abundance is consistent with WMAP and can represent the (cold) dark matter however large $T_{\rm RH}$ may be! We show an example in Fig. 3 for $T_{\rm RH}$ as large as 10^{12} GeV in the case of $\mathbf{5}_M + \bar{\mathbf{5}}_M$ of SU(5) and with the first operator of $K_{\rm hol} \supset \mathbf{5}_M \bar{\mathbf{5}}_F$ given in Sec. 4 for illustration. One sees that the details of the messenger/spurion sectors can have an important effect on the viability of the DM scenario. For instance the small red-hatched area in the left-hand panel of Fig. 3 corresponds to gravitino DM solutions in the scenario of [33] where $\langle F_S \rangle \simeq \langle F_{\rm TOT} \rangle$. However it corresponds to a spurion much heavier than the LMP in a parameter space region (above the dashed black line) where spurion mediated LMP annihilation into gravitinos violates perturbative unitarity, thus theoretically unreliable. In contrast, viable solutions exist when the spurion is lighter than the LMP, as shown by the green/yellow region on the right-hand panel. A systematic study including other possible operators has been carried out in [31]. A more promising case is the $\mathbf{16}_M + \overline{\mathbf{16}}_M$ of SO(10). The LMP being an MSSM singlet in this case, its interaction with the thermal bath is loop suppressed leading to a much higher Y_M than in the SU(5) case for comparable M_{s_-}. Taking into account the decay induced by $K_{\rm hol}$ or $K_{\rm non-hol}$, one finds gravitino DM solutions when the spurion is much heavier than the LMP, but this time in regions where perturbative unitarity remains reliable [35], [31, 32]. By the same token, one can justify here not considering explicitly the SBGS and messenger sectors (left-hand part of Fig. 1) by assuming them to be much heavier than the LMP, thus playing a role similar to that of the spurion (i.e. essentially gravitational contributions to LMP annihilation into gravitinos).[e]

Finally, let us briefly discuss the issue of primordial nucleosynthesis of the light elements which constitutes an important observational probe of the earliest epochs of the thermal history. A late decaying particle from physics beyond the SM (with a lifetime $\tau \sim \mathcal{O}(1\,sec)$) can affect the successful standard big bang nucleosynthesis (SBBN) predictions through ei-

[e]Obviously these sectors could offer DM candidates, or lead to cosmological closer problems on their own. In this case they can be treated along similar lines quite symmetrical to the ones considered in the present study.

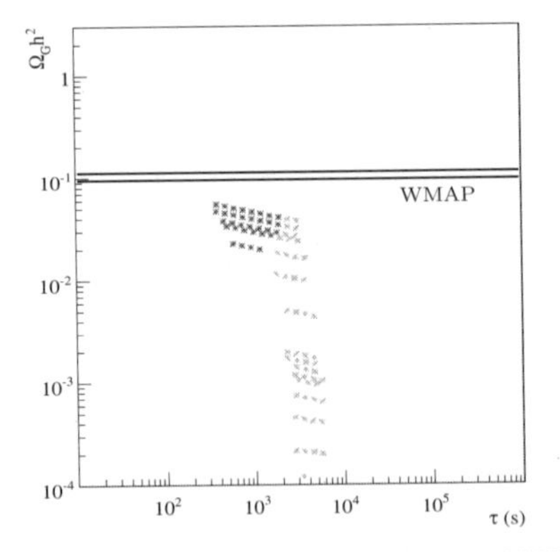

Fig. 2. The non-thermal stau-NLSP decay contribution $\Omega_{\tilde{G}}^{\text{non-th}}h^2$ to the gravitino abundance versus the NLSP lifetime, with $N_{\text{mess}} = 2$, $M_{s_-} = 5 \times 10^6\text{GeV}$, $\tan\beta = 10$, $10\text{MeV} \leq m_{\tilde{G}} \leq 100\text{GeV}$ and a scan over Λ, taken from [39]. The horizontal band corresponds to the $0.095 < \Omega_{\text{CDM}}h^2 < 0.136$ WMAP consistent region (see Fig. 4 for the green/red color code.)

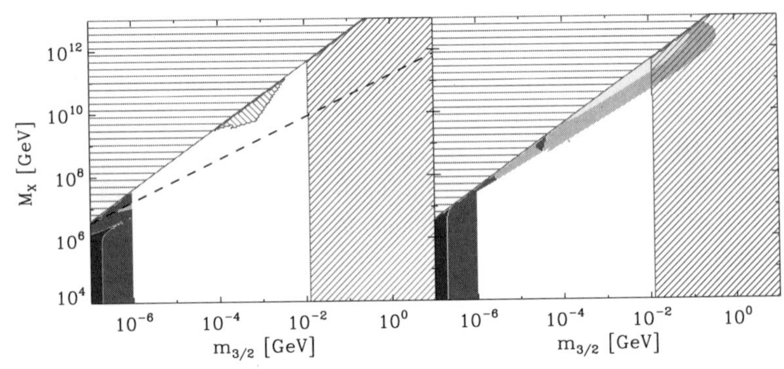

Fig. 3. sneutrino-like LMP versus gravitino masses. The spurion is heavier (lighter) than the LMP in the left-hand (right-hand) panel. Green/yellow region corresponds to gravitino cold DM with $\Omega_{\tilde{G}}h^2 < 0.3$; $T_{\text{RH}} = 10^{12}\text{GeV}$. (red/blue correspond to warm/hot gravitinos); the NLSP is assumed to be a 150 GeV Neutralino, decaying mainly into a photon (or a Z-boson) and a gravitino. The red-hatched bands to the right of each panel indicate the $m_{\tilde{G}}$ regions where this decay occurs after ~ 1 sec, thus potentially affecting primordial nucleosynthesis. Taken from [31] to which we refer for further details.

ther electromagnetic injections or hadronically induced nuclear reactions [f]. This possibility has become particularly interesting in the perspective of solving a problematic deviation from SBBN of the ^{7}Li and ^{6}Li inferred observational abundancies in low metalicity stars. Moreover, a very efficient

[f]See for instance [40], [41], [42] and references therein and thereout.

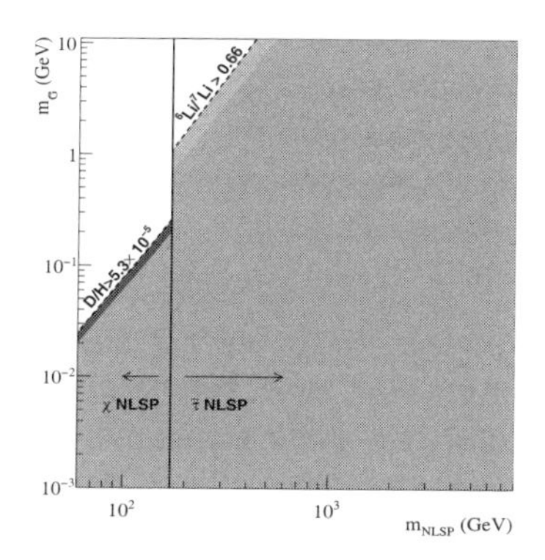

Fig. 4. Gravitino versus NLSP masses. same GMSB parameters as in Fig. 2; ^{7}Li/H$<$ 2.5×10^{-10} (red); $0.015 <$ ^{6}Li/^{7}Li$<$ 0.66 (green); SBBN: $Y_p \leq 0.258$, $1.2 \times 10^{-5} \leq$ D/H $\leq 5.3 \times 10^{-5}$, ^{3}He/D$\leq$ 1.72 (light blue); taken from [44].

catalyses of the ^{6}Li–producing reaction can occur if the decaying particle is electrically charged and sufficiently long lived [43]. Constraints on physics beyond the SM are thus of two types: conservative (consistency with SBBN) or speculative (solving the Lithium problems). We illustrate these two features in Fig. 4 within the GMSB context [44], showing the effect of the nature of the NLSP on the lithium yields.

In this respect, it is to be noted that the LMP decays typically at temperatures $\mathcal{O}$(100MeV) if $M_{s_-} \geq 10^3$TeV, thus rendering the gravitino DM scenarios we have described here quite safe from the BBN perspective. Nonetheless, one should keep in mind that it remains exclusively a task for the colliders to ultimately favor or disprove GMSB scenarios.

Acknowledgments

My thanks go to the organizers of DARK 2009 for the very enjoyable atmosphere and the quite diversified topics of the conference. This work was supported in part by ANR under contract NT05-1_43598/ANR-05-BLAN-0193-03.

References

1. G. Jungman, M. Kamionkowski and K. Griest, *Phys. Rept.* **267**, 195 (1996).
2. E. Komatsu *et al.*, *Astrophys. J. Suppl.* **180**, 330 (2009).
3. A. H. Chamseddine, R. L. Arnowitt and P. Nath, *Phys. Rev. Lett.* **49**, p. 970 (1982).
4. R. Barbieri, S. Ferrara and C. A. Savoy, *Phys. Lett.* **B119**, p. 343 (1982).
5. L. J. Hall, J. D. Lykken and S. Weinberg, *Phys. Rev.* **D27**, 2359 (1983).

6. H. Goldberg, *Phys. Rev. Lett.* **50**, p. 1419 (1983).
7. J. R. Ellis, J. S. Hagelin, D. V. Nanopoulos, K. A. Olive and M. Srednicki, *Nucl. Phys.* **B238**, 453 (1984).
8. K. A. Olive, *Eur. Phys. J.* **C59**, 269 (2009).
9. P. Fayet, *Phys. Lett.* **B78**, p. 417 (1978).
10. M. Dine, W. Fischler and M. Srednicki, *Nucl. Phys.* **B189**, 575 (1981).
11. S. Dimopoulos and S. Raby, *Nucl. Phys.* **B192**, p. 353 (1981).
12. M. Dine and W. Fischler, *Phys. Lett.* **B110**, p. 227 (1982).
13. M. Dine and M. Srednicki, *Nucl. Phys.* **B202**, p. 238 (1982).
14. M. Dine and W. Fischler, *Nucl. Phys.* **B204**, p. 346 (1982).
15. L. Alvarez-Gaume, M. Claudson and M. B. Wise, *Nucl. Phys.* **B207**, p. 96 (1982).
16. C. R. Nappi and B. A. Ovrut, *Phys. Lett.* **B113**, p. 175 (1982).
17. S. Dimopoulos and S. Raby, *Nucl. Phys.* **B219**, p. 479 (1983).
18. M. Dine and A. E. Nelson, *Phys. Rev.* **D48**, 1277 (1993).
19. M. Dine, A. E. Nelson and Y. Shirman, *Phys. Rev.* **D51**, 1362 (1995).
20. M. Dine, A. E. Nelson, Y. Nir and Y. Shirman, *Phys. Rev.* **D53**, 2658 (1996).
21. S. Shirai, F. Takahashi, T. T. Yanagida and K. Yonekura, *Phys. Rev.* **D78**, p. 075003 (2008).
22. H. P. Nilles, *Phys. Rept.* **110**, 1 (1984).
23. G. F. Giudice and R. Rattazzi, *Phys. Rept.* **322**, 419 (1999).
24. K. A. Intriligator, N. Seiberg and D. Shih, *JHEP* **04**, p. 021 (2006).
25. S. Dimopoulos, G. F. Giudice and A. Pomarol, *Phys. Lett.* **B389**, 37 (1996).
26. T. Han and R. Hempfling, *Phys. Lett.* **B415**, 161 (1997).
27. J. Wess and J. Bagger Princeton, USA: Univ. Pr. (1992) 259 p.
28. J. Bagger, E. Poppitz and L. Randall, *Nucl. Phys.* **B426**, 3 (1994).
29. A. D. Dolgov, M. V. Sazhin and Y. B. Zeldovich, *Basics of modern cosmology* Ed. Frontieres (1991).
30. L. M. Krauss and F. Wilczek, *Phys. Rev. Lett.* **62**, p. 1221 (1989).
31. K. Jedamzik, M. Lemoine and G. Moultaka, *Phys. Rev.* **D73**, p. 043514 (2006).
32. M. Kuroda and G. Moultaka, *in preparation* .
33. M. Fujii and T. Yanagida, *Phys. Lett.* **B549**, 273 (2002).
34. E. A. Baltz and H. Murayama, *JHEP* **05**, p. 067 (2003).
35. M. Lemoine, G. Moultaka and K. Jedamzik, *Phys. Lett.* **B645**, 222 (2007).
36. M. Bolz, A. Brandenburg and W. Buchmuller, *Nucl. Phys.* **B606**, 518 (2001).
37. J. Pradler and F. D. Steffen, *Phys. Lett.* **B648**, 224 (2007).
38. V. S. Rychkov and A. Strumia, *Phys. Rev.* **D75**, p. 075011 (2007).
39. S. Bailly (Ph.D. thesis (in french), 2008).
40. S. Dimopoulos, R. Esmailzadeh, L. J. Hall and G. D. Starkman, *Astrophys. J.* **330**, p. 545 (1988).
41. K. Jedamzik, *Phys. Rev.* **D70**, p. 063524 (2004).
42. M. Kawasaki, K. Kohri and T. Moroi, *Phys. Lett.* **B625**, 7 (2005).
43. M. Pospelov, *Phys. Rev. Lett.* **98**, p. 231301 (2007).
44. S. Bailly, K. Jedamzik and G. Moultaka, *arXiv:0812.0788 [hep-ph]* .

THE PHENOMENOLOGY OF GRAVITINO DARK MATTER SCENARIOS IN SUPERGRAVITY MODELS

Y. SANTOSO

Institute for Particle Physics Phenomenology,
Department of Physics, University of Durham,
Durham, DH1 3LE, United Kingdom
E-mail: yudi.santoso@durham.ac.uk

We review the phenomenology of gravitino dark matter within supergravity framework. Gravitino can be dark matter if it is the lightest supersymmetric particle, which is stable if R-parity is conserved. There are several distinct scenarios depending on what the next to lightest supersymmetric particle (NLSP) is. We discuss the constraints and summarize the phenomenology of neutralino, stau, stop and sneutrino NLSPs.

Keywords: Dark matter; Supergravity phenomenology.

1. Introduction

The identity of dark matter has not yet been resolved at the time this article was written. One interesting possibility is to have a neutral lightest supersymmetric particle (LSP), which is stable via R-parity conservation, as the dark matter.[1] Neutralino is a popular candidate for dark matter and sneutrino is another possibility [a]. Here, we focus our attention on yet another hypothesis within supergravity, i.e. gravitino as cold dark matter.[3,4]

In supergravity models (i.e. models with gravity mediated supersymmetry breaking),[5] the gravitino mass is close to the other sparticle masses. However, it is not a priori whether the gravitino is lighter or heavier than the others. Note that this is different from gauge mediated models in which the gravitino mass can be naturally very light.[6] We assume supergravity models here, with supersymmetric masses of $\sim 1\,\text{GeV} - 1$ TeV and the gravitino is the LSP. In this framework, the coupling between gravitino and mat-

[a] Left-sneutrino is excluded by direct detection,[2] but right-sneutrino is still viable.

ter fields is very small, $\sim 1/M_{\rm Pl}$. Because of this, gravitino is practically undetectable (aside from its gravitational effect). Also, the next lightest supersymmetric particle (NLSP) could be long lived, with a lifetime of typically O(1 s) or longer. In this case, the NLSP decay affects the primordial light element abundances.[7] The phenomenology of this scenario depends largely on what the NLSP is. We will discuss below various possibilities, each with its own distinct phenomenology.

2. Gravitino Dark Matter in Supergravity Models

The biggest theoretical uncertainty in supersymmetric models arises from the fact that we do not know how supersymmetry, if it does exist, is broken in nature. Because of this, the values of the soft couplings are uncertain. We can take them as free parameters. However, because of the large number of parameters, we need to make some simplifying assumptions. Motivated by the Grand Unified Theory (GUT), the usual assumption is that parameters of the same type are unified at the GUT scale. Their values at weak scale are then derived by employing the renormalization group equation (RGE). The simplest model of this kind is the CMSSM (Constrained Minimal Supersymmetric Standard Model) [b], in which we have universal gaugino mass $m_{1/2}$, universal sfermion mass m_0, and universal trilinear coupling A_0 at the GUT scale. In addition, we have two parameters from the Higgs sector, i.e. the ratio of the two Higgs vevs $\tan\beta \equiv \langle H_1\rangle/\langle H_2\rangle$, and the sign of μ where μ is the Higgs mixing parameter in the superpotential. Note however that in GUT theories, e.g. $SU(5)$ or $SO(10)$, the Higgs fields are contained in different multiplets as compared to the matter multiplets. This motivates a generalization of the CMSSM, in which the Higgs soft masses $m_{1,2}$ are not necessarily equal to m_0 at the GUT scale.[8] Furthermore, we can trade $m_{1,2}$ with μ and the CP-odd Higgs mass m_A as our free parameters through the electroweak symmetry breaking condition. The resulting model is called Non-Universal Higgs Masses (NUHM) model.[9] Thus the NUHM parameters are $m_{1/2}$, m_0, A_0, $\tan\beta$, μ and m_A.

In the usual scenario with neutralino LSP, we implicitly assume that gravitino is sufficiently heavy such that it decouples from the low energy theory. In the gravitino dark matter (GDM) scenario, on contrary, we assume that the gravitino mass $m_{\tilde{G}} = m_{3/2}$ is sufficiently small such that the gravitino is the LSP. For our purposes, we can take $m_{3/2}$ as another free parameter. Within CMSSM, with gravitino LSP, there are three possible

[b] Also known as mSUGRA, depending on your preference.

NLSP, i.e. neutralino, stau and stop particles.[3,10] For NUHM, in addition, we can have selectron or sneutrino as the NLSP.[11] Of course for a more general model of MSSM we can have more possibilities.

3. Phenomenological Constraints

3.1. *Dark matter relic density constraint*

Being a very weakly interacting particle, gravitino decoupled very quickly from the thermal plasma in the early universe. This leads to a concern that the gravitino could be over-abundance. However, inflation can solve this problem.[12] In inflationary models, the early gravitino density together with other densities are diluted by the inflation. Gravitino is then reproduced by reheating after the inflation,[13] although with a smaller yield that can still satisfy the relic density constraint [c].

Gravitino relic density consists of two parts, the thermal relic $\Omega^{\rm T}_{\widetilde{G}}$ which is produced by reheating, and the non-thermal relic $\Omega^{\rm NT}_{\widetilde{G}}$ coming from the decay of the NLSP.

$$\Omega_{\widetilde{G}}h^2 = \Omega^{\rm T}_{\widetilde{G}}h^2 + \Omega^{\rm NT}_{\widetilde{G}}h^2 \tag{1}$$

The thermal relic is related to the reheating temperature T_R through the following relation[15]

$$\Omega^{\rm T}_{\widetilde{G}}h^2 \simeq 0.27\left(\frac{T_R}{10^{10}\,{\rm GeV}}\right)\left(\frac{100\,{\rm GeV}}{m_{\widetilde{G}}}\right)\left(\frac{m_{\tilde{g}}}{1\,{\rm TeV}}\right)^2 \tag{2}$$

where $m_{\tilde{g}}$ is the gluino mass. We can see that for $m_{\widetilde{G}} = 100$ GeV and $m_{\tilde{g}} = 1$ TeV, to get $\Omega^{\rm T}_{\widetilde{G}}h^2 \lesssim 0.1$ we need $T_R \lesssim 10^{10}$ GeV. The value of T_R depends on the inflation model. For our purpose, we take T_R as a free parameter. For one to one decays of NLSP to gravitino, which is generally the case, the gravitino non-thermal relic can be written as:

$$\Omega^{\rm NT}_{\widetilde{G}}h^2 = \frac{m_{\widetilde{G}}}{m_{\rm NLSP}}\Omega_{\rm NLSP}h^2 \tag{3}$$

where $\Omega_{\rm NLSP}h^2$ is the NLSP density before the decay. Due to the long lifetime, the NLSP density is frozen out long before its decay, and this can be calculated by the usual method of solving the Boltzmann equation in the expanding universe. Note that even if the NLSP density is larger than the WMAP value, we might still satisfy the dark matter relic density constraint because of the rescaling by the mass ratio $m_{\widetilde{G}}/m_{\rm NLSP}$. The NLSP density

[c] This can still impose a strong constraint on the inflationary theories.[14] However, this topic is beyond the scope of this article.

can also be written in term of the yield, $Y_{\rm NLSP} = n_{\rm NLSP}/s$, where n is the number density and s is the entropy density. This is related to $\Omega_{\rm NLSP}h^2$ by

$$Y_{\rm NLSP}M_{\rm NLSP} = \Omega_{\rm NLSP}h^2 \times (3.65 \times 10^{-9}\ {\rm GeV}) \tag{4}$$

The total relic density of the gravitino must not exceed the upper limit of the dark matter relic density range as suggested by WMAP:[16]

$$\Omega_{\rm DM}^{\rm WMAP}h^2 \simeq 0.113 \pm 0.004 \tag{5}$$

Taking 2σ, this means that $\Omega_{\widetilde{G}}h^2 \lesssim 0.121$. The percentage of the thermal versus non-thermal relic density depends on the strength of the NLSP interactions which determine $\Omega_{\rm NLSP}h^2$. If we can measure these interactions at colliders we can deduce the reheating temperature.[17]

3.2. *The BBN constraints*

Big bang nucleosynthesis (BBN) is often cited as the greatest success of the big bang theory. By using simple assumptions that the early universe is in thermal equilibrium and expanding one can calculate the primordial light element abundances (using standard nuclear cross sections) and gets results which agree very well with the observations. If there is a metastable particle that decays during or after the BBN era, the light element abundances can be altered by the participation of the energetic decay products in the nucleosynthesis processes. Thus, BBN provides a stringent constraint for gravitino dark matter. On the other hand, the prediction of the standard BBN (sBBN) is not perfectly in agreement with the observational data. There seems to be discrepancy between the observed lithium abundances and the predicted values as shown in Table 1. This is known as the lithium problem. The discrepancy on ^{6}Li is particularly difficult to be solved within the standard theory. There is no known astrophysical process that can produce large amount of ^{6}Li. Moreover ^{6}Li is fragile. Therefore we should expect less rather than more ^{6}Li compared to the prediction. The lithium problem could be an indication of a new physics beyond the standard model. There are two proposed solution to this problem through

Table 1. Comparison of lithium abundances from standard model prediction and observations.

	^{7}Li/H	^{6}Li/^{7}Li
Observation	$(1-2)\times 10^{-10}$	$\sim 0.01 - 0.15$
Standard BBN (with CMB)	$\sim 4\times 10^{-10}$	$< 10^{-4}$

a hypothesized metastable particle. The first one is through catalytic effect.[18,19] The process $d+{}^4\text{He} \to {}^6\text{Li} + \gamma$ is suppressed by parity. If there is a massive negatively charged particle X^- that is bound to ^{4}He by Coulomb interaction it can absorb the emitted photon, hence the process is no longer parity suppressed. Simultaneously, the X^- particle is freed from the bound by the energy released and can subsequently be attached to another ^{4}He, thus acting as a catalysis for ^{6}Li production. This catalytic effect can also effect other light element abundances such as beryllium.[20]

Another proposed solution to the lithium problem is through hadronic decay of a metastable particle.[21] The decay produces energetic n, p and also T, ^{3}He (through spallation of ^{4}He), which then interact with the ambient nuclei, e.g. $n + p \to$D, T + ${}^3\text{He} \to {}^6$Li (producing more ^{6}Li), and ${}^7\text{Be}(n,p){}^7\text{Li}(p,{}^4\text{He}){}^4\text{He}$ (reducing ^{7}Li). Note that deuterium is also produced, hence put some constraints on this scenario.

3.3. *Astrophysical constraints*

If the NLSP decays at a time later than the BBN, the photons produced by the decay might not be fully thermalised by the time of recombination. This can cause distortion on the cosmic microwave background radiation black body spectrum.[22] The size of the distortion depends on the amount of the energy injected into photons. This is represented by a chemical potential μ for photon. CMB spectrum measurement by the FIRAS instrument onboard of COBE satellite sets an upper limit on μ:[23]

$$|\mu| \lesssim 9 \times 10^{-5} \tag{6}$$

This limit is only important for lifetime $\gtrsim 10^6$ s since photons produced earlier should have enough time to thermalize before recombination.

Gravitinos from the NLSP decay at a late time have larger velocities compared to the primordial gravitino. This leads to a longer free-streaming length, smoothing out small scale density perturbation. If the dark matter relic density is dominated by non-thermal relic, the structure formation would be affected. This scenario is proposed as a solution to the small scale problem.[24]

3.4. *Collider constraints*

At colliders, heavier supersymmetric particles can still be produced provided that there is enough energy in the collisions. These sparticle would quickly decay, cascading down to the NLSP. Due to its long lifetime the NLSP itself would escape from the detector before eventually decays to gravitino, hence appear as a stable particle with respect to the detectors. There have been searches for stable massive particles (SMP) at colliders.[25]

A particularly interesting signal would be produced if the NLSP is electromagnetically charged. In this case the NLSP should traverse the calorimeter and subsequently be detected by the muon detector. The first obvious step of the data analysis is of course to discover this charged NLSP. The CDF collaboration, based on 1.0 fb^{-1} of data at $\sqrt{s} = 1.96$ TeV, sets a lower bound on (meta)stable stop particle at 249 GeV;[26] while the D0 collaboration, using 1.1 fb^{-1} of data, sets upper limits for stable stau pair production cross section from 0.31 pb to 0.04 pb for stau masses between 60 GeV and 300 GeV[27] and lower mass limits of 206 GeV and 171 GeV for pair produced stable charged gauginos and higgsinos respectively.

However, not all possible NLSP are charged. If neutralino or sneutrino is the NLSP, they would not be detected. Only neutralino with a very short lifetime ($\lesssim$ few ns) can be detected through its decay product. The CDF sets a lower limit on the neutralino mass at 101 GeV for lifetime 5 ns.[28] Similar to the familiar case with neutralino LSP, there are also various signatures from the cascade decays. The same methods of analysis can be applied to the case with stable neutral NLSP. For long-lived neutralino NLSP the signatures would be indistinguishable from that of neutralino LSP scenario. For sneutrino NLSP, however, the signatures would in general be different.[29]

4. Phenomenology of GDM with Various NLSP

In this section we look at each scenario of neutralino, stau, stop and sneutrino NLSP. We do not include chargino NLSP[30] here.

4.1. *Neutralino NLSP*

For neutralino mass of 1 TeV and gravitino mass of 1 GeV the neutralino lifetime is about $O(1)$ s. The lifetime is longer for a smaller mass gap. Thus the neutralino in this scenario would escape the collider detectors and trigger large missing energy signatures. Assuming that all primordial neutralinos has eventually decayed to gravitino, only gravitino is floating around today. Therefore WIMP direct detection experiments would not see any signal. There would be no indirect astrophysical signal from dark matter annihilation in the halo either. In this case it would be difficult to proof the identity of the dark matter, i.e. whether it is gravitino, axino or whether something else. We will also need to check whether R-parity is really conserved.

The neutralino NLSP in the CMSSM is much constrained by the BBN, especially when the neutralino lifetime is $\gtrsim 10^4$ s where the electromagnetic shower effect on the light element abundances becomes important. The main reason is because neutralino has relatively large freeze-out density for most of the parameter space.

4.2. *Stau NLSP*

If produced, a stau NLSP would be seen as a massive stable charged particle at colliders. It would leave a clean track in the inner detector and then reach the muon detector, hence it would look like a slow/heavy muon. Because of its electromagnetic charge, the stau can be slowed down by making it go through a bulky medium. Thus it can be trapped and stored until it decays.[31,32] In this way one can hope to measure its lifetime.

Within the CMSSM, stau NLSP has the largest allowed region of parameter space, hence thought as the natural candidate. Stau NLSP would yield catalytic effect on BBN. This has attract much attention and many papers are devoted in the study of this topic, in particular regarding the lithium problem solution.[33]

4.3. *Stop NLSP*

A long lived stop would hadronize once it is produced. By taking analogy with heavy quark hadrons one can deduce the lightest hadron states and their lifetimes. The light stop sbaryons are $\Lambda^+_{\tilde{T}} \equiv \tilde{t}_1 ud$ (which is the lightest), $\Sigma^{++,+,0}_{\tilde{T}} \equiv \tilde{t}_1(uu, ud, dd)$ (which decays through strong interaction), and $\Xi^{+,0}_{\tilde{T}} \equiv \tilde{t}_1 s(u,d)$ (which decays semileptonically with lifetime $\tau \lesssim 10^{-2}$ s). The light stop mesinos are $\widetilde{T}^0 \equiv \tilde{t}_1\bar{u}$ (which is the lightest), $\widetilde{T}^+ \equiv \tilde{t}_1\bar{d}$ (with lifetime $\tau \simeq 1.2$ s), and $\widetilde{T}_s \equiv \tilde{t}_1\bar{s}$ (with lifetime $\tau \simeq 2\times 10^{-6}$ s). The antistop would hadronize into the corresponding antisbaryons and antimesinos. In the early universe, being the lighter one, the neutral $\widetilde{T}^0$ is more abundance than the charged $\Lambda^{\pm}_{\tilde{T}}$. This reduces the catalytic effect on BBN. Moreover, due to its strong interaction, the freeze out density of stop is generally small. Therefore this scenario can generally satisfy the BBN constraint.[10]

On the other hand, it was shown[34] that stop NLSP scenario is fit to solve the lithium problem through hadronic decay. Note that further annihilation of stop occurs after the hadronization, with annihilation rate $\Gamma_{\rm ann} = \langle \sigma v \rangle n_{\tilde{t}}$ where $\sigma \sim R^2_{\rm had}$ and $v \simeq \sqrt{3T/m_{\tilde{t}}}$. The final stop abundance before its decay can be written as

$$m_{\tilde{t}} Y_{\tilde{t}} = 0.87 \times 10^{-14}\,\text{GeV} \left(\frac{f_\sigma}{0.1}\right)^{-2} \left(\frac{g_*}{17.25}\right)^{-1/2} \times \left(\frac{T_{\text{QCD}}}{150\,\text{MeV}}\right)^{-3/2} \left(\frac{m_{\tilde{t}}}{10^2\,\text{GeV}}\right)^{3/2} \quad (7)$$

It was found that, with $f_\sigma = 0.1$, the lithium problem solution prefers $m_{\tilde{t}} = 400 - 600$ GeV and $m_{3/2} = 2 - 10$ GeV.

4.4. *Sneutrino NLSP*

Since sneutrino does not interact strongly nor electromagnetically the effect of sneutrino decays on BBN can be guessed to be small, but nonzero. The BBN effect comes through energy transfer (elastic/inelastic) from energetic neutrino (produced by the decay $\tilde{\nu} \to \tilde{G} + \nu$) to the background particles, and from 3/4-body decay modes[35] ($\tilde{\nu} \to \tilde{G} + \nu + (\gamma, Z)$, $\tilde{\nu} \to \tilde{G} + \ell + W$) and 4-body ($\tilde{\nu} \to \tilde{G} + \nu + f + \bar{f}$, $\tilde{\nu} \to \tilde{G} + \ell + f + \bar{f}'$). Although the 3 and 4 body decay branching ratios are small, they can still be important since they produced particles that can directly involve in the nucleosynthesis processes. For $M_{\tilde{\nu}} \sim O(100$ GeV$)$, the BBN constraint can be satisfied if the sneutrino density before the decay is

$$Y_{\tilde{\nu}} M_{\tilde{\nu}} \lesssim \mathcal{O}(10^{-11})\ \text{GeV} \qquad \text{for} \quad B_h = 10^{-3} \quad (8)$$

$$Y_{\tilde{\nu}} M_{\tilde{\nu}} \lesssim \mathcal{O}(10^{-8})\ \text{GeV} \qquad \text{for} \quad B_h = 10^{-6} \quad (9)$$

where B_h is the hadronic branching ratio. The sneutrino NLSP and gravitino LSP scenario was explored within NUHM models, and it was found that there are large regions of parameter space still allowed.[11]

At colliders, similar to the neutralino (N)LSP case, sneutrino NLSP would yield a missing energy signature. We can study this scenario through cascade decays of heavier supersymmetric particles. In general, the signatures are different from those in neutralino case. Signatures that are thought to be the best for neutralino LSP might not be suitable for this sneutrino NLSP case. A preliminary study of collider phenomenology with sneutrino NLSP has been done in Ref. 29. However, a more detail study might reveal more information.

5. Concluding Remarks

Gravitino is a feasible and interesting candidate for dark matter. There are many possible phenomenology with gravitino dark matter depending on the choice for the NLSP, which we have summarized in this article. Future and upcoming collider experiments, such as the LHC, might be able to unveil some hints on the identity of dark matter. Progresses in direct and indirect detection experiments are also looked promising. The next few years would be an interesting time to find out more about dark matter, and whether gravitino can still stand up as a candidate for dark matter.

Acknowledgments

My participation in Dark2009 conference was made possible by the support of British Royal Society. I thank my collaborators on the subject of this paper: John Ellis, Lorenzo Diaz-Cruz, Terrance Figy, Kazunori Kohri, Keith Olive, Krzysztof Rolbiecki, and Vassilis Spanos.

References

1. J. R. Ellis, J. S. Hagelin, D. V. Nanopoulos, K. A. Olive and M. Srednicki, Nucl. Phys. B **238**, 453 (1984).
2. T. Falk, K. A. Olive and M. Srednicki, Phys. Lett. B **339**, 248 (1994).
3. J. R. Ellis, K. A. Olive, Y. Santoso and V. C. Spanos, Phys. Lett. B **588**, 7 (2004).
4. J. L. Feng, A. Rajaraman and F. Takayama, Phys. Rev. Lett. **91**, 011302 (2003); Phys. Rev. D **68**, 063504 (2003); J. L. Feng, S. F. Su and F. Takayama, Phys. Rev. D **70**, 063514 (2004); Phys. Rev. D **70**, 075019 (2004); L. Roszkowski, R. Ruiz de Austri and K. Y. Choi, JHEP **0508**, 080 (2005).
5. A. H. Chamseddine, R. L. Arnowitt and P. Nath, Phys. Rev. Lett. **49**, 970 (1982); R. Barbieri, S. Ferrara and C. A. Savoy, Phys. Lett. B **119**, 343 (1982); L. J. Hall, J. D. Lykken and S. Weinberg, Phys. Rev. D **27**, 2359 (1983).
6. G. F. Giudice and R. Rattazzi, Phys. Rept. **322**, 419 (1999).
7. T. Moroi, arXiv:hep-ph/9503210.
8. D. Matalliotakis and H. P. Nilles, Nucl. Phys. B **435**, 115 (1995); M. Olechowski and S. Pokorski, Phys. Lett. B **344**, 201 (1995); V. Berezinsky, A. Bottino, J. R. Ellis, N. Fornengo, G. Mignola and S. Scopel, Astropart. Phys. **5**, 1 (1996); M. Drees, M. M. Nojiri, D. P. Roy and Y. Yamada, Phys. Rev. D **56**, 276 (1997) [Erratum-ibid. D **64**, 039901 (2001)]; P. Nath and R. L. Arnowitt, Phys. Rev. D **56**, 2820 (1997); A. Bottino, F. Donato, N. Fornengo and S. Scopel, Phys. Rev. D **63**, 125003 (2001); S. Profumo, Phys. Rev. D **68**, 015006 (2003); D. G. Cerdeno and C. Munoz, JHEP **0410**, 015 (2004); H. Baer, A. Mustafayev, S. Profumo, A. Belyaev and X. Tata, JHEP **0507**, 065 (2005); J. R. Ellis, K. A. Olive and P. Sandick, Phys. Rev. D **78**, 075012 (2008); L. Roszkowski, R. R. de Austri, R. Trotta, Y. L. Tsai and T. A. Varley, arXiv:0903.1279 [hep-ph].
9. J. R. Ellis, K. A. Olive and Y. Santoso, Phys. Lett. B **539**, 107 (2002); J. R. Ellis, T. Falk, K. A. Olive and Y. Santoso, Nucl. Phys. B **652**, 259 (2003).

10. J. L. Diaz-Cruz, J. R. Ellis, K. A. Olive and Y. Santoso, JHEP **0705**, 003 (2007).
11. J. R. Ellis, K. A. Olive and Y. Santoso, JHEP **0810**, 005 (2008).
12. J. R. Ellis, A. D. Linde and D. V. Nanopoulos, Phys. Lett. B **118**, 59 (1982).
13. M. Y. Khlopov and A. D. Linde, Phys. Lett. B **138**, 265 (1984).
14. M. Kawasaki, F. Takahashi and T. T. Yanagida, Phys. Lett. B **638**, 8 (2006).
15. M. Bolz, A. Brandenburg and W. Buchmuller, Nucl. Phys. B **606**, 518 (2001) [Erratum-ibid. B **790**, 336 (2008)].
16. E. Komatsu *et al.* [WMAP Collaboration], Astrophys. J. Suppl. **180**, 330 (2009).
17. J. Pradler and F. D. Steffen, Phys. Lett. B **648**, 224 (2007); K. Y. Choi, L. Roszkowski and R. Ruiz de Austri, JHEP **0804**, 016 (2008).
18. M. Pospelov, Phys. Rev. Lett. **98**, 231301 (2007); C. Bird, K. Koopmans and M. Pospelov, Phys. Rev. D **78**, 083010 (2008).
19. K. Kohri and F. Takayama, Phys. Rev. D **76**, 063507 (2007); M. Kaplinghat and A. Rajaraman, Phys. Rev. D **74**, 103004 (2006).
20. M. Pospelov, J. Pradler and F. D. Steffen, JCAP **0811**, 020 (2008).
21. K. Jedamzik, Phys. Rev. D **70**, 063524 (2004); D. Cumberbatch, K. Ichikawa, M. Kawasaki, K. Kohri, J. Silk and G. D. Starkman, Phys. Rev. D **76**, 123005 (2007).
22. R. Lamon and R. Durrer, Phys. Rev. D **73**, 023507 (2006).
23. D. J. Fixsen, E. S. Cheng, J. M. Gales, J. C. Mather, R. A. Shafer and E. L. Wright, Astrophys. J. **473**, 576 (1996).
24. M. Kaplinghat, Phys. Rev. D **72**, 063510 (2005); J. A. R. Cembranos, J. L. Feng, A. Rajaraman and F. Takayama, Phys. Rev. Lett. **95**, 181301 (2005); F. D. Steffen, JCAP **0609**, 001 (2006).
25. M. Fairbairn *et al.*, Phys. Rept. **438**, 1 (2007).
26. T. Aaltonen *et al.* [CDF Collaboration], arXiv:0902.1266 [hep-ex].
27. V. M. Abazov *et al.* [D0 Collaboration], arXiv:0809.4472 [hep-ex].
28. T. Aaltonen *et al.* [CDF Collaboration], Phys. Rev. D **78**, 032015 (2008).
29. L. Covi and S. Kraml, JHEP **0708**, 015 (2007).
30. G. D. Kribs, A. Martin and T. S. Roy, JHEP **0901**, 023 (2009).
31. J. L. Feng and B. T. Smith, Phys. Rev. D **71**, 015004 (2005) [Erratum-ibid. D **71**, 019904 (2005)].
32. K. Hamaguchi, M. M. Nojiri and A. de Roeck, JHEP **0703**, 046 (2007).
33. D. G. Cerdeno, K. Y. Choi, K. Jedamzik, L. Roszkowski and R. Ruiz de Austri, JCAP **0606**, 005 (2006); R. H. Cyburt, J. R. Ellis, B. D. Fields, K. A. Olive and V. C. Spanos, JCAP **0611**, 014 (2006); J. Kersten and K. Schmidt-Hoberg, JCAP **0801**, 011 (2008); J. Pradler and F. D. Steffen, Phys. Lett. B **666**, 181 (2008); M. Kawasaki, K. Kohri, T. Moroi and A. Yotsuyanagi, Phys. Rev. D **78**, 065011 (2008); G. Panotopoulos, Phys. Lett. B **671** 327 (2009); S. Bailly, K. Jedamzik and G. Moultaka, arXiv:0812.0788 [hep-ph].
34. K. Kohri and Y. Santoso, arXiv:0811.1119 [hep-ph].
35. T. Kanzaki, M. Kawasaki, K. Kohri and T. Moroi, Phys. Rev. D **76**, 105017 (2007).

LIKELIHOOD ANALYSIS OF THE NEXT-TO-MINIMAL SUPERGRAVITY MOTIVATED MODEL

C. BALÁZS[1] and D. CARTER[2]

School of Physics, Monash University,
Melbourne Victoria 3800, Australia
[1]*E-mail: csaba.balazs@sci.monash.edu.au*
[2]*E-mail: daniel.carter@sci.monash.edu.au*
www.physics.monash.edu.au

Will supersymmetry be found at the CERN Large Hadron Collider (LHC)? If it is, which supersymmetric model is chosen by Nature? Will the available data be enough to sort these puzzles out?

Addressing these questions in this work, we show that the next-to-minimal version of the popular supergravity motivated model (NmSuGra) has a good chance to be observed at the LHC. We also demonstrate that regions of the NmSuGra parameter space which the LHC cannot reach will be detectable at upgraded versions of WIMP direct detection experiments, such as super-CDMS.

Keywords: Supersymmetry; Collider phenomenology; Dark matter; Rare decays; Electroweak precision experiments.

1. Introduction

Supersymmetry is one of the most robust theories that can solve outstanding problems of the standard model (SM) of elementary particles. Supersymmetry naturally explains the dynamics of electroweak symmetry breaking while preserving the hierarchy of fundamental energy scales. It readily accommodates dark matter, the asymmetry between baryons and anti-baryons, the unification of gauge forces, gravity, and more.

If supersymmetry is the solution to these problems, then its natural scale is the electroweak scale, and it is expected to be observed in upcoming experiments, such as the CERN Large Hadron Collider (LHC). But could supersymmetry actually be detected at the LHC? In this work, we will attempt to give a quantitative answer to this question. To simplify the problem, we will examine a constrained supersymmetric model.

One of the main motivations for supersymmetry is that it can naturally bridge the hierarchy between the weak and Planck scales. Unfortunately, the presence of the superpotential μ term in the minimal supersymmetric extension of the standard model (MSSM) undermines this very aim [1]. Experimental data have also squeezed the MSSM into fine-tuned regions, creating the supersymmetric little hierarchy problem[a]. Extensions of the MSSM by gauge singlet superfields not only resolve the μ problem, but can also reduce the little hierarchy [3–5]. In the next-to-minimal MSSM (NMSSM), the μ term is dynamically generated and no dimensionful parameters are introduced in the superpotential (other than the vacuum expectation values that are all naturally weak scale), making the NMSSM a truly natural model [6–23].

Over the last two decades, due to its simplicity and elegance, the constrained MSSM (CMSSM) and the minimal supergravity-motivated (mSuGra) model became a standard in supersymmetry phenomenology. Guided by this, within the NMSSM, we impose the universality of sparticle masses, gaugino masses, and tri-linear couplings at the grand unification theory (GUT) scale, thereby defining the next-to-minimal supergravity-motivated (NmSuGra) model. This approach ensures that all dimensionful parameters of the NMSSM scalar potential also naturally arise from supersymmetry breaking in a minimal fashion. NmSuGra also reduces the electroweak and dark matter fine-tunings of mSuGra.

Using a Bayesian likelihood analysis, we identify the parameter regions of the NmSuGra model that are preferred by the present experimental limits from collider, astrophysical, and various low energy measurements. We combine theoretical exclusions with experimental limits from the CERN Large Electron-Positron (LEP) collider, the Fermilab Tevatron, NASA's Wilkinson Microwave Anisotropy Probe (WMAP) satellite (and other related astrophysical measurements), the Soudan Cryogenic Dark Matter Search (CDMS), the Brookhaven Muon g−2 Experiment, and various b-physics measurements including the rare decay branching fractions $b \to s\gamma$ and $B_s \to l^+l^-$. Finally, we show that the favored parameter space can be detected by a combination of the LHC and an upgraded CDMS at the 95 % confidence level.

[a]For a review see Ref. [2].

2. The NmSuGra Model

In this work, we adopt the superpotential

$$W_{NMSSM} = W_{MSSM}|_{\mu=0} + \lambda \hat{S}\hat{H}_u \cdot \hat{H}_d + \frac{\kappa}{3}\hat{S}^3, \tag{1}$$

where $W_{MSSM}|_{\mu=0}$ is the MSSM superpotential containing only Yukawa type terms but with μ set to zero, $\hat{S}$ ($\hat{H}_{u,d}$) is a standard gauge singlet ($SU(2)_L$ doublet) chiral superfield, λ and κ are dimensionless couplings, and $\hat{H}_u \cdot \hat{H}_d = \epsilon_{\alpha\beta}\hat{H}_u^\alpha \hat{H}_d^\beta$ with the fully antisymmetric tensor normalized as $\epsilon_{11} = 1$. The corresponding soft supersymmetry breaking terms are

$$\begin{aligned}\mathcal{L}^{soft}_{NMSSM} &= \mathcal{L}^{soft}_{MSSM}|_{B=0} + m_S^2|S|^2 + \\ &\quad (\lambda A_\lambda S H_u \cdot H_d + \frac{\kappa A_\kappa}{3} S^3 + h.c.),\end{aligned} \tag{2}$$

where $\mathcal{L}^{soft}_{MSSM}|_{B=0}$ contains the mass and Yukawa terms but not the $B\mu$ term.

The superpotential (1) possesses a global Z_3 symmetry which is broken during the electroweak phase transition in the early universe. The resulting domain walls should disappear before nucleosynthesis; however Z_3 breaking (via singlet tadpoles) leads to a vacuum expectation value (vev) for the singlet that is much larger than the electroweak scale. Thus the requirement of the fast disappearance of the domain walls appears to destabilize the hierarchy of vevs in the NMSSM. Fortunately, in Ref. [24, 25] is was shown that, by imposing a Z_2 R-symmetry, both the domain wall and the stability problems can be eliminated. Following [24], we assume that tadpoles are induced, but they are small and their effect on the phenomenology is negligible.

We also assume that the soft masses of the gauginos unify to $M_{1/2}$, those of the sfermions and Higgses to M_0, and all the tri-linear couplings (including A_κ and A_λ) to A_0 at the GUT scale. This leaves nine free parameters: M_0, $M_{1/2}$, A_0, $\langle H_u \rangle$, $\langle H_d \rangle$, $\langle S \rangle$ (the Higgs and singlet vevs), m_S, λ and κ. The three minimization equations for the Higgs potential [26] and $\langle H_u \rangle^2 + \langle H_d \rangle^2 = v^2$ (here $v = \sqrt{2}/g_2$ is the standard Higgs vev) eliminate four parameters. With the introduction of $\mu = \lambda\langle S \rangle$, and $\tan\beta = \langle H_u \rangle / \langle H_d \rangle$, our remaining free parameters are

$$P = \{M_0, M_{1/2}, A_0, \tan\beta, \lambda, \text{sign}(\mu)\}. \tag{3}$$

Constrained versions of the NMSSM have been studied in the recent literature. The most constrained version is the cNMSSM [27] with $m_S = M_0$. In other cases the $A_\kappa = A_\lambda$ relation is relaxed [26], and/or κ is taken as a free parameter [28, 29], or the soft Higgs masses are allowed to deviate from M_0 [30] giving less constrained models. In the spirit of the CMSSM/mSuGra, we adhere to universality and use only λ to parametrize the singlet sector. This way, we keep all the attractive features of the CMSSM/mSuGra while the minimal extension alleviates problems rooted in the MSSM.

3. Bayesian Likelihood Analysis

In this section, we recapture the interpretation of the problem of theoretical parameter extraction within the context of probability theory. Given a theoretical model T and a set of experimental data D, we introduce the probability $\mathcal{P}(P|T;D)$ that the theoretical model describes the data with the theory parameters set to values P.

In the theory of probabilities $\mathcal{P}(P|T;D)$ is known as the conditional probability of the occurrence of P provided that T is assumed while D holds. According to Bayes' theorem, this probability can be simply calculated as

$$\mathcal{P}(P|T;D) = \mathcal{P}(D|T;P)\frac{\mathcal{P}(T;P)}{\mathcal{P}(D)}. \tag{4}$$

Here $\mathcal{P}(D|T;P)$ is the likelihood that a certain set of data is predicted by a theoretical model with a specified set of theory parameters. The probability $\mathcal{P}(T;P)$ gives the a-priori distribution of the parameters within the theory, fixed by only theoretical considerations independently from the data. Finally, the evidence $\mathcal{P}(D)$ gives the integrated likelihood of the theory T in terms of the data alone. From unitarity it follows that

$$\mathcal{P}(D) = \int \mathcal{P}(D|T;P)\mathcal{P}(T;P)dP, \tag{5}$$

where the integral extends over the whole parameter space of the theory.

The likelihood function $\mathcal{P}(D|T;P)$ maps from the theory space to the signature space. If the data under consideration are independent, as in our case, the likelihood $\mathcal{P}(D|T;P)$ factorizes:

$$\mathcal{P}(D|T;P) = \prod_i \mathcal{L}_i(D|T;P), \tag{6}$$

where each term is formed as a convolution

$$\mathcal{L}_i(D|T;P) = \mathcal{L}_{i,e}(D) \otimes \mathcal{L}_{i,t}(T;P). \tag{7}$$

If the experimental and theoretical likelihoods, $\mathcal{L}_{i,e}$ and $\mathcal{L}_{i,t}$, are normally distributed, then

$$\mathcal{L}_i(D|T;P) = \frac{1}{\sqrt{2\pi}\sigma_i} \exp(\chi_i^2(D,T,P)/2). \tag{8}$$

In this case the exponents

$$\chi_i^2(D,T,P)/2 = (d_i - t_i(P))^2/2\sigma_i^2 \tag{9}$$

are defined in terms of the experimental data $D = \{d_i \pm \sigma_{i,e}\}$ and theoretical predictions for these measurables $\{t_i \pm \sigma_{i,t}\}$. Independent experimental and theoretical uncertainties combine into $\sigma_i^2 = \sigma_{i,e}^2 + \sigma_{i,t}^2$.

In cases when the experimental data only specify a lower (or upper) limit, the likelihood function can be written in terms of the error function

$$\mathcal{P}_i(D|T;P) = \frac{1}{2}\text{erf}(\sqrt{\chi_i^2(D,T,P)/2}). \tag{10}$$

In the stochastic context, the task of theoretical parameter extraction boils down to the determination of the posterior probability distribution $\mathcal{P}(P|T;D)$ describing the probability that the parameters of a given theory have certain values in light of a certain set of data. This can be calculated by evaluating the right hand side of Bayes' theorem. A parameter is well determined by the data if the variance of the $\mathcal{P}(P|T;D)$ distribution is small. This means that the problem of theoretical parameter extraction can be quantified in terms of the likelihood function $\mathcal{P}(D|T;P)$, the theoretical prior and the experimental evidence.

While the posterior probability $\mathcal{P}(P|T;D)$ depends on all the parameters $P = \{p_i\}$ of the theory, sometimes it is useful to know the probability distribution of a given single parameter p_i. This latter is referred to as the marginalized probability and given by

$$\mathcal{P}(D|T;p_i) = \int \mathcal{P}(D|T;P)dp_1...dp_{i-1}dp_{i+1}...dp_n. \tag{11}$$

Similarly to the above, marginalization can be performed to two variables resulting in a two dimensional distribution

$$\mathcal{P}(D|T;p_i,p_j) = \int \mathcal{P}(D|T;P)dp_1...dp_{i-1}dp_{i+1}...dp_{j-1}dp_{j+1}...dp_n. \tag{12}$$

Our main goal is to evaluate these marginalized probability distributions for the five theoretical parameters of the NmSuGra model.

4. Detectability of NmSuGra

Our intent is to show that the experimentally favored region of the NmSuGra model can be discovered by nascent experiments in the near future. To this end, we use the publicly available computer code NMSPEC [31] to calculate the spectrum of the superpartner masses and their physical couplings from the model parameters given in Eq. (3). Then, we use NMSSMTools 2.1.0 and micrOMEGAs 2.2 [32] to calculate the abundance of neutralinos (Ωh^2), the spin-independent neutralino-proton elastic scattering cross section (σ_{SI}), the NmSuGra contribution to the anomalous magnetic moment of the muon (Δa_μ), and various b-physics related quantities.

Table 1. Observables used in the calculation of the posterior probability distribution $\mathcal{P}(P|T;D)$. Experimental data are listed under column $d_i \pm \sigma_{i,e}$, and typical uncertainties related to the NmSuGra calculations are under $\sigma_{i,t}$.

observable	limit type	$d_i \pm \sigma_{i,e}$	$\sigma_{i,t}$
m_h	lower limit	up to 114.4 GeV [33]	3.0 GeV [34]
$m_{\tilde{W}_1}$	lower limit	103.5 GeV [35]	1.0 GeV [36]
Δa_μ	central value	$(29.0 \pm 9.0)\times 10^{-10}$ [37]	negligible
Δm_d	central value	$(5.07 \pm 0.04)\times 10^{11}$ ps^{-1} [38]	1 % [32]
Δm_s	central value	$17.31^{+0.33}_{-0.18} \pm 0.07$ ps^{-1} [39]	1 % [32]
$B(b \to s\gamma)$	central value	$(3.50 \pm 0.17)\times 10^{-4}$ [38]	10 % [40]
$B(B^+ \to \tau + \nu_\tau)$	central value	$(1.73 \pm 0.35) \times 10^{-4}$ [39]	10 % [32]
$B(B_s \to \mu^+\mu^-)$	upper limit	4.7 $\times 10^{-8}$ [39]	10 % [41]
Ωh^2	upper limit	0.1143 [42]	10 % [32]
σ_{SI}	upper limit	CDMS 2008 [43]	20 % [44]

For each NmSuGra model parameters p_i, we quantify the experimental preference in terms of the marginalized posterior probability distributions $\mathcal{P}(D|T;p_i)$. The data D used in our analysis are listed in Table 1. Uncertainties arising in the supersymmetric calculation of (Δa_μ) and the b-physics related quantities are calculated using NMSSMTools. Among the standard input parameters, $m_b(m_b) = 4.214$ GeV and $m_t^{pole} = 171.4$ GeV are used.

Our main results, the marginalized posterior probability distributions for the NmSuGra input parameters are shown in Fig. 1. A glance at these distributions reveals a significant statistical preference for relatively narrow intervals of M_0, $M_{1/2}$, A_0, and $\tan\beta$. From Fig. 1, we conclude that present data favors low M_0 regions while having a slight preference for a focus point type region. The gaugino mass parameter $M_{1/2}$ is limited below 2 TeV. The universal tri-linear coupling $|A_0|$ is most favoured around 1 TeV, and its value is well constrained between -3 and 5 TeV. There are two preferred

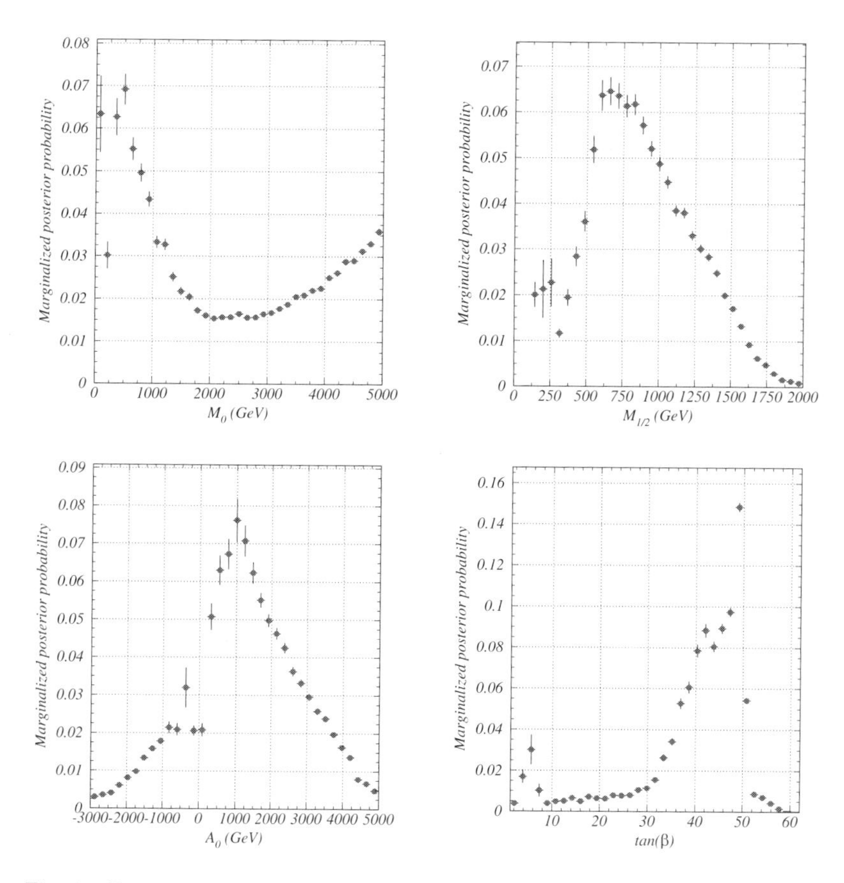

Fig. 1. Posterior probability distributions marginalized to the input parameters M_0, $M_{1/2}$, A_0, and $\tan\beta$.

regions for $\tan\beta$, one around 40-50 and a less preferred one around 5. The data do not show any preference over λ (not shown), but limit its value to the $10^{-6} - 0.5$ region.

Fig. 2 shows the posterior probability distribution marginalized to the common scalar and gaugino masses, M_0 vs. $M_{1/2}$. The maximum likelihood region at low M_0 and $M_{1/2}$ corresponds to slepton-neutralino coannihilation and Higgs resonance annihilation, while the moderate likelihood tail at high M_0 is the familiar focus point region. These high probability regions also occur in mSuGra [45]. Black contour lines mark regions above which the gluino mass and the SUSY scale $\sqrt{m_{\tilde{t}_1} m_{\tilde{t}_2}}$ are higher than 2 (solid), 1.5 (dashed), and 1.0 (dotted) TeV. These regions should be accessible by the

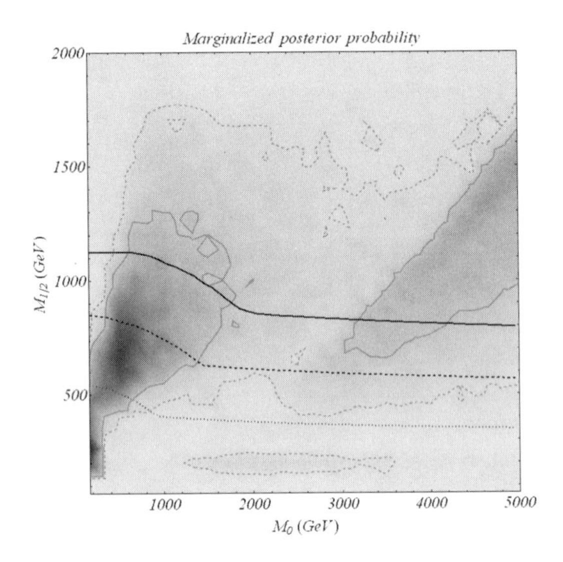

Fig. 2. Posterior probability distribution marginalized to the common scalar and gaugino masses, M_0 and $M_{1/2}$. Confidence level contours are shown for 68 (solid red) and 95 (dashed red) %. The reach of the LHC is estimated by showing contours above which the gluino mass and the SUSY scale are higher than 2 (solid black), 1.5 (dashed black), and 1.0 (dotted black) TeV.

LHC for 10 (solid), 1 (dashed) and 0.1 fb^{-1} [46, 47]. According to this, most of the 68 % confidence level region (delineated by solid red lines) at moderate M_0 and $M_{1/2}$ will be covered, while the focus point region will hardly be probed by the LHC.

Fig. 3 shows the posterior probability distribution marginalized to the plane of the lightest neutralino mass $m_{\tilde{Z}_1}$ and the $\tilde{Z}_1$-proton elastic recoil cross section σ_{SI}. The projected reach of the upgraded CDMS experiment is shown in black for a 25 (solid), 100 (dashed), and a 1000 (dotted) kg detector [48]. This plot clearly shows that direct detection experiments might play a pivotal role in discovering or ruling out the simplest constrained supersymmetric scenarios. Unlike the LHC, an upgraded CDMS is able to explore the whole model parameter space of NmSuGra. This underlines the complementarity of collider and direct dark matter searches.

Acknowledgments

We thank the organizers of the DARK09 conference for their kind invitation and hospitality during the meeting. We also thank D. Kahawala, F.

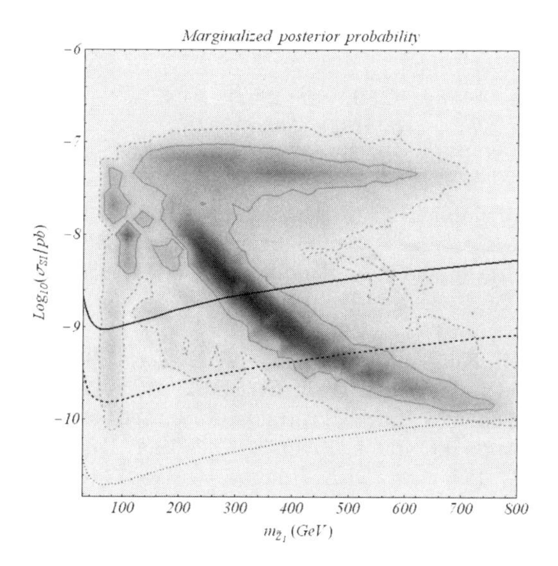

Fig. 3. Posterior probability distribution marginalized to the lightest neutralino mass $m_{\tilde{Z}_1}$ and $\tilde{Z}_1$-proton elastic recoil cross section σ_{SI}. Confidence level contours are shown for 68 (solid red) and 95 (dashed red) %. The projected reach of the upgraded CDMS experiment is shown for a 25 (solid black), 100 (dashed black), and a 1000 (dotted black) kg detector.

Wang and M. White for invaluable discussions on various aspects of the NMSSM and the likelihood analysis. This research was funded in part by the Australian Research Council under Project ID DP0877916.

References

1. J. E. Kim and H. P. Nilles, *Phys. Lett.* **B138**, p. 150 (1984).
2. G. F. Giudice, *In Kane, Gordon (ed.), Pierce, Aaron (ed.): Perspectives on LHC physics, 155-178* (2008).
3. R. Dermisek and J. F. Gunion, *Phys. Rev. Lett.* **95**, p. 041801 (2005).
4. M. Bastero-Gil, C. Hugonie, S. F. King, D. P. Roy and S. Vempati, *Phys. Lett.* **B489**, 359 (2000).
5. J. F. Gunion, *AIP Conf. Proc.* **1030**, 94 (2008).
6. P. Fayet, *Nucl. Phys.* **B90**, 104 (1975).
7. H. P. Nilles, M. Srednicki and D. Wyler, *Phys. Lett.* **B120**, p. 346 (1983).
8. J. M. Frere, D. R. T. Jones and S. Raby, *Nucl. Phys.* **B222**, p. 11 (1983).
9. J. P. Derendinger and C. A. Savoy, *Nucl. Phys.* **B237**, p. 307 (1984).
10. B. R. Greene and P. J. Miron, *Phys. Lett.* **B168**, p. 226 (1986).
11. J. R. Ellis *et al.*, *Phys. Lett.* **B176**, p. 403 (1986).
12. L. Durand and J. L. Lopez, *Phys. Lett.* **B217**, p. 463 (1989).
13. M. Drees, *Int. J. Mod. Phys.* **A4**, p. 3635 (1989).

14. J. R. Ellis, J. F. Gunion, H. E. Haber, L. Roszkowski and F. Zwirner, *Phys. Rev.* **D39**, p. 844 (1989).
15. P. N. Pandita, *Z. Phys.* **C59**, 575 (1993).
16. P. N. Pandita, *Phys. Lett.* **B318**, 338 (1993).
17. U. Ellwanger, M. Rausch de Traubenberg and C. A. Savoy, *Phys. Lett.* **B315**, 331 (1993).
18. B. Ananthanarayan and P. N. Pandita, *Phys. Lett.* **B371**, 245 (1996).
19. B. Ananthanarayan and P. N. Pandita, *Phys. Lett.* **B353**, 70 (1995).
20. B. Ananthanarayan and P. N. Pandita, *Int. J. Mod. Phys.* **A12**, 2321 (1997).
21. U. Ellwanger, M. Rausch de Traubenberg and C. A. Savoy, *Nucl. Phys.* **B492**, 21 (1997).
22. T. Elliott, S. F. King and P. L. White, *Phys. Lett.* **B351**, 213 (1995).
23. S. F. King and P. L. White, *Phys. Rev.* **D52**, 4183 (1995).
24. C. Panagiotakopoulos and K. Tamvakis, *Phys. Lett.* **B446**, 224 (1999).
25. C. Panagiotakopoulos and A. Pilaftsis, *Phys. Rev.* **D63**, p. 055003 (2001).
26. C. Hugonie, G. Belanger and A. Pukhov, *JCAP* **0711**, p. 009 (2007).
27. A. Djouadi, U. Ellwanger and A. M. Teixeira, *Phys. Rev. Lett.* **101**, p. 101802 (2008).
28. G. Belanger, F. Boudjema, C. Hugonie, A. Pukhov and A. Semenov, *JCAP* **0509**, p. 001 (2005).
29. D. G. Cerdeno, E. Gabrielli, D. E. Lopez-Fogliani, C. Munoz and A. M. Teixeira, *JCAP* **0706**, p. 008 (2007).
30. A. Djouadi *et al.*, *JHEP* **07**, p. 002 (2008).
31. U. Ellwanger and C. Hugonie, *Comput. Phys. Commun.* **177**, 399 (2007).
32. G. Belanger, F. Boudjema, A. Pukhov and A. Semenov, *Comput. Phys. Commun.* **176**, 367 (2007).
33. S. Schael *et al.*, *Eur. Phys. J.* **C47**, 547 (2006).
34. M. Frank *et al.*, *JHEP* **02**, p. 047 (2007).
35. G. Abbiendi *et al.*, *Eur. Phys. J.* **C35**, 1 (2004).
36. F. E. Paige, S. D. Protopopescu, H. Baer and X. Tata, *hep-ph/0312045* (2003).
37. F. Jegerlehner and A. Nyffeler, *arXiv:0902.3360* (2009).
38. E. Barberio *et al.*, *arXiv:0808.1297* (2008).
39. M. Artuso, E. Barberio and S. Stone, *PMC Phys.* **A3**, p. 3 (2009).
40. M. Misiak *et al.*, *Phys. Rev. Lett.* **98**, p. 022002 (2007).
41. A. J. Buras, P. H. Chankowski, J. Rosiek and L. Slawianowska, *Nucl. Phys.* **B659**, p. 3 (2003).
42. E. Komatsu *et al.*, *Astrophys. J. Suppl.* **180**, 330 (2009).
43. Z. Ahmed *et al.*, *Phys. Rev. Lett.* **102**, p. 011301 (2009).
44. J. R. Ellis, K. A. Olive and C. Savage, *Phys. Rev.* **D77**, p. 065026 (2008).
45. H. Baer and C. Balazs, *JCAP* **0305**, p. 006 (2003).
46. H. Baer, C. Balazs, A. Belyaev, T. Krupovnickas and X. Tata, *JHEP* **06**, p. 054 (2003).
47. O. Buchmueller *et al.*, *JHEP* **09**, p. 117 (2008).
48. D. S. Akerib *et al.*, *Nucl. Instrum. Meth.* **A559**, 411 (2006).

HIERARCHY PROBLEM AND DILATONIC DARK MATTER

Y.M. CHO

Center for Theoretical Physics and School of Physics
College of Natural Sciences, Seoul National University
Seoul 151-742, Korea
and
School of Basic Sciences
Ulsan National Institute of Science and Technology
Ulsan 689-805, Korea
ymcho@yongmin.snu.ac.kr

We discuss the physical implications of the extra space in higher-dimensional unification, and show that a large extra space requires an extremely small fine structure constant (of the order of 10^{-42}) for the Kaluza-Klein gauge boson, or else a huge violation of equivalence principle in 4-dimensional Einstein gravity. More importantly, the extra space generates the dilaton which represents the volume element of the extra space. We show that the scalar curvature of the extra space generates the dilaton mass, and that the dilaton mass determines the size of the extra space. Moreover, as a massive scalar graviton, the dilaton can be an excellent candidate of dark matter. Assuming that the dilaton is the dark matter, we estimate the lower bound of the scale of the extra space to be of 10^{-9} m.

Keywords: dilatonic fifth force, violation of equivalence principle, hierarchy problem, mass of dilaton, origin of mass, dilatonic dark matter, size of extra space

1. Introduction

All modern unified theories, Kaluza-Klein theory, superstring, and supergravity, are based on higher dimensional space-time.[1–4] This makes the experimental verification of the extra dimension a very important issue. There are different perceptions, but basically two contending views, on the extra space. The popular view is that the extra space is real (i.e., physical), but the 4-dimensional world that we are living in is some kind of approximation which can be described by a low energy effective theory of the full higher-dimensional theory.[3,4] The other view is that the universe is higher-dimensional, but a symmetry (an isometry) of the unified space makes the extra space physically inaccessible.[1,5] So far there has been no experimental evidence of the extra space that can determine which view is realistic. But new experiments from Large Hadronic Collider (LHC) could change this and produce decisive results on the extra space. Under this circumstance a pedagogical discussion on the extra space and its physical implications appears to be necessary.

Independent of how one perceives the extra space, the higher-dimensional unification has unmistakable consequences in the 4-dimensional physics. A most important consequence is the appearance of gravitationally interacting massive particles (GIMPs) or massive scalar gravitons (MSGs), the dilaton and other internal gravitons which originate from the gravitons of the extra space, in the 4-dimensional effective theory.[1,2] But traditionally they have been regarded to be unimportant because the extra space has been thought to be very small, of the order of the Planck scale.[1,3,4] Because of this these GIMPs have been assumed to be extremely heavy. In this case they decay quickly, so that they leave almost no trace in the present universe.[2,6] Moreover, one can not produce them easily because it costs too much energy to do that. This makes them totally irrelevant in 4-dimensional physics. Unfortunately this is not exactly true, because even a small extra space can make these scalar gravitons massless when the extra space is flat.[1,2] Moreover, there is the possibility that the extra space may actually not be so small as has been thought.[7–9] Indeed it has been asserted that this (a large extra space) might be exactly what we need to resolve the hierarchy problem.[10,11] But in this case these GIMPs become very light, and can have deep impacts in the 4-dimensional effective theory. This certainly makes the discussion of the extra space an important issue.

As we have remarked, one of the big puzzles in theoretical physics is the hierarchy problem. We need to understand why the gravitational force is so weak compared to other forces. To explain this Dirac long time ago conjectured that the Newton's constant may not be a constant but a time-dependent parameter, and that in early universe the gravity could have been as strong as other forces but has become so weak in the present universe precisely because the universe has become so old.[12] The dilaton plays a crucial role to resolve the hierarchy problem. Notice that, at the classical level, the hierarchy problem is the problem of the mass hierarchy. Assuming the Planck mass to be the only natural mass scale of the underlying unified theory, one has to express all other masses in terms of the Planck mass. The problem is that the Planck mass is so large compared to the masses of elementary particles, so that it is extremely difficult to obtain the mass of elementary particles from the Planck mass. The dilaton, in particular the dilaton mass, plays a crucial role to explain this.

The dilaton (also recently dubbed as the radion[10] or gravexiton[13]), is well known to have a deep impact in physics as the scalar graviton which couple to all matter fields. First, it generates the fifth force which can compromize the equivalence principle in Einstein's gravity.[8,14] And the dilaton mass determines the range of the fifth force. Second, as a GIMP it can be an excellent candidate of the dark matter.[6,8,15] And the dilaton mass determines the dilatonic dark matter density. Moreover, it plays a central role to resolve the hierarchy problem. And here the dilaton mass is determined by the size of the extra space, so that we can infer the size of the extra space from the dilaton mass.[6,15] This confirms that the dilaton and the dilaton mass is closely related to the extra space and the hierarchy problem.

In this talk we first discuss the implications the large extra space and show that a large extra space can explain the huge gap in the mass hierarchy, but requires an extremely small fine structure constant for the Kaluza-Klein (KK) gauge boson. Moreover, we show that the cosmology puts a strong constraint on dilaton massand thus the size of the extra space. In particular, we show that the cosmological constraint puts a lower limit on the scale of the extra space to be of $10^{-9}\ m$.

The paper is organized as follows. In Section II we review the KK dilaton in higher-dimensional unification, and show that a large extra space requires an extremely small fine structure constant for the KK gauge boson. In Section III we discuss how the dilaton mass originates from the curvature of the extra space, and show how the dilaton can explain the huge gap in the mass hierarchy. In Section IV we discuss the potential problems of the dimensional reduction by zero mode approximation, and show how the dimensional reduction by isometry can avoid the potential problems of light KK excited modes in the popular dimensional reduction by zero mode approximation. In Section V we discuss the experimental and theoretical constraints of dilaton mass, and estimate the size of the extra space from the known constraints on dilaton mass. In particular, we set the lower limit on the size of the extra space. Finally in Section VI we discuss the physical implications of our analysis, and propose experiments to confirm the existence of the dilaton.

2. Kaluza-Klein Dilaton: A Review

All known interactions in nature are mediated by spin-one or spin-two fields. However, the unification of all interactions inevitably requires the existence of a fundamental spin-zero field. In fact, all modern unified theories (Kaluza-Klein theory, supergravity, and superstring) contain a fundamental scalar field called the dilaton, or more precisely the KK dilaton.[1,2] What makes this scalar field unique is that unlike other scalar fields such as the Higgs field, it couples directly to the (trace of the) energy-momentum tensor of the matter fields. As such it behaves like a scalar graviton, and thus generates a new force which modifies Einstein's gravitation in a fundamental way.

The simplest unified theory which contains the dilaton is the Jordan-Brans-Dicke theory.[16] It is the simplest theory which contains a scalar graviton, the Brans-Dicke dilaton. As such it predicts a composition-dependent fifth force which violates the equivalence principle.[14,17] And it is the prototype theory which provides a possible resolution of the hierarchy problem. It makes Newton's constant space-time dependent, and thus reduces the hierarchy problem to a time-dependent artifact. So it naturally implements Dirac's conjecture.[12] Because of this, historically Jordan-Brans-Dicke theory has played a very important role in theoretical physics. On the other hand the Brans-Dicke dilaton is massless, so that it creates a long-range fifth force which has not been observed in nature.[18,19] This rules out the Brans-Dicke theory as unphysical.

But the KK dilaton can easily acquire a mass, and this makes the KK dilaton very important. The importance of the KK dilaton is based on the following facts. First, the dilaton can generate a fifth force which modifies the Newton's gravity.[14,17] Second, the dilaton mass can play a crucial role to resolve the hierarchy problem. Third, the dilaton with a proper mass could play the role of dark matter of the universe.[6,8] And the dilaton mass can determine the size of the extra space. To discuss the dilaton physics in higher dimensional unification, however, we must first make the dimensional reduction and obtain the 4-dimensional effective theory.

Since all higher dimensional unified theories contain $(4+n)$-dimensional gravitation, we start from the Kaluza-Klein theory and adopt the zero mode approximation to obtain the effective 4-dimensional theory. As we have pointed out, a simple and elegant way to select the zero modes is to impose a proper isometry. In this zero mode approximation by isometry the isometry reduces the $(4+n)$-dimensional unified space to a principal fiber bundle P(M,G) made of the 4-dimensional base manifold M and the n-dimensional extra space (the group manifold) G. Thus it naturally compactifies the extra space and separates the extra coordinates from the 4-dimensional space-time coordinates. Moreover, it automatically selects the zero modes. So, in this zero mode approximation by isometry the popular dimensional reduction becomes dimensional reduction by isometry.

Let ∂_a $(a = 1, 2, ...n)$ be the (vertical) Killing vectors of G which describes the n-dimensional isometry and ∂_μ be a coordinate basis on M. And let $D_\mu = \partial_\mu + e\kappa A_\mu{}^a \partial_a$ $(\mu = 0, 1, 2, 3)$ be the horizontal lift of ∂_μ to P which are orthogonal to ∂_a, where e is a coupling constant of the KK gauge fields $A_\mu{}^a$ and κ is the scale of the extra space. Now, let $\gamma_{\mu\nu}$ and $\tilde{\phi}_{ab}$ be the 4-dimensional metric on M and the n-dimensional metric on G in the block diagonal basis, γ and $\tilde{\phi}$ be the determinants of $\gamma_{\mu\nu}$ and $\tilde{\phi}_{ab}$, and $\rho_{ab} = \tilde{\phi}^{-1/n}\ \tilde{\phi}_{ab}$ $(|\det\rho_{ab}| = 1)$ be the normalized metric on G. In this setting the (4+n)-dimensional Einstein-Hilbert action on P leads, in the zero mode approximation, to the following 4-dimensional Lagrangian in the Jordan frame[1,7]

$$I_{CF} = -\frac{1}{16\pi G_0}\int \sqrt{-\gamma}\sqrt{\tilde{\phi}}\ \Big[R_M - \frac{n-1}{4n}\gamma^{\mu\nu}\frac{(\partial_\mu\tilde{\phi})(\partial_\nu\tilde{\phi})}{\tilde{\phi}^2}$$

$$+\frac{\gamma^{\mu\nu}}{4}(D_\mu\rho^{ab})(D_\nu\rho_{ab}) + \frac{1}{\kappa^2\sqrt[n]{\tilde{\phi}}}\ \hat{R}_G(\rho_{ab}) + \frac{\kappa^2}{4}\sqrt[n]{\tilde{\phi}}\ \rho_{ab}\gamma^{\mu\alpha}\gamma^{\nu\beta}F^a_{\mu\nu}F^b_{\alpha\beta}$$

$$+\Lambda_0 + \lambda(|\det\rho_{ab}| - 1)\ \Big]d^4x d^n\theta, \qquad (1)$$

where G_0 is the (4+n)-dimensional Newton's constant, R_M is the scalar curvature of M fixed by $\gamma_{\mu\nu}$, $\hat{R}_G(\rho_{ab})$ is the (dimensionless) normalized curvature of the extra space fixed by ρ_{ab}, $F^a_{\mu\nu}$ is the gauge field of the isometry group G, Λ_0 is a (4+n)-dimensional cosmological constant, and λ is a Lagrange multiplier. There are two things to be noticed. First, the extra coordinate dependence of the zero modes has been completely taken care of by the isometry, so that all fields now can be treated as the functions of the 4-dimensional space-time coordinates. Secondly, the scale $1/\kappa^2$ sets the scale of the curvature $\hat{R}_G(\rho_{ab})$ of the extra space.

To proceed further, it is important for us to keep two things in mind. First, one must choose the correct conformal frame in which the metric describes the massless spin-two graviton. Second, one must make sure that the dimensional reduction does not compromise the equivalence principle. For this reason we first introduce the Pauli metric $g_{\mu\nu}$ and the Kaluza-Klein dilaton σ by[1,2]

$$g_{\mu\nu} = \sqrt{\tilde{\phi}}\ \gamma_{\mu\nu}, \qquad \tilde{\phi} = \langle\tilde{\phi}\rangle\ \exp\Big(2\sqrt{\frac{n}{n+2}}\ \sigma\Big) \quad (\langle\sigma\rangle = 0). \qquad (2)$$

With this we can reduce (1) to the following 4-dimensional Lagrangian in the Pauli frame (up to a total divergence),[1,7]

$$\mathcal{L}_{CF} = -\frac{\hat{V}_G}{16\pi G_0}\sqrt{-g}\Big[R + \frac{1}{2}(\partial_\mu\sigma)^2 - \frac{1}{4}(D_\mu\rho^{ab})(D^\mu\rho_{ab})$$
$$+\frac{\hat{R}_G(\rho_{ab})}{\kappa^2\ \langle\tilde{\phi}\rangle^{(n+2)/2n}}\exp\Big(-\sqrt{\frac{n+2}{n}}\,\sigma\Big) + \frac{\Lambda_0}{\sqrt{\langle\tilde{\phi}\rangle}}\exp\Big(-\sqrt{\frac{n}{n+2}}\,\sigma\Big)$$
$$+\frac{e^2\ \kappa^2}{4}\exp\Big(\sqrt{\frac{n+2}{n}}\,\sigma\Big)\rho_{ab}F^a_{\mu\nu}F^{\mu\nu b}\Big], \tag{3}$$

where $\hat{V}_G = \kappa^n/\sqrt{\langle\tilde{\phi}\rangle}$ is the volume of the extra space. In this form the Pauli metric now describes the massless spin-two graviton.

Clearly $\langle\tilde{\phi}\rangle$ describes the vacuum value of the volume (and thus the size) of the extra space, since $\tilde{\phi} = |\det\,\tilde{\phi}_{ab}|^{1/n}$. And we can make the size of the extra space arbitrary by leaving $\langle\tilde{\phi}\rangle$ arbitrary. On the other hand there is another quantity κ which also represents the scale (and thus the size) of the extra space, which is *a priori* arbitrary. So we can always normalize $\langle\tilde{\phi}\rangle = 1$, without loss of generality by adjusting the scale κ. Actually there are two ways to make the size of extra space arbitrary, to fix $\langle\tilde{\phi}\rangle = 1$ and leave κ arbitrary or to fix the scale κ (with $\kappa = l_p$ as usual) and leave $\langle\tilde{\phi}\rangle$ arbitrary. Obviously the two approaches are identical and will not change the outcome. For this reason we choose the first approach in this paper and adopt $\langle\tilde{\phi}\rangle = 1$.

With this understanding, we now impose the following conditions

$$\frac{\hat{V}_G}{16\pi G_0} = \frac{\kappa^n}{16\pi G_0} = \frac{1}{16\pi G}, \qquad \frac{e^2\kappa^2}{16\pi G} = 1, \tag{4}$$

and find that the Lagrangian can be written as

$$\mathcal{L}_{CF} = -\frac{\sqrt{-g}}{16\pi G}\Big[R + \frac{1}{2}(\partial_\mu\sigma)^2 - \frac{1}{4}(D_\mu\rho^{ab})(D^\mu\rho_{ab})$$
$$+m^2\hat{R}_G(\rho_{ab})\exp\Big(-\sqrt{\frac{n+2}{n}}\,\sigma\Big) + \Lambda_0\exp\Big(-\sqrt{\frac{n}{n+2}}\,\sigma\Big)$$
$$+\frac{1}{4}\exp\Big(\sqrt{\frac{n+2}{n}}\,\sigma\Big)\rho_{ab}F_{\mu\nu}{}^{a}F^{\mu\nu b}, \tag{5}$$

where $m^2 = 1/\kappa^2 = \alpha_{KK}m_p^2/4$, $\alpha_{KK} = e^2/4\pi$ is the fine structure constant of the Kaluza-Klein gauge field, and m_p is the Planck mass.

The justification of (4) must be clear. The first equation $V_G/16\pi G_0 = 1/16\pi G$ assures that Newton's constant describes the 4-dimensional gravitational coupling after the dimensional reduction. This (with $V_G = \kappa^n$) means that the scale of the higher dimensional gravitational constant l_0 is given by[2,8]

$$l_0 = {G_0}^{1/(n+2)} \simeq \kappa^{n/(n+2)} G^{1/(n+2)} = \left(\frac{\kappa}{l_p}\right)^{n/(n+2)} l_p. \tag{6}$$

This clearly shows that, if κ becomes large compared to l_p, the higher-dimensional gravity becomes much stronger than Einstein's gravity.[10,11] In fact, this has been the basis of the ADD proposal to resolve the hierarchy problem with a large extra space.[10,11] Now, the second equation $e^2\kappa^2 = 16\pi G$ is to make sure that the Pauli metric has the minimal coupling to the KK gauge field in the absence of dilaton. This is needed to guarantee the equivalence principle in Einstein's gravity. Notice, however, that the second equation requires

$$\kappa^2 = \frac{4}{\alpha_{KK}}\, l_p^2, \qquad l_0 = \left(\frac{\kappa}{l_p}\right)^{n/(n+2)} l_p = \left(\frac{4}{\alpha_{KK}}\right)^{n/2(n+2)} l_p. \tag{7}$$

This means that to have a large extra space we need an extremeny small α_{KK}. In other words, unless α_{KK} is extremely small, the higher-dimensional gravitatioal constant as well as the size of the extra space must be of the order of the Planck scale. In fact (7) tells that, to have the ADD proposal which assumes $\kappa \simeq 10^{-5}\ nm$, we need to have $\alpha_{KK} \simeq 10^{-42}$ for $n = 6$ (For $n = \infty$ we have $\alpha_{KK} \simeq 4 \times 10^{-32}$, which sets the upper limit of α_{KK}). This is a very important point which has been completely overlooked in the ADD proposal.

One might think that such a small α_{KK} is absurd, and try to overcome this absurdity. An obvious way to do this is to modify the second equation of (4) and allow the violation of the equivalence principle for the KK gauge field. Suppose we do that, and let $\eta^2 = e^2\kappa^2/16\pi G$. In this case (7) will be replaced by

$$l_0 = \left(\frac{\kappa}{l_p}\right)^{n/(n+2)} l_p = \left(\frac{4\eta}{\alpha_{KK}}\right)^{n/2(n+2)} l_p, \tag{8}$$

so that one could have a large extra space making η large, without requiring α_{KK} extremely small. Of course, this will make the gravitational coupling to the KK gauge field η^2 times stronger than the minimal coupling in (5).

So, if one is willing to accept the violation of the weak equivalence principle by the KK gauge boson, it appears that one could have a large extra space without having very small α_{KK}.

But this appearance need clarification, because the above conclusion depends on which quantum field we identify as the physical particle (i.e., the quantum) of KK gauge field. To see this, let $\eta^2 = e^2\kappa^2/16\pi G$ and introduce the renormalized the gauge potential $\bar{A}_\mu{}^a$ by $\bar{A}_\mu{}^a = \eta\ A_\mu{}^a$. With this the Lagrangian (5) acquires the identical form, except that $A_\mu\ a$ and e are replaced by $\bar{A}_\mu{}^a$ and $\bar{e} = e/\eta$. So, in terms of the renormalized gauge field and the renormalized gauge coupling, the size of the extra space is determined exactly as before, by the fine structure constant of the KK gauge field. This leads us to the important question: Which field, $A_\mu{}^a$ or $\bar{A}_\mu{}^a$, do we identify as the the quantum field (the physical particle) of KK gauge boson? In quantum field theory where the quantization is done in the Minkwski space-time in the absence of gravity, this question is totally irrelevant because the overall normalization of the quantum field is irrelevant. But here it becomes important, because this overall normalization of the fields determines the gravitational coupling strength (and the gravitational mass) of the quantum field. And the weak equivalence principle tells that all quantum fields must have the standard normalization such that the theory automatically guarantees the equivalence principle. But *a priori* (and logically) there is no reason why all quantum fields (and physical particles) must obey the equivalence principle.

So, we have two options: to keep the equivalence principle but allow an extremely small α_{KK}, or else to accept a huge ($\eta^2 \simeq 10^{20}$ for $n = 6$) violation of equivalence principle by the KK gauge field and have a resonable α_{KK}. And we emphasize that this violation of equivalence principle is not the violation by dilatonic fifth force we discussed before. It is the violation by the Einstein's gravity itself which one must have even in the absence of dilaton, which is much more serious. And this is the price to pay for a large extra space. This shows that one has to accept the ADD proposal with a grain of salt.

The popular wisdom in the higher-dimensional unification has been that the extra space is extremely small (of the order of the Planck scale).[1,3,4] Our analysis confirms that this is true if we accept the weak equivalence principle and a resonable (not so small) coupling strength for the KK gauge boson. If we are willing to accept an extremely small fine structure constant for the KK gauge boson, however, the extra space can be made large. As we will see, a large extra space has deep implications.

3. Dilaton Mass and Hierarchy Problem

An intimately related issue to large extra space and hierarchy problem is the dilaton mass. Now, we discuss how a large extra space makes the dilaton mass very small and thus explain the huge gap in the mass hierarchy, in spite of the fact that the Planck mass is the only mass scale of the theory. In doing so we show that the dilaton mass originates from the curvature of the extra space, which implies that mass could originate from the curvature of space-time.

As we have remarked, the essence of the hierarchy problem is to understand the huge gap between the Planck scale and the mass scale of the ordinary elementary particles, or equivalently to produce the ordinary mass scale from the Planck scale. The Planck mass is supposed to be the fundamental scale of the universe which sets the mass scale of any unified theory, yet it is so different from the mass scale we have in elementary particles. And this is the origin of the hierarchy problem. To see how the dilaton can explain the mass gap, notice first that in cosmology the classical dilaton field may not be in vacuum yet but, just like the metric, may actually evolve in time.[8,17] Moreover, just as in Brans-Dicke theory, the classical value of dilaton determines Newton's constant. This means that in the Kaluza-Klein cosmology Newton's constant becomes a time-dependent parameter in general. So the Kaluza-Klein dilaton can naturally implement the Dirac's conjecture proposed to resolve the hierarchy problem. This indicates that the dilaton could indeed play a crucial role in the hierarchy problem.

Now we show tha the dilaton mass tells us how the large Planck mass can be reduced to the small mass scale of elementary particles, and thus explain the huge gap between them. To do this, suppose the Lagrangian (5) has a unique vacuum at $\langle g_{\mu\nu}\rangle = \eta_{\mu\nu}$, $\langle\sigma\rangle = 0$, $\langle\rho_{ab}\rangle = \delta_{ab}$, $\langle A_\mu{}^a\rangle = 0$. With this we have the following dilatonic potential $V(\sigma)$ from (5)[2,8]

$$\begin{aligned} V(\sigma) &= \frac{1}{16\pi G}\Big[m^2\langle\hat{R}_G\rangle\ \exp(-\sqrt{\frac{n+2}{n}}\ \sigma) + \Lambda_0\ \exp(-\sqrt{\frac{n}{n+2}}\ \sigma)\Big] + V_0 \\ &= \frac{m^2\langle\hat{R}_G\rangle}{16\pi G}\Big[\exp(-\sqrt{\frac{n+2}{n}}\sigma) - \frac{n+2}{n}\exp(-\sqrt{\frac{n}{n+2}}\sigma) + \frac{2}{n}\Big], \end{aligned} \tag{9}$$

where $m^2 = (\alpha_{KK}/4)m_p^2$ is the mass scale we have in (5), $\langle\hat{R}_G\rangle = \hat{R}_G(\langle\rho_{ab}\rangle) \simeq O(1)$ is the dimensionless vacuum curvature of the extra space obtained by the bi-invariant Cartan-Killing metric δ_{ab}, and V_0 is a constant which we introduce to assure that (9) does not create a non-vanishing 4-dimensional cosmological constant (vacuum energy). An important point here is that Λ_0 and V_0 are not free parameters, because they are completely fixed by the vacuum condition $dV(0)/d\sigma = 0$ and $V(0) = 0$. Notice that Λ_0 and V_0 are necessary for the existence of a stable physical vacuum. Without Λ_0 we have no vacuum at all. This would mean that the dilaton becomes massless and must generate a long-range fifth force, which is excluded by experiments. And without V_0 we have a negative $V(0)$ and thus a negative 4-dimensional cosmological constant, which is also unacceptable.

Clearly the potential (9) determines the dilaton mass. To find the dilaton mass from the potential, it is important to realize that σ is not exactly the physical dilaton field. The physical dilaton field must have the dimension of mass, but σ here is dimensionless. Now the Lagrangian (5) tells that, if the dilaton couples minimally to gravity and obeys the equivalence principle, the physical dilaton $\hat{\sigma}$ must be given by $\hat{\sigma} = \sigma/\sqrt{16\pi G}$. With this normalization we have

$$V(\sigma) = \frac{\mu^2}{2}\hat{\sigma}^2 + \text{higher order interactions},$$
$$\mu^2 = (16\pi G)\,\frac{d^2V(0)}{d\sigma^2} = -\frac{2\langle \hat{R}_G\rangle}{n}\,m^2 = -\frac{\langle \hat{R}_G\rangle}{2n}\,\alpha_{KK}\,m_p^2. \tag{10}$$

This shows that the dilaton mass μ, with $\langle \hat{R}_G\rangle \simeq O(1)$, becomes very small compared to Planck mass when α_{KK} becomes small. This confirms that the dilaton mass and the size of the extra space is closely related. A large extra space does make the dilaton mass small even though it is determined by the Planck mass (and thus resolve the hierarchy problem), when the fine structure constant of KK gauge boson becomes small.

Of course, even when the dilaton becomes massive, the dilatonic fifth force can compromise the equivalence principle. But this violation is due to the dilaton, which is absent in Einstein's gravity. Moreover, this violation can easily be made compatible with experiments assigning a proper dilaton mass and making the fifth force short-ranged as we have remarked.

Our analysis teaches us the follwings. First, a large extra space (together with the equivalence prnciple) requires a small α_{KK}, and this in turn reduces the dilaton mass from the Planck scale to a resonable scale. Actually it reduces all mass scales of the theory and thus resolve the hierarchy problem, as we will see later. On the other hand, if one is willig to compromise the weak equivalence principle, one can have large extra space without making the gauge coupling of the KK gauge boson very small.

Secondly, the dilaton mass is fixed by the size of the extra space. From (10) we see that the scale of the extra space κ is determined by the dilaton mass, $\kappa = 1/m = \sqrt{-2\langle \hat{R}_G \rangle / n} \times 1/\mu \simeq 1/\mu$. So, for example, for the S^3 compactification of the 3-dimensional extra space in $(4+3)$-dimensional unification we have G=$SU(2)$ and $\langle \hat{R}_G \rangle = -3/2$, so that $\kappa = 1/\mu$. In general we always have $\kappa \simeq 1/\mu$ (unless the dimension of the extra space n becomes extremely large), as far as $\langle \hat{R}_G \rangle \neq 0$. This shows that we can estimate the size of the extra space from the dilaton mass.

Thirdly, the curvature of the extra space $\langle \hat{R}_G \rangle$ plays the crucial role in dilaton mass. In fact we have $\mu = 0$ when $\langle \hat{R}_G \rangle = 0$, independent of the dimension and the size of the extra space. This tells that the dilaton mass is created by the scalar curvature of the extra space. If the extra space is flat, the dilaton becomes massless. And since a massless dilaton is excluded by the fifth force experiments, we can conclude that the extra space must have a non-vanishing scalar curvature.

Another important point of our analysis is that we have other scalar particles ρ_{ab} known as the internal gravitons which are very similar to the dilaton in the higher dimensional unification.[1,7] And (5) tells that the mass of internal gravitons is given by $m = \langle \hat{R}_G \rangle / \kappa$, so that they acquire mass of the order of the dilaton mass. In particular, a small α_{KK} makes the mass of the internal gravitons (as well as dilaton) small. This implies that the same mechanism which reduces the dilaton mass reduces all masses of the unified theory.

The fact that the extra space has to be curved to make the dilaton massive has a very important implication in string theory. As is well-known, the string theory favors a Ricci-flat extra space (the Calabi-Yau manifold or the orbifold).[4,20] In this case the gravitons of the extra space (just like the Einstein's graviton) should remain massless, and this will make the dilaton and internal gravitons massless. This is incompatible with the known experiments, because the massless dilaton and internal gravitons should generate a long-range fifth force which does not exist in nature. This implies that the string vacuum may not be based on the Calabi-Yau manifold or the orbifold. Otherwise the string theory must provide a new mechanism which can make these massless scalar gravitons massive, or else make the effect of these massless scalar gravitons extremely small.

Our analysis strongly indicates that there is another mass generation mechanism other than the Higgs mechanism, a geometric mass generation through the curvature of space-time.[2,8] Understanding the origin of mass has been a fundamental problem in physics. The interpretation of Einstein's equation $R_{\mu\nu} - \frac{1}{2}R\, g_{\mu\nu} = -8\pi G\, T_{\mu\nu}$ has always been one way, that the energy-momentum tensor is the source of the curvature of space-time and thus creates the curvature. Now, the dilaton mass suggests that we can interpret it the other way, that the curvature creates the energy-momentum and thus mass. This provides a natural explanation of the origin of mass, and shows that the hierarchy problem is closely related to the problem of the origin of mass.

So far we have discussed only the higher dimensional gravity, and shown how a large extra space reduces all masses in the theory. But in superstring and supergravity one has higher-dimensional matter fields, and it is not clear how the dimensional reduction will fare when we include them. A detailed inspection shows that the dimensional reduction can make some of the gauge bosons massive.[2,7] Moerover, a geometric symmetry breaking which can create fermion mass through the geometric Higgs mechanism may take place in these theories.[2,7] In general, however, all masses becomes of the order of the dilaton mass we had in (5). This confirms that a large extra space does make all masses in the unified theory small, even though they all are expressed by the Planck mass.

4. Kaluza-Klein Excited Modes and Dimensional Reduction by Isometry

As we have seen, a large extra space can resolve the hierarchy problem as the ADD proposal suggests, but with a stiff price. One must accept an extremely small fine structure constant for the KK gauge boson. This implies that a large extra space does not really resolve the hierarchy problem, but actually shift the problem to another absurdly small number which need to be explained. Now, the large extra space can cause another serious problem, the problem with the KK excited modes. And this is the problem we face in the zero mode approximation.

The ADD proposal is based on a large extra space of the order of $10^{-5}\ nm$ for $n = 6$. Although the extra space of this size may appear to be inaccessible at present energy, it can actually be detected indirectly. This is because such extra space should produce massive tower of KK excited modes in 4-dimension, whose masses start from 9.2 MeV (Notice that with $n = 3$ the mass of the lightest KK mode becomes of 42.5 eV). This is because $1/\kappa$, not $1/l_0$, determines the mass scale of KK excited modes. The problem is that no such light KK modes have been detected yet. Even with a unrealistically large n (infinite-dimensional extra space) one must have massive KK modes of the order of TeV, which is well within the energy range of LHC. So these excited modes, if exist, should be detectable at LHC. Of course, LHC could produce such excited modes. But if not, the zero mode approximation based on the large extra space should be discarded. Clearly this problem with the KK modes can not be avoided as far as the extra dimension is treated as physical (real), unless of course one assumes the extra space to be extremely small.

Under such circumstances, one may wonder if there is any way to get rid of the troublesome KK excited modes. Actually there is a simple way to get rid of the problem. To understand this it is important to remember that this problem arises because we have treated the extra space as physical in the dimensional reduction. As we have mentioned the popular dimensional reduction by zero mode approximation has many problems.[2,8] To have the zero mode approximation we first have to obtain a compact extra space as the vacuum of the unified theory. But this is very difficult to implement in any unified theory which has the higher-dimensional general invariance. Indeed this is the main problem in string theory.[4,20] To circumbent this one assumes the spontaneos compactification of the extra space. But even with this *ad hoc* assumption, it is not simple to obtain the zero mode approximation when the extra space has a non-trivial topology.[2,8]

A straightforward way to remove this problem is to adopt the dimensional reduction by isometry.[1,7] So far we have adopted the dimensional reduction by isometry as a mathematical way to obtain the zero mode approximation, assuming the extra space is real and treating the isometry as an approximate symmetry. The isometry naturally decomposes the unified space to a principal fiber buldle P(M,G), and allows us to obtain the zero modes. But as far as one treats the extra space real, the isometry should be interperted as an approximation, a trick to separate the extra space and to select the zero modes. Now, suppose that the isometry is exact. In this case the zero mode approximation no longer becomes an approximation but provides the full dimensional reduction. The exact isometry makes the extra space "unphysical", or more precisely inaccessible, independent of how large it is. Moreover, it prevents the KK excited modes.

This does not mean that we can not confirm the existence of the extra space. As we have shown, we still have plenty of evidences of extra space in the 4-dimensional effective theory, the existence of the dilaton and the internal gravitons. We can even tell the curvature of the extra space through the dilaton mass.

Of course, at the moment we have no concrete experimental evidence on whether we have the KK excited modes or not, so that it is still premature to decide which dimensional reduction we should adopt. But LHC soon will tell which is realistic. If the popular dimensional reduction is correct, LHC must produce the KK excited modes. If not, we may have no choice but to accept the dimensional reduction by isometry. The logical advantage of the dimensional reduction by isometry has also recently been emphasized by Penrose.[5]

5. Dilatonic Dark Matter and Size of Extra Space

Now, an interesting question to ask is how large the extra space is supposed to be. As we have shown, we can infer the size of the extra space from the dilaton mass. Although we do not know the dilaton mass we have two important constraints on it, the constraint from the fifth force experiments[18,19] and the cosmological constraint on the dilatonic dark matter which is theoretical.[6,15]

Consider the fifth force constraint first, and let F_g and F_5 be the gravitational and fifth force between the two baryonic point particles separated by a distance r. From the dimensional argument, one may express the total force between two test particles in Newtonian limit as

$$\begin{aligned} F = F_g + F_5 &\simeq \frac{\alpha_g}{r^2} + \frac{\alpha_5}{r^2}(1+\mu r)\ \exp\big(-\mu r\big) \\ &= \frac{\alpha_g}{r^2}\Big\{1+\beta(1+\mu r)\ \exp\big(-\mu r\big)\Big\}, \end{aligned} \tag{11}$$

where α_g and α_5 are the fine structure constants of the gravitation and fifth force, $\beta = \alpha_5/\alpha_g$ is the ratio between them, and μ is the dilaton mass. In terms of Feynman diagrams the first term represents one graviton exchange but the second term represents one dilaton exchange in the zero momentum transfer limit. In the Kaluza-Klein unification we have $\beta = n/(n+2)$.[2,8] But in general one may assume $\beta \simeq 1$ treating the dilaton as a GIMP. On the other hand, it is important to keep in mind that in principle there is no *a priori* reason why the dilaton interaction has to be gravitational. So in principle we can treat α_g and α_5 independent coupling constants in the fifth force experiments.

With this in mind one may try to measure the coupling constant and the range of the fifth force experimentally. A recent torsion-balance fifth force experiment puts (with $\beta \simeq 1$) the upper bound of the range of the fifth force to be around 56 μm with 95% confidence level.[18,19] This tells that the dilaton mass has to be larger than 10^{-2} eV. In general the fifth force experiments tell us the allowed region of the dilaton mass versus the coupling constant β, which can be used for us to estimate the size of the extra space.

Cosmology also puts a theoretical constraint on the dilaton mass. As a GIMP the dilaton could be a natural candidate for the dark matter of the universe.[6,8] It starts off in thermal equilibrium at the beginning and decouples from other sources very early near the Planck time. Moreover, since its coupling to matter fields is very weak, it may easily survive in the present universe and comprise a significant amount of dark matter. So we can estimate the mass of the dilaton, assuming that the energy density of the relic dilaton provides the dark matter density. A typical decay process of the dilaton is the two-photon process and the fermion-antifermion pair production process. The Lagrangian (5) implies that, in the linear approximation where σ is assumed small enough, dilaton decay may be described by the following interaction Lagrangian,[6,15]

$$\mathcal{L}_{int} \simeq -\sqrt{16\pi G}\,\Big[\,\frac{g_1}{4}\;\hat{\sigma}\;F_{\mu\nu}F^{\mu\nu} + g_2\;m_f\;\hat{\sigma}\;\bar{\psi}\psi\,\Big], \tag{12}$$

where g_1 and g_2 are dimensionless coupling constants, m_f is the mass of the fermion, and $\hat{\sigma}$ is the physical dilaton field. From this we can calculate the two-photon decay rate and the fermion-antifermion decay rate of the dilaton at tree level, and estimate the dilaton energy density $\rho(\mu)$ at present universe.[6,15]

According to recent cosmological observations, dark matter accounts for about 23% of the critical density $\rho_c = 3H_0^2/(8\pi G) \simeq 5.2$ keV cm^{-3}. So, for the dilaton to be the dark matter, we must have the following constraints for the dilaton mass. First, the dilaton energy density $\rho(\mu)$ must be equal to the dark matter density $\rho_0(\mu) \simeq 1.2$ keV cm^{-3}.[6,15] Secondly, the energy density of the daughter particles of dilaton decay (photons and light fermions) $\tilde{\rho}(\mu)$ must be negligibly small compared to the dark matter density, $\tilde{\rho}(\mu) \ll \rho_0(\mu)$. We can find the dilaton mass which satisfies the above constraints, assuming only $g_1 \simeq g_2$.[6,15] Our calculation shows that when $\mu < 160$ eV or $\mu > 276$ MeV, the dilaton undercloses the universe, but when 160 eV $< \mu <$ 276 MeV it overcloses the universe. Moreover, the 160 eV dilaton has the lifetime $\tau_1 \simeq 8.1 \times 10^{17}\; t_0$, so that the contribution of the daughter particles to the energy density becomes negligible. However, the 276 MeV dilaton has the lifetime $\tau_2 \simeq 6.9 \times 10^{-2}\; t_0$, so that the contribution of decays into photons and light fermions becomes huge, much more than the dark matter density. This rules out 276 MeV as a possible mass for the dilaton.[6,15] As importantly, our result shows that, even when the dilaton

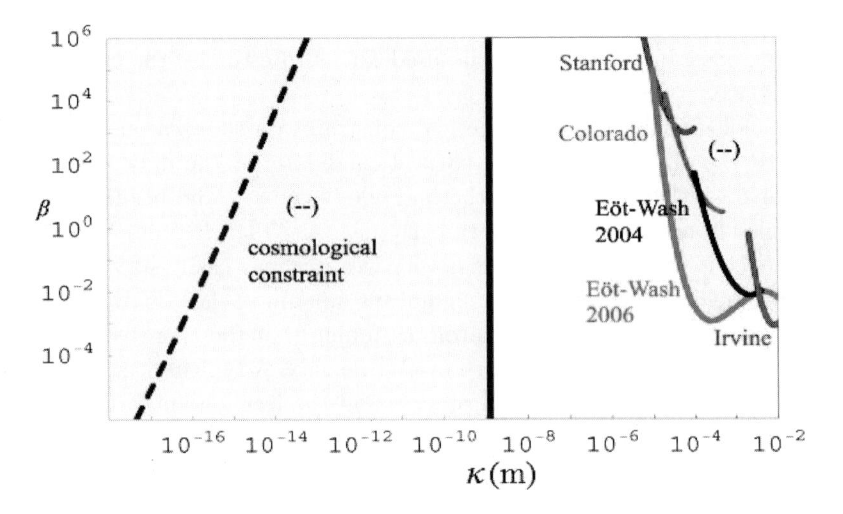

Fig. 1. The allowed range κ of the extra space versus the relative fine structure constant $\beta = \alpha_5/\alpha_g$ of the fifth force in $(4+3)$-dimensional Kaluza-Klein unification with S^3 compactification of the extra space. The colored region marked by (–) is the excluded region. The dotted line represents the constraint of the heavy dilaton whose daughter particles overclose the universe.

does not turn out to be the dark matter, it can not have mass between 160 eV and 276 MeV.

Clearly these constraints on the dilaton mass restrict the allowed size of the extra space. Putting the two constraints together we obtain Fig. 1, which shows the allowed regions of the scale of the extra space versus the relative fine structure constant $\beta = \alpha_5/\alpha_g$ of the fifth force. Notice that the dotted line represents the heavy dilaton which is excluded by the constraint $\tilde{\rho}(\mu) \ll \rho_0(\mu)$. From these constraints on the dilaton mass we can determine the scale of the extra space. The fifth force experiments require that the scale of the extra space cannot be larger than $10^{-5}\ m$, but the cosmological constraint excludes the scale of the extra space smaller than $10^{-9}\ m$. Notice that the cosmological constraint plays a crucial role in restricting the scale of the extra space. It would be extremely difficult to obtain such a short-range constraint by fifth force experiments alone, or by any type of laboratory experiment.

6. Discussion

In this talk we have shown that a large extra space makes the fine structure constant of the KK gauge boson extremely small, and creates very light KK excited modes which could be inconsistent with the known experiments. On the other hand, we showed that a large extra space does explain the huge gap between the Planck mass and ordinary masses of elementary particles, and thus could resolve the hierarchy problem. In particular we have shown that the small fine structure constant of the KK gauge boson plays the key role to reduce the Planck mass to ordinary mass scale.

The dimensional reduction by isometry makes the extra space inaccessible. This, however, does not mean that we can not confirm the existence of the extra space. There are ample evidence of the extra space, a straightforward evidence being the existence of the dilaton and internal gravitons. And we can even estimate the size and curvature of the extra space from the dilaton mass.

In particular, the dilaton plays a crucial role for us to understand the mass hierarchy in physics. A large extra space makes the dilaton mass small, and can reduce the Planck mass to the mass scale of ordinary elementary particles, and thus explain the huge gap between them. We emphasize, however, that for the dilaton to acquire a mass the extra space must be curved. This is because the curvature of the extra space creates the dilaton mass. And a flat extra space makes the dilaton massless, independent of its size.

Our discussion also implies that mass can originate from the curvature of space-time. Understanding the origin of mass has been a fundamental problem in physics. Indeed a primary purpose of LHC is to confirm the origin of mass based on the Higgs mechanism. The dilaton tells that this problem is closely connected to the hierarchy problem, and solves the hierarchy problem by providing a geometric mass generation mechanism which is different from the Higgs mechanism.

It should be emphasized that our results apply essentially to all unified theories which contain the higher dimensional gravity, including the superstring and supergravity. Of course, in superstring or supergravity the situation is more complicated. For example, here one has an extra higher dimensional dilaton (the string dilaton) in addition to the Kaluza-Klein dilaton which remains massless to all orders in perturbation theory.[4] Moreover, the potential of the Kaluza-Klein dilaton becomes more complicated in the presence of other higher-dimensional matter fields.[2,6] Nevertheless the generic feature of our results could still apply to all higher dimensional unified theories.

This tells two things. First, the string vacuum must be made of a curved extra space. In superstring the popular string vacuum has often been assumed to be a Ricci flat Calabi-Yau manifold or orbifold. Our result implies that any Ricci flat extra space will make the Kaluza-Klein dilaton (as well as the internal gravitons) massless, and thus make the theory inconsistent with the fifth force experiments. So a realistic string vacuum must consist of a compact extra space which has a non-vanishing curvature. Otherwise one has to invent a new mechanism which can make the massless scalar gravitons consistent with the known experiments. Secondly, the string dilaton must acquire a mass not to contradict with the fifth force experiments. The question is how. Unfortunately there is no simple way to make it massive. This is a serious problem in string theory. Otherwise one should have to invent a mechanism that makes the string dilaton coupling to matter fields extremely weak (much weaker than gravitational coupling).[21]

The theoretical imperative for the existence of the dilaton in any higher dimensional unification makes the dilaton a fundamental scalar field of nature which one cannot ignore, at least from a theoretical point of view. This makes the experimental confirmation of it an urgent issue. A natural way to confirm the existence of the dilaton would be through a fifth force experiment. But our result shows that, if indeed the dilaton becomes the dark matter of the universe, it would be very difficult to detect it by a fifth force experiment since the fifth force would be too short-ranged to be detected. We certainly need a totally different type of experiment.

One can think of two types of experiments, both based on the dilaton-photon interaction.[6,15] First, one can consider the dilaton-photon conversion experiment in a strong electromagnetic resonant cavity, similar to the axion detection experiment.[22] Secondly, one can try to detect the two photon decay of the relic dilaton directly from the sky, assuming that the dilaton makes up the dark matter of the universe. Of course, even in these experiments it would be difficult to detect the dilaton, because the dilatonic coupling to the photon is extremely weak. This makes the experimental detection of the dilaton more challenging.[15]

Acknowledgments

The work is supported in part by the BSR Program (Grant KRF-2007-314-C00055) of the Korea Research Foundation.

References

1. Y.M. Cho, J. Math. Phys. **16**, 2029 (1975); Y.M. Cho and P.G.O. Freund, Phys. Rev. **D12**, 1711 (1975); Y.M. Cho and P.S. Jang, Phys. Rev. **D12**, 3138 (1975).

2. Y.M. Cho, Phys. Rev. **D35**, 2628 (1987).
3. See e.g., M. Duff, B. Nilsson, and C. Pope, Phys. Rep. **130**, 1 (1986).
4. See e.g., E. Witten, Phys. Lett. **B 155**, 151, (1985). See also M. Green, J. Schwartz, and E. Witten, *Superstring theory*, Vol. 2 (Cambridge University Press) 1987.
5. R. Penrose, *The Road to Reality : A Complete Guide to The Laws of The Universe* (A.A. Knopf) 2005.
6. Y.M. Cho and Y.Y. Keum, Class. Quant. Grav. **15**, 907 (1998).
7. Y.M. Cho, Phys. Lett. **B199**, 358 (1987).
8. Y.M. Cho, Phys. Rev. **D41**, 2462 (1990); Y.M. Cho and J.H. Yoon, Phys. Rev. **D47**, 3465 (1993). See also, Y.M. Cho, in *Proceedings of XXth Yamada Conference on Big Bang, Active Galactic Nuclei, and Supernovae*, edited by S. Hayakawa and K. Sato (Universal Academy Press, Tokyo) 1988.
9. I. Antoniadis, Phys. Lett. **B246**, 377 (1990).
10. N. Arkani-Hamed, S. Dimopoulos, and G. Dvali, Phys. Lett. **B429**, 263 (1998); N. Arkani-Hamed, I. Antoniadis, S. Dimopoulos, and G. Dvali, Phys. Lett. **B436**, 257 (1998).
11. I. Antoniadis, S. Dimopoulos, and G. Dvali, Nucl. Phys. **B516**, 70 (1998); Phys. Rev. Lett **84**, **D62**, 105002 (2000).
12. P.A.M. Dirac, Nature (London) **136**, 323 (1937).
13. U. Gunther, A. Starobinsky, and A. Zhuk, Phys. Rev. **D 69**, 044003, (2004).
14. Y.M. Cho and D.H. Park, Gen. Rel. Grav. **23**, 741 (1991).
15. Y.M. Cho and J.H. Kim, Phys. Rev. **D 79**, 023504 (2009).
16. P. Jordan, Ann. Phys. (Leipzig) **1**, 218 (1947); W. Pauli, in *Schwerkraft und Weltall* (F. Vieweg und Sohn, Braunschweig) 1955; C. Brans and R. Dicke, Phys. Rev. **124**, 921 (1961).
17. Y.M. Cho, Phys. Rev. Lett. **68**, 21 (1992).
18. E. Fishbach et al., Phys. Rev. Lett. **56**, 3 (1986); C. Hoyle et al., Phys. Rev. Lett. **86**, 1418 (2001).
19. D. Kapner et al., Phys. Rev. Lett. **98**, 021101 (2007); E. Adelberger et al., Phys. Rev. Lett. **98**, 131104 (2007).
20. See, for example, M. Kaku *Introduction to Superstrings* (Springer-Verlag New York Inc.) 1988.
21. T. Damour and A. Polyakov, Nucl. Phys. **B423**, 532 (1994).
22. P. Sikivie, Phys. Rev. Lett. **51**, 16 (1983); Phys. Rev. **D32**, 11 (1985); P. Sikivie, D. Tanner, and Y. Wang, Phys. Rev. **D50**, 4744 (1994).

DARK ENERGY AND DARK MATTER IN MODELS WITH WARPED EXTRA DIMENSIONS

ISHWAREE. P. NEUPANE

Department of Physics and Astronomy, University of Canterbury,
Private Bag 4800, Christchurch 8041, New Zealand
**E-mail: ishwaree.neupane@canterbury.ac.nz*

In this note I discuss about the recent attempts, difficulties and prospects of explaining the current acceleration of the universe (attributed to dark energy) and also dark matter using cosmological warped compactifications of higher-dimensional gravity models, including five-dimensional braneworld models.

Keywords: Warped extra dimensions; cosmic acceleration; dark energy; dark matter.

1. Introduction

The observed current acceleration of our universe[1] is among the most puzzling discoveries in modern cosmology. This somewhat unexpected result together with an inflationary epoch required to solve the horizon and flatness problems of the big bang cosmology need to be understood in the framework of fundamental theories. While the basic principles of the early inflation are rather well established, with many of its predictions being supported by observational data from WMAP,[2] it still remains a paradigm in search of a concrete theoretical model. Efforts are still underway to explain dark energy and dark matter from 10d or 11d supergravity models which are the low-energy effective theories of superstrings.

In light of observational evidences for both an inflationary epoch in the distant past and a recent cosmic acceleration, it is of importance to construct explicit de Sitter solutions from 10d or 11d supergravity models which are the low-energy effective theories of superstrings. In the last two decades, various low energy versions of string theory have been widely used to understand and to probe the mathematical structures of quantum field theories and cosmologies at a microscopic scale. String theory has been a source of new ideas and has greatly inspired novel scenarios of cosmology, but it has yet to confront data and make predictions. Examples of predictions which have received considerable interests include mechanism of supersymmetry breaking, techniques of generating a small positive cosmological constant[3] or quintessence[4] and general statements about the scale and properties of inflation.[5] Motivated by this success and observation driven cosmology, it seems worth studying string theory models in cosmological backgrounds.[6–8]

There are multiple ambitious experimental and theoretical efforts currently planned to find answer to the origin of cosmic acceleration, including the primordial inflation. One of the most intriguing ideas to this problem centers on the role of extra dimensions. Randall and Sundrum in a theory referred to as RS1[9] realized that a five-dimensional braneworld model with a brane can reasonably address the mass hierarchy in particle physics if there is a second brane some distance away from the first, which perhaps mimics the physical 3+1 spacetime that we live in. For this simple and elegant proposal to work one needs a five-dimensional anti de Sitter spacetime, i.e., a background geometry which is negatively curved. Once the five-dimensional cosmological constant is assumed to be zero then the solution would be lost. In a sense the RS solution is very much fine-tuned. It is therefore of natural importance to find nontrivial solutions that exist in flat spacetimes as well. It would be equally interesting to see if cosmic acceleration (attributed to dark energy) can arise as purely an outcome of the vacuum Einstein equations in higher dimensions.

2. Extra Dimensions and Warped Compactification

One of the most remarkable achievements in physics around the mid 1990s is a series of new understanding about string theory dualities, roles of "branes" of different dimensionality and extended objects such as orientifold planes and anti-branes. These new ingredients in string/M theory have helped physicists to learn some fabulous ideas, such as localization of gravity in braneworld models, gravity -gauge theory duality in type IIb string theory and inflationary cosmologies in models of warped flux compactifications.

Following,[9,10] one may find interest in warped metrics that maintain the usual four-dimensional Poincaré symmetry:

$$ds_D^2 = W(y)^2\, \hat{g}_{\mu\nu} dX^\mu dX^\nu + W(y)^\gamma \tilde{g}_{mn}(y)\, dy^m dy^n, \tag{1}$$

where X^μ are the usual spacetime coordinates ($\mu, \nu = 0, 1, 2, 3$), $W(y)$ is the warp factor as a function of one of the internal coordinates, y, and γ is a constant. This form of the metric is more general than the usual product metric for which $W(y) = 1$. The internal m-dimensional metric is $ds_m^2 = \tilde{g}_{mn}(y)dy^m dy^n$. Non-factorizable metrics as above, where the metric of the four familiar dimensions is dependent of coordinate in the extra dimensions, are not only phenomenologically motivated as in RS models they are quite natural in string theory.[11]

For many years, it was assumed that cosmic acceleration was ruled out for classical 10d or 11d supergravity models on the basis of a "no-go" theorem,[12,13] which forbids accelerating cosmologies in the presence of static and warped extra dimensions. Recent studies 10d string theory and 11d supergravity models[6,7] showed that it is possible to circumvent earlier no-go theorems in a time-dependent background using compact extra dimensions that are negatively curved and also time-dependent. In fact, time-dependent solutions would be acceptable only if the their growth rate is not too large or not big enough to affect the big-bang nucleosynthesis bounds imposed on various matter and gauge coupling constants.

Here I only consider standard warped metrics as defined in (1) and present explicit cosmological solutions for which not only the warp factor is nontrivial but also the expansion rate of our part of the spacetime can undergo an inflationary de Sitter expansion. An intriguing feature of these solutions is that the scale factor of the universe becomes a constant only in the limit where the warp factor $W(y)$ also becomes a constant. This feature is quite remarkable and also distinctive from that in the original RS models.

Recent cosmological observations appear to fit reasonably well with the standard four-dimensional Einstein general relativity supplemented by a cosmological constant term. It might therefore be important to know whether this picture can be realized by considering compactified versions of 10d string theory. A recent novel understanding in this direction is that inflationary de Sitter solutions are possible in the presence of static and warped extra dimensions, without violating the four- and higher-dimensional null energy condition or any other energy conditions other than the four-dimensional strong energy condition. In order to realize such an outcome one should write the higher dimensional metric ansatz in a general form rather than specializing to a particular class of metrics.

The full spacetime dimensions is D, which may be split as $D \equiv 4+m \equiv 4+1+q$. It is natural to suppose that q of the extra dimensions are physically compact. The radial direction y may be closed or even infinitely large, as in RS2 model.[10] In principle, one can write

$$\tilde{g}_{mn}(y)\, dy^m dy^n \equiv dy^2 + f(y) g_{pq} dy^p dy^q \tag{2}$$

and absorb the conformal factor W^γ inside dy^2 by defining $W^{\gamma/2}dy \equiv d\tilde{y}$ and $W^\gamma f(y) \equiv \mathcal{F}(\tilde{y})$. But there is no advantage in doing this. In fact, only some specific values of γ become relevant when one writes an explicit metric for the internal space, say Y_m. The choices made by Gibbons,[12] Maldacena and Nuñez[13] and Giddings et al.,[14] with respect to braneworld no-go theorems, are all different. These are, respectively, $\gamma = 0$, $\gamma = 2$ and $\gamma = -2$. This difference may not be much relevant in $D = 5$ dimensions but in dimensions $D \geq 6$ the choice of γ can be associated with the spatial curvature of the internal space Y_m, as explicitly shown in.[15]

To strengthen the above point, let me assume that the internal space metric $\tilde{g}_{mn}dy^m dy^n$ is Ricci flat. This is the case for instance if $f(y) = y^2$ and the base space X_m is an m-dimensional sphere or some other Einstein spaces having positive spatial curvature. Then Einstein's field equations will show that there exists a specific value of γ for which there is an inflationary de Sitter solution in four-dimensions. Such a solution can easily go unnoticed have I started with a choice like $\gamma = 0$.

If one starts with $\gamma = 0$ and also imposes the condition that the internal space is Ricci flat, then one easily ends up without finding an inflationary de Sitter solution at least in the vacuum case. The story is similar for some other choices of γ (see below). So, here I shall keep the coefficient γ and also the spatial curvature of the internal spaces unspecified, so Y_m can have the positive, zero or negative scalar curvature.

3. An Explicit Model in Five Dimensions

One may assume that some of the extra spatial dimensions (denoting X_q) are too tiny or curled up in a compact way, and the real world looks like a five-dimensional universe described by warped geometry of the form

$$ds_5^2 = W(y)^2\, \hat{g}_{\mu\nu} dX^\mu dX^\nu + e^{2Q}\, W(y)^\gamma\, dy^2, \tag{3}$$

where Q is an arbitrary constant. The classical action describing this geometry is given by

$$S = \frac{M_5^3}{2} \int d^5x \sqrt{-g_5}\, R_{(5)}, \tag{4}$$

where M_5 is the fundamental 5D Planck scale. My starting point is different from that in the RS braneworld models only in that I will take the metric of the usual four-dimensional spacetime in a general form

$$ds_4^2 \equiv \hat{g}_{\mu\nu} dX^\mu dX^\nu = -dt^2 + a^2(t) \left[\frac{r_0^2\, dr^2}{r_0^2 - kr^2} + r^2(d\theta^2 + \sin^2\theta d\phi^2) \right], \tag{5}$$

where $a(t)$ is the scale factor of the universe and r_0 is an arbitrary constant. I have allowed all three possibilities for the physical 3d spatial curvature: flat ($k = 0$), open ($k = -1$) and closed ($k = +1$). Models similar to the one here were studied before, see for example,[16] but a new observation here is that for the existence of inflationary de Sitter solutions we do not necessarily require a (higher-dimensional) bulk cosmological constant term.

Equations of motion following from the metrics (3) and (5) are given by

$$W'^2 - e^{2Q}\left(\frac{\dot{a}^2}{a^2} + \frac{k}{a^2 r_0^2}\right) W^\gamma = 0, \tag{6a}$$

$$2WW'' - \gamma W'^2 = 0, \tag{6b}$$

$$\frac{\ddot{a}}{a} - \frac{\dot{a}^2}{a^2} - \frac{k}{a^2 r_0^2} = 0, \tag{6c}$$

where $\dot{}$ and $'$ denote respectively $\partial/\partial t$ and $\partial/\partial y$. From Eq. (6c) we get

$$a(t) = \frac{1}{2}\exp\left(\frac{\mu(t - t_0)}{e^Q}\right) + \frac{k\, e^{2Q}}{2r_0^2\mu^2}\exp\left(\frac{\mu(t_0 - t)}{e^Q}\right), \tag{7}$$

where μ and t_0 are integration constants. In the $\gamma \neq 2$ case, from Eqs. (6a) and (6b), we obtain

$$[W(y)]^{2-\gamma} = \frac{1}{4}\,(2-\gamma)^2\,\mu^2 (y + c)^2. \tag{8}$$

The bulk singularity at $y = -c$ is a coordinate artifact, which would be absent in some other coordinate system. To quantify this one can either introduce a new coordinate z satisfying $W(y)^{\gamma/2}dy \equiv W(z)\,dz$ or solve Einstein's equations by setting $\gamma = 2$ in (3). We then obtain

$$ds_5^2 = e^{-2\mu(z+z_0)}\left(\hat{g}_{\mu\nu}dX^\mu dX^\nu + e^{2Q}\,dz^2\right). \tag{9}$$

In the $k = -1$ case, there is a big-bang type singularity at $\mu(t - t_0) = (Q - \ln(r_0\mu))\,e^Q$, while in the $k = 0$ and $k = +1$ cases, the scale factor is always non zero. One could in principle set $Q = 0$ in Eq. (3), but an essential point here is that the scale factor and warp factor can have different slopes. The physical scale factor $a(t)$ grows exponentially while the radial coordinate z is static, so the model is cosmologically viable.

I have not specified yet whether the radial direction y is compact or not, but solutions to 5D Einstein equations are the same for both compact and non-compact extra space. To draw a distinction between these two cases, which become relevant especially when one wants to solve some other problems in physics, e.g., the mass hierarchy, one may introduce two branes of opposite brane tensions at orbifold fixed points, as in RS1 model.

Let us introduce a bulk cosmological term Λ and also specify boundary conditions such that the warp factor is regular at $y = y_0$ where we place a 3-brane with brane tension T_3 and well-behaved at infinite distances from the brane. The classical action describing this set up is

$$S = \frac{M_5^3}{2} \int_B d^5x \sqrt{-g_5}\,(R - 2\Lambda) + M_5^3 \int_{\partial B} d^4x \sqrt{-g_{\rm b}}(-T_3), \tag{10}$$

where g_b is the determinant of the metric g_{ab} evaluated at $y = y_0$. Einstein's equations are given by

$$G_{AB} = -T_3 \frac{\sqrt{-g_b}}{\sqrt{-g}} g^b_{\mu\nu} \delta^\mu_A \delta^\nu_B \delta(y - y_0) - \Lambda g_{AB}. \tag{11}$$

Eqs. (6a) and (6b) are now modified as

$$W'^2 - e^{2Q}\left(\frac{\dot{a}^2}{a^2} + \frac{k}{a^2 r_0^2}\right) W^\gamma = -\frac{\hat{\Lambda}}{6} W^{2+\gamma}, \tag{12a}$$

$$2WW'' - \gamma W'^2 = -\frac{\hat{\Lambda}}{3} W^{2+\gamma} - \frac{\tau_3\,\delta(y - y_0)}{3} W^{2-\gamma/2}, \tag{12b}$$

where $\hat{\Lambda} \equiv \Lambda\, e^{2Q}$ and $\tau_3 \equiv T_3\, e^Q$, while Eq. (6c) is the same, which is not modified by a bulk cosmological term. One replaces $\delta(y - y_0)$ by $\delta(z)$ in the $\gamma = 2$ case. With the widely used choice that $\gamma = 0$, we find[17]

$$W(y) = \frac{\sqrt{6}\,\mu}{\sqrt{-\hat{\Lambda}}} \sinh\left[\frac{\sqrt{-\hat{\Lambda}}\,(y + c)}{\sqrt{6}}\right]. \tag{13}$$

By defining $W(y)^{\gamma/2} dy \equiv W(z)\, dz$, we find

$$W(z) = \frac{24\mu^2\, e^{\mu(z_0 - z)}}{24\mu^2\, e^{2\mu z_0} + \hat{\Lambda}\, e^{-2\mu z}}, \tag{14}$$

which has a smooth $\Lambda \to 0$ limit. This result is an exact solution of 5D Einstein equations obtainable by setting $\gamma = 2$ in Eq. (3).

Let us consider only the $\Lambda = 0$ case. We then find

$$W(z) = e^{-\mu|z|}, \quad T_3 = 12\mu\, e^{-Q}. \tag{15}$$

The brane tension vanishes in the limit $\mu \to 0$ and in the same limit $a(t) \to$ const and $W(y) =$ const. This defines a flat Minkowski vacuum. Although the details and the motivations are different, the above solution inherits certain features of a 5D braneworld model discussed by Dvali et al. and Deffayet[18] where acceleration is supported by a 4D scalar curvature.

The four-dimensional effective theory follows by substituting Eq. (3) into the classical action (4) or (10). Here I focus on the 5D curvature term from which we can derive the scale of gravitational interactions:

$$S_{\rm eff} \supset \frac{M_5^3}{2}\int d^4x\sqrt{-\hat{g}_4}\int e^Q dy\, W^{2+\gamma/2}\left(\hat{R}_4 - \mathcal{L}_0\right), \tag{16}$$

where $\mathcal{L}_0 \equiv e^{-2Q}\, W^{-\gamma}\left(12W'^2 + 8WW'' - 4\gamma W'^2\right)$. From Eq. (9) we get $\mathcal{L}_0 = 12\mu^2 e^{-2Q} \equiv \Lambda_4$. This result holds even if $\gamma \neq 2$. In the $\gamma = 2$ case, one replaces dy by dz. The range of the extra dimensional coordinate is then $0 \le z \le \pi r_c$, where r_c is a compactification radius. From Eq. (15), we find the relation between four- and five-dimensional effective Planck masses:

$$M_{\rm Pl}^2 = M_5^3\, e^Q \int_0^{\pi r_c} dz\, e^{-3\mu|z|} = \frac{2M_5^3 e^Q}{3\mu}\left[1 - e^{-3\mu r_c \pi}\right]. \tag{17}$$

As in,[10] there is a well-defined value for $M_{\rm Pl}$, even in the $r_c \to \infty$ limit.

4. Revisiting Braneworld No-Go Theorems

One may ask: what prevented theorists from discovering a large class of inflationary solutions given above? To answer this question we may have to examine the conditions imposed in earlier no-go theorems. Here I again focus on the case of a 5D Minkowski bulk.

We can be little more precise with regard the braneworld no-go theorems. In the early 1980s, Gary Gibbons came up with a sort of negative result about the implication of 10d or 11d warped supergravity models to a realistic four-dimensional cosmology. The braneworld "no-go" theorem forbids cosmic acceleration in cosmological solution arising from compactification of pure SUGRA where the internal space is time-dependent, non-singular compact manifold without boundary. The message of this no-go theorem was something like that "10 into 4 wont's go". Here I have to emphasize that the no-go theorems discussed so far only apply to models with physically compact dimensions but perhaps not to the models where the extra dimensions are only geometrically compact.

Acceleration requires a violation of 4d strong energy condition (SEC):

$$R^4_{00} \quad \text{or} \quad T_{AB}\xi^A\xi^B < 0, \tag{18}$$

where ξ^A and ξ^B are some time-like four vectors. Indeed, the SEC is known to hold both in ten- and eleven-dimensional supergravity models, i.e. $R_{00}^{(D)} \ge 0$. The common lore of several versions of braneworld no-go theorems is that if extra dimensions are warped and static, then in a compactified theory, one normally get

$$R_{00}^{(4)} \geq 0. \tag{19}$$

On the other hand, we also know that in order obtain a cosmic acceleration one is required to have $R_{00}^{(4)} < 0$. The easiest way to see this is to consider the conventional four-dimensional Friedman-Robertson-Walker metric, i.e. Eq. (5). The time-time component of the 4d Ricci tensor of this metric gives

$$R_{00}^{(4)} = -\frac{3\ddot{a}}{a}. \tag{20}$$

Clearly, acceleration requires $\ddot{a}/a > 0$ and hence $R_{00}^{(4)} < 0$. To this end, one should also note that the Einstein field equations give

$$R_{00}^{(4)} = T_{00} + g^{ii}T_{ii} = \rho + 3p. \tag{21}$$

Obviously, the SEC on the matter stress tensor requires the right hand side of this equation to be non-negative. Thus an accelerated expansion is possible in a universe governed by Einstein's gravity only if the matter in it or the source of cosmic acceleration (or dark energy) violates the SEC at least in four-dimensions.

The most celebrated version of "no-go" theorem can be explained by considering the following $(4+m)$-dimensional metric

$$ds_D^2 = W(y)^2 ds_4^2(x) + g_{mn}(y)dy^m dy^n. \tag{22}$$

This gives

$$R_{00}^{(D)}(x,y) = R_{00}^{(4)} - \frac{1}{4W(y)^2}\nabla_y^2 W(y)^4. \tag{23}$$

Multiplying by $W(y)^2$ and performing an integration, we obtain

$$\left[W(y)^2\right] R_{00}^{(4)} = \int W(y)^2\, R_{00}^{(D)} + \frac{1}{4}\int \nabla_y^2 W(y)^4. \tag{24}$$

The second term on the right hand vanishes, especially, when evaluated in a closed cycle or compact extra spaces without boundary. If so, $R_{00}^{(D)} \geq 0$ only if $R_{00}^{(4)} \geq 0$. This seems to rule out de Sitter type compactifications if the extra dimensions are warped and static. This was a kind of setback to many who believe that a four-dimensional accelerating cosmology should arise from some sort of compactifications of fundamental theories of gravity, including string theory. The result eventually led to some simple suggestions, such as, one may have to either modify Einstein's theory with higher curvature terms[19] or some other less well understood non-perturbative effects arising in a higher-dimensional framework.[14]

Nearly 25 years since the first proposition of no-go theorem for a de Sitter solution in many classes of 10d or 11d supergravity models,[12,13] in recent papers,[15,20] I have shown that the limitation with warped models studied to date has arisen partly from the oversimplification of higher dimensional metric ansatzë, other than the fact that some of the conditions imposed in earlier no-go theorems may not apply to cosmological solutions, especially, when the extra dimensions are only geometrically compact. Recent work in[15] provides an explicit example of a non-singular warped compactification on de Sitter space dS_4 with finite four-dimensional Newton's constant, which I will briefly discuss below.

For the metric (3), the basic equations reduce to

$$ {}^{(5)}R_{\mu\nu} = {}^{(4)}R_{\mu\nu} - \frac{\hat{g}_{\mu\nu}}{4W^\gamma}\left[\frac{(W^4)''}{W^2} - 2\gamma W'^2\right], \tag{25a} $$

$$ R_{55} = -\frac{4}{W}W'' + \frac{2\gamma}{W^2}W'^2. \tag{25b} $$

Here I will momentarily set $Q = 0$, without loss of generality. One can rewrite the above two equations as follows

$$ R_g = R_{\hat{g}}\, W^{-2} - 2(6-\gamma)W'^2 W^{-2-\gamma} - 4W''W^{-1-\gamma}, \tag{26a} $$

$$ R_5{}^5 = -4W''W^{-1-\gamma} + 2\gamma W'^2 W^{-2-\gamma}, \tag{26b} $$

where $R_g \equiv {}^{(5)}R_\mu{}^\mu$ and $R_{\hat{g}} \equiv {}^{(4)}R_\mu{}^\mu$ are, respectively, the curvature scalars of the 5- and 4-dimensional spacetimes with the metric tensors $g_{\mu\nu}$ and $\hat{g}_{\mu\nu}$. One can easily check that a linear combination of $(1-n)W^{n+\gamma}\times$ Eq. (26a) and $(n-4)W^{n+\gamma}\times$ Eq. (26b) gives (for any positive n)

$$ \frac{(W^n)''}{n} - \frac{\gamma}{2}W'^2 W^{n-2} = W^{n+\gamma}\left[\frac{1-n}{12}\left(R_g - R_{\hat{g}}W^{-2}\right) + \frac{n-4}{12}R_5{}^5\right]. \tag{27} $$

From Eq. (27) and the equations $R_{AB} = 8\pi G_5(T_{AB} - \frac{1}{3}g_{AB}T_C{}^C)$, we get

$$ \frac{2\pi G_5}{3}(T_g + (2n-4)T_5{}^5)e^{(n+\gamma)A} = \frac{1-n}{12}\, e^{(n+\gamma-2)A}R_{\hat{g}} - (A'e^{nA})' + \frac{\gamma}{2}A'^2 e^{nA}, \tag{28} $$

where $e^{A(y)} \equiv W(y)$ and $T_g \equiv {}^{(5)}T_\mu{}^\mu$. As is evident, with $\gamma = 0$, one recovers the braneworld sum rule discussed in.[21]

One of the common assumptions of the work by Gibbons, Maldacena and Nuñez, and many other authors, with respect to braneworld no-go theorem, is that the extra-dimensional space is closed, i.e., that without boundary. They also assumed that it is compact, though the underlying principle here is that one should make sure that the 4D Newton's constant in finite, as is required to get a sensible 4D effective theory. In fact, the boundedness conditions like $\oint \nabla^2 W^4 = 0$ and $\oint \nabla(W^{n-1}\nabla W) = 0$ discussed in[12,22] are "strict", which are not essentially satisfied by cosmological solutions, especially, when the extra dimensions are only geometrically compact.

With $W(z) = e^{-\mu|z|}$, which was obtained above by taking $\gamma = 2$ in Eq. (3), one finds that $\oint \nabla^2 W^4 \equiv \oint (W^4)'' - 4\oint W'^2 W^2 = -4\mu^2 \oint W^4$, which vanishes only if $\mu = 0$. In the limit $\mu \to 0$, we get $M_{\rm Pl}^2 \to M_5^3\, \pi r_c$, $T_3 \to 0$ and $W(z) \to$ const. Here I consider the cosmologically relevant case for which $\mu \neq 0$ or $\dot{a} \neq 0$. The 3-brane tension is thus non-vanishing.

Assuming that there always exists a class of solutions for which $\oint (A' e^{nA})' = 0$, we find

$$\oint e^{(n+\gamma)A} \left(T_g + (2n-4){T_5}^5\right) = \frac{1-n}{8\pi G_5} R_{\tilde{g}} \oint e^{(n+\gamma-2)A} + \frac{3\gamma}{4\pi G_5} \oint A'^2\, e^{nA}. \tag{29}$$

As is evident, we can get $R_{\hat{g}} > 0$ by appropriately choosing n or γ, even if the term on the left hand side vanishes. This result is in agreement with the explicit cosmological solutions of 5D vacuum Einstein equations presented above. Similar results hold in dimensions $D \geq 6$.

5. An Explicit Model in Ten Dimensions

The above discussion can be extended to higher dimensions. With this motivation, one may assume that ten-dimensional supergravity is the relevant framework. The internal six-dimensional space Y_6 can have an arbitrary scalar curvature, which is a generalized 6D space with the metric

$$ds_6^2 \equiv \tilde{g}_{mn}(y) dy^m dy^n \equiv \lambda^2 \left(dy^2 + \beta f(y)\, ds_{X_5}^2\right), \tag{30}$$

where λ is a scale associated with the size of Y_6 or compactification scale. Here X_5 can be some compact manifolds, such as S^5 or Einstein-Sasaki space $T^{1,1} = (S^2 \times S^2) \rtimes S^1$. The metric of the latter example is

$$\begin{aligned} ds_{X_5}^2 &= g_{ii} d\theta^i d\theta^j = \frac{1}{9}\, e_\psi^2 + \frac{1}{6}\left(e_{\theta_1}^2 + e_{\phi_1}^2 + e_{\theta_2}^2 + e_{\phi_2}^2\right) \\ &= \frac{1}{9}\left(d\psi + \cos\theta_1 d\phi_1 + \cos\theta_2 d\phi_2\right)^2 + \frac{1}{6}\sum_{q=1}^{2}\left(d\theta_q^2 + \sin\theta_q^2 d\phi_q^2\right), \end{aligned} \tag{31}$$

where (θ_1, ϕ_1) and (θ_2, ϕ_2) are coordinates on each S^2 and ψ is the coordinate of a $U(1)$ fiber. The solutions discussed below would be valid even when $T^{1,1}$ is replaced by S^5. The 6D scalar curvature is

$$R_{(6)} = \frac{5}{\lambda^2 f(y)} \left(\frac{4}{\beta} - f'' + \frac{f'^2}{2f} \right). \tag{32}$$

With $f(y) \equiv y^2$, we find $R_6 = 20(1-\beta)/(\beta y^2)$, which implies that Y_6 becomes Ricci flat only when $\beta = 1$. Although it is not difficult to generalize the metric (30) such that Y_6 is a deformed conifold without any conical singularities,[8,24] here I only consider the simplest example where $f(y) = y^2$.

Inflationary de Sitter solutions be obtained also from string compactifications on general conifolds or Ricci flat 6d spaces, provided one writes the 10d metric ansatz in an appropriate form. To this end, we make the ansatz

$$ds_{10}^2 = e^{2A(y)} \left(-dt^2 + a(t)^2\, d\mathbf{x} \cdot d\mathbf{x}\right) + e^{\gamma A(y)} \lambda^2 \left(dy^2 + \beta\, y^2 ds_{X_5}^2\right), \tag{33}$$

where β and γ are some numerical constants. This gives

$$\begin{aligned} \sqrt{-g_{10}} &= \frac{\lambda^6}{108} \beta^{5/2} y^5 e^{(4+3\gamma)A} \sin\theta_1 \sin\theta_2 \sqrt{-\hat{g}_4}, \\ \mathcal{R}_{10} &= e^{-2A}\left(\hat{R}_4 - \mathcal{L}_\Lambda\right), \end{aligned} \tag{34}$$

where $\hat{R}_4 = 6(\ddot{a}/a + \dot{a}^2/a^2)$ and

$$\mathcal{L}_\Lambda = \frac{e^{(2-\gamma)A}}{\lambda^2} \left[(8+5\gamma)\left(A'' + \frac{5}{y} A'\right) + (20 + 16\gamma + 5\gamma^2) A'^2 + \frac{20}{y^2} \frac{(\beta - 1)}{\beta} \right], \tag{35}$$

where $A' = dA/dy$. The 6d scalar curvature is $R_{(6)} = 20(1-\beta)/(\beta\lambda^2 y^2)$. The effective 4d theory can have a nonzero cosmological constant-like term, even when the 6d space is Ricci flat, which corresponds to the choice $\beta = 1$. To be precise, one solves the 10d Einstein equations, whose solution is

$$\begin{aligned} e^{(2-\gamma)A} &= \frac{3(2-\gamma)^2 y^2}{32 L^2}, \quad \beta = \frac{(2-\gamma)^2}{8}, \\ a(t) &\propto e^{Ht}, \quad \left(H^2 = \frac{1}{\lambda^2 L^2}\right). \end{aligned} \tag{36}$$

From Eqs. (35) and (36), we get $\mathcal{L}_\Lambda = 12/(\lambda^2 L^2) \equiv 2\Lambda$. In the case of a Ricci-flat 6d space, we take $\gamma = 2 \pm 2\sqrt{2}$, which gives a de Sitter solution. The authors of[14] made the choice $\gamma = -2$ and $R_6(\tilde{g}) = 0$, which already belongs to a restrictive class of metrics, for which the solution of 10d vacuum

Einstein equations is trivial: $a(t) = \text{const}$, $A(y) = \text{const}$. With $\gamma = -2$, a de Sitter solution is obtained by taking $\beta = 2$, i.e. when $\tilde{R}_6 = \tilde{R}_{mn}\tilde{g}^{mn} < 0$.

With the choice that $\gamma = 0$, the above solution looks like RS model, except the presence of additional compact dimensions:

$$ds_{10}^2 = e^{2A(y)} ds_4^2(x) + r_c^2 \left(dy^2 + \frac{y^2}{2}\, ds_{X_5}^2 \right), \tag{37a}$$

$$e^{2A} = \frac{3}{8}\frac{y^2}{L^2}, \quad a(t) \propto e^{Ht}, \quad H = \sqrt{\frac{1}{r_c^2 L^2}}. \tag{37b}$$

The metric is different from RS models in one more aspect: the 10d spacetime here looks like $dS_4 \times Y_6$, since both the usual 4d spacetime and the internal 6d space have positive curvatures $\hat{R}_{(4)} = 12/(\lambda^2 L^2)$ and $\tilde{R}_{(6)} = 20/(\lambda^2 y^2)$. Of course, 10d Ricci scalar curvature is vanishing, $R_{10} = e^{-2A}\hat{R}_{(4)} + \tilde{R}_{(6)} - \frac{4}{\lambda^2}(2A'' + 5A'^2 + 10A'/y) = 0$.

The $\gamma = 2$ case must be treated separately. This is also the choice made by Maldacena and Nuñez[13] in reference to braneworld no-go theorem. In this particular case, one could actually introduce a new coordinate z, satisfying $z \propto \ln y$. We then find that the following metric ansatz

$$ds_{10}^2 = e^{2A(z)} \left(-dt^2 + a(t)^2\, d\mathbf{x} \cdot d\mathbf{x} \right) + e^{-2A(z)} ds_6^2, \tag{38}$$

solves the 10d vacuum Einstein equations when

$$A(z) = -\frac{k(z + z_0)}{4}, \quad a(t) = a_0 \exp\sqrt{\frac{4t^2 e^{-kz_0}}{27\, r_c^2}}, \tag{39a}$$

$$ds_6^2 = r_c^2\, e^{kz} \left(\frac{9k^2}{8}\, dz^2 + e_\psi^2 + \frac{3}{2}\sum_{i=1}^{4} e_i^2 \right), \tag{39b}$$

where r_c is a compactification scale. This gives an explicit example of a non-singular warped compactification on de Sitter space dS_4. The no-go theorem discussed by Maldacena and Nuñez[13] is clearly avoided, or at least that it may not apply to cosmological warped solutions. This argument may be extended to some other forms of no-go theorem.[12] For example the model considered by Gibbons corresponds, in the present model, to the choice $e^{A(y)} = W(y)$ and $\gamma = 0$, but in this case we have already seen that there is a 4d de Sitter space solution with $\beta = 1/2$ or when $\tilde{R}_6 = 10/(\lambda^2 y^2) > 0$. The message is simple and clear. If one does not restrict the curvature of the internal space or allow a general 6d metric $ds_6^2 = \tilde{g}_{mn}(y) dy^m dy^n$ having positive, zero or negative scalar curvature, then one finds that inflationary cosmology is always possible with a specific choice of γ.

One may have already noted that the solutions given in Eq. (36) contain a conical singularity at $y = 0$. This is a singularity where the warp factor $e^{A(y)}$ vanishes, which may have a physical interpretation in terms of four-dimensional effective field theory, perhaps with some fields supported near the singularity. The framework of low energy effective four-dimensional theory is actually provided by the solutions given Eqs. (39a)-(39b), which are non-singular and also related to the solutions in Eqs. (33)-(36) by the transformation $e^{\gamma A(y)/2}dy = e^{A(z)}dz$. In the discussion below we assume the existence of a positive tension 3-brane, imposing Z_2 symmetry at $z = 0$.

The product $\mathcal{K}z$ must be positive, which is required to make the warped volume of the 6d space finite. With a particular value of $\mathcal{K}$, i.e. $\mathcal{K} = -2/3$, the 6d space (39a) is related to asymptotic form of the metric

$$ds_6^2 \equiv 3 \times 2^{2/3} \times r_c^2 \left[\frac{1}{3K^2}\left(\frac{dz^2}{2} + e_\psi^2\right) + \frac{g_+^2 K}{2} \cosh^2\left(\frac{z}{2}\right) + \frac{g_-^2 K}{2} \sinh^2\left(\frac{z}{2}\right)\right], \tag{40}$$

where $e_1 = -\sin\theta_1 d\phi_1$, $e_2 = d\theta_1$, $e_3 = \cos\psi \sin\theta_2 d\phi_2 - \sin\psi d\theta_2$ and $e_4 = \sin\psi \sin\theta_2 d\phi_2 + \cos\psi d\theta_2$, $g_\pm^2 \equiv (e_1 \pm e_3)^2 + (e_2 \pm e_4)^2$, and

$$K(z) \equiv \frac{(c + \sinh(2z) - 2z)^{1/3}}{2^{1/3} \sinh z}. \tag{41}$$

Here c is a constant. The asymptotic form ($z \to \infty$) of the metric (40),

$$ds_6^2 \to r_c^2\, e^{2z/3} \left(\frac{dz^2}{2} + e_\psi^2 + \frac{3}{2}\sum_{i=1}^{4} e_i^2\right), \tag{42}$$

is different from the metric of a warped deformed conifold given in[24] by a factor of 1/2 in the element dz^2, but in the present model, this is the metric that explicitly solves the 10d vacuum Einstein equations.

6. Cosmological Solutions and Null Energy Condition

With $\gamma = 2$, the following metric solves all of the 10D Einstein equations

$$ds_{10}^2 = e^{-2\mathcal{K}|z|}\, ds_4^2 + \frac{4L^2}{3}\, e^{-2\mathcal{K}|z|} \left(2\mathcal{K}^2\, dz^2 + ds_{X_5}^2\right), \tag{43}$$

with the scalar factor $a(t)$ satisfying

$$a(t) = \frac{1}{2} \exp\left(\frac{t - t_0}{L}\right) + \frac{kL^2}{2r_0^2} \exp\left(\frac{t_0 - t}{L}\right), \tag{44}$$

where L is an integration constant. From this solution we derive

$$M_{Pl}^2 = \frac{M_{10}^8}{(2\pi)^6} \times \frac{256\sqrt{2}\,\pi^3\, L^6 \mathcal{K}}{729} \int_0^{\pi r_c} dz\, e^{-10\mathcal{K}|z|}. \tag{45}$$

The 4d Planck mass is finite even in the $r_c \to \infty$ limit. Inflationary de Sitter solutions exist with other choices of γ, i.e. $\gamma \neq 2$.[15] Note that the warping becomes stronger away from the brane. In other words $e^{-10\mathcal{K}|z|}$ becomes exponentially small as $|z| \to \infty$. This behavior is similar to that in RS models but opposite to that in the Klebanov-Strassler model.[24]

In the presence of bulk matter fields, we may write an ansatz for the stress-energy tensor of the form

$$T_{AB} = -T_3 P(g_{AB})\, \delta(y) + \mathcal{T}_{AB}, \tag{46}$$

where $P(g_{AB})$ is the pull-back of the spacetime to the world volume of a 3-brane and the last term $\mathcal{T}_{AB}$ represents the contribution of bulk matter fields. In recent papers,[23] it was argued that to get a four-dimensional de Sitter space solution one may have to violate the five- and higher-dimensional null energy conditions or allow a time-dependent Newton's constant or even both. But this is not what we find here. To quantify this one considers

$$R_5{}^5 - R_0{}^0 = \left[3\left(1+\frac{\gamma}{2}\right)\frac{W'^2}{W^2} - \frac{3W''}{W} - e^{2Q}\,\frac{3\ddot{a}}{a}\right]\frac{e^{-2Q}}{W^\gamma}.$$

In order not to violate the 5D null energy condition (NEC) we require $R_5{}^5 - R_0{}^0 \geq 0$. In the simplest case that $\mathcal{T}_{AB} = 0$, we find $R_5{}^5 - R_0{}^0 = 6\mu\, e^{-2Q} > 0$ on the brane and $R_5{}^5 - R_0{}^0 = 0$ in the bulk. Similarly, for the 10D metric solution given above, Eq. (43), we find

$$\tilde{R}_{mn}\tilde{g}^{mn} - R_0{}^0 = \left[15 - \frac{57}{8\mathcal{K}^2}\frac{W'^2}{W^2} - \frac{39}{8\mathcal{K}^2}\frac{W''}{W} - L^2\,\frac{3\ddot{a}}{a}\right]\frac{1}{W(y)^2 L^2}, \tag{47}$$

where $W(z) = e^{-\mathcal{K}|z|}$. Again $\tilde{R}_{mn}\tilde{g}^{mn} - R_0{}^0 = 0$ in the bulk, while $\tilde{R}_{mn}\tilde{g}^{mn} - R_0{}^0 > 0$ on the brane. There are no violations of the null energy condition (NEC), both in four- and ten-dimensional spactimes. This result is conceivable from the viewpoint that the NEC can be violated only by introducing non-standard bulk matter fields (i.e. $\mathcal{T}_0{}^0 < 0$).

7. Dark Matter with Warped Extra Dimensions

I only briefly comment on the prospects of understanding dark matter in models with warped extra dimensions. Except some phenomenologically motivated discussions, especially, in the models of RS1 type, very little is known about dark matter phenomenology in higher dimensional models.

In simplest RS1 type braneworld scenarios, in,[25] Agashe and Servant have argued that the lightest supersymmetric particle in such models could be stable and a well-motivated dark matter candidate. They further argued that solving the problem of baryon number violation in non supersymmetric grand unified theories (GUTs) in warped higher-dimensional spacetime may lead to a stable Kaluza-Klein particle. An exciting aspect of such models is that the entire parameter space could be tested at near future dark matter direct detection experiments.

In a recent work,[26] Pacino et al. have argued that a discrete exchange symmetry may give rise to realistic dark matter candidates in models with warped extra dimensions. In their model, a realistic pattern of electroweak symmetry breaking typically occurs in a region of parameter space in which the Higgs is heavier than the experimental bound and new light quark resonances are predicted. This scenario involves some level of fine-tuning.

In ref.,[27] Mukohyama made a specific suggestion with respect to a possibility of characterizing anti-branes as the standard dark matter particles. He considered the following ten-dimensional metric ansatz

$$ds_{10}^2 = e^{2A(z)} ds_4^2 + e^{-2A(z)} ds_6^2, \tag{48a}$$

$$ds_6^2 \propto \left[\frac{1}{3K^2}(dz^2 + e_\psi^2) + K \cosh^2\left(\frac{z}{2}\right) e_+^2 + K \sinh^2\left(\frac{z}{2}\right) e_-^2 \right], \tag{48b}$$

where

$$K \equiv \frac{(\sinh(2z) - 2z)^{1/3}}{2^{1/3} \sinh z}. \tag{49}$$

The equation of motion for a normalized scalar field Φ (or radial modulus), which is defined via $\Phi \equiv T_3 z$ (with T_3 being the 3-brane tension as introduced above) and the total energy density have following forms

$$\ddot{\Phi} + 3H\dot{\Phi} + \frac{\partial V}{\partial \Phi} = 0, \tag{50a}$$

$$\rho = \frac{1}{2}\dot{\Phi}^2 + V(\Phi) + \rho_6 + \rho_0. \tag{50b}$$

The potential $V(\Phi)$ and density parameters ρ_6 and ρ_0 are defined as

$$V(\Phi) = \frac{1}{2} m^2 \Phi^2 + \frac{J_2^2}{2a^6 \Phi^2}, \quad \rho_6 \propto \frac{J_3^2}{T_3 a^6}, \quad \rho_0 \equiv 2T_3 \left[e^{4A}\right]_{z=0}, \tag{51}$$

where J_2 and J_3 are related to the angular momenta along the S^2 and S^3 parts of the base space $X_5 = T^{1,1}$. Here ρ_3 could behave as dark matter density. This is realized by considering a particular solution $\Phi = \sqrt{|J_2/m|}a^{-3/2}$. With $H^2 \propto a^3$, one gets

$$\rho \simeq \left(1 + \frac{9H^2}{8m^2}\right)\rho_3 + \rho_6 + \rho_0, \tag{52}$$

where $\rho_3 \equiv |mJ_2|/a^3$. The present-day Hubble scale is much smaller than the mass related to the modulus field Φ, i.e. $H^2 \ll m^2$, so $\rho \simeq \rho_0 + \rho_3$, where ρ_0 is associated with a 4d cosmological constant. If the extra dimensions are Ricci flat and/or not warped, which imply that $J_2 = 0 = J_3$, then $\rho \simeq \rho_0$.

8. Conclusion

I conclude the paper with a short summary of the results, with an emphasis on prospects of understanding cosmic acceleration problem (attributed to dark energy of the present universe) in some specific models with warped extra dimensions. One of the novel observations is that de Sitter solutions can arise with an arbitrary curvature of the internal 6d space ($\tilde{R}_{(6)} = 0$, > 0 or < 0), provided one writes the 10d metric in an appropriate form. A feature common to all inflationary de Sitter solutions given above is that upon the dimensional reduction from 10d supergravity one generates in the 4d effective theory a cosmological constant like term. Explicitly,

$$S = \frac{M_{10}^8}{2(2\pi)^6}\int d^{10}x\sqrt{-g_{10}}\,\mathcal{R}_{10} = \frac{M_{Pl}^2}{2}\int d^4x\sqrt{-\hat{g}_4}\left(\hat{R}_4 - 2\Lambda_4\right), \tag{53}$$

where $\Lambda_4 = \frac{6}{\lambda^2 L^2}$ and $\Lambda_4 = \frac{8}{9}r_c^{-2}e^{-kz_0}$, respectively, for the metrics (33) and (38). The relation between four- and ten-dimensional effective Planck masses, $M_{\rm Pl}$ and M_{10} is given by $M_{Pl}^2 \equiv (M_{10}^8 \times V_6^{\rm w})/(2\pi)^6$, where the warped volume $V_6^{\rm w} = \int d^6y\sqrt{\tilde{g}_6}\,e^{(2+3\gamma)A(y)}$. In the particular case $\gamma = 2$ in (33), which is also the choice widely used in the literature, e.g.,[13] we find

$$V_6^{\rm w} \sim \frac{54|k|\pi^3 r_c^6 e^{\mathcal{K}z_0}}{\sqrt{2}}\int_0^{z_1} e^{-2|\mathcal{K}|z}. \tag{54}$$

Λ_4 may be tuned to be the present value of dark energy $\Lambda_0 \sim e^{-\mathcal{K}z_0}/r_c^2 \sim 10^{-120}\,M_{\rm Pl}^2$, which then implies that $M_{\rm Pl} \simeq 3.13\times 10^{29}\,M_{10}^2 \times r_c$. If one does not require the constant Λ_4 to be close to present value of dark energy, $e^{\mathcal{K}z_0}$ or the ratio $e^{\mathcal{K}z_0}/r_c^2$ does not have to be fine-tuned so precisely. The above discussions show that de Sitter space background solutions such as given in (36) or (39a) could serve as models of currently accelerating universe.

Acknowledgments

I would like to thank Kei-ichi Maeda, Juan Maldacena, Shinji Mukhoyama, Nobu Ohta, Gary Shiu, Paul Steinhardt, David Wiltshire for valuable conversations on topics related to the present work. I am also grateful to the organizers of the DARK09 for the opportunity to talk as a key speaker. This research was supported by the New Zealand Foundation for Research, Science and Technology Grant No. E5229 and also by Elizabeth EE Dalton Grant No. 5393.

References

1. A. G. Riess *et al.* Astron. J. **116**, 1009 (1998); S. Perlmutter *et al.* Astrophys. J. **517**, 565 (1999).
2. D. N. Spergel et al. [WMAP Collaboration], Astrophysical Journal Suppl. 148 (2003) 175; *ibid*, **170**, 377 (2007).
3. S. Kachru, R. Kallosh, A. Linde and S. P. Trivedi, Phys. Rev. D **68**, 046005 (2003); C. P. Burgess, J. M. Cline, H. Stoica and F. Quevedo, JHEP **0409**, 033 (2004).
4. I. P. Neupane, Class. Quant. Grav. **21**, 4383 (2004).
5. M. Becker, L. Leblond and S. E. Shandera, Phys. Rev. D **76**, 123516 (2007).
6. P. K. Townsend and M. N. R. Wohlfarth, Phys. Rev. Lett. **91**, 061302 (2003); N. Ohta, Phys. Rev. Lett. **91**, 061303 (2003); Prog. Theor. Phys. **110**, 269 (2003).
7. C. M. Chen, P. M. Ho, I. P. Neupane and J. E. Wang, JHEP **0307**, 017 (2003); JHEP **0310**, 058 (2003); I. P. Neupane and D. L. Wiltshire, Phys. Lett. B **619**, 201 (2005); Phys. Rev. D **72**, 083509 (2005).
8. I. P. Neupane, Phys. Rev. Lett. **98**, 061301 (2007).
9. L. Randall and R. Sundrum, Phys. Rev. Lett. **83**, 3370 (1999).
10. L. Randall and R. Sundrum, Phys. Rev. Lett. **83**, 4690 (1999).
11. J. Polchinski, Phys. Rev. Lett. **75**, 4724 (1995).
12. G. W. Gibbons, in *Supersymmetry, Supergravity and Related Tolics*, edited by F. del Aguila, J. A. de Azcarraga, and L. E. Ibanz (World Scientific, 1985), pp. 123-146.
13. J. M. Maldacena and C. Nunez, Int. J. Mod. Phys. A **16**, 822 (2001).
14. S. B. Giddings, S. Kachru and J. Polchinski, Phys. Rev. D **66**, 106006 (2002).
15. I. P. Neupane, arXiv:0901.2568 [hep-th].
16. N. Kaloper, Phys. Rev. D **60**, 123506 (1999); S. Forste, Z. Lalak, S. Lavignac and H. P. Nilles, JHEP **0009**, 034 (2000).
17. J. Garriga and M. Sasaki, Phys. Rev. D **62**, 043523 (2000).
18. G. R. Dvali, G. Gabadadze and M. Porrati, Phys. Lett. B **485**, 208 (2000); C. Deffayet, Phys. Lett. B **502**, 199 (2001).
19. Y. M. Cho and I. P. Neupane, Int. J. Mod. Phys. A **18**, 2703 (2003) [arXiv:hep-th/0112227].
20. I. P. Neupane, arXiv:0903.4190 [hep-th].
21. G. W. Gibbons, R. Kallosh and A. D. Linde, JHEP **0101**, 022 (2001).

22. F. Leblond, R. C. Myers and D. J. Winters, JHEP **0107**, 031 (2001).
23. D. H. Wesley, JCAP **0901**, 041 (2009); P. J. Steinhardt and D. H. Wesley, arXiv:0811.1614 [hep-th].
24. I. R. Klebanov and M. J. Strassler, JHEP **0008**, 052 (2000).
25. K. Agashe and G. Servant, Phys. Rev. Lett. **93**, 231805 (2004) [arXiv:hep-ph/0403143].
26. G. Panico, E. Ponton, J. Santiago and M. Serone, Phys. Rev. D **77**, 115012 (2008) [arXiv:0801.1645 [hep-ph]].
27. S. Mukohyama, Phys. Rev. D **72**, 061901 (2005) [arXiv:hep-th/0505042].

PART III

Hot Dark Matter, Neutrino Mass and Dark Energy

NUCLEAR DOUBLE BETA DECAY, FUNDAMENTAL PARTICLE PHYSICS, HOT DARK MATTER, AND DARK ENERGY

HANS V. KLAPDOR-KLEINGROTHAUS

University of Heidelberg, Germany
Postal address: Stahlbergweg 12, 74931 Lobbach, Germany
prof.klapdor-kleingrothaus@hotmail.de, http : //www.klapdor − k.de

IRINA V. KRIVOSHEINA

Radiophysical Research Institute (NIRFI), Nishnij-Novgorod, Russia,
irinakv57@mail.ru, http : //www.klapdor − k.de

Nuclear double beta decay, an extremely rare radioactive decay process, is - in one of its variants - one of the most exciting means of research into particle physics beyond the standard model. The large progress in sensitivity of experiments searching for neutrinoless double beta decay in the last two decades - based largely on the use of large amounts of enriched source material in "active source experiments" - has lead to the observation of the occurrence of this process in nature (on a 6.4 sigma level), with the largest half-life ever observed for a nuclear decay process (2.2×10^{25} y). This has fundamental consequences for particle physics - violation of lepton number, Majorana nature of the neutrino. These results *are independent* of any information on nuclear matrix elements (NME) *. It further leads to sharp restrictions for SUSY theories, sneutrino mass, right-handed W-boson mass, superheavy neutrino masses, compositeness, leptoquarks, violation of Lorentz invariance and equivalence principle in the neutrino sector.
The masses of light-neutrinos are found to be degenerate, and to be at least 0.22 ± 0.02 eV. This fixes the contribution of neutrinos as hot dark matter to $\geq$4.7% of the total observed dark matter. The neutrino mass determined might solve also the dark energy puzzle.

Keywords: Heidelberg-Moscow Experiment, lepton number violation, Majorana neutrino, double beta decay, neutrino mass, R-parity violating SUSY, sneutrino mass, superheavy neutrinos, composite neutrinos, right-handed W boson mass, hot dark matter, dark energy.

---

*It is briefly discussed *how* important NME for $0\nu\beta\beta$ decay *really* are.

1. Double Beta Decay

In the Nuclidic Chart a special and intriguing role for the research into fundamental particle physics, astrophysics and cosmology is offered by neutron-rich nuclei more or less far from the stability line. In the 5. Edition in 1981 the editors of the Karlsruhe Nuclidic Chart were pioneering in including the first microscopic calculations of beta decay half lives of nuclei far from stability which later were published in Nuclear Data Tables,[1] and which lead to new insights into element synthesis by the astrophysical r-process, the age of the universe and cosmology (the cosmological constant).[2,3]

There are about 35 neutron-rich nuclei, which can undergo the socalled nuclear double beta decay - a second-order weak decay mode.[3,4]

The weak interaction is the most universal interaction after gravitation and operates on at least all fermions. It is the *only* interaction, which can alter the charge of the fermions (the most famous example is beta decay) and their flavours.

From a perturbation theory point of view nuclear beta decay is understood as first order effect of the classical theory. In the Glashow-Weinberg-Salam (GWS) theory, in which the point-like current-current interaction is replaced by a boson exchange interaction nuclear beta decay is a second order effect. The very heavy bosons W^{+-} and Z are responsible for the extremely short range of the weak interaction. Double beta decay is a second order effect of the classical theory (and fourth order in the GWS theory). That into the expression for the $\beta\beta$ decay rate the weak coupling constant (Fermi constant) which is of the order $G_\beta = 1.008.10^{-5} \times m_p^{-2}$ enters in the fourth power, leads to extremely small half lives. Double beta decay can become observable for nuclei for which no other decay process (in particular β decay) is possible. This is the case for several even-even nuclei (even number of protons and of neutrons) which because of the pairing energy have lower energy ground states than their odd-odd neighbours (odd number of protons and of neutrons). They may be converted into a more stable isotope then only under double beta decay (see Fig. 1). This process may be understood as simultaneous β decay of two neutrons (for $\beta^-\beta^-$ decay) or of two protons (for $\beta^+\beta^+$ decay).

There are essentially two modes of double beta decay, the neutrino-accompanied ($2\nu\beta\beta$) mode, which is allowed in the Standard Model of Particle Physics and has been observed by geochemical experiments already since 1950, and by direct detection first in 1987,[21] and the neutrinoless mode ($0\nu\beta\beta$), which is *not* allowed in the Standard Model. In $2\nu\beta^-\beta^-$ decay two electrons are emitted together with two electron-antineutrinos

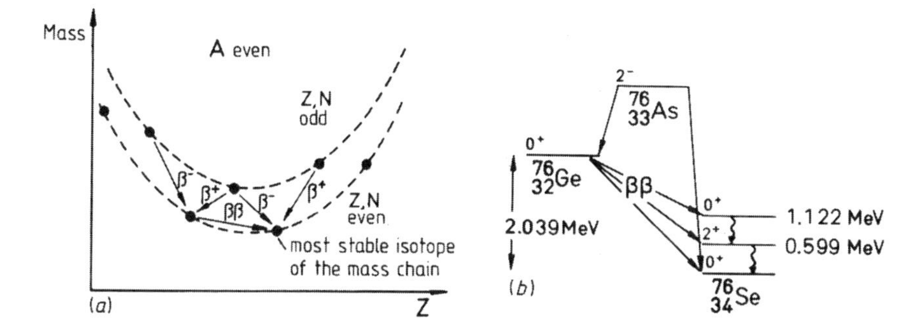

Fig. 1. (a) Energetic situation of potential double beta emitters. Because of the pairing energy, nuclei with an even number of protons and an even number of neutrons are energetically depressed in comparison with neighbouring nuclei. Thus many nuclei are stable against single β decay, but may be converted into a more stable isotope under double beta decay. (b) Schematic diagram of double beta decay of ^{76}Ge.

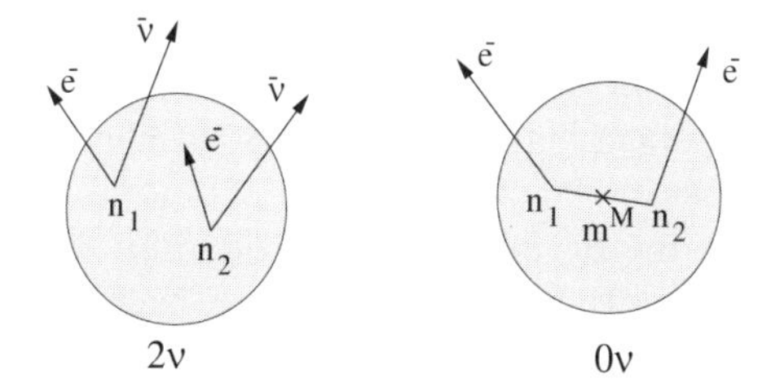

Fig. 2. 0ν and 2ν double β decay pioneered by E. Majorana and M. Goeppert-Mayer.

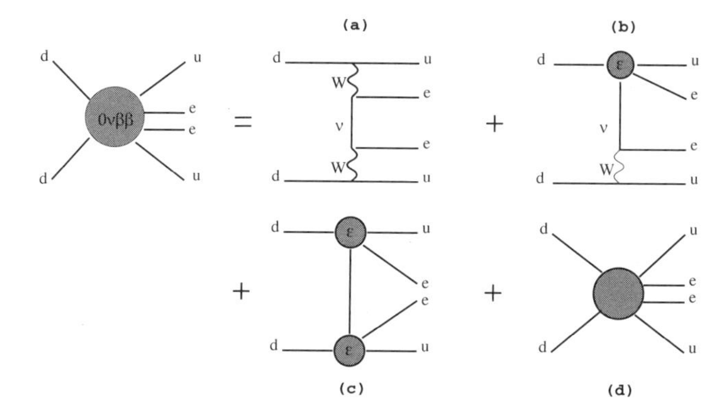

Fig. 3. Feynman graphs of the general double beta decay rate, with long range (a-c) and short range interactions (d) (from[7]).

$\bar{\nu}_e$. (see Fig.2), so that lepton number L is conserved:

$$
\begin{array}{lcccccc}
{}_A^Z X & \longrightarrow & {}_A^{Z+2}X & + 2e^- & + 2\bar{\nu}_e & & \\
L:0 & & 0 & +2 & -2 & \Longrightarrow & \Delta L = 0.
\end{array} \quad (1)
$$

Much more interesting than $2\nu\beta\beta$ decay, which has been observed meanwhile for about ten nuclides, with half lives of the order of 10^{19}-10^{24} years, is the socalled neutrinoless double beta decay ($0\nu\beta\beta$), which may be viewed as an exchange of a neutrino between the two decaying nucleons (see Figs. 2, 3):

$$
\begin{array}{lcccc}
{}_A^Z X & \longrightarrow & {}_A^{Z+2}X & + 2e^- & \\
L:0 & & 0 & +2 & \Longrightarrow \quad \Delta L = 2
\end{array} \quad (2)
$$

In this case the (total) lepton number L [a)] is *not* conserved. Such a process is only possible if neutrino and antineutrino are identical (i.e. the neutrino is a Majorana particle) [b], *and* if either the neutrino has a non-vanishing mass, or there exists a right-handed weak interaction. In Grand Unified Theories (GUT's) the latter two conditions are not independent.[32] A right-handed component here is only effective in simultaneous association with a Majorana mass.

If we go beyond the Standard Model, there are further mechanisms for $0\nu\beta\beta$ decay, such as Higgs boson exchange, exchange of a SUSY particle (gluino, photino, ..), of leptoquarks, composite neutrinos, etc. (see, e.g.,[6,8]). The process (2) therefore yields broad access to many topics of particle physics beyond the standard model at the TeV range, on which new physics is expected to manifest itself. It can provide an absolute scale of the neutrino mass, and yields sharp restrictions for SUSY models, leptoquarks, compositeness, left-right symmetric models, test of special relativity and equivalence principle in the neutrino sector and others (see section 4). For details, we refer to.[6,8,11,12]

The history of $\beta\beta$ decay using the nucleus as a complicated laboratory for a wide range of particle physics started about 70 years ago. This history is connected with fundamental discoveries of particle physics, such as parity non-conservation and of gauge theories, and double beta research has

---

[a]Total lepton number (L) is defined as the sum of the family lepton numbers L_e,L_μ, L_τ ($L = L_e + L_\mu + L_\tau$). Non-conservation of family lepton number (but *not* of L) has been observed by socalled neutrino oscillations. L was found in all experiments up to now, as also the baryon number, to be a conserved quantity in particle physics.

[b]All other fermions known today are Dirac particles, where each particle has its defined antiparticle.

become one of the most important fields of non-accelerator particle physics. Concerning neutrino physics, without $0\nu\beta\beta$ decay there is no way to decide the nature of the neutrino (Dirac or Majorana particle), and of the structure of the ν mass matrix, since neutrino oscillation experiments measure only differences of neutrino mass eigenstates. Only investigation of neutrino oscillations *and* double beta decay together can lead to an absolute mass scale.

2. History of Experimental Development of Double Beta Decay

The long and close association between the phenomenon of nuclear double beta decay, the violation of lepton number conservation and the nature and mass of the neutrino began shortly after the "discovery" of the neutrino by W. Pauli in 1930. The motivation of M. Goeppert-Mayer, however, when performing the first calculations of the half life of $\beta\beta$ decay (in 1935) was not the nature of the neutrino, not conservation of leptons, but the stability of even-even nuclei over geological times.

In 1939 W.H. Furry showed that the "symmetrical" theory of neutrino and antineutrino by E. Majorana (1937) could give rise to the process of neutrinoless double beta decay.

The first experiments on double beta decay were undertaken, before the existence of neutrinos was proved *directly* by Cowan and Reines (in 1955). While most of the very first experiments in the period 1948–1952 were looking for the decay electrons, a remarkable exception was the experiment performed by M.G. Inghram and J.H. Reynolds (1949, 1950). They looked for the daughter nucleus and exploited the fact, that measurable amounts of the daughter might accumulate over geological times in ores, which are rich in the corresponding parent nucleus. They analyzed a tellurium ore from Boliden, Sweden, which was about 1.5 billion years old, and reported evidence for the transition $^{130}Te \longrightarrow ^{130} Xe$ with a half-life of 1.4×10^{21} years, which they attributed to $2\nu\beta\beta$ decay of ^{130}Te.[38] Another early approach was to look for radioactive daughter nuclei which in principle are detectable in much smaller quantities than stable rare gases. The experiment of Inghram and Reynolds was the forerunner of a series of geochemical experiments which definitely proved the occurrence of $2\nu\beta\beta$ decay, and confirmed their value within a factor of about 2.[39,40] The first observation in "direct" experiments (not geochemical and radiochemical) was claimed in 1987, for ^{82}Se.[21] The first "active source experiment" (in which the detector material is at the same time the $\beta\beta$ emitter) was the one by E. der Mateosian and M. Goldhaber using $CaFe_2$, in 1966.[41]

A particularly favorable case is presented by the $\beta\beta$ candidate ^{76}Ge. This germanium isotope occurs with an abundance of 7.8% in natural germanium, from which large high resolution detectors can be manufactured. Thus Ge can be used simultaneously as source and detector allowing for large source strength without spoiling the high energy resolution of such detectors. The most sensitive experiment using detectors from natural Ge was over many years the one by D. Caldwell in California,[34–36] until in the early nineties the first Ge experiments using Ge *enriched* in the isotope ^{76}Ge (to 86%) were started[5,37] . The use of enriched Ge drastically increased the sensitivity, and started a new era of $\beta\beta$ experiments. The largest experiment of this type (with 11 kg of enriched detectors) and the first one using high-purity Ge detectors, and the most sensitive experiment since 1993 until now is the Heidelberg–Moscow experiment,[5] which was operated in the Gran Sasso underground laboratory from 1990 to 2003.

3. Nuclear Matrix Elements - Some Necessary Comments

The half-life for neutrinoless double beta decay is connected with particle physics parameters. If we consider only two mechanisms for triggering the decay, exchange of a massive Majorana neutrino and right-handed weak currents, we have[79–81]

$$[T^{0\nu}_{1/2}(0^+_i \to 0^+_f)]^{-1}$$

$$= C_{mm}\frac{\langle m\rangle^2}{m_e^2} + C_{\eta\eta}\langle\eta\rangle^2 + C_{\lambda\lambda}\langle\lambda\rangle^2 + C_{m\eta}\langle\eta\rangle\frac{\langle m\rangle}{m_e} + C_{m\lambda}\langle\lambda\rangle\frac{\langle m_\nu\rangle}{m_e} + C_{\eta\lambda}\langle\eta\rangle\langle\lambda\rangle,$$

$$\langle m\rangle = |m^{(1)}_{ee}| + e^{i\phi_2}|m^{(2)}_{ee}| + e^{i\phi_3}|m^{(3)}_{ee}| \;, \tag{3}$$

where $m^{(i)}_{ee} \equiv |m^{(i)}_{ee}| \exp{(i\phi_i)}$ $(i = 1, 2, 3)$ are the contributions to the effective mass $\langle m\rangle$ from individual mass eigenstates, with ϕ_i denoting relative Majorana phases connected with CP violation, and $C_{mm}, C_{\eta\eta}, ...$ denote nuclear matrix elements squared, multiplied by a phase factor. Ignoring contributions from right-handed weak currents on the right-hand side of the above equation, only the first term remains. $\langle m_\nu\rangle$ is the effective neutrino mass, η and λ are right-handed weak current parameters.

The mere occurrence of the process eq.(2) proves violation of total lepton number, and proves according to the fundamental paper by Schechter and Valle[32] that the neutrino is a Majorana particle. For these most fundamental conclusions from an observed $0\nu\beta\beta$ signal obviously *no* knowledge of nuclear matrix elements is needed at all.

We *have* to calculate the nuclear matrix elements (NME), when we want to determine the contributions of neutrino mass, right-handed weak currents and others, such as contributions by SUSY, leptoquarks, etc. (see [6]) to the process, or if we want to deduce information on the neutrino mass etc. from an experiment.

Such calculations have been made since more than three decades now, and large progress has been made in improving the precision of these calculations - with fundamental steps of progress in the understanding of the field by the inclusion of the effects of spin-isospin and quadrupole-quadrupole ground state correlations in the wave functions, by Grotz and Klapdor around 1984, [3,31] which have triggered the in some sense partly equivalent inclusion of the pp force in QRPA (quasiparticle random phase) calculations, 1986 - 1989, 2000, 2001[79–81,103,104,106,107] . In the last 15 years various kinds of refinements have been tried, such as 'second QRPA', 'RQRPA', 'full-RQRPA' etc. (see, e.g.[6]). They have led, however, except perhaps for the first of them to rather limited progress and to more or less oscillating (in time and in sign) corrections.

The reason for the latter is simply that our knowledge of nuclear theory just does not allow to pin down the matrix elements to something better than the order of a factor of two or so in good cases (see, e.g.[108]). So, the results by e.g.[102] agree (as the results of these authors did earlier already for short time in 2004 Ref.[109]), after many "back and forth" finally with the results obtained twenty years ago by,[79–81] which were used in the analysis given in sections 4, 5.

In spite of this it seems that sometimes the importance of and the potential for calculations of NME is systematically exaggerated - from reasons, which may not always be justified by its real scientific importance and by the real potential of improvement. NME calculations seem to have somehow become an end in itself - or a kind of 'perpetuum mobile'. Let us illustrate the kind of sometimes unproper argumentation in NME work by some examples from recent papers.

1. How Many $0\nu\beta\beta$ Signals Needed ? In an otherwise very useful paper on NME by[102] it is stated in the Introduction that 'it is widely recognized that a convincing case for $0\nu\beta\beta$ decay must involve independent signals in three or more nuclei'. This is simply wrong. As stated above, such a discovery and the mere occurrence of this process is enough to conclude the fact of violation of lepton number, and the fact that the neutrino is a Majorana particle - with all its consequences, for SUSY and the structure of space-time - independent of any knowledge about NME for $0\nu\beta\beta$ decay.

'Widely recognized' according to their list of references, is the opinion of a representative of *one* earlier not successful (see Ref.[111]) experiment (IGEX) which had to be given up in 1999 already (their[102] reference 1) (for more details see[6]). Second the opinion of a representative (their reference 5) of an experiment in preparation, which still is ten orders of magnitude away from the sensitivity required to become useful for double beta decay research (for the latter see[14]).

2. Could several $0\nu\beta^-\beta^-$ signals allow to determine different contributions to the $0\nu\beta\beta$ amplitude ? In the same paper[102] the authors 'promised' in the Introduction, that measurement of several $0\nu\beta^-\beta^-$ signals would allow, with sufficient improvement of the calculation of NME, to identify other $0\nu\beta\beta$ mechanisms in addition to the mass mechanism. These, and other authors, however, are running into the same deadlock. It is certainly true - and trivial - that $0\nu\beta\beta$ signals from several nuclei could help to teach us about better calculations of nuclear matrix elements. However, there is, almost *no hope* (in contrast to what is said in the Introduction of [102]), that we could learn from several $0\nu\beta\beta$ experiments about the various contributions to the $0\nu\beta\beta$ amplitude - *as long as* we restrict ourselves to $\beta^-\beta^-$ decay. This has been pointed out already in[110] (see also[13,14,123]), where it has been shown, that only $\beta^-\beta^-$ experiments *alone cannot* give a solution to the different contributions to the decay amplitude such as mass mechanism, right-handed weak currents and others. (This can by the way also somehow be concluded from section IV,C of,[102] - in contradiction to the hopes raised in the Introduction of their paper.) Some details to this principle limitation of $\beta^-\beta^-$ experiments are given in section 6 of this paper, where also a way out of this problem is outlined.

3. Uncertainties of NME - "Theoretical errors": It has been tried in,[102] and earlier in,[112] and other papers to determine uncertainties of theoretical calculations of NME (socalled theoretical errors). This in principle very useful effort has been done perhaps in some biassed way. *First* the authors *limit* themselves to one method, QRPA. *Second*, they perform calculations with different entrance nuclear parameters sets - which are limited by a priori choice - and then average over the results, and determine some kind of mean error - which naturally depends on the chosen range of entrance parameters. Since they stick to *one* theoretical approach (of many possible), and within this one approach only to *their own* calculations, their error can hardly be generally representative and probably will tend to be too small. In a finally unpublished preprint to the paper[112] even *very unrealistic* small errors were presented. The fig. 2 in[102] compares their results with QRPA calculations of other authors and with shell model calculations. The figure demonstrates that what they call conservative estimate of error originating from known parameters in QRPA (where they *ignore* the known strong effects of deformation) may be indeed too low. They state that the independent calculations of other authors fall within their own 3 sigma error ranges, although from their figure 2 obviously this is not true for 4 of the shown 7 cases (nuclei).

4. Nuclear deformation: It has been stated by the authors[102] themselves (on their page 4), that further parametric uncertainties exist, among them those related to nuclear deformation, and to low-lying β^+ strength (which are *ignored* in their work). They ignore also that serious efforts to include deformation in NME calculations *exist for long time* (e.g.[113]), and have shown that the effects can be remarkably large. A good example for the latter is ^{150}Nd (see Ref.[113]).

5. 'New' old developments: In the paper [102] it has been noted that 'the uncertainties can be largely kept under control by systematically fixing, in each nucleus, the strength parameter g_{pp} via two neutrino $\beta\beta$ decay rates'.

It has probably been *'forgotten'*, however, to mention that this method has been *used* already in 2000 and 2001 by Stoica and Klapdor [106,107] and that much earlier already Muto and Klapdor [79] fixed g_{pp} by experimental β^+ decay data. This is another example for how an impression of *new developments* can be given by just *repeating* long known methods.

6. Several $0\nu\beta\beta$ signals improve precision of neutrino mass? In Ref. [102] (page 9) it is shown that regardless of how many accurate experimental signals of $0\nu\beta\beta$ decay are combined, the *final accuracy* of the deduced Majorana neutrino mass *will not be improved* relevantly compared to a single experiment, as result of correlations of theoretical errors. This also *contradicts* to the request made in point 1, that many experimental $0\nu\beta\beta$ signals are required.

7. The limits of $0\nu\beta^-\beta^-$ experiments: It is visible from section IV C of [102], but not expressively stated there, that as mentioned in point 2 above, from several positive $0\nu\beta^-\beta^-$ experiments there is almost *no chance* to get information on other $\beta\beta$ mechanisms beyond the mass mechanism. Again the authors, have *'forgotten'* that such conclusion in a *much more* direct and transparent way has been drawn already in[110] and discussed in Refs. e.g. [13,14,123]. It is also not stated that there are ways to overcome this problem and that they have been published already in the above papers (see also section 6 of this paper).

8. Minor effects of Fermi $0\nu\beta\beta$ NME: Another example for some overweighting of the importance of NME calculations is the paper by[114] . In very short: After the hope-raising title (Can one measure nuclear matrix elements of neutrinoless double beta decay ?) it is discussed how one might measure the Fermi $0\nu\beta\beta$ NME – which is known to give only a minor contribution to the total $0\nu\beta\beta$ NME. The difficulties and limitations of measuring this cross section by the suggested (p,n) experiments (see, e.g. Ref. [25]) are overlooked. Anyway, improvement in the errors of $M_F{}^{0\nu}$ if really obtainable, would have only a minor effect on the total uncertainties of the $0\nu\beta\beta$ NME, and thus will hardly be of major practical importance.

9. $0\nu\beta\beta$ and CP phase: In [102] the authors conclude from their NME calculations, and analysis, preference of the CP-conserving case with $e^{i\phi}$ =+1 over the other, $e^{i\phi}$ =-1, with ϕ denoting the Majorana phase (see eq. 3). Again the authors *'forget' to mention* that already in 2002 [18] it has been pointed out, that as consequence of the HEIDELBERG-MOSCOW experimental result together with the result of WMAP, the neutrino mass eigenvalues have the same CP parity (see also[19] , and [15]) - as required the theoretical work of Ref.[115] , who showed that assuming the quark and lepton mixing to be identical at or near the GUT scale, the large solar and athmospheric neutrino mixing angles can be understood purely as result of renormalization group evolution, *if* neutrino masses are quasi-degenerate (with same CP parity). The common neutrino masses according to this work of[115] then must be larger than 0.1 eV.

10. $0\nu\beta\beta$ and WMAP rules out old NME: It has been shown also in[15,18] that the results of the HEIDELBERG-MOSCOW experiment together with the results by WMAP *rule out* completely a 20 years old old-fashioned nuclear matrix element of neutrinoless double beta decay by Vogel et al. (still basing on *unrealistic* nuclear forces), used in an analysis of WMAP (see Ref.[116]).

11. Many experimental $0\nu\beta\beta$ signals for better NME?: In some papers Ref.[105] it is stated in some way, which seems rather far from any practicability that observation of $0\nu\beta\beta$ decay of three (or more) nuclei would allow to differentiate between good and less good models for NME calculations - although, according to the authors, would *not necessarily* lead to the correct effective neutrino masses. Not regarding the limited usefulness mentioned even by the authors, such a procedure in view of the *extreme* difficulty performing several *such* experiments corresponds to turning the problem upside down.

Concluding this section: *No major progress in calculations of NME has been made in the last decade or so, at least in the various QRPA approaches.* Most 'refined' present calculations give results agreeing with what has been obtained twenty years ago. The reason is that a serious improvement in the accuracy of calculated NME (beyond a factor of two or so) simply cannot be achieved as result of our limited knowledge of nuclear forces. The reached *accuracy is on the other hand sufficient* to deduce important particle physics information from measured $0\nu\beta\beta$ decay - in the cases, where NME are needed. The from time to time stated urgent importance of improving NME calculations for *planning* of $0\nu\beta\beta$ experiments (e.g.,[117,118] and others) is far from feasibility - is it sometimes done on *another* purpose ? The serious experimentalist makes the experiment sensitive enough to be independent of the unavoidable uncertainties in the calculations.

The somewhat artificial way, in which sometimes further (practically impossible) improvement of NME calculations is demanded as indispensable pre-requisite for any further progress in double beta research is scientifically not justified.

For more details about nuclear matrix elements for double beta decay we refer to[6] .

4. The Result of the HEIDELBERG-MOSCOW Experiment - and the Neutrino Mass

The result from the HEIDELBERG-MOSCOW experiment (see[13,17]) is shown in Figs. 4, 5. The background around $Q_{\beta\beta}$ is (with pulse shape analysis) around 5×10^{-3} counts/kg y keV, i.e. close to the level which had been planned in the GENIUS project.[22] The signal at $Q_{\beta\beta}$, where a $0\nu\beta\beta$ signal should occur, has a confidence level of 6.4σ (7.05 ± 1.11 events).

This is the *first* and up to now *only* indication for the occurrence of this process.

The intensity of the observed signal[13] corresponds to a half-life for $0\nu\beta\beta$ decay of $T_{1/2}^{0\nu\beta\beta} = (2.23^{+0.44}_{-0.31}\times10^{25})$ y. From the half-life one can derive information on the effective neutrino mass $\langle m\rangle$ and the right-handed weak current parameters $\langle\eta\rangle$, $\langle\lambda\rangle$. Under *the assumption* that only *one* of the three terms (the effective mass term and the two right-handed weak current terms) contributes to the decay process, and ignoring potential other processes connected with SUSY theories, leptoquarks, compositeness, etc. (see[6]), we find:

$$\langle m\rangle = (0.32^{+0.03}_{-0.03})\,\text{eV, or}$$
$$\langle\eta\rangle = (3.05^{+0.26}_{-0.25}) \times 10^{-9}\text{, or}$$
$$\langle\lambda\rangle = (6.92^{+0.58}_{-0.56})\times10^{-7}.$$

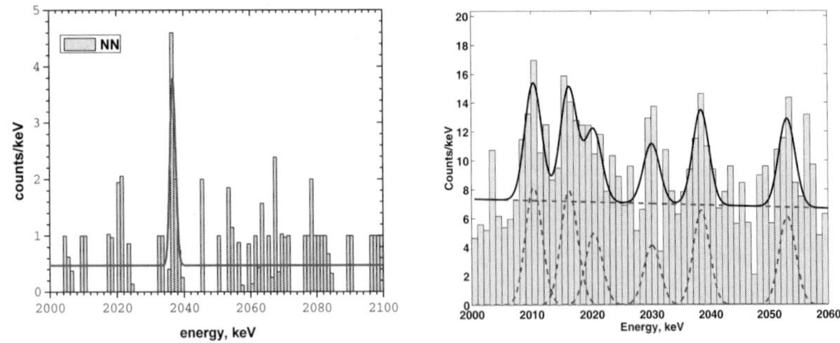

Fig. 4. The pulse shape selected spectrum (basing on the analysis of the measured *time structure* of the individual pulses[13] in the range 2000-2100 taken with detectors 2,3,4,5 (51.39 kg y) (left) and the corresponding full spectrum of all five detectors in the range 2000-2060 keV (56.66 kg y) (right), in the period 1995-2003 (see[13]). The $Q_{\beta\beta}$ value of $\beta\beta$ decay of ^{76}Ge is known to be 2039.006 ± 0.050 keV. Fig 4 (left) is a zoom of Fig. 5 (upper part).

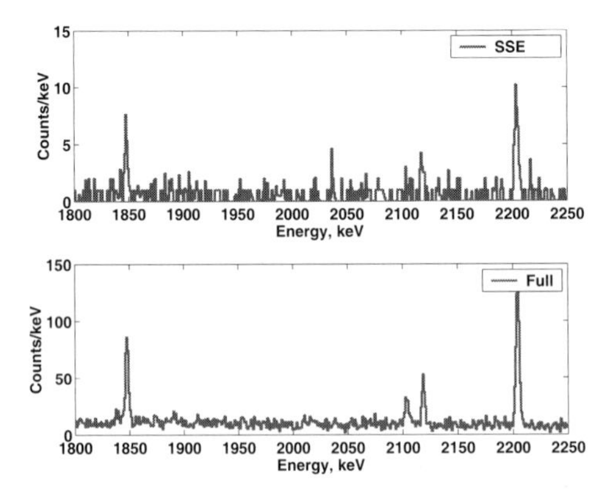

Fig. 5. Top: The pulse-shape selected spectrum of single site events measured with detectors 2,3,4,5 from 1995-2003. Below: The full spectrum measured with detectors 2,3,4,5 from 1995-2003 (from[13]).

These are the upper limits for these quantities. When 'calibrating' the corresponding matrix element from the *measured* rate of $2\nu\beta\beta$ decay of ^{76}Ge (see[13]) the effective neutrino mass becomes lower, down to (0.22 ± 0.02) eV.

The effective electron neutrino mass is $\langle m_\nu \rangle = \left|\sum (U_{ei}^*)^2 m_i\right|$, where $U_{ei}^* = \langle \nu_e | \nu_i \rangle = \langle \nu_i | \nu_e \rangle^*$, and $|\nu_e\rangle = \sum_i U_{ei} |\nu_i\rangle$. From neutrino oscillation experiments [c] one can determine the mixing parameters and thus the effec-

[c]From observation of solar, atmospheric and accelerator neutrinos, neutrino oscillations have been observed, i.e. transitions from one neutrino flavour into another, e.g. $\nu_e \longrightarrow \nu_\mu$. These experiments allow to determine the difference of the ν mass eigenstates (which are

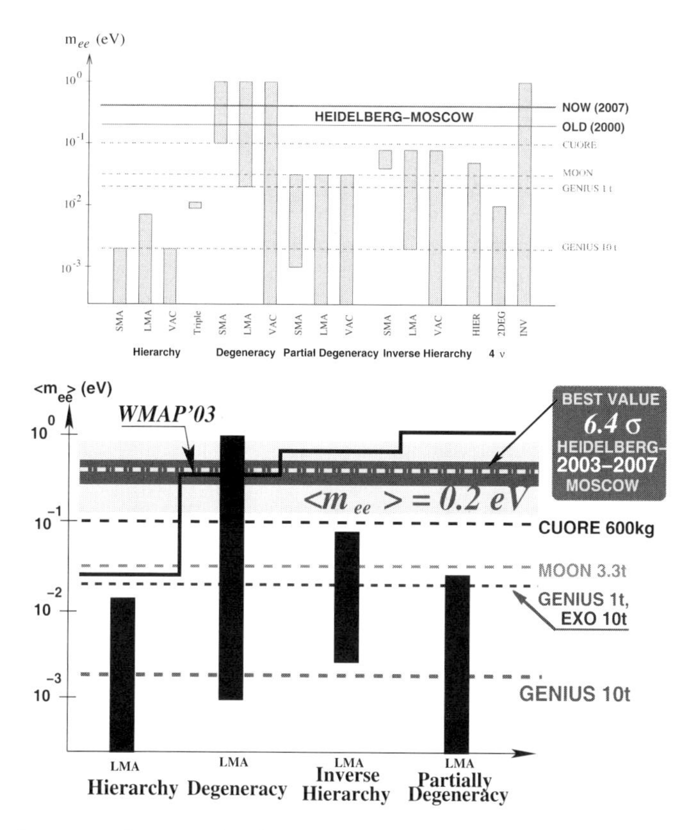

Fig. 6. Summary of **expected values** for the *effective neutrino mass* $\langle m_{ee} \rangle$ in different neutrino mass schemes and the result of the HEIDELBERG-MOSCOW experiment. The bars denote allowed ranges of $\langle m \rangle$ in different neutrino mass scenarios, still allowed by neutrino oscillation experiments (see[13,20]). All models except the degenerate one and the sterile neutrino scenario with inverted hierarchy are excluded by the $0\nu\beta\beta$ decay result. Also shown is (lower part) the exclusion line from WMAP, plotted for $\sum m_\nu < 1.0\, eV$, (which is according to[26], too strict). WMAP does not rule out *any* of the neutrino mass schemes. Further shown are for history the expected sensitivities expected earlier for the planned double beta experiments CUORE, MOON, EXO and the 1 ton and 10 ton project of GENIUS[6,22].

tive mass $\langle \mathrm{m} \rangle$ to be expected in $0\nu\beta\beta$ experiments *for different ν mass scenarios* (different ν mass models). Fig. 6 shows these expectations together with the value of $\langle \mathrm{m} \rangle$ determined from the HEIDELBERG-MOSCOW experiment (under the assumption of dominating mass mechanism). It can

found to be of the order of 0.008 and 0.05 eV),[29,30] and their flavour composition. The *absolute* neutrino masses *cannot* be determined by neutrino oscillation experiments.

be seen that in a scenario of three neutrino flavours only the solution of *degenerate* (i.e. essentially identical, except for the small differences determined by neutrino oscillation experiments, see footnote [c]) masses remains. If allowing for sterile neutrinos (4ν scenarios), also the sterile neutrino scenario with inverted hierarchy remains (see also[122]). All other mass models (hierarchical, inverse hierarchy, partially degenerate) are excluded by the HEIDELBERG-MOSCOW $\beta\beta$ experiment. The common mass eigenvalue corresponding to this effective mass can be determined with recent mixing angles from solar neutrinos to be 0.22-0.48 eV [6,33] .

The analysis of the cosmological experiments SDSS and WMAP together[26] yields an upper limit for the sum of the neutrino masses of $\sum m_\nu$ <1.7 eV (95% c.l.). This would correspond to $\leq$ 12% of hot dark matter of the total dark matter observed in the universe. The above double beta result yields a *lower* limit of 4.7.%. The SDSS result means that the individual neutrino mass should be smaller than $\sim$ 0.6 eV, which is consistent with the above value from double beta decay [d]. The Fig. 6 shows as example the *limits* set by the cosmic microwave experiment WMAP (assuming $\sum$ m$_i$ = 1.0 eV). The Planck mission - the new generation cosmic microwave background experiment just started, is expected to reach a sensitivity down to $\sum m_\nu = 0.2\,eV$.[28] This means it will be able to test the HEIDELBERG-MOSCOW - result of $\sum m_\nu \geq 0.6\,eV$ (derived under the assumption of a dominating ν mass mechanism) earlier than any other $\beta\beta$ experiment (see below, section 6). PLANCK thus can decide whether the double beta decay process is triggered by the neutrino mass mechanism or another one.

5. The Results for Other Beyond Standard Model Physics

Assuming *other* mechanisms to dominate the $0\nu\beta\beta$ decay amplitude, which have been studied extensively in our group, and other groups, in recent years, the result allows to set stringent limits on parameters of SUSY models, leptoquarks, compositeness, masses of heavy neutrinos, the right-handed W boson and possible violation of Lorentz invariance and equivalence principle in the neutrino sector and others. Figs. 7,8,9 and 10, show as examples some of the relevant graphs which can in principle contribute to the $0\nu\beta\beta$ amplitude and from which bounds on the corresponding pa-

---

[d]Other evaluations of WMAP [27] yield systematically lower values, down to $\sum$ m$_i$ < 0.6 eV (also still consistent with the $0\nu\beta\beta$ result), but according to the convincing argumentation of[26] as result of less generally justified priors used by Ref.[27]

rameters can be deduced assuming conservatively the measured half-life as upper limit for the individual processes. Figs.11,13 and 12 show some results deduced from the experimental $0\nu\beta\beta$ result for the underlying theories.

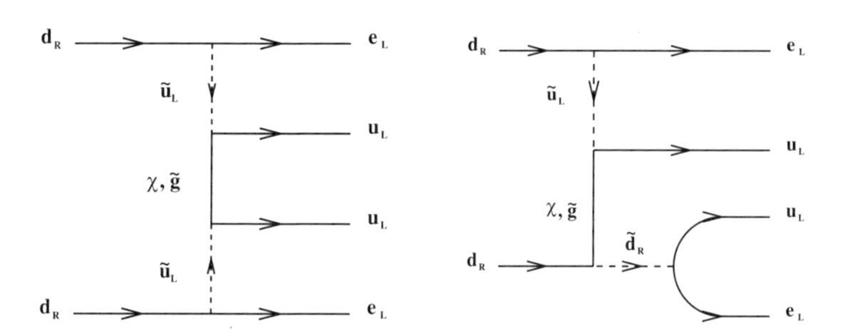

Fig. 7. Examples of Feynman graphs for $0\nu\beta\beta$ decay within R–parity violating supersymmetric models (see, e.g.[11]).

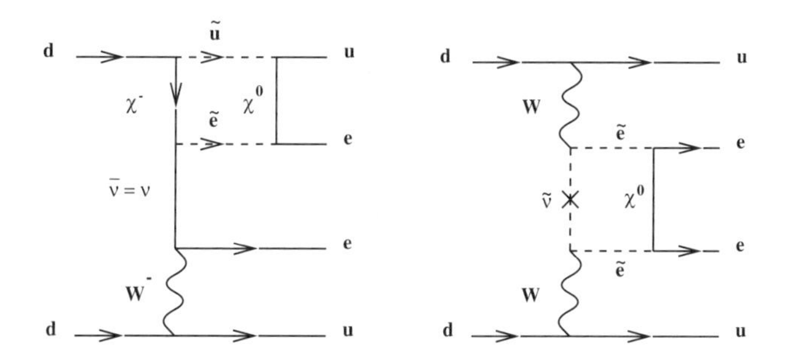

Fig. 8. Examples of R_P conserving SUSY contributions to $0\nu\beta\beta$ decay (see, e.g.[11]).

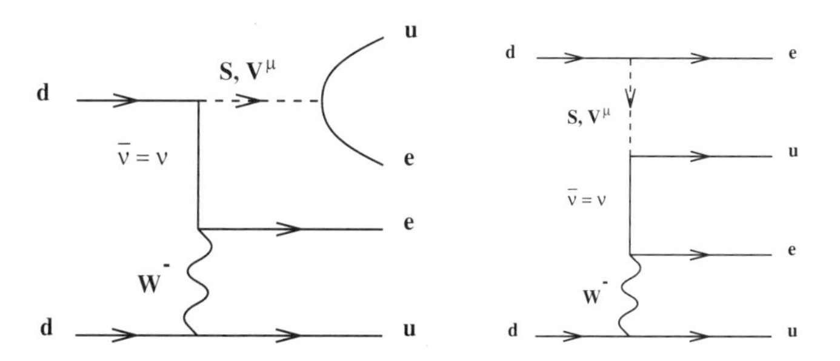

Fig. 9. Examples of Feynman graphs for $0\nu\beta\beta$ decay within LQ models. S and V^{μ} stand for scalar and vector LQs, respectively (see, e.g.[11]).

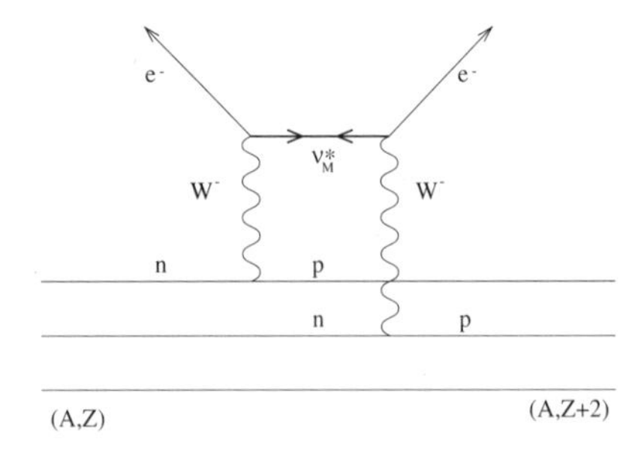

Fig. 10. Neutrinoless double beta decay ($\Delta L = +2$ process) mediated by a composite heavy Majorana neutrino.

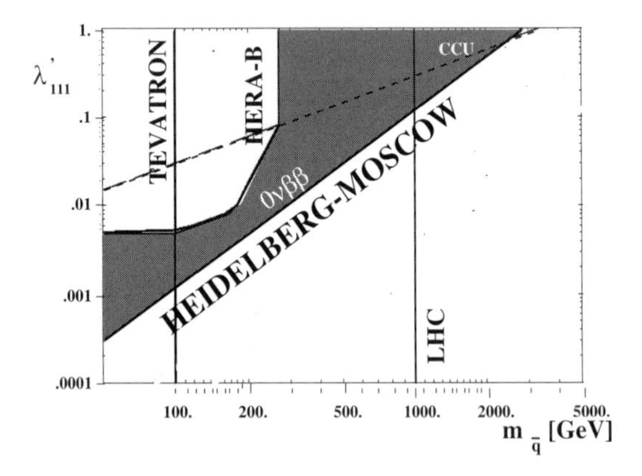

Fig. 11. Comparison of sensitivities of existing and future experiments on R_p-violating SUSY models in the plane λ'_{111}-$m_{\tilde{q}}$. *Note the double logarithmic scale!* Shown are the areas currently excluded by the experiments at the TEVATRON and HERA-B, the limit from charged-current universality, denoted by CCU, and the limit from $0\nu\beta\beta$-decay from the HEIDELBERG-MOSCOW Experiment. *The area beyond (or left of) the lines is excluded.* The estimated sensitivity of LHC is also given (from[11]).

SUSY With R–Parity Breaking

The constraints on the parameters of the minimal supersymmetric standard model with explicit R–parity violation deduced[49–51] from the $0\nu\beta\beta$ half–life limit are more stringent than those from other low–energy processes and from the largest high energy accelerators. The limit for the R-parity breaking Yukawa coupling λ'_{111} (see Fig. 11) is

$$\lambda'_{111} \leq 3.9 \cdot 10^{-4} \Big(\frac{m_{\tilde{q}}}{100GeV}\Big)^2 \Big(\frac{m_{\tilde{g}}}{100GeV}\Big)^{\frac{1}{2}} \tag{4}$$

with $m_{\tilde{q}}$ and $m_{\tilde{g}}$ denoting squark and gluino masses, respectively, and with the assumption $m_{\tilde{d}_R} \simeq m_{\tilde{u}_L}$. This result was important for the discussion of new physics in the connection with the high–Q^2 events seen at HERA. It excluded the possibility of squarks of first generation (of R–parity violating SUSY) being produced in the high–Q^2 events.[59–62]

We find further[51,52]

$$\lambda'_{113}\lambda'_{131} \leq 3 \cdot 10^{-8} \tag{5}$$

$$\lambda'_{112}\lambda'_{121} \leq 1 \cdot 10^{-6}. \tag{6}$$

SUSY with R–Parity Conservation

Also *R–parity* conserving softly broken supersymmetry has been found (by the Heidelberg group,[64,66]) to give contributions to neutrinoless double beta decay, via the *(B–L)* violating sneutrino mass term being a generic ingredient of any weak–scale *SUSY* model with a Majorana neutrino mass. These contributions are realized on the level of sneutrino box diagrams (Fig. 8).

For the $(B-L)$ violating sneutrino mass $\tilde{m}_M$ the following limits are obtained[64]

$$\tilde{m}_M \leq 2\left(\frac{m_{SUSY}}{100GeV}\right)^{\frac{3}{2}} GeV, \qquad \chi \simeq \tilde{B} \tag{7}$$

$$\tilde{m}_M \leq 11\left(\frac{m_{SUSY}}{100GeV}\right)^{\frac{7}{2}} GeV, \qquad \chi \simeq \tilde{H} \tag{8}$$

for the limiting cases that the lightest neutralino is a pure Bino $\tilde{B}$, as suggested by the SUSY solution of the dark matter problem,[65] or a pure Higgsino. Actual values for $\tilde{m}_M$ for other choices of the neutralino composition should lie in between these two values.

Another way to deduce a limit on the 'Majorana' sneutrino mass $\tilde{m}_M$ is to start from the experimental neutrino mass limit, since the sneutrino contributes to the Majorana neutrino mass m_M^{ν} at the 1–loop level proportional to $\tilde{m}_M^2$. This yields under some assumptions[64]

$$\tilde{m}_{M_{(i)}} \leq (60-125)\left(\frac{m_{\nu(i)}^{exp}}{1eV}\right)^{1/2} MeV \tag{9}$$

Starting from the mass limit determined for the electron neutrino by $0\nu\beta\beta$ decay this leads to

$$\tilde{m}_{M_{(e)}} \leq 40MeV. \tag{10}$$

This result is somewhat dependent on neutralino masses and mixings. A non–vanishing 'Majorana' sneutrino mass would result in new processes at future colliders, like sneutrino–antisneutrino oscillations. Reactions at the Next Linear Collider (NLC) like the SUSY analog to inverse neutrinoless double beta decay $e^-e^- \to \chi^-\chi^-$ (where χ^- denote charginos) or single sneutrino production, e.g. by $e^-\gamma \to \tilde{\nu}_e\chi^-$ could give information on the Majorana sneutrino mass, also. This is discussed by.[64,66,67] A conclusion is that future accelerators can give information on second and third generation sneutrino Majorana masses, but for first generation sneutrinos cannot compete with $0\nu\beta\beta$ decay.

Leptoquarks

Assuming that either scalar or vector leptoquarks contribute to $0\nu\beta\beta$ decay (see Fig. 9), the following constraints on the effective LQ parameters can be derived:[69]

$$\epsilon_I \leq 2.8 \times 10^{-9} \left(\frac{M_I}{100\text{GeV}}\right)^2, \tag{11}$$

$$\alpha_I^{(L)} \leq 3.5 \times 10^{-10} \left(\frac{M_I}{100\text{GeV}}\right)^2, \tag{12}$$

$$\alpha_I^{(R)} \leq 7.9 \times 10^{-8} \left(\frac{M_I}{100\text{GeV}}\right)^2. \tag{13}$$

Since the LQ mass matrices appearing in $0\nu\beta\beta$ decay are (4×4) matrices,[69] it is difficult to solve their diagonalization in full generality algebraically. However, if one assumes that only one LQ-Higgs coupling is present at a time, the (mathematical) problem is simplified greatly and one can deduce that either the LQ-Higgs coupling must be smaller than $\sim 10^{-(4-5)}$ or there can not be any LQ with e.g. couplings of electromagnetic strength with masses below $\sim 250GeV$. These bounds from $\beta\beta$ decay were of interest in connection with recently discussed evidence for new physics from HERA.[59,70–72] Assuming that actually leptoquarks have been produced at HERA, double beta decay (the HEIDELBERG-MOSCOW experiment) would allow to fix the leptoquark–Higgs coupling to a few 10^{-6}.[62]

Compositeness

Evaluation of the $0\nu\beta\beta$ half–life assuming exchange of excited composite Majorana neutrinos ν^* yields for the mass of the excited neutrino a lower bound of[45,68]

$$m_N \geq 3.4 m_W \tag{14}$$

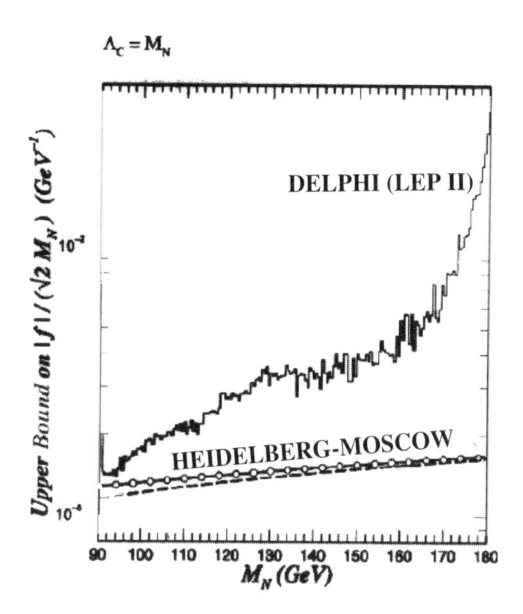

Fig. 12. Comparison between the $\beta\beta_{0\nu}$ HEIDELBERG-MOSCOW experiment and the LEP II upper bound on the quantity $|f|/(\sqrt{2}M_N)$ as a function of the heavy composite neutrino mass M_N, with the choice $\Lambda_C = M_N$. *Regions above the curves are excluded.* The dashed and solid circle curves are the $\beta\beta_{0\nu}$ bounds from the HEIDELBERG-MOSCOW experiment (for details see[45]).

for a coupling of order O(1) and $\Lambda_c \simeq m_N$. Here, m_W is the W–boson mass. The constraints concerning composite excited neutrinos of mass m_N deduced from $\beta\beta$ decay are more strict than the results of LEPII, as shown in Fig. 12.

Superheavy Neutrinos, Right-Handed W-Boson

It was discussed already since *1980* by Mohapatra[53,54] that not only exchange of a left–handed neutrino, but also of a heavy *right–handed neutrino*, which naturally occurs in left–right symmetric models, could induce neutrinoless double beta decay. This process which was discussed in more detail later by the Heidelberg group,[57,58] yields at present the *most restrictive lower bound* on the mass of a right–handed W boson, of $m_{W_R} \geq 1.4\,\mathrm{TeV}$[11,52,57] .

In the case of the exchange of heavy or superheavy *left–handed* neutrinos one can exploit the mass dependence of the matrix element (see e.g.[79]) to obtain *lower* limits on the latter (see in[6] [HM95]).

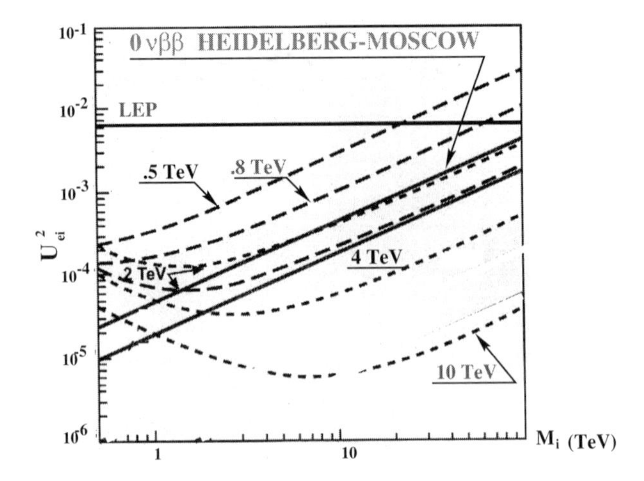

Fig. 13. Discovery limit for $e^-e^- \longrightarrow W^-W^-$ at a linear collider as function of the mass M_i of a heavy left-handed neutrino, and of U^2_{ei} for $\sqrt{s}$ between 500 GeV and 10 TeV. In all cases the parameter space above the line corresponds to observable events. The limits from the HEIDELBERG-MOSCOW $0\nu\beta\beta$ experiment are shown also, the areas above the $0\nu\beta\beta$ contour line are excluded. The horizontal line denotes the limit on neutrino mixing, U^2_{ei}, from LEP (from[43]).

The deduced lower limit[10,11,52] is $\langle m_H \rangle \geq 9 \times 10^7$ GeV. Assuming the bound on the mixing matrix[42–44] $U^2_{ei} = 5 \times 10^{-3}$ and assuming no cancellation between the involved states, the limit implies a bound on the mass eigenstate $M_i > 5 \times 10^5$ GeV. To obtain the same information by inverse double beta decay $e^-e^- \longrightarrow W^-W^-$ at a Next Linear Collider, the latter should have a center of mass energy of two TeV[42,43] (Fig. 13), which would be very far future.

Extra Dimensions

Neutrinoless double beta decay in the presence of extra dimensions has been discussed by R. Mohapatra and A. Pérez-Lorenzana.[55] They show that the higher Kaluza-Klein modes of the right-handed W-boson provide new contributions to this process. In this way correlated limits on m_{W_R} and the inverse size of the extra dimensions can be obtained from double beta decay.

The model building conditions under which a $0\nu\beta\beta$ signal of the observed level can be obtained due to Kaluza-Klein singlet neutrinos in theories with large extra dimensions we have investigated in[56] (see chapter 1.5.5 of [6]).

Special Relativity, Equivalence Principle

Violation of Lorentz invariance (VLI): The bound obtained from the HEIDELBERG-MOSCOW experiment is

$$\delta v < 4 \times 10^{-16} \quad \text{for} \quad \theta_v = \theta_m = 0 \tag{15}$$

where $\delta v = v_1 - v_2$ is the measure of VLI in the neutrino sector. θ_v and θ_m denote the velocity mixing angle and the weak mixing angle, respectively. In Fig. 14 (from[74]) the bound implied by double beta decay is presented for the entire range of $sin^2(2\theta_v)$, and compared with bounds obtained from neutrino oscillation experiments (see[73]).

Violation of equivalence principle (VEP): Assuming only violation of the weak equivalence principle, there does not exist any bound on the amount of VEP. It is this region of the parameter space which is most restrictively bounded by neutrinoless double beta decay. In a linearized theory the gravitational part of the Lagrangian to first order in a weak gravitational field $g_{\mu\nu} = \eta_{\mu\nu} + h_{\mu\nu}$ ($h_{\mu\nu} = 2\frac{\phi}{c^2} diag(1,1,1,1)$) can be written as $\mathcal{L} = -\frac{1}{2}(1+g_i)h_{\mu\nu}T^{\mu\nu}$, where $T^{\mu\nu}$ is the stress-energy in the gravitational

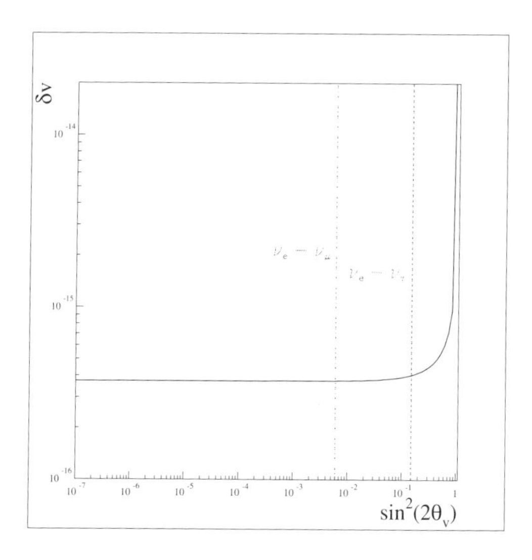

Fig. 14. Double beta decay bound (solid line) on violation of Lorentz invariance in the neutrino sector, excluding the region to the upper left. Shown is a double logarithmic plot in the δv–$\sin^2(2\theta)$ parameter space. The bound becomes most stringent for the small mixing region, which has not been constrained from any other experiments. For comparison the bounds obtained from neutrino oscillation experiments (from[73]) in the $\nu_e - \nu_\tau$ (dashed lines) and in the $\nu_e - \nu_\mu$ (dashed-dotted lines) channel, excluding the region to the right, are shown (from[74]).

eigenbasis. In the presence of VEP the g_i may differ. We obtain[74] the following bound from the Heidelberg–Moscow experiment, for $\theta_v = \theta_m = 0$:

$$\begin{aligned} \phi\delta g &< 4 \times 10^{-16} \text{ (for } \bar{m} < 13\text{eV)} \\ \phi\delta g &< 2 \times 10^{-18} \text{ (for } \bar{m} < 0.08\text{eV)}. \end{aligned} \tag{16}$$

Here $\bar{g} = \frac{g_1+g_2}{2}$ can be considered as the standard gravitational coupling, for which the equivalence principle applies. $\delta g = g_1 - g_2$. The bound on the VEP thus, unlike the one for VLI, will depend on the choice for the Newtonian potential ϕ.

Dark Matter and Dark Energy

A degenerate neutrino mass of $\geq 0.22\,\text{eV}$ corresponds to a contribution of neutrinos as hot dark matter of $\geq 4.7\%$ to the total observed dark matter in the Universe (see Fig. 15). According to Goldman et al.[121] there is a possible correlation between the mass of the neutrino and dark energy. The conclusion is that a neutrino of mass around $0.3\,\text{eV}$ could solve the problem of dark energy.

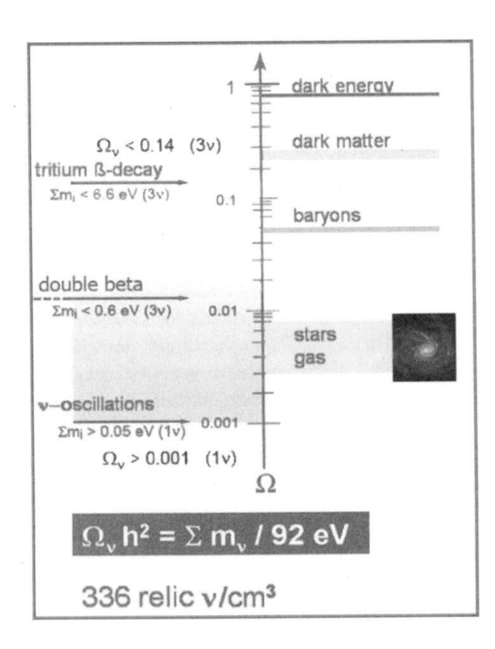

Fig. 15. Contribution of hat dark matter to the mass distribution in the Universe.

"In a cloud of massive fermions interacting by exchange of a light scalar field, the effective mass and the total energy density eventually increases with decreasing density. In this regime, the pressure density relation can approximate that required for dark matter energy. Applying this to the expansion of the Universe with a very light scalar field leads to the conclusion that Majorana neutrinos of a mass of $\sim 0.3\,\mathrm{eV}$" (as observed in $0\nu\beta\beta$ decay) "may be consistent with current observation of dark energy".[121]

6. Future $\beta\beta$ Experiments and Some Problems

The main problem is that present and future 'confirmation' experiments usually are not sensitive enough (see[14] and more in[6]). A good example is the NEMO III experiment. The half-life limits reached (at a 1.5σ level) of $T^{0\nu}_{1/2}$ $=1.0\times10^{23}$ and 4.6×10^{23} years for ^{100}Mo and ^{82}Se (see[84]) after 389 days of effective measurement are a factor 20 away from the half-lives required to check the HEIDELBERG-MOSCOW result on a 1.5σ level. Since the half-life is connected with the measuring time by $T^{0\nu}_{1/2} \sim \sqrt{\frac{t\cdot M}{\delta E\cdot B}}$, this means that NEMO III would have to measure *more than 400 years, to see the signal on a 1.5σ level*, and correspondingly longer, to see it on a higher c.l.[14,76] CUORICINO - which has the general problem, that it *cannot* distinguish between β and γ-events, and because of its high background cannot see the $2\nu\beta\beta$ spectrum of ^{130}Te, could see the HEIDELBERG-MOSCOW signal assuming an uncertainty in the knowledge of the nuclear matrix element[77] of a factor of only 2, within 1 and 30 years - on a 1.5σ c.l.[14,76] It can thus *never disprove* the HEIDELBERG-MOSCOW result. The large version CUORE with a by a factor of 16 larger mass also would need many years for a statement on a 6σ level. EXO - the main problem is that *no tracks* are visible in a liquid ^{136}Xe experiment.[87] This kills the main idea of the experiment to separate $\beta\beta$ from γ events, and just reduces it to complicated calorimeter. Since the other main idea - laser identification of the daughter nucleus, is not (yet) working, the present rather modest aim is to reach a background level as reached in the HEIDELBERG-MOSCOW experiment, instead of the factor of 1000 less, projected earlier[88] . They claim to reach this goal now around 2001 (see Ref.[88]). GERDA - (the 'copied' GENIUS project proposed in 1997[22] , planning to operate naked ^{76}Ge crystals in liquid nitrogen). From our earlier Monte Carlo calculations we expected to get a large potential for $\beta\beta$ research. The only long-term experience with naked detectors in liquid nitrogen has been collected since then with our GENIUS-Test-Facility[23] . Why any GENIUS-like project will not be able to confirm our evidence in a reasonably short time see[24] .

SNO$^+$ - The only experiment having a good chance to perform a *high-statistics $0\nu\beta\beta$ experiment*, is *perhaps* the SNO$^+$ project, sometimes in its application for double beta decay called SNO^{++}[86] . The idea is to fill SNO, the Sudbury (Solar) Neutrino Observatory, with liquid scintillator. This liquid scintillator then serves as the medium, in which double beta decay candidate materials would be deployed. The use of ^{150}Nd (in form of Nd_2O_3 nanoparticles) would take advantage of the relatively large phase space and nuclear matrix element of ^{150}Nd (see[85]). For 50% enriched ^{150}Nd and 0.1% loading in SNO$^+$ a 5σ sensitivity for an effective neutrino mass of 0.04 eV is expected to be reached after 3 years of running. Since the start of the ^{150}Nd experiment is expected for the year 2010, SNO$^+$ could be *the near- or medium-time scale future* of double beta decay.

Concerning expected information on the ν mass, there is another problem in present experimental approaches. Even if one of these $\beta^-\beta^-$ experiments would be able to confirm the HEIDELBERG-MOSCOW result, no *new* information would be obtained.

It is known for 20 years - but surprisingly often overlooked - that a $\beta^-\beta^-$ experiment can give information on the effective neutrino mass *only* under some *assumption* on the contribution of right-handed weak currents (parameters η, λ) or others like SUSY, ... to the $\beta\beta$-amplitude (see e.g.[6]). *In general* one obtains *only an upper limit on $\langle m \rangle$*. So if neutrino masses are deduced from $0\nu\beta\beta$ experiments, this is always done under the assumption of vanishing η, λ etc. *In that sense* it is premature to compare as often done such number with numbers deduced e.g. from WMAP or other experiments, or to use it as a landmark for future tritium experiments.

It is unfortunate that even an additional high-sensitive $\beta^-\beta^-$ experiment (e.g. ^{136}Xe) *together* with the ^{76}Ge HEIDELBERG-MOSCOW result can give *no* information to decide the individual contribution of $\langle m \rangle$, $\langle \eta \rangle, \langle \lambda \rangle$ to the $0\nu\beta\beta$ decay rate. This has been shown already in 1994[78] , and investigated in more detail in[123] .

Proposed way out

In the same paper[78] it has been shown that the only realistic way to get this information on the individual contributions of m, η, λ is to combine the $\beta^-\beta^-$ result from ^{76}Ge (HEIDELBERG-MOSCOW), with a very high-sensitivity (level of 10^{27} y) mixed mode β^+/EC decay experiment (e.g. of ^{124}Xe), (see also[14,123] and Fig. 16.)

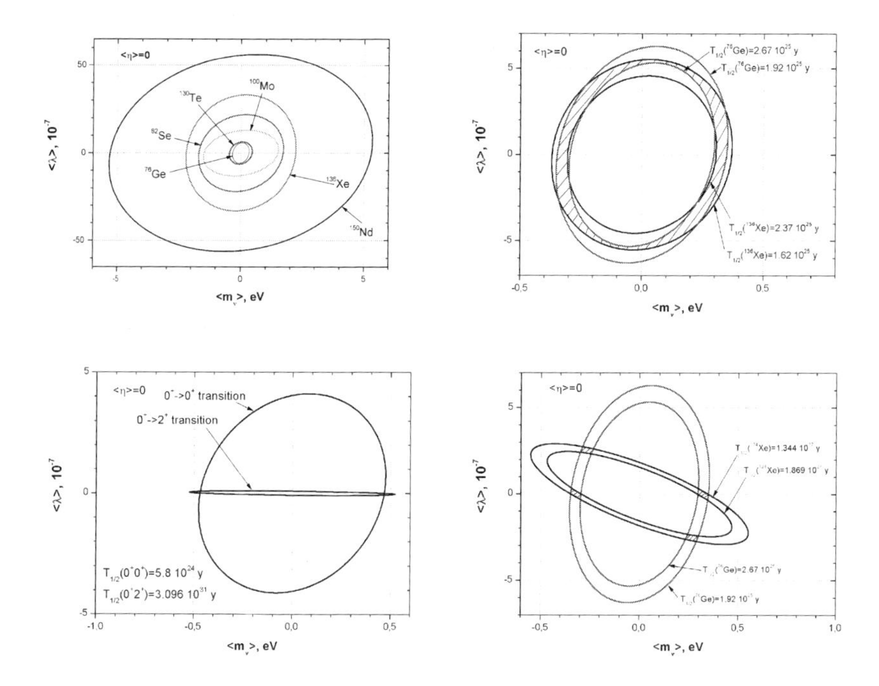

Fig. 16. Upper left: The allowed regions in the $\langle\lambda\rangle$, $\langle m\rangle$ parameter space for different $\beta\beta$ parent isotopes using experimental limits. Upper right: Allowed region for the $\langle m\rangle - \langle\lambda\rangle$ parameter space for ^{76}Ge $(T_{1/2}^{0\nu\beta\beta} = 2.23^{+0.44}_{-0.31}\times 10^{25})$ y and assuming ^{136}Xe $(T_{1/2}^{0\nu\beta\beta} = 1.93^{+0.44}_{-0.31}\times 10^{25})$ y. Bottom left: Allowed region for ^{100}Mo for the $0^+ \to 2^+$ and $0^+ \to 0^+$ transition (assuming half lives, respectively, 3.1×10^{31} y and 5.8×10^{24} y. Bottom right: The case of neutrinoless mixed β^+/EC mode of ^{124}Xe for the assumed half-life (1.34-1.87)$\times 10^{27}$ y compared to the ^{76}Ge measured half-life ($\langle m_\nu\rangle$- $\langle\lambda\rangle$ plane). The dashed areas are consistent with both experiments. (From Ref.[123])

So it might be wise to combine future efforts to confirm the HEIDELBERG-MOSCOW result with a possibility to pin down some of the various contributions to the $0\nu\beta\beta$ decay amplitude.

7. Summary and Outlook

We reached with the HEIDELBERG-MOSCOW experiment,[5,13,16,75,76] what we wanted to learn from our large GENIUS project, proposed in 1997[22] at a time where a signal was not yet seen - namely observation of $0\nu\beta\beta$ decay. There is now a $>6\sigma$ signal for $0\nu\beta\beta$ decay. The neutrino is a Majorana particle. Total lepton number is violated. Presently running and planned experiments are usually - with one possible exception - not sensi-

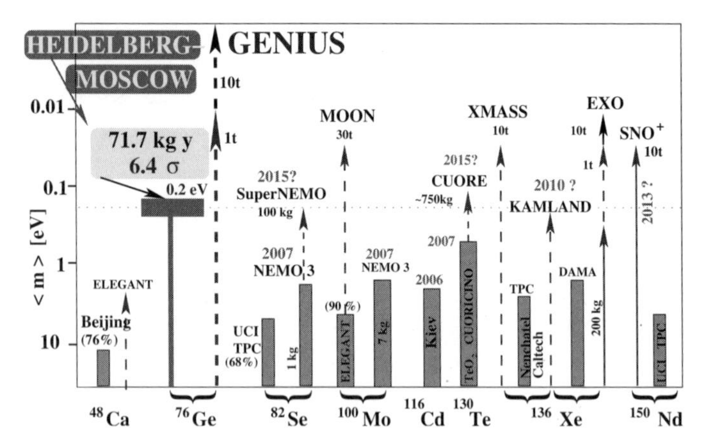

Fig. 17. Present sensitivity, and expectation for the future, of the most promising $\beta\beta$ experiments. Given are limits for $\langle m \rangle$, except for the HEIDELBERG-MOSCOW experiment where the measured *value* is given. Framed parts of the *bars:* present status for running experiments; solid and dashed lines: experiments under construction or proposed, respectively - dashed lines: far from realization. For references see[6,13]

tive enough to check the HEIDELBERG-MOSCOW result on a reasonable time scale (see Fig. 17) [e].

Nuclear double beta decay has developed in the last decades to one of the most exciting means of the research into physics beyond the standard model. That for the first time, an evidence has been found, that the process of $0\nu\beta\beta$ decay is occurring in nature is, because of its fundamental consequences for particle physics, a huge challenge for future experiments. For a better understanding of the various mechanisms contributing to the $0\nu\beta\beta$ amplitude, we will have to go to *completely different* types of $\beta\beta$ experiments, than presently persued (see Ref.[6,13,14]).

Recent information from many *independent* sides is consistent with a neutrino mass of the order of the value found by the HEIDELBERG-MOSCOW experiment. This is the case for the results from CMB and LSS, neutrino oscillations, particle theory and cosmology (for a detailed discussion see[6]). To mention a few examples: Neutrino oscillations require in the case of degenerate neutrinos common mass eigenvalues of $m > 0.04\,\text{eV}$. An

[e] Also the authors of Ref.[102] have to confirm (see their Figs. 3 and 5), that also according to their own calculations of nuclear matrix elements no present experiment probes the 90% c.l. signal range of KK-K[13] , and none of them can exclude a fraction of the range given by HEIDELBERG-MOSCOW at a comparable confidence level. They state, that more optimistic claims by[119] were based on a larger favoured range for the ^{130}Te half-life, and thus were wrong.

analysis of CMB measurements by SDSS and large scale structure yields $\sum m_\nu < 1.7\,\text{eV}$[26] (see footnote (d) on page 14). Theoretical papers require degenerate neutrinos with m > 0.1, or 0.2 eV or 0.3 eV,[89,91,92,99,100] and the recent alternative cosmological concordance model requires relic neutrinos with mass of order of eV.[101] As mentioned already earlier[18,75] the results of double beta decay and CMB measurements together indicate that the neutrino mass eigenvalues have the same CP parity, as required by the model of.[92] Also the approach of[120] comes to the conclusion of a Majorana neutrino. The Z-burst scenario for ultra-high energy cosmic rays requires $m_\nu \sim 0.4\,\text{eV}$,[93,94] and also a non-standard model (g-2) has been connected with degenerate neutrino masses >0.2 eV.[90] The neutrino mass determined from $0\nu\beta\beta$ decay is consistent also with present models of leptogenesis in the early Universe.[98]

Finally we have pointed out that the probably fasted test and confirmation of the result of our ^{76}Ge $\beta\beta$ result will be delivered by the new PLANCK satellite mission launched in May 2009. It will investigate the cosmic microwave background with unprecedented precision and in this way also the neutrino mass. It will either confirm the neutrino mass deduced by the ^{76}Ge experiment (under the assumption of a dominating mass mechanism), or will decide that the process of $\beta\beta$ decay is triggered by another mechanism. It thus will yield information which can hardly be obtained ever by any future $\beta\beta$ experiments.

Acknowledgments

The authors thank Dr. Irina Titkova for efficient and pleasant collaboration. The authors acknowledge the invaluable support from DFG and BMBF, and LNGS for this project.

References

1. H.V. Klapdor et al., *Atomic Data and Nuclear Data Tables* **31** (1984) 81-111;
2. H.V. Klapdor, *Progr. in Part. and Nucl. Phys.* **10** (1983) 131-225; **17** (1986) 419-455; *Sterne und Weltraum* **1985/3** 132-139; *Fortschritte der Physik* **33, Heft 1** (1985) 1-55; *The Astrophysical J.* **301** (1986) L39-43.
3. K. Grotz and H.V. Klapdor, "Die schwache Wechselwirkung in Kern-, Teilchen- und Astrophysik", Teubner, Stuttgart 1989; Adam Hilger – IOP Bristol 1990; Mir, Moskva, 1992; Shandong Science and Technology Press, Jinan, China 1996.
4. H.V. Klapdor-Kleingrothaus and A. Staudt, "Teilchenphysik ohne Beschle-

uniger", Teubner, Stuttgart 1995; IOP, Bristol 1995, sec. ed. 1998; Nauka Fizmatlit, Moscow, 1997.

5. H.V. Klapdor, *Proposal, Internal Report*, MPI-1987-**V17**, September 1987.
6. H.V. Klapdor-Kleingrothaus, "70 Years of Double Beta Decay - From Nuclear Physics to Beyond the Standard Model", WS, Singapore (2009); "60 Years of Double Beta Decay - From Nuclear Physics to Beyond the Standard Model", WS, Singapore (2001).
7. H. Päs, M. Hirsch, H.V. Klapdor-Kleingrothaus and S.G. Kovalenko *Phys. Lett.* **B453** (1999) pp. 194-198.
8. H.V. Klapdor-Kleingrothaus, *Int. Journ. of Mod. Phys.* **D13** (2004) 2107.
9. K. Muto and H.V. Klapdor "Neutrinos", "Graduate texts in contemporary physics", ed. *H.V. Klapdor Berlin, Germany: Springer* (1988) pp. 183-238.
10. H.V. Klapdor-Kleingrothaus in Proc. "Lepton and Baryon Number Violation in Particle Physics, Astrophysics and Cosmology", eds. *H.V. Klapdor-Kleingrothaus and I.V. Krivosheina, International Workshop at ECT,* Trento, Italy, 20-25 April, 1998, *World Scientific* (1998) pp. 251-301.
11. H.V. Klapdor-Kleingrothaus in the Proc. "Symmetries in Intermediate High Energy Physics", eds. *A. Faessler et al., , Springer-Verlag, Berlin, Heidelberg* (2000), *Tracts in Modern Physics* **63** pp. 69-104.
12. H.V. Klapdor-Kleingrothaus, *Int. J. Mod. Phys.* **A 13** (1998) 3953.
13. H.V. Klapdor-Kleingrothaus, I.V. Krivosheina et al., *Mod. Phys. Lett.* **A21** (2006) 1547; *Phys. Lett.* **B586** (2004) 198-212 and *Nucl. Instr. & Methods* **A522** (2004) 371-406; *Mod. Phys. Lett.* **A16** (2001) 2409-2420, *Phys. Lett.* **B632** (2006) 623-631; *Phys. Rev.* **D73** (2006) 013010; *Phys. Scripta* **T127** (2006) 40-42; Proc. of DARK2007, Sydney, Sept. 2007, eds. H.V. Klapdor-Kleingrothaus et al., pp. 442-467; *Int. J. Mod. Physics* **E17** (2008) 505-517.
14. H.V. Klapdor-Kleingrothaus *Int. J. Mod. Physics* **E17** (2008) 505-517.
15. H.V. Klapdor-Kleingrothaus et al., *Phys. Lett.* **B578** (2004) 54-62.
16. H.V. Klapdor-Kleingrothaus, in Proc. of BEYOND03, Castle Ringberg, Germany, 9-14 June 2003, *Springer* (2004), ed. H.V. Klapdor-Kleingrothaus, 307-364.
17. K.Ya. Gromov et al., *J. Part. Nucl. Lett.* **3** (2006) 30-41.
18. H.V. Klapdor-Kleingrothaus, in Proc. of Intern. Conf. BEYOND'02, Oulu, Finland, 2-7 Jun. 2002, IOP, Bristol, 2003 ed. H.V. Klapdor-Kleingrothaus, 215-240 pp.
19. in Proc. of Neutrinos and Implications for Physics Beyond the Standard Model, Stony Brook, New York, 11-13 Oct 2002. *Int. J. Mod. Phys.* **A18** (2003) 4113-4128; *hep-ph/* **0303217**.
20. H.V. Klapdor-Kleingrothaus, H. Päs and A.Yu. Smirnov, *Phys. Rev.* **D63** (2001) 073005.
21. M. Moe et al., *Phys. Rev. Lett.* **59** (1987) 989.
22. H.V. Klapdor-Kleingrothaus, J. Hellmig and M. Hirsch, *GENIUS-Proposal,* 20 Nov. 1997; J. Hellmig, HVKK, *Z. Phys.* **A359** (1997) 351-359; H.V. Klapdor-Kleingrothaus, M. Hirsch, *Z. Phys.* **A359** (1997) 361-372; H.V. Klapdor-Kleingrothaus, *CERN Courier,* **Nov. 1997**, 16- 18; H.V. Klapdor-Kleingrothaus, J. Hellmig, M. Hirsch, *J. Phys.* **G24** (1998) 483-

516; H.V. Klapdor-Kleingrothaus et al., Proposal Aug. 1999 sec. draft, *hep-ph/***9910205**, Proc. of "Beyond the Desert99" eds. H.V. Klapdor-Kleingrothaus, I.V. Krivosheina (IOP, 2000), pp. 915-1015.

23. H.V. Klapdor-Kleingrothaus, *CERN Courier* **43 Nr.6** (2003) 9; H.V. Klapdor-Kleingrothaus et al., *Nucl. Instr. Meth.* **A 511** (2003) 341; H.V. Klapdor-Kleingrothaus et al., *Nucl. Instr. Meth.* **A 530** (2004) 410-418; *Nucl. Instrum. Meth.* **A 481** (2002) 149-159.
24. H.V. Klapdor-Kleingrothaus and I.V. Krivosheina, *Nucl. Instr. Meth.* **A566** (2006) 472; *Phys. Scripta* **T127** (2006) 52.
25. B.S. Flanders, R. Madey, B.D. Anderson, A.R. Baldwin, J.W. Watson, C.C. Foster, H.V. Klapdor and K. Grotz *Phys. Rev.* **C40** (1989) pp. 1985-1992; R. Madey, B.S. Flanders, B.D. Anderson, A.R. Baldwin, J.W. Watson, S.M. Austin, C.C. Foster, H.V. Klapdor and K. Grotz, *Phys. Rev.* **C40** (1989) pp. 540-552.
26. SDSS Collaboration (M. Tegmark et al.), *Phys. Rev.***D 69** (2004) 103501, *astro-ph/***0310723**.
27. WMAP Collaboration, *Astrophys. J. Suppl.* **148** (2003) 213-233; **170** (2007) 377.
28. "Planck-The Scientific programme", *ESA-SCI* (2005) **1**, *astro-ph/* **0604069** v1 and http://www.rssd.esa.int/SA/PLANCK/docs/Bluebook-ESA-SCI-(2005)1_V2.pdf.
29. SNO Collaboration *Phys. Rev. Lett.* **92** (2004) 181-301; J.N. Bahcall et al., *J. High Energy Phys.* **0311** (2003) 1-48.
30. Super-Kamiokande Coll., *Phys. Rev. Lett.* **93** (2004) 101801; *Phys. Rev.* **D71** (2005) 112005.
31. H.V. Klapdor and K. Grotz *Phys. Lett.* **B142** (1984) pp. 323-328; and K. Grotz and H.V. Klapdor *Nucl. Phys.* **A460** (1986) pp. 395-436.
32. J. Schechter and J.W.F. Valle, *Phys. Rev.* **D25** (1982) 2951-2954.
33. H. Sugiyama, Proc. BEYOND'02, Oulu, Finland, June 2002, ed. H.V. Klapdor- Kleingrothaus (IOP 2003), pp. 409-415.
34. D.O. Caldwell, in Proc. 12th International Conference on "Neutrino Physics and Astrophysics", Sendai, Japan, June 3-8, 1986, *eds.* T. Kitagaki and H. Yuta, *World Scientific, Singapore* (1986) pp. 77-92.
35. D.O. Caldwell *Int. J. Mod. Phys.* **A4** (1989) 1851-1869.
36. D.O. Caldwell in Proc. of 14th Europhysics Conference on Nuclear Physics: Rare Nuclear Decays and Fundamental Physics, Bratislava, Czechoslovakia, 22-26 October, 1990, ed *P. Povinec, J. Phys.* **G17** (1991) *Suppl.* S137-S144.
37. I. Kirpichnikov, in Proc. of International Conference "Underground Physics", Baksan, Russia, August 1987, *eds.* E.N. Alekseev et.al.; A.A. Vasenko, I.V. Kirpichnikov et al., *Mod. Phys. Lett.* **A5** (1990) pp. 1299-1306.
38. M.G. Inghram and J.H. Reynolds, *Phys. Rev.* **78** (1950) pp. 822-823.
39. N. Takaoka and G. Ogata, *Z. Naturforsch* **A21** (1966) pp. 84-90.
40. T. Kirsten, W. Gentner and O. Schaeffer, *Z. Phys.* **202** (1967) pp. 273-292.
41. E. der Mateosian and M. Goldhaber *Phys. Rev.* **146** (1966) pp. 810-815.
42. G. Belanger, F. Boudjema, D.London and H. Nadeau, *Phys. Rev.* **D 53** (1996) 6292.

43. G. Belanger et al., *Phys. Rev.* **D53** (1996) 6292 and in Proc. of Lepton-Baryon Int. Conf., April 1998, Trento, IOP, Bristol, (1999), eds. H.V. Klapdor-Kleingrothaus and I.V. Krivosheina.
44. E. Nardi, E. Roulet and D. Tommasini *Phys. Lett.* **B344** (1995) pp. 225-232.
45. O. Panella et al., *Phys. Rev.* **D 62** (2000) 015013.
46. M. Hirsch, H.V. Klapdor–Kleingrothaus and S.G. Kovalenko, *Phys. Lett.* **B 398** (1997) 311 and **403** (1997) 291.
47. H.V. Klapdor–Kleingrothaus and H. Päs; in: Proc. of the 6th Symp. on Particles, Strings and Cosmology (PASCOS'98), Boston/USA, 1998.
48. H.V. Klapdor-Kleingrothaus and U. Sarkar, *hep-ph/***0302237**.
49. M. Hirsch, H.V. Klapdor–Kleingrothaus and S.G. Kovalenko, *Phys. Rev. Lett.***75** (1995) 17.
50. M. Hirsch, H.V. Klapdor–Kleingrothaus and S. Kovalenko, *Phys. Rev.* **D 53** (1996) 1329.
51. M. Hirsch, H.V. Klapdor–Kleingrothaus and S.G. Kovalenko, *Phys. Lett.* **B 372** (1996) 181, Erratum: *Phys. Lett.* **B381** (1996) 488.
52. H.V. Klapdor-Kleingrothaus and H. Päs *"Neutrinoless Double Beta Decay and New Physics in the Neutrino Sector"*, in Proc. COSMO 99: 3rd International Conference on Particle Physics and the Early Universe, Trieste, Italy, 27 Sept. - 3 Oct. 1999, *eds* U. Cotti, R. Jeannerot, G. Senjanovic and A. Smirnov. *Singapore, World Scientific* (2000), and in *H.V. Klapdor-Kleingrothaus (ed.)*: Sixty years of double beta decay *Singapore, World Scientific* (2001) pp. 755-762, and *hep-ph/*0002109.
53. R.N. Mohapatra and G. Senjanović *Phys. Rev. Lett.* **44** (1980) pp. 912-915.
54. R.N. Mohapatra *Phys. Rev.* **D34** (1986) pp. 909-910.
55. R.N. Mohapatra and A. Pérez-Lorenzana *hep-ph/*9909389 (1999) pp. 1-10 and *Phys. Lett.* **B468** (1999) pp. 195-200.
56. G. Bhattacharyya, H.V. Klapdor-Kleingrothaus, H. Päs and A. Pilaftsis *Phys. Rev.* **D67** (2003) 113001-1-17 and *hep-ph/*0212169.
57. M. Hirsch, H. V. Klapdor-Kleingrothaus and O. Panella *Phys. Lett.* **B374** (1996) pp. 7-12.
58. M. Hirsch and H.V. Klapdor-Kleingrothaus in Proc. International Workshop "Double Beta Decay and Related Topics", Trento, Italy, April 24 - May 5, 1995, eds. *H.V. Klapdor-Kleingrothaus and S. Stoica, Singapore: World Scientific* (1996) pp. 175-191.
59. D. Choudhury and S. Raychaudhuri, preprint *hep-ph/* **9702392**.
60. D. Cline, in: Proceedings of the *International Workshop Dark Matter in Astro- and Particle Physics* (DARK96), Eds. H.V. Klapdor–Kleingrothaus and Y. Ramachers, World Scientific 1997, p. 479.
61. G. Altarelli, J. Ellis, G.F. Guidice, S. Lola and M.L. Mangano, preprint *hep-ph/* **9703276**.
62. M. Hirsch, H.V. Klapdor–Kleingrothaus, S. Kovalenko, in.[63]
63. H.V. Klapdor-Kleingrothaus in Proc. of BEYOND'97, Castle Ringberg, Germany, 8-14 June 1997, ed. by H.V. Klapdor-Kleingrothaus and H. Päs, IOP Bristol (1998) 485-531, and *Int. J. Mod. Phys.* **A13** (1998) 3953.
64. M. Hirsch, H.V. Klapdor–Kleingrothaus and S.G. Kovalenko, *Phys. Rev.* **D**

57 (1998) 1947.
65. G. Jungmann, M. Kamionkowski and K. Griest, *Phys. Rep.* **267** (1996) 195.
66. M. Hirsch, H.V. Klapdor–Kleingrothaus and S.G. Kovalenko, *Phys. Lett.* **B 398** (1997) 311 and **403** (1997) 291.
67. M. Hirsch, H.V. Klapdor–Kleingrothaus, St. Kolb and S.G. Kovalenko, *Phys. Rev.* **D 57** (1998) 2020.
68. E. Takasugi *"Double Beta Decay Constraint on Composite Neutrinos"*, in Proc. "Beyond the Desert'97": Accelerator and Nonaccelerator Approaches, eds. *H.V. Klapdor-Kleingrothaus and H. Päs*, "Conference on Physics Beyond the Standard Model", Tegernsee, Germany, 8-14 June, 1997, *IOP* (1998) p. 360.
69. M. Hirsch, H.V. Klapdor–Kleingrothaus and S.G. Kovalenko, *Phys. Lett.* **B 378** (1996) 17 and *Phys. Rev.* **D 54** (1996) R4207.
70. J.L. Hewett and T.G. Rizzo, *preprint hep-ph/***9703337v3** (May1997).
71. K.S. Babu et al., *preprint hep-ph/***9703299** (March 1997).
72. J. Kalinowski et al., *preprint hep-ph/***9703288v2** (March 1997).
73. A. Halprin, C.N. Leung and J. Pantalone, *Phys. Rev.* **D 53** (1996) 5365.
74. H.V. Klapdor-Kleingrothaus, H. Päs and U. Sarkar *Eur. Phys. J.* **A5** (1999) pp. 3-6 and *hep-ph/***9809396**.
75. H.V. Klapdor-Kleingrothaus, I.V. Krivosheina et al., *Phys. Lett.* **B586** (2004) 198-212 and H.V. Klapdor-Kleingrothaus, A. Dietz, I.V. Krivosheina et al., *Nucl. Instr. & Methods* **A522** (2004) 371-406; H.V. Klapdor-Kleingrothaus, I.V. Krivosheina et al., *Mod. Phys. Lett.* **A21** (2006) 1547.
76. H.V. Klapdor-Kleingrothaus, in Proc. of Int. Conf. "Neutrino Telescopes", Febr. 2005, Venice, Italy, ed. M. Baldo-Ceolin., p. 215, *hep-ph/***0512263**.
77. A. Staudt, K. Muto, H.V. Klapdor-Kleingrothaus, *Eur. Lett.* **13** (1990) 31.
78. M. Hirsch, K. Muto, T. Oda and H.V. Klapdor-Kleingrothaus, *Z. Phys.* **A 347** (1994) 151.
79. K. Muto and H.V. Klapdor "Neutrinos", In "Graduate texts in contemporary physics", ed. *H.V. Klapdor Berlin, Germany: Springer* (1988) pp. 183-238.
80. A. Staudt and H.V. Klapdor-Kleingrothaus *Nucl. Phys.* **A549** (1992) pp. 254-264.
81. K. Muto, E. Bender and H.V. Klapdor *Z. Phys.* **A334** (1989) pp. 177-186 and pp. 187-194.
82. M. Hirsch, A. Staudt, K. Muto and H.V. Klapdor-Kleingrothaus *At. Data Nucl. Data Tables* **53** pp. 165-193.
83. M. Hirsch, K. Muto, T. Oda and H.V. Klapdor-Kleingrothaus *Z. Phys.* **A347** (1994) pp. 151-160.
84. R. Arnold et. al. (NEMO Collaboration) *"First Results of the Search of Neutrinoless Double Beta Decay With the NEMO 3 Detector"*, *Phys. Rev. Lett.* **95** (2005) pp. 182302-1–4, *hep-ex/*0507083.
85. M. Moe in Proc. of the 16th Int. Conf. on "Neutrino Physics and Astrophysics, NEUTRINO'94", Eilat, Israel, 29 May-3 June, 1994, eds. *A. Dar, G. Eilam and M. Gronau, Nucl. Phys.* **B38** *Proc. Suppl.* (1995) pp. 36-44.
86. M. Boulay, M. Chen, M. Di Marco et al., *"A Letter Experessing Interest in Starting an Experiment at SNOLAB Involving Filling SNO with Liquid*

Scintillator Plus Double Beta Decay Candidate Isotopes", April 9, (20040), *http : //snoplus.phy.queensu.ca/*

87. J. Vuilleumier for the EXO coll., Proc. idm2004, Edinburg, Scotland 2004, WS, Singapore (2005) 635.
88. A. Piepke, (Talk) at Heidelberg, ν Workshop, 2005. and in Proc. of XXImes Rencontres de Blois "Windows on the Universe", Chateau Royal de Blois, France, 21- 26 June, 2009, http://confs.obspm.fr/Blois2009/.
89. K.S. Babu, E. Ma and J.W.F. Valle, *Phys. Lett.* **B552** (2003) 207-213 and *hep-ph/***0206292**.
90. E. Ma and M. Raidal, *Phys. Rev. Lett.* **87** (2001) 011802; *Erratum-ibid.* **87** (2001) 159901 and *hep-ph/***0102255**.
91. E. Ma in Proc. of Intern. Conf. BEYOND'02, Oulu, Finland, 2-7 Jun. 2002, IOP, Bristol, 2003, and BEYOND 2003, Ringberg Castle, Tegernsee, Germany, 9-14 Juni 2003, Springer, Heidelberg, Germany, 2004, ed. H.V. Klapdor-Kleingrothaus.
92. R.N. Mohapatra, M.K. Parida and G. Rajasekaran, (2003) *hep-ph/***0301234**.
93. D. Fargion et al., in Proc. of DARK2000, Heidelberg, Germany, July 10-15, 2000, Ed. H.V. Klapdor-Kleingrothaus, *Springer*, (2001) 455-468 and in Proc. of Beyond the Desert 2002, BEYOND02, Oulu, Finland, June 2002, IOP 2003, and BEYOND03, Ringberg Castle, Tegernsee, Germany, 9-14 Juni 2003, Springer, Heidelberg, Germany, 2003, ed. H.V. Klapdor-Kleingrothaus.
94. Z. Fodor, S.D. Katz, A. Ringwald, *Phys. Rev. Lett.***88** (2002) 171101 and Z. Fodor et al., *JHEP* (2002) 0206:046 or *hep-ph/***0203198**, and in Proc. of Intern. Conf. Beyond the Desert 02, BEYOND'02, Oulu, Finland, 2-7 Jun 2002, IOP, Bristol, 2003, ed. H V Klapdor-Kleingrothaus and *hep-ph/***0210123**.
95. H. V. Klapdor-Kleingrothaus and U. Sarkar, *Mod. Phys. Letter.* **A18** (2003) 2243-2254.
96. H. Minakata and O. Yasuda, *Phys. Rev.* **D 56** (1997) 1692.
97. O. Yasuda, in Proc. of Int. Conf. BEYOND'99, Ringberg Castle, Germany, June 6-12, 1999, IOP, Bristol (2000), eds. H.V. Klapdor-Kleingrothaus and I.V. Krivosheina, p.223.
98. M.N. Rebelo, Proc. of BEYOND'2003, Castle Ringberg, Germany, July 2003, ed. H.V. Klapdor-Kleingrothaus, Springer, Heidelberg (2004) 267.
99. K.S. Babu, E. Ma and J.W.F. Valle, *Phys. Lett.* **B 552** (2003) 207-213.
100. M. Hirsch, J. C. Romao, S. Skadhauge, J. W. F. Valle and A. Villanova del Moral *Phys. Rev.* **D69** (2004) 093006.
101. A. Blanchard, M. Douspis, M. Rowan-Robinson and S. Sarkar, *astro-ph/***0304237**.
102. A. Faessler et al., *Phys. Rev.* **D79** (2009) 053001 and *hep-ph/* **0810.5733**, March, 2009.
103. J. Engel and P. Vogel *Phys. Rev.* **C69** (2004) 034304 and nucl-th/0311072.
104. T. Tomoda and A. Faessler *Phys. Lett.* **B191** (1987) pp. 475-481.
105. S.M. Bilenky and J. A. Grifols, *Phys. Lett.* **B550** (2002) 154 and hep-

ph/0211101; S.M. Bilenky and S. T. Petcov, hep-ph/0405237; S.M. Bilenky, Proc. 11 Intern Work. on Neutrino telescopes, Venezia, Italy, Febr. 22-25, 2005, ed. M. Baldo-Ceolin (2005) page 247-256.
106. S. Stoica and H.V. Klapdor-Kleingrothaus *Eur. Phys. J.* **A9** (2000) pp. 345-352, and *nucl-th/*0010106.
107. S. Stoica and H.V. Klapdor-Kleingrothaus *Phys. Rev.* **C63** (2001) pp. 064304-1–6.
108. F. Simkovic et al., *Phys. Rev.* **C79** (2009) pp. 055501.
109. A. Faessler, in Proc. of Neutrino Oscillation Worksh., Otranto, Italy, 2004, eds. G.L. Fogli et al., *Nucl. Phys. Proc. Suppl.* **145** (2005) pp. 213-218.
110. M. Hirsch, K. Muto, T. Oda and H.V. Klapdor-Kleingrothaus *Z. Phys.* **A347** (1994) pp. 151-160.
111. H.V. Klapdor-Kleingrothaus, A. Dietz and I.V. Krivosheina *Phys. Rev.* **D70** (2004) pp. 078301-1–4, *hep-ph/*0403056.
112. V.A. Rodin et al., *Nucl. Phys.* **A766** (2006) 107-131, *Erratum-ibid.* **A793** (2007) 213-215, *nucl-th* **0706.4304** and *nucl-th* **0503063**.
113. P.K. Rath, R. Chandra, K. Chaturvedi, P.K. Raina, J.G. Hirsch *nucl-th* **0906.4476**; K. Chaturvedi et al., *Phys. Rev.* **C78** (2008) 054302., and earlier papers.
114. V. Rodin and A. Faessler, *nucl-th* **0906.1759 v.1**, 9 June 2009.
115. R.N. Mohapatra, M.K. Parida, G. Rajasekaran *Phys. Rev.* **D69** (2004) 053007-1–8, and *hep-ph/*0301234.
116. A. Pierce and H. Murayama *Phys. Lett.* **B581** (2004) 218-223. *hep-ph/*0302131; S. Hannestad, *astro-ph/***0303076**; P. Vogel in Particle Data Group (ed. K. Hagiwara et al.) *Phys. Rev* **D66** (2002) 010001.
117. [Bro05] C. Brofferio et al., (CUORICINO and CUORE Collaborations) in Proc. 11th International Workshop on Neutrino Telescopes, Venice, Italy, 22-25 Feb 2005, *ed* M. Baldo-Ceolin, Padova, Papergraf (2005) pp. 239-245.
118. F. Simkovic in *http://www1.jinr.ru/News/News_1_2009.pdf*, page 20.
119. C. Arnaboldi et al. (CUORICINO Collaboration, *Phys. Rev.* **C78** (2008) 035502, *hep-ex* **0802.3439**.
120. R. Hofmann, *hep-ph/***0401017 v.1**.
121. T. Goldman, G.J. Stephenson Jr., P.M. Alsing and B.H.J. McKellar, in Proc. of DARK2009, Christchurch, New Zealand, January 15-23, 2009, World Scientific (2009), eds. H.V. Klapdor-Kleingrothaus and I.V. Krivosheina.
122. G. Karagiorgi et al., *hep-ph/***0906.1997v1**; C. Athanassopoulos et al. (LSND Collaboration), *Phys. Rev. Lett.* **77** (1996) 3082 and *nucl-ex/***9605003**; C. Athanassopoulos et al. (LSND Collaboration), *Phys. Rev.* **C 58** (1998) 2489 and *nucl-ex/***9706006**; A. Aguilar et al. (LSND Collaboration), *Phys. Rev.* **D 64** (2001) 112007 *hep-ex/***0104049**.
123. H. V. Klapdor-Kleingrothaus, I. V. Krivosheina and I. V. Titkova, to be published 2009.

NEUTRINO MASS, DARK MATTER AND BARYON ASYMMETRY VIA TEV-SCALE PHYSICS

MAYUMI AOKI [1*], SHINYA KANEMURA [2], and OSAMU SETO [3]

[1] *Department of Physics, Tohoku University, Sendai 980-8578, Japan*
[2] *Department of Physics, University of Toyama, 3190 Gofuku, Toyama 930-8555, Japan*
[3] *William I. Fine Theoretical Physics Institute, University of Minnesota, Minneapolis, MN 55455, USA*
[*] *E-mail: mayumi@tuhep.phys.tohoku.ac.jp*

We discuss a TeV-scale model, in which neutrino masses and mixings, dark matter, and baryon asymmetry of the Universe can be simultaneously explained without assuming large hierarchy among the mass scales. Imposing the exact Z_2 symmetry, tiny neutrino masses are generated at the three loop level and stability of the dark matter candidate is guaranteed. Moreover the necessary conditions for the electroweak baryogenesis are satisfied via the extended Higgs sector. The model provides a distinctive experimental signature especially in Higgs phenomenology and dark matter physics.

1. Introduction

There are at present definite reasons to consider a model beyond the standard model (SM). First of all, it is established from a variety of experiments that neutrinos have small masses below the eV scale and mix among themselves.[1] It would be a natural picture for the origin of neutrino masses that new particles are introduced to form the well-known dimension-five operator for Majorana neutrino masses, $(f_{ij}/\Lambda)(\nu_i\nu_j\phi^0\phi^0)$, where Λ is the cut off scale, $L^i = (\nu_i, l_i)$ is the i-th generation left-handed lepton doublet and ϕ^0 is the SM(-like) Higgs boson. Second, recent cosmological observations indicate that 23% of the present energy density consists of the dark matter (DM).[2] To obtain the candidate of the DM, we need in general some new particles which are odd under an exact discrete symmetry. Finally, asymmetry of matter and anti-matter in our Universe has been measured as

$(n_B - n_{\overline{B}})/n_\gamma \sim 6.1 \times 10^{-10}$,[3] where n_B ($n_{\overline{B}}$) is the (anti-)baryon density and n_γ is the primordial photon density. The asymmetry may be understood within the framework of physics at the electroweak scale, since all the necessary ingredients for the baryogenesis[4] could be available there. Although the SM cannot account for the asymmetry quantitatively, certain extensions of the Higgs sector would be able to overcome the difficulties which the SM encounters. Explanation of all of above the phenomena, tiny neutrino mass and existence of DM and baryon asymmetry of the Universe, is an extremely important issue for the particle physics.

One of the simplest scenarios to generate tiny neutrino masses would be via the seesaw mechanism.[5] Introducing the right-handed (RH) neutrinos, which are gauge singlets under the SM, with a large Majorana mass M_R to the SM, a small mass for the left-handed neutrinos would be obtained as $\sim m_D^2/M_R$, where m_D is the Dirac mass of the electroweak scale.This leads the neutrino scale in natural if $M_R \sim 10^{13-16}$ GeV. This scenario would be compatible with the framework with large mass scales like grand unification. However, such a large scale causes a problem of hierarchy and is far from experimental reach.

On the other hand, another attractive possibility for generating neutrino masses has been proposed, where neutrino masses are explained at TeV-scale physics via radiative effects. Original idea of such approach was given by Zee.[6] A Majorana mass for the light neutrinos arises at the one-loop level in his model. A new interesting model with a TeV-scale RH neutrino was proposed by Krauss, Nasri and Trodden,[7] where the neutrino masses are generated at the three-loop level due to the exact discrete Z_2 symmetry, and the Z_2-odd RH neutrino is a candidate of DM. This has been extended with two RH neutrinos to describe the neutrino data.[8] Several models with adding baryogenesis have been considered in Ref.[9]

In this talk, we discuss a new TeV-scale model which would explain neutrino masses and mixings, origin of DM and baryon asymmetry of the Universe *simultaneously* by an extended Higgs sector with RH neutrinos.[10] In order to avoid large hierarchy, masses of the RH neutrinos are to be at most TeV scales. Tiny neutrino masses are then generated at the three loop level due to an exact discrete Z_2 symmetry, by which tree-level Yukawa couplings of neutrinos are prohibited. The lightest neutral odd state under the discrete symmetry is a candidate of DM. Baryon asymmetry can be generated at the electroweak phase transition (EWPT) by additional CP violating phases in the Higgs sector.[11,12] In this framework, a successful model can be built without contradiction of the current data.

Table 1. Particle properties under the discrete symmetries.

	Q^i	u_R^i	d_R^i	L^i	e_R^i	Φ_1	Φ_2	$S^\pm$	η	N_R^α
Z_2 (exact)	+	+	+	+	+	+	+	−	−	−
$\widetilde{Z}_2$ (softly broken)	+	−	−	+	+	+	−	+	−	+

2. Model

In addition to the *known* SM fields, particle entries are two scalar isospin doublets with hypercharge 1/2 (Φ_1 and Φ_2), charged singlets ($S^\pm$), a real scalar singlet (η) and two generation isospin-singlet RH neutrinos (N_R^α with $\alpha = 1, 2$). In order to generate tiny neutrino masses at the three-loop level, we impose an exact Z_2 symmetry, which we refer to as Z_2. We assign the Z_2 odd charge to $S^\pm$, η and N_R^α, while ordinary gauge fields, quarks and leptons and Higgs doublets are Z_2 even. Introduction of two Higgs doublets would cause a dangerous flavor changing neutral current. To avoid this in a natural way, we impose another discrete symmetry ($\widetilde{Z}_2$) that is softly broken.[13,14] From a phenomenological reason discussed later, we assign $\widetilde{Z}_2$ charges such that only Φ_1 couples to leptons whereas Φ_2 does to quarks;

$$\mathcal{L}_Y = -y_{e_i} \overline{L}^i \Phi_1 e_R^i - y_{u_i} \overline{Q}^i \tilde{\Phi}_2 u_R^i - y_{d_i} \overline{Q}^i \Phi_2 d_R^i + \text{h.c.}, \tag{1}$$

where Q^i is the ordinary i-th generation left-handed quark doublet, and u_R^i and d_R^i (e_R^i) are RH-singlet up- and down-type quarks (charged leptons), respectively. We summarize the particle properties under Z_2 and $\widetilde{Z}_2$ in Table 1. Notice that the Yukawa coupling in Eq. (1) is different from that in the minimal supersymmetric SM (MSSM).[15] Consequently, the Higgs boson phenomenology in our model shows discriminative features as compared to that in the MSSM.[14,16,17] The scalar potential is given by

$$\begin{aligned} V = &\sum_{a=1}^{2} \left(-\mu_a^2 |\Phi_a|^2 + \lambda_a |\Phi_a|^4\right) - \left(\mu_{12}^2 \Phi_1^\dagger \Phi_2 + \text{h.c.}\right) \\ &+\lambda_3 |\Phi_1|^2 |\Phi_2|^2 + \lambda_4 |\Phi_1^\dagger \Phi_2|^2 + \left\{\frac{\lambda_5}{2} (\Phi_1^\dagger \Phi_2)^2 + \text{h.c.}\right\} \\ &+\sum_{a=1}^{2} \left(\rho_a |\Phi_a|^2 |S|^2 + \sigma_a |\Phi_a|^2 \frac{\eta^2}{2}\right) + \sum_{a,b=1}^{2} \left\{\kappa \, \epsilon_{ab} (\Phi_a^c)^\dagger \Phi_b S^- \eta + \text{h.c.}\right\} \\ &+\mu_s^2 |S|^2 + \lambda_s |S|^4 + \mu_\eta^2 \eta^2/2 + \lambda_\eta \eta^4 + \xi |S|^2 \eta^2/2, \end{aligned} \tag{2}$$

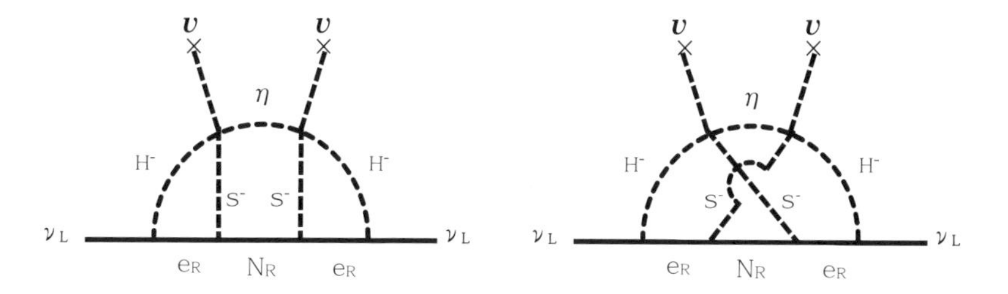

Fig. 1. Three-loop diagrams for tiny neutrino masses.

where ϵ_{ab} is the anti-symmetric tensor with $\epsilon_{12} = 1$. The mass term and the interaction for N_R^α are given by

$$\mathcal{L}_Y = \sum_{\alpha=1}^{2} \left\{ \frac{1}{2} m_{N_R^\alpha} \overline{N_R^{\alpha c}} N_R^\alpha - h_i^\alpha \overline{(e_R^i)^c} N_R^\alpha S^- + \text{h.c.} \right\}. \tag{3}$$

In general, μ_{12}^2, λ_5 and κ (as well as h_i^α) can be complex. The phases of λ_5 and κ can be eliminated by rephasing $S^\pm$ and Φ_1. The remaining phase of μ_{12}^2 causes CP violation in the Higgs sector. Although the phase is crucial for successful baryogenesis at the EWPT,[11] it does not much affect the following discussions. Thus, we neglect it for simplicity. We later give a comment on the case with the non-zero CP-violating phase.

As Z_2 is exact, the even and odd fields cannot mix. Mass matrices for the Z_2 even scalars are diagonalized as in the usual two Higgs doublet model (THDM) by the mixing angles α and β, where α diagonalizes the CP-even states, and $\tan\beta = \langle \Phi_2^0 \rangle / \langle \Phi_1^0 \rangle$.[15] The Z_2 even physical states are two CP-even (h and H), a CP-odd (A) and charged ($H^\pm$) states. We here define h and H such that h is always the SM-like Higgs boson when $\sin(\beta - \alpha) = 1$.

3. Neutrino Mass

The left-handed neutrino mass matrix M_{ij} is generated by the three-loop diagrams in Fig. 1. The absence of lower order loop contributions is guaranteed by Z_2. $H^\pm$ and e_R^i play a crucial role to connect left-handed neutrinos with the one-loop sub-diagram by the Z_2-odd states. We obtain

$$M_{ij} = \sum_{\alpha=1}^{2} C_{ij}^\alpha F(m_{H^\pm}, m_{S^\pm}, m_{N_R^\alpha}, m_\eta), \tag{4}$$

where $C_{ij}^{\alpha} = 4\kappa^2 \tan^2\beta (y_{e_i}^{\rm SM} h_i^{\alpha})(y_{e_j}^{\rm SM} h_j^{\alpha})$ and

$$F(m_{H^\pm}, m_{S^\pm}, m_{N_R}, m_\eta) = \left(\frac{1}{16\pi^2}\right)^3 \frac{(-m_{N_R} v^2)}{m_{N_R}^2 - m_\eta^2}$$
$$\times \int_0^\infty dx \left[x \left\{ \frac{B_1(-x, m_{H^\pm}, m_{S^\pm}) - B_1(-x, 0, m_{S^\pm})}{m_{H^\pm}^2} \right\}^2 \right.$$
$$\left. \times \left(\frac{m_{N_R}^2}{x + m_{N_R}^2} - \frac{m_\eta^2}{x + m_\eta^2} \right) \right], \quad (m_{S^\pm}^2 \gg m_{e_i}^2), \tag{5}$$

with m_f representing the mass of the field f, $y_{e_i}^{\rm SM} = \sqrt{2} m_{e_i}/v$, $v \simeq 246$ GeV and B_1 being the tensor coefficient function in Ref.[18] Magnitudes of $\kappa \tan\beta$ as well as F determine the universal scale of M_{ij}, whereas variation of h_i^{α} ($i = e, \mu, \tau$) reproduces the mixing pattern indicated by the neutrino data.[1] M_{ij} is related to the data by $M_{ij} = U_{is}(M_\nu^{\rm diag})_{st}(U^T)_{tj}$, where U_{is} is the unitary matrix and $M_\nu^{\rm diag} = {\rm diag}(m_1, m_2, m_3)$. Under the *natural* requirement $h_e^{\alpha} \sim \mathcal{O}(1)$, and taking the $\mu \to e\gamma$ search results into account,[19] we find that $m_{N_R^\alpha} \sim \mathcal{O}(1)$ TeV, $m_{H^\pm} \lesssim \mathcal{O}(100)$ GeV, $\kappa \tan\beta \gtrsim \mathcal{O}(10)$, and $m_{S^\pm}$ being several times 100 GeV. On the other hand, the LEP direct search results indicate $m_{H^\pm}$ (and $m_{S^\pm}$) $\gtrsim 100$ GeV.[1] In addition, with the LEP precision measurement for the ρ parameter, possible values uniquely turn out to be $m_{H^\pm} \simeq m_H$ (or m_A) $\simeq 100$ GeV for $\sin(\beta - \alpha) \simeq 1$. It is notable that such a light $H^\pm$ is not excluded by the $b \to s\gamma$ data,[20] thanks to the Yukawa coupling in Eq. (1).[14,17] Since we cannot avoid to include the hierarchy among $y_i^{\rm SM}$, we only require $h_i^{\alpha} y_i \sim \mathcal{O}(y_e) \sim 10^{-5}$ for values of h_i^{α}. Several sets for h_i^{α} are shown in Table 2 with the predictions on the branching ratio of $\mu \to e\gamma$ assuming the normal hierarchy, $m_1 \simeq m_2 \ll m_3$ with $m_1 = 0$. For the inverted hierarchy ($m_3 \ll m_1 \simeq m_2$ with $m_3 = 0$), $\kappa \tan\beta$ is required to be larger. Our model turns out to prefer the normal hierarchy scenario.[21]

Table 2. Values of h_i^{α} for $m_{H^\pm}(m_{S^\pm}) = 100(400)$GeV $m_\eta = 50$ GeV, $m_{N_R^1} = m_{N_R^2} = 3.0$ TeV for the normal hierarchy. For Set A (B), $\kappa \tan\beta = 28$ (32) and $U_{e3} = 0$ (0.18). Predictions on the branching ratio of $\mu \to e\gamma$ are also shown.

Set	h_e^1	h_e^2	h_μ^1	h_μ^2	h_τ^1	h_τ^2	$B(\mu \to e\gamma)$
A	2.0	2.0	-0.019	0.042	-0.0025	0.0012	6.9×10^{-12}
B	2.2	2.2	0.0085	0.038	-0.0012	0.0021	6.1×10^{-12}

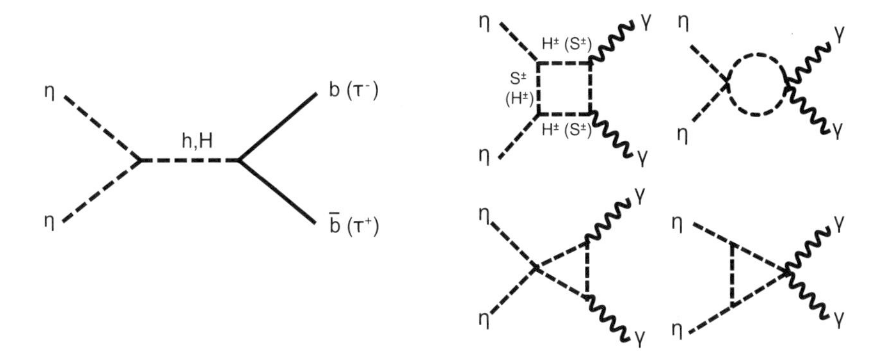

Fig. 2. Feynman diagrams for pair annihilation of dark matter into $b\bar{b}$ and $\tau^+\tau^-$ through s-channel neutral Higgs bosons (h and H) exchange diagrams and into $\gamma\gamma$ through one-loop diagrams by $H^\pm$ and $S^\pm$.

4. Cold Dark Matter

The lightest Z_2-odd particle is stable and can be a candidate of DM if it is neutral. In our model, N_R^α must be heavy, so that the DM candidate is identified as η. When η is lighter than the W boson, η dominantly annihilates into $b\bar{b}$ and $\tau^+\tau^-$ via tree-level s-channel Higgs (h and H) exchange diagrams, and into $\gamma\gamma$ via one-loop diagrams. From their summed thermal averaged annihilation rate $\langle\sigma v\rangle$, the relic mass density $\Omega_\eta h^2$ is evaluated as

$$\Omega_\eta h^2 = 1.1 \times 10^9 \frac{(m_\eta/T_d)}{\sqrt{g_*} M_P \langle\sigma v\rangle} \ \text{GeV}^{-1}, \tag{6}$$

where M_P is the Planck scale, g_* is the total number of relativistic degrees of freedom in the thermal bath, and T_d is the decoupling temperature.[22] Fig. 3 shows $\Omega_\eta h^2$ as a function of m_η. Strong annihilation can be seen near 50 GeV $\simeq m_H/2$ (60 GeV $\simeq m_h/2$) due to the resonance of H (h) mediation. The data[2] ($\Omega_{\text{DM}} h^2 \simeq 0.11$) indicate that m_η is around 40-65 GeV.

5. Baryon Asymmetry

The model satisfies the necessary conditions for baryogenesis.[4] Especially, departure from thermal equilibrium can be realized by the strong first order

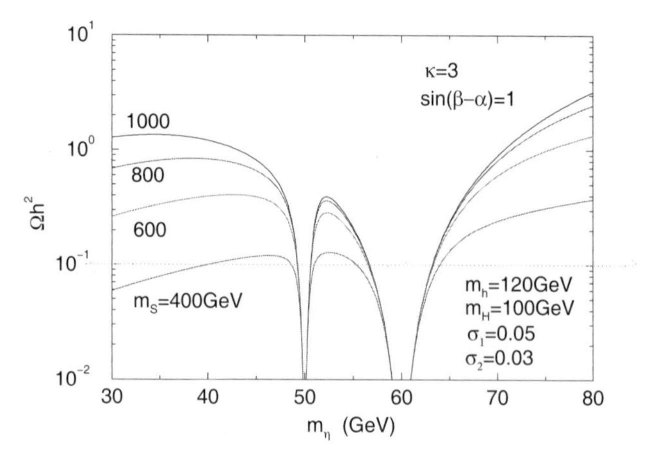

Fig. 3. The relic abundance of η.

EWPT. The free energy is given at a high temperature T by[23]

$$V_{eff}[\varphi, T] = D(T^2 - T_0^2)\varphi^2 - ET\varphi^3 + \frac{\lambda_T}{4}\varphi^4 + ..., \tag{7}$$

where φ is the order parameter, and

$$E \simeq \frac{1}{12\pi v^3}(6m_W^3 + 3m_Z^3 + m_A^3 + 2m_{S^\pm}^3), \tag{8}$$

with $D \simeq (6m_W^2 + 3m_Z^2 + 6m_t^2 + m_A^2 + 2m_{S^\pm}^2)/(24v^2)$, $T_0^2 \sim m_h^2/(4D)$ and $\lambda_T \sim m_h^2/(2v^2)$. A large value of the coefficient E is crucial for the strong first order EWPT.[12] In Eq. (8), quantum effects by h, H and $H^\pm$ are neglected since they are unimportant for $\sin(\beta - \alpha) \simeq 1$ and $m_{H^\pm} \simeq m_H \simeq M$ ($\equiv \sqrt{2\mu_{12}^2/\sin 2\beta}$) [the soft $\tilde{Z}_2$ breaking scale[24]]. For sufficient sphaleron decoupling in the broken phase, it is required that[25]

$$\frac{\varphi_c}{T_c}\left(\simeq \frac{2E}{\lambda_{T_c}}\right) \gtrsim 1, \tag{9}$$

where φ_c ($\neq 0$) and T_c are the critical values of φ and T at the EWPT. In Fig. 4, the allowed region under the condition of Eq. (9) is shown. The condition is satisfied when $m_{S^\pm} \gtrsim 350$ GeV for $m_A \gtrsim 100$ GeV, $m_h \simeq 120$ GeV, $m_H \simeq m_{H^\pm}(\simeq M) \simeq 100$ GeV, $\mu_S \simeq 200$ GeV and $\sin(\beta - \alpha) \simeq 1$. Unitarity bounds are also satisfied unless m_A (m_S) is too larger than M (μ_S).[12,26]

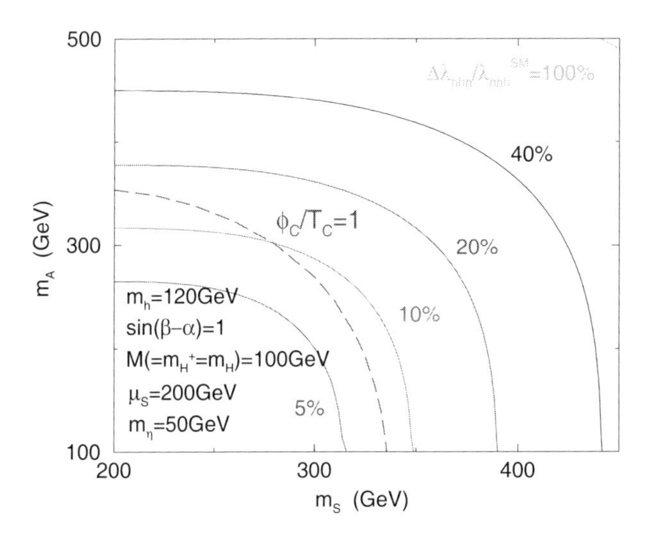

Fig. 4. The region of strong first order EWPT. Deviations from the SM value in the hhh coupling are also shown.

6. Phenomenology

A successful scenario which can simultaneously solve the above three issues under the data[1,19,20] would be

$$\begin{array}{ll}\sin(\beta-\alpha)\simeq 1, & \kappa\tan\beta\simeq 30, \\ m_h = 120\text{ GeV}, & m_H\simeq m_{H^\pm}(\simeq M)\simeq 100\text{ GeV}, \\ m_A \gtrsim 100\text{ GeV}, & m_{S^\pm}\sim 400\text{ GeV}, \\ m_\eta\simeq 40-65\text{ GeV}, & m_{N_R^1}\simeq m_{N_R^2}\simeq 3\text{ TeV}.\end{array} \qquad (10)$$

This is realized without assuming unnatural hierarchy among the couplings. All the masses are between $\mathcal{O}(100)$ GeV and $\mathcal{O}(1)$ TeV. As they are required by the data, the model has a predictive power. We note that the masses of A and H can be exchanged with each other.

We outline phenomenological predictions in the scenario in (10) in order. The detailed analysis is shown elsewhere.[26] (I) h is the SM-like Higgs boson, but decays into $\eta\eta$ when $m_\eta < m_h/2$. The branching ratio is about 36% (25%) for $m_\eta \simeq 45$ (55) GeV. This is related to the DM abundance, so that our DM scenario is testable at the CERN Large Hadron Collider (LHC). (II) η is potentially detectable by direct DM searches,[27] because η can scatter with nuclei via the scalar exchange.[28] (III) For successful baryogenesis, the hhh coupling has to deviate from the SM value by more than 10-20 % (see Fig. 4),[12] which can be tested at the International Linear Collider (ILC).[29] (IV) H (or A) can predominantly decay into $\tau^+\tau^-$ instead of $b\bar{b}$ for $\tan\beta \gtrsim 3$.[17] When A (or H) is relatively heavy it can decay into $H^\pm W^\mp$ and HZ (or AZ). (V) the scenario with light $H^\pm$ and H (or A) can be directly tested at the LHC via $pp \to W^* \to HH^\pm$ and $AH^\pm$.[30] (VI) $S^\pm$ can be produced in pair at the LHC (the ILC),[31] and decay into $\tau^\pm \nu\eta$. The signal would be a hard hadron pair[32] with a large missing energy. (VII) The couplings h_i^α cause lepton flavor violation such as $\mu \to e\gamma$ which would provide information on $m_{N_R^\alpha}$ at future experiments.

Finally, we comment on the case with the CP violating phases. Our model includes the THDM, so that the same discussion can be applied in evaluation of baryon number at the EWPT.[11] The mass spectrum would be changed to some extent, but most of the features discussed above should be conserved with a little modification.

7. Summary

We have proposed an explicit model with the extended Higgs sector and TeV-scale right handed neutrinos, which would simultaneously explain neutrino masses and mixing, dark matter, and baryon asymmetry of the Universe via the TeV scale physics. As a result of the imposed exact Z_2 symmetry, neutrino masses are generated at the three loop level and the lightest Z_2-odd particle (a real scalar singlet) becomes the candidate of the dark matter. Baryon asymmetry of the Universe would be explained within the framework of the electroweak baryogenesis scenario due to the modification of the Higgs potential of the standard model. Our model gives specific predictions in Higgs phenomenology, dark matter physics and flavor physics, so that it is testable at current and future experiments.

References

1. W. M. Yao, et al., J. Phys. G **33** (2006) 1.
2. E. Komatsu, et al., arXiv:0803.0547 [astro-ph].
3. Particle Data Group, C. Amsler et al., Phys. Lett. B **667**, 1 (2008)
4. A. D. Sakharov, Pisma Zh. Eksp. Teor. Fiz. **5**, 32 (1967).
5. T. Yanagida, in Proceedings of Workshop on *the Unified Theory and the Baryon Number in the Universe*, p.95 KEK Tsukuba, Japan (1979); M. Gell-Mann, P. Ramond and R. Slansky, in Proceedings of Workshop *Supergravity*, p.315, Stony Brook, New York, 1979.
6. A. Zee, Phys. Lett. B **93**, 389 (1980) [Erratum-ibid. B **95**, 461 (1980)]; A. Zee, Phys. Lett. B **161**, 141 (1985).
7. L. M. Krauss, S. Nasri and M. Trodden, Phys. Rev. D **67**, 085002 (2003).
8. K. Cheung and O. Seto, Phys. Rev. D **69**, 113009 (2004).

9. E. Ma, Phys. Rev. D **73**, 077301 (2006); J. Kubo, E. Ma and D. Suematsu, Phys. Lett. B **642**, 18 (2006); T. Hambye, et al., Phys. Rev. D **75**, 095003 (2007); K. S. Babu and E. Ma, arXiv:0708.3790 [hep-ph]; N. Sahu and U. Sarkar, arXiv:0804.2072 [hep-ph].
10. M. Aoki, S. Kanemura, O. Seto, Phys. Rev. Lett. **102**, 051805 (2009).
11. J. M. Cline, K. Kainulainen and A. P. Vischer, Phys. Rev. D **54**, 2451 (1996); L. Fromme, S. J. Huber and M. Seniuch, JHEP **0611**, 038 (2006).
12. S. Kanemura, Y. Okada and E. Senaha, Phys. Lett. B **606**, 361 (2005).
13. S. L. Glashow and S. Weinberg, Phys. Rev. D **15**, 1958 (1977);
14. V. D. Barger, J. L. Hewett and R. J. N. Phillips, Phys. Rev. D **41**, 3421 (1990).
15. J. F. Gunion, et al., "*The Higgs Hunters's Guide*" (Addison Wesley, 1990).
16. Y. Grossman, Nucl. Phys. B **426**, 355 (1994).
17. M. Aoki, S. Kanemura, K. Tsumura and K. Yagyu, arXiv:0902.4665 [hep-ph].
18. G. Passarino and M. J. G. Veltman, Nucl. Phys. B **160**, 151 (1979).
19. A. Baldini, Nucl. Phys. Proc. Suppl. **168**, 334 (2007).
20. E. Barberio *et al.* [Heavy Flavor Averaging Group], arXiv:0808.1297 [hep-ex].
21. The unitarity bound for the scattering process $h\eta \to H^+S^-$ turns out to give $\kappa \lesssim 25$, whereas $\tan\beta$ should not be too large for successful electroweak baryogenesis.[11]
22. E. W. Kolb and M. S. Turner, *The Early Universe* (Addison-Wesley, 1990).
23. G. W. Anderson and L. J. Hall, Phys. Rev. D **45**, 2685 (1992); M. Dine, et al., Phys. Rev. D **46**, 550 (1992).
24. S. Kanemura, et al., Phys. Lett. B **558**, 157 (2003); Phys. Rev. D **70**, 115002 (2004).
25. G. D. Moore, Phys. Lett. B **439**, 357 (1998); Phys. Rev. D **59**, 014503 (1998).
26. M. Aoki, S. Kanemura and O. Seto, in preparation.
27. Y. D. Kim, Phys. Atom. Nucl. **69**, 1970 (2006); D. S. Akerib, et al., Phys. Rev. Lett. **96**, 011302 (2006).
28. J. McDonald, Phys. Rev. D **50**, 3637 (1994); for a recent study, see *e.g.*, H. Sung Cheon, S. K. Kang and C. S. Kim, J. Cosmol. Astropart. Phys. 05 (2008) 004.
29. M. Battaglia, E. Boos and W. M. Yao, arXiv:hep-ph/0111276; Y. Yasui, et al., arXiv:hep-ph/0211047.
30. S. Kanemura and C. P. Yuan, Phys. Lett. B **530**, 188 (2002); Q. H. Cao, S. Kanemura and C. P. Yuan, Phys. Rev. D **69**, 075008 (2004).
31. S. Kanemura, et al., Phys. Rev. D **64**, 053007 (2001).
32. B. K. Bullock, K. Hagiwara and A. D. Martin, Phys. Rev. Lett. **67**, 3055 (1991).

A POSSIBLE CONNECTION BETWEEN MASSIVE FERMIONS AND DARK ENERGY

T. GOLDMAN

Theoretical Division, Los Alamos National Laboratory,
Los Alamos, New Mexico, 87545, USA
E-mail: tgoldman@lanl.gov

G. J. STEPHENSON, JR.* and P. M. ALSING†

Dept. of Physics & Astronomy, University of New Mexico,
Albuquerque, New Mexico 87131, USA
**E-mail: gjs@phys.unm.edu* †*E-mail: alsingpm@hotmail.com*

B. H. J. MCKELLAR

School of Physics, University of Melbourne,
Victoria 3010, Australia
E-mail: bhjm@unimelb.edu.au

In a dense cloud of massive fermions interacting by exchange of a light scalar field, the effective mass of the fermion can become negligibly small. As the cloud expands, the effective mass and the total energy density eventually increase with decreasing density. In this regime, the pressure-density relation can approximate that required for dark energy. We apply this phenomenon to the expansion of the Universe with a very light scalar field and infer relations between the parameters available and cosmological observations. Majorana neutrinos at a mass that may have been recently determined, and fermions such as the Lightest Supersymmetric Particle (LSP) may both be consistent with current observations of dark energy.

Keywords: dark energy; dark matter; massive fermions.

1. Introduction

Several years ago it was suggested that neutrinos might interact weakly among themselves through the exchange of a very light scalar particle,[1,2] with possible consequences for the evolution of the Universe and for the propagation of neutrinos from distant events. We examined such a system for scalars with astrophysical ranges to explore the possibility of neutrino clustering[3] and noted at the time that the neutrino clouds thus formed could seed structure formation in the Early Universe. More generally, in such clouds of massive fermions interacting by exchange of a light scalar field, the effective mass of the fermion can become negligibly small. We found that, as a consequence, when the cloud expands, the effective mass and the total energy density must eventually increase with decreasing density. We studied this system in 1996,[3] well before the discovery of Dark Energy, in connection with experimental problems encountered in the search for the mass of the (electron) neutrino. Those anomalies have since disappeared, but provided us with the technology to describe dark energy in a well understood dynamical system.

In the following, we first review our previous work on the theory of massive fermions interacting via exchange of a scalar field. This is carried out with scaled variables so the regime of applicability is not constrained. We next recall the relation between Dark Energy and equations of state and define the w parameter used therein. After this, we apply our model results for w and discuss the numerical, analytical and scaling properties relevant to the accuracy of our results. Penultimately, we extract a rough mass value from applying those results to describe Dark Energy assuming the currently accepted value for its energy density in the epoch corresponding to $z = 1$. Finally, we present our conclusions and discuss some open questions.

2. Summary of a Theory of Massive Fermions Interacting via Light Scalar Field Exchange

The effective Lagrangian for a Dirac field, ψ, interacting with a scalar field, ϕ, is:

$$\mathcal{L} = \bar{\psi}(i\partial\!\!\!/ - m_\nu^{(0)})\psi + \frac{1}{2}\left[\phi(\partial^2 - m_s^2)\phi\right] + g\bar{\psi}\psi\phi \tag{1}$$

which gives as the equations of motion

$$\left[\partial^2 + m_s^2\right]\phi = g\bar{\psi}\psi \tag{2}$$

$$\left[i\partial\!\!\!/ - m_\nu^{(0)}\right]\psi = -g\phi\psi. \tag{3}$$

As usual, we set $\hbar = c = 1$. We have omitted nonlinear scalar selfcouplings here, even though they are required to exist by field theoretic selfconsistency,[4] as they may consistently be assumed to be sufficiently weak as to be totally irrelevant. The parameter $m_f^{(0)}$ is the renormalized vacuum mass that the fermion would have in isolation, and takes into account any contributions from all other interactions, as well as contributions from the vacuum expectation value of the new scalar field, ϕ.

We look for solutions of these equations in infinite matter which are static and translationally invariant. Eq.(2) then gives

$$\phi = \frac{g}{m_s^2}\bar{\psi}\psi, \tag{4}$$

which, when substituted in Eq.(3) gives an effective mass for the fermion of

$$m_f^* = m_f^{(0)} - \frac{g^2}{m_s^2}\bar{\psi}\psi. \tag{5}$$

These equations are operator equations. We next act with each of these equations on a state $|\Omega\rangle$ defined as a filled Fermi sea, with a number density n per fermion state, and Fermi momentum k_F, related as usual by $n = {k_F}^3/(6\pi^2)$. The operator $\bar{\psi}\psi$ acting on this state gives

$$\bar{\psi}\psi|\Omega\rangle = \frac{\zeta}{(2\pi)^3}\int_{|\vec{k}|<k_F} d^3k \left| \frac{m_f^*}{\sqrt{(m_f^*)^2 + k^2}} \right| |\Omega\rangle, \tag{6}$$

where ζ is the number of fermion states which contribute — $\zeta = 2$ for Majorana fermions and $\zeta = 4$ for Dirac fermions. Thus the effective mass is determined by an integral equation

$$m_f^* = m_f^{(0)} - \frac{g^2\zeta}{2\pi^2 m_s^2}\int_0^{k_F} k^2\, dk \frac{m_f^*}{\sqrt{(m_f^*)^2 + k^2.}} \tag{7}$$

To discuss the solutions of this equation, we reduce it to dimensionless form, dividing by $m_f^{(0)}$, and introducing the parameter

$$K_0 \equiv \zeta\frac{g^2(m_f^{(0)})^2}{2\pi^2 m_s^2}, \tag{8}$$

and the variables $y = \frac{m_f^*}{m_f^{(0)}}, x = \frac{k}{m_f^{(0)}}, x_F = \frac{k_F}{m_f^{(0)}}$. Then Eq.(7) becomes

$$y = 1 - yK_0\int_0^{x_F}\frac{x^2\, dx}{\sqrt{y^2 + x^2}} \tag{9}$$

$$= 1 - \frac{yK_0}{2}\left[e_F x_F - y^2 \ln\left(\frac{e_F + x_F}{y}\right)\right], \tag{10}$$

with $e_F \equiv \sqrt{x_F^2 + y^2}$. This choice of scaled variables gives all energies (and momenta) in units of the vacuum fermion mass. For consistency, we define the dimensionless scalar mass as $\mu = \frac{m_s}{m_f^{(0)}}$ in these same units. One can regard Eq.(10) as a non-linear equation for y as a function of either e_F or x_F. As a function of e_F, y is multiple valued (when a solution exists at all), whereas y is a single valued function of x_F.

The total energy of the system is a sum of the energy of the fermions, $E_f = e_f\, m_f^{(0)}\, \zeta N$, and the energy in the scalar field, $E_s = e_s\, m_f^{(0)}\, \zeta N$, where N is the total number of neutrinos in each contributing state. These expressions serve to define the per fermion quantities e_f and e_s. Also, $E_s = \mathcal{E}_s\, V$, where $\mathcal{E}_s = \frac{1}{2} m_s^2 \phi^2$ is the energy density of the (here uniform) scalar field.

One finds that

$$
\begin{aligned}
e_f &= \frac{3}{x_F^3} \int_0^{x_F} x^2\, dx \sqrt{x^2 + y^2} \\
&= \frac{3}{x_F^3} \left\{ \frac{e_F x_F^3}{4} + \frac{e_F x_F y^2}{8} - \frac{y^4}{8} \ln\left(\frac{e_F + x_F}{y} \right) \right\}
\end{aligned}
\tag{11}
$$

and

$$
\begin{aligned}
e_s &= \frac{K_0}{2} \frac{3}{x_F^3} y^2 \left(\int_0^{x_F} \frac{x^2\, dx}{\sqrt{x^2 + y^2}} \right)^2 \\
&= \frac{1}{2K_0} \frac{3}{x_F^3} (1 - y)^2.
\end{aligned}
\tag{12}
$$

and the total energy density per fermion is just the sum,

$$
< e >= e_f + e_s. \tag{13}
$$

Notice that for large values of x_F,

$$
\begin{aligned}
y &\rightarrow \frac{2}{K_0 x_F^2} \\
e_f &\rightarrow \frac{3 e_F}{4} \\
&\rightarrow \frac{3 x_F}{4} \\
e_s &\rightarrow \frac{3}{2K_0} \frac{1}{x_F^3}.
\end{aligned}
\tag{14}
$$

It is also useful to note that, for small x_F,

$$
\begin{aligned}
y &\to 1 - \frac{K_0 x_F^3}{3} \\
e_f &\to 1 + \frac{3x_F^2}{10} \\
e_s &\to \frac{K_0}{2}\frac{x_F^3}{3}.
\end{aligned}
\tag{15}
$$

For the fermion system to be bound, the minimum of $e = e_f + e_s$ as a function of density (or x_F) must be less than 1, its value in the zero density limit. Fig.(1) shows the variation of e and y as a function of x_F for several values of K_0. Note that for sufficiently large K_0, there is a minimum relative to both the large and small x_F regimes, that is, relative to regions of both large and small fermion density.

Thinking of this in terms of expansion of the Universe, early times correspond to the large x_F region on the right and late times, including presumably the present, are to the left. Thus we see that at some intermediate

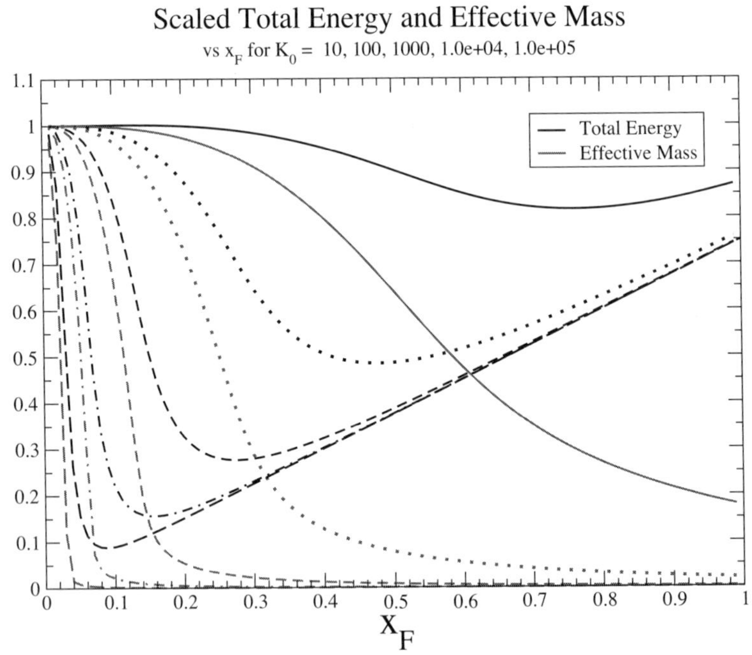

Fig. 1. Total energy density per fermion e and effective mass y vs. x_F for several values of K_0.

period, the energy density passes through a minimum and as the system approaches the present, it passes through a regime in which the energy density is increasing as the number density decreases – characteristic of a regime of negative pressure. Fig.(2) gives an advance peek at the value of

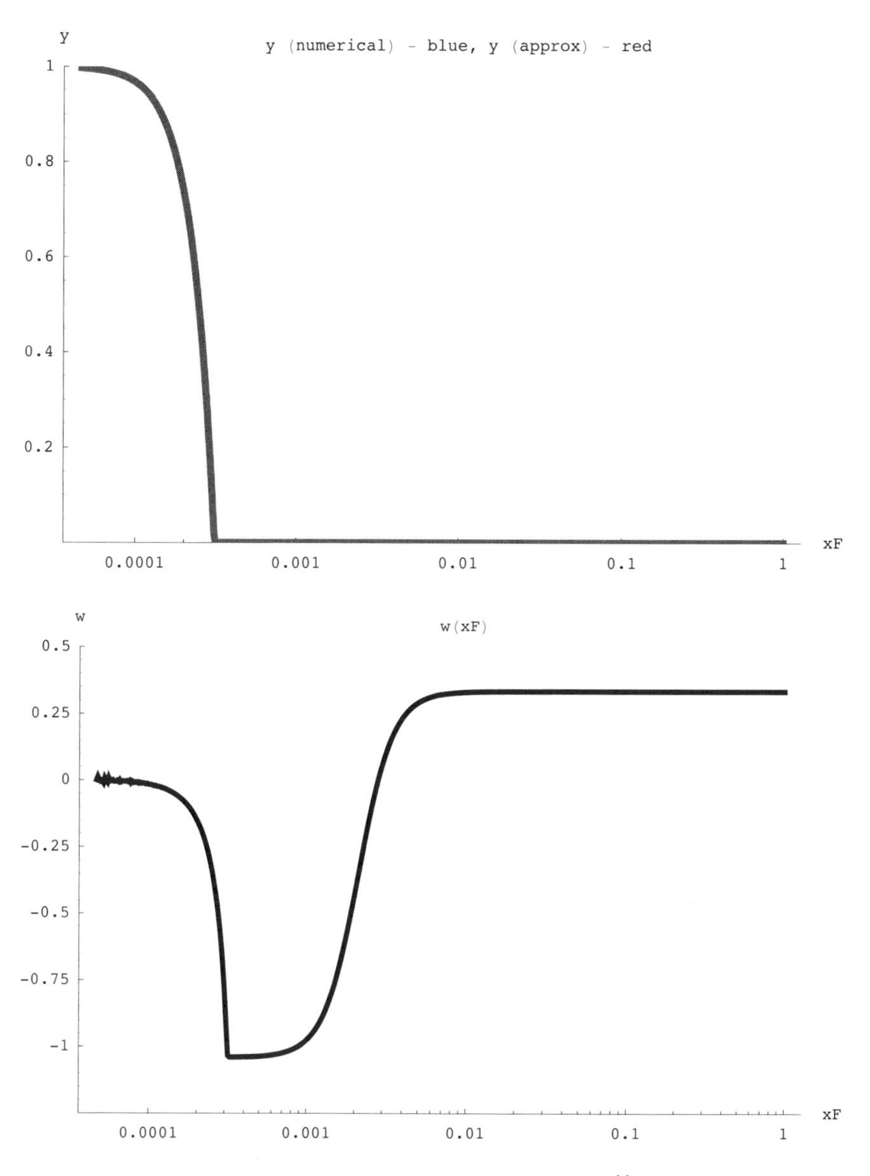

Fig. 2. y and w vs. $\log(x_F)$ for $K_0 = 10^{11}$.

the equation of state parameter, w, that we derive from the dependence of e vs. x_F. We will demonstrate later how we do this numerically, but it should be noted that there are still some numerical difficulties evidenced by the "hash" in the low x_F limit where we know that $w \to 0$ as the cold, now non-interacting fermion "dust" turns effectively into new "dark matter".

3. Einstein, FRW and Equations of State

In a Friedmann-LeMaître-Robertson-Walker Universe, Einstein's equations produce the second order time derivatiive equation of motion that relates the expansion (size) scale parameter, a, to Newton's constant, G, the matter density, ρ, pressure, P, and a cosmological constant, Λ:

$$\frac{\ddot{a}}{a} = -\frac{4\pi G}{3}(\rho + 3P) + \frac{\Lambda}{3} \tag{16}$$

Therefore, it is necessary to know the relevant equation of state (EoS) before the time development of the scale factor can be determined. For dust, which has no pressure, $P = 0$, while for a relativistic gas, $P = \frac{1}{3}\rho$. Note that acceleration of the expansion parameter occurs, even in the absence of a cosmological constant, when $P < -\frac{1}{3}\rho$. Finally, for a spatially and temporally homogenous scalar field, $P < -\rho$, which is more than enough to produce acceleration of the expansion. More generally, we can parametrize equations of state in this regime by a constant, w, as

$$P = w\rho \tag{17}$$

There are a great number of models for this Dark Energy phenomenon. They go by names such as "quintessence" for $-1 < w < 0$, or "phantom energy" for $w < -1$. Among others, this issue has been addressed by Fardon, Nelson and Weiner,[6] Peccei,[7] Barshay and Kreyerhoff,[8] Baushev[9] and Mukhopadhyay, Ray and Choudhoury.[10]

3.1. *Our EoS*

For the system under consideration, the total (matter plus field) energy is given by the product of the total energy density per fermion $< e >$ and the total number of fermions, which in turn is determined by the number density n times the volume, V:

$$U = \rho V = m_f^{(0)} < e > nV, \tag{18}$$

where we recall that

$$n = \zeta \frac{(m_f^{(0)})^3}{6\pi^2} x_F^3 \tag{19}$$

is the fermion number density. The pressure is defined by

$$P = -\frac{\partial U}{\partial V}, \tag{20}$$

where U is the internal energy given above. Since nV is a constant,

$$\begin{aligned} P &= -m_f^{(0)} nV \frac{\partial < e >}{\partial V} \\ &= -m_f^{(0)} nV \frac{\partial n}{\partial V} \frac{\partial < e >}{\partial n} \\ &= m_f^{(0)} n^2 \frac{\partial < e >}{\partial n} \\ &= \frac{\rho}{3} \frac{x_F}{< e >} \frac{\partial < e >}{\partial x_F} \\ &= \frac{\rho}{3} \frac{\partial ln(< e >)}{\partial ln(x_F)} \end{aligned} \tag{21}$$

From this and Eq.(17), we can identify

$$w = \frac{1}{3} \frac{\partial ln(< e >)}{\partial ln(x_F)}. \tag{22}$$

This is our central model result.

4. General Character of Model Results

In Fig.(3), we shows the value of w as computed numerically from Eq.(22) for 8 values of K_0 on a log scale for $x_F < 0.4$ and in FIg.(4) for $x_F < 0.1$ on a linear scale. Note that it approaches close to -1 as the density decreases (as the Universe expands and the scale factor a increases from right to left) and then departs sharply towards zero, as also indicated earlier in Fig.(2).

At large x_F, it is clear that w approaches $+1/3$ as it should for a relativistic gas of fermions. It is perhaps less clear, due to numerical fluctuations, that the value goes to zero at zero density. This can be checked by considering the small x_F expansions of the energy densities shown at Eq.(15).

Finally, we performed a number of numerical scaling checks to examine whether w can fall below -1. The are shown in Figs.(5,6,7).

We are continuing our efforts to ensure numerical stability primarily by seeking analytic formulae and approximations, especially to avoid taking derivatives numerically. We will report on these improvements elsewhere,

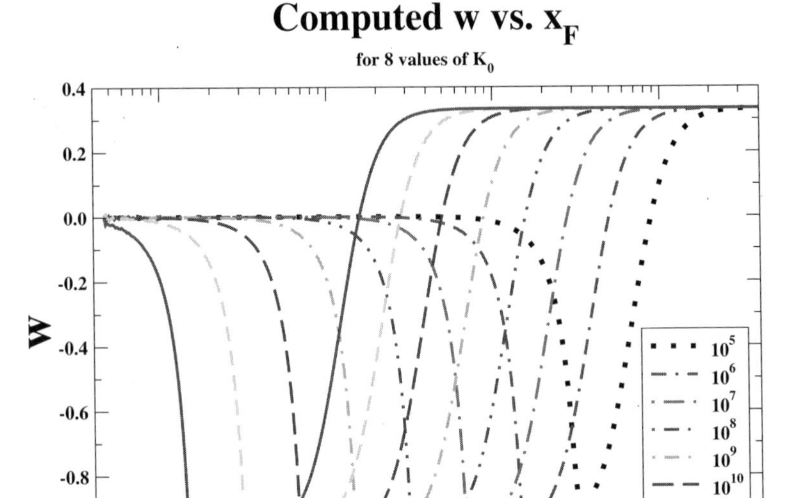

Fig. 3. w vs. $\log(x_F)$ for 8 values of K_0.

but suffice it to say that we have found no contradiction to the results obtained here by purely numerical means, and confirmed that the difficulties encountered at very small x_F are indeed due to numerical noise.

5. Numerical Values

All of the above is carried out with scaled variables. However, was shown in Ref.[3], there are only a few, weak constraints on the actual parameter values. Very large values of K_0 are possible even for very small values of g^2 if the range of the scalar is very large, corresponding to very small values of m_s. Even if long-ranged, such weak interactions between fermions, especially neutrinos or those outside of the Standard Model altogether (such as the LSP) are exceptionally difficult to constrain by any laboratory experiments.

The energy density of Dark Energy is quoted[5] as $(3.20 \pm 0.4) \times 10^{-47}\text{Gev}^4$. In more manageable units, this is given as $(2.3meV)^4$. What does this imply for the allowed value of m_0? If we set this energy density

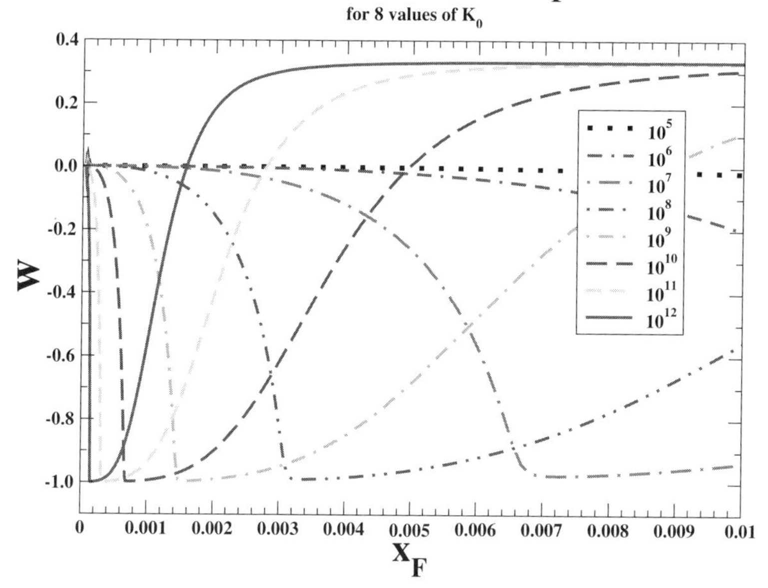

Fig. 4. w vs. x_F for 8 values of K_0.

equal to that of this system at $w = -1$,

$$\rho_\Lambda = \frac{(m_f^{(0)})^4 < e > x_F^3}{6\pi^2} \tag{23}$$

then solving for $m_f^{(0)}$ gives

$$m_f^{(0)} = \{\frac{6\pi^2 \rho_\Lambda}{< e > x_F^3}\}^{1/4} \tag{24}$$

If we further suppose that this occurs at cosmological $z \sim 1$, then the range of the scalar field must be comparable to the size of the Universe at that time. That is,

$$m_s \sim 7 \times 10^9 \,\text{lightyears} \sim 3 \times 10^{-30} \,\text{meV}. \tag{25}$$

For the relatively modest value of $K_0 \sim 8 \times 10^6$, this implies that $m_f^{(0)} \sim 300\text{meV}$ and $g^2/(4\pi) \sim 6 \times 10^{-58}$. These values emphasize the virtual impossibility of constraining this physics by means of laboratory experiments.

We note with interest that following the curve for y from $z = 1$ to "now", i.e., $z = 0$, tells us that the effective mass of this fermion would now

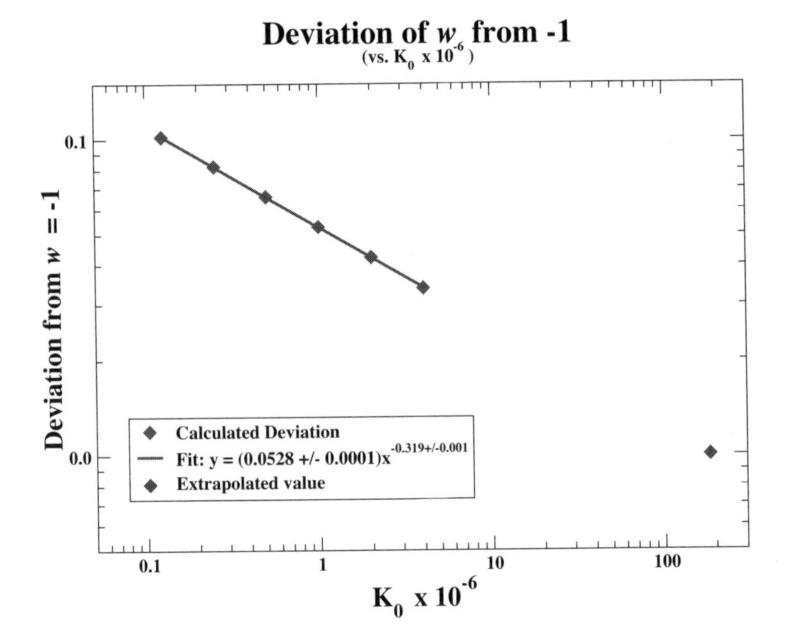

Fig. 5. Deviation of minimum value of w from -1 vs. K_0.

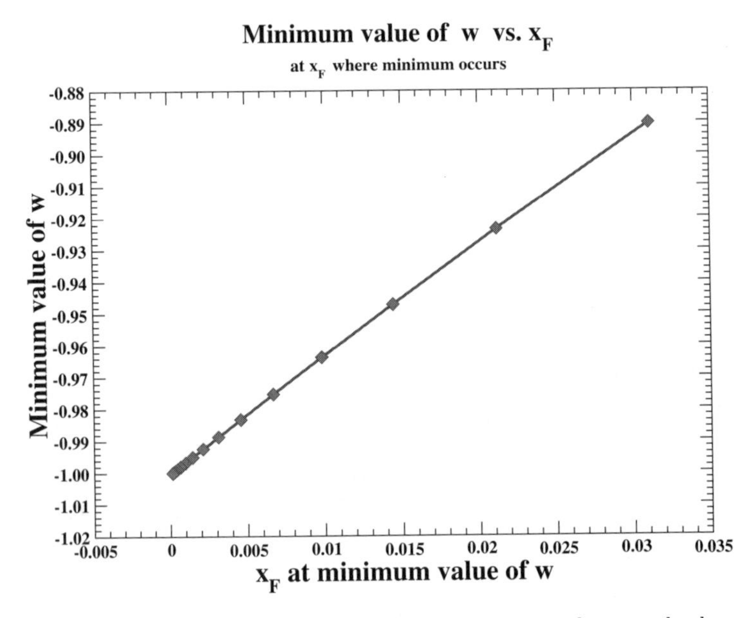

Fig. 6. Minimum value of w vs. x_F at which minimum occurs, for several values of K_0.

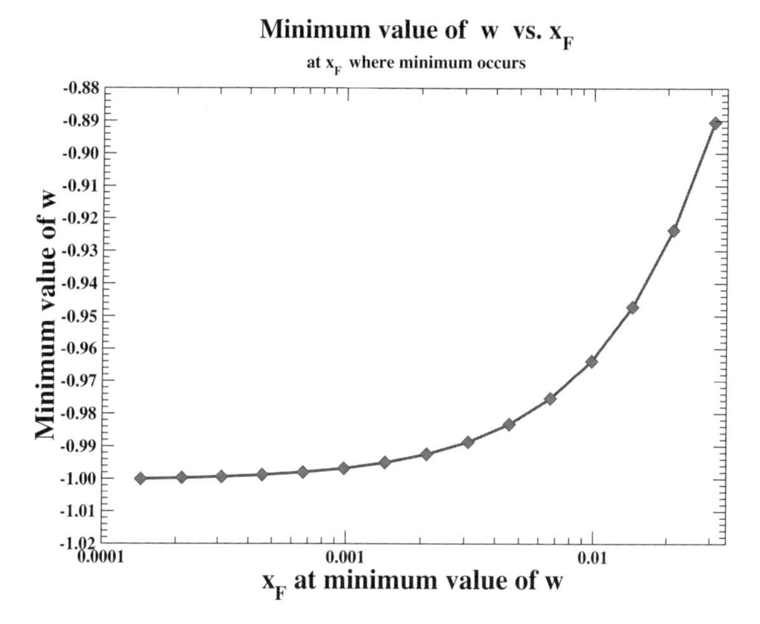

Fig. 7. Minimum value of w vs. logarithm of x_F at which minimum occurs, for several values of K_0.

be measured to be approximately $(7/8)m_f^{(0)}$, a value tantalizingly close to the Majorana neutrino mass that Prof. Klapdor has reported from his experiments.[11]

Other solutions are possible, and Fig.(8) shows how our results scale very accurately (for sufficiently large K_0) with the 4th root of K_0, both numerically and under one of our analytic approximations to the region where w is a minimum. In particular, if one has the LSP at ~ 1 TeV in mind, K_0 becomes very large, $\sim 10^{57}$ and $g^2/(4\pi)$ also increases, $\sim 10^{-32}$, but these values are not ruled out by anything known.

6. Conclusions and Questions

We have displayed an explicit and calculable dynamical mechanism that describes Dark Energy and connects it to what in the current epoch becomes a kind of dark matter. Although Dark Matter is usually thought of as existing at early epochs in the life of the Universe, the system described here turns (hot) relativistic fermions, that interact weakly with a very light scalar field, into Dark Energy which lasts for a limited time during the expansion of the Universe which then morphs into new cold dark matter

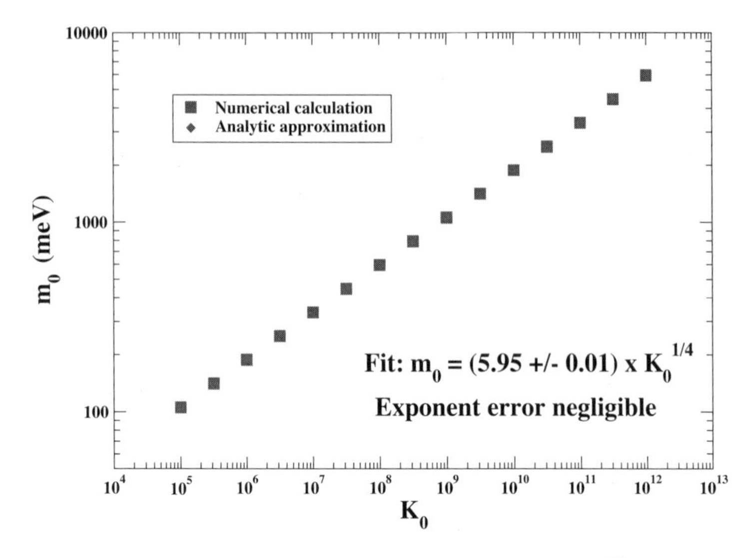

Fig. 8. Scaling variation of the vacuum fermion mass, $m_f^{(0)}$, vs. K_0.

components. Thus, neutrinos can contribute to both. Nor need there be only one time scale or one species for which this applies. (See, e.g., Ref.[12].) If there are many sterile fermions with sufficiently long decay lifetimes, the acceleration/deceleration history of the Universe could be much more complicated that presently envisioned.

We may also ask generally why $w > -1$, but it is fairly clear in this model: A scalar field strength uniform in space and time produces[13] exactly $w = -1$, but here the source for the scalar field is the density of massive fermions. Relativistically, their strength for producing scalar field is severely reduced at high momentum (in the rest frame of the Universe) but as they slow, the scalar field strength grows nonlinearly until, due to the expansion of the Universe, the fermions separate so much (greater than the Yukawa range for exchange of the scalar field) that they cannot act collectively and the scalar field strength declines again.

Finally, we note that our model has definitive if difficult tests, as it predicts specific variations of w, slowly approaching -1 from above as z decreases through 1, and rising rapidly towards zero as z approaches the present. We hope this encourages observationalists in their efforts to discern variation of w with z.

Acknowledgments

This work was carried out in part under the auspices of the National Nuclear Security Administration of the U.S. Department of Energy at Los Alamos National Laboratory under Contract No. DE-AC52-06NA25396 and supported in part by the Australian Research Council.

References

1. M. Kawasaki, H. Murayama and T. Yanagida, *Mod. Phys. Lett. A* **7**, 563 (1992).
2. R. A. Malaney, G. D. Starkman and S. Tremaine, *Phys. Rev. D* **51**, 324 (1995).
3. G. J. Stephenson, Jr., T. Goldman and B. H. J. McKellar, *Int. J. Mod. Phys. A* **13**, 2765 (1998); arXiv:hep-ph/9603392.
4. B. H. J. McKellar, M. Garbutt, T. Goldman and G. J. Stephenson, Jr., *Mod. Phys. Lett. A* **19**, 1155 (2004).
5. J. Dunkley *et al.*, *Astrophys. J. Suppl.* **180**, 306 (2009); arXiv:0803.0586.
6. R. Fardon, A. E. Nelson and N. Weiner, *J. Cosmol. Astropart. Phys.* **10**. 005 (2004); arXiv:astro-ph/0309800; see also, R. Takahashi and M. Tanimoto, *JHEP* **0605** 021 (2006); arXiv:astro-ph/0601119.
7. R. D. Peccei, *Phys. Rev. D* **71**, 023527 (2005); arXiv:hep-ph/0411137.
8. S. Barshay and G. Kreyerhoff, *Mod. Phys. Lett. A* **23**, 2897 (2008).
9. A. Baushev, arXiv:0809.0235.
10. U. Mukhopadhyay, S. Ray and A. A. Usmani, arXiv:0811.0782.
11. H. V. Klapdor-Kleingrothaus, I.V. Krivosheina, *Mod. Phys. Lett. A* **21**, 1547 (2006); H. V. Klapdor-Kleingrothaus, *et al.*, *Phys. Lett. B* **586**, 198 (2004); arXiv:hep-ph/0404088; H. V. Klapdor-Kleingrothaus, *et al.*, *Mod. Phys. Lett. A* **16**, 2409 (2001); arXiv:hep-ph/0201231.
12. G. J. Stephenson, Jr., T. Goldman and B. H. J. McKellar, *Mod. Phys. Lett. A* **12**, 2391 (1997); arXiv:hep-ph/9610317.
13. S. M. Carroll, M. Hoffman and M. Trodden, arXiv:astro-ph/0301273.

NONEXTENSIVE STATISTICS IN ASTRO-PARTICLE PHYSICS: STATUS AND IMPACT FOR DARK MATTER/DARK ENERGY THEORY

M. P. LEUBNER*

Institute for Astro- and Particle Physics, University of Innsbruck, Innsbruck, A-6020, Austria
**E-mail: manfred.leubner@uibk.ac.at*

Nature appears non-local and nonlinear on all observable levels resulting in a large variety of complex phenomena in different scientific fields. In this situation the classical Boltzmann-Gibbs extensive thermo-statistics, applicable whenever microscopic interactions and memory are short ranged and the environment is a continuous and differentiable manifold, fails. We are dealing with systems generally subject to spatial or temporal non-local interactions, evolving in a non-Euclidean/multi-fractal space-time, making their behavior nonextensive. An appropriate generalization of the entropy functional yields upon entropy maximization naturally power-law distributions as manifestation of long-range interactions and correlations in the system, controlled by a single and physically interpretable parameter. Moreover, the nonextensive context is per se subject to entropy bifurcation, generating a tandem character of structures, where higher order stationary states of reduced entropy reside besides lower order stationary states of increased entropy. After reviewing the fundamental theoretical concepts of nonextensive statistics, we focus on the significance and present status with particular attention to the problem of dark matter density distributions in relaxed large scale astrophysical structures as well as the dark energy domain, associating dark energy with self-interacting scalar fields, subject to highest degree of correlations, and appearing as natural content within the nonextensive statistical landscape.

Keywords: Nonextensive statistics; Dark matter; Dark energy.

1. Introduction

Power-law behavior as manifestation of fractal or multi-fractal structures is found in a large variety of complex phenomena in different scientific fields. A novel context of description is based on entropy generalization and maximization of the corresponding particular entropy function. Such a generalization is an intrinsic nonlinear process where the resulting power-law distributions follow in a natural way. Nature is per se nonextensive and complex since any member of an ensemble of particles, e.g. of the interplanetary medium, star clusters or clusters of galaxies, is subject to electromagnetic and/or gravitational interactions with the environment. With regard to the two limits, the crystal as system of maximum order describable by simple geometry and, vice versa, a thermalized gas of independently moving particles accessible by standard Boltzmann-Gibbs-Shannon (BGS) statistics, nature appears somewhere between. In extensive systems no interactions or correlations are present and the BGS logarithmic entropy measure yields the Maxwellian distribution. On the other hand, the members of nonextensive systems are subject to long-range interactions and couplings, controlled by a generalized entropy functional, where the corresponding power-law distributions depend on one specific parameter, the entropic index.

Leptokurtic, long-tailed probability distribution functions (PDFs) subject to a non-Maxwellian core and pronounced halo are a persistent feature in a variety of different astrophysical environments. Those include the thermo-statistical properties of the interplanetary medium where the electron and proton velocity space distributions show ubiquitously suprathermal halo patterns, well described by the empirical family of κ-distributions, a power law in particle speed.[1–3] Theoretically, significant progress was provided on the basis of a novel kinetic approach, demonstrating that power-law velocity distributions are a particular thermodynamic equilibrium state.[4] Finally, the empirical family of κ-distributions was linked to power-law distributions derived in the context of nonextensive statistics,[5–7] thus providing the hitherto missing theoretical foundation of non-thermal equilibrium PDFs.

Moreover, also the PDFs of the turbulent fluctuations of the magnetic field strength, density and velocity field differences in astrophysical plasmas show pronounced leptokurtic cores and extended tails on small scales. Based on high resolution in situ satellite observations and detailed studies of the multi-scale statistical properties in fully developed turbulence it was verified that the nonextensive κ-distribution family represents also correctly the characteristics of intermittent turbulence on small scales. Reflecting a decoupled state, the PDFs approach a Gaussian on on large scales, thus confirming fundamentally that the probability of rare events is raised on small scales.[8–10]

Remarkably, we have to add to this diversity also scale invariant power-law distributions relying on self-organized criticality (SOC).[11–13] Moreover, gravitationally bound large scale astrophysical structures are subject to long-range interactions, motivating a generalization of the nonextensive formalism to self-gravitating systems as well. In particular, the radial dark matter (DM) and plasma density distributions in galaxies and galaxy clusters are commonly fitted by simple empirical models with lack of any physical justification.[14–17] In the context of entropy generalization in gravitational fields this shortcoming was fundamentally improved, where a nonextensive theory was developed representing precisely both, the hot plasma and DM density profiles observed in galaxies and galaxy clusters.[18,19] The significance and accuracy of nonextensive statistics, modeling density distributions of astrophysical bound structures, was confirmed by N-body DM and hydrodynamic simulations as well as by observations.[20,21] Finally, strongly interacting chaotic scalar fields were introduced as suitable models for dark energy in the universe where the fluctuating field momenta obey a nonextensive power-law distribution.[22] In this formalism the chaotic fields generating dark energy are restricted by one specific negative value of the entropic index, contrary to the negative pressure dark energy domain, subject to highest degree of correlations and appearing as fundamental sector within the entire nonextensive landscape, see below.

In all physically different situations nonextensive statistics, accounting for long-range interactions and correlations, serves as a highly successful context of description whereas standard BGS statistics does not apply. An analysis and summary of the entire nonextensive landscape in section 5 provides finally clear insight into the distinct nonextensive domains and corresponding physics, as depending on the sign and value of the entropic index.

2. Nonextensive Entropy Generalization

The classical BGS extensive thermo-statistics applies when microscopic interactions are short ranged and the environment is a continuous and differentiable manifold. Astrophysical systems, however, are generally subject to long-range interactions in a non-Euclidean, for instance fractal or multifractal environment. A proper nonextensive generalization of the BGS entropy for statistical equilibrium was recognized five decades ago[23] and later revived,[24] suitably extending the standard additivity of entropies to the nonextensive, nonlinear situation. A variety of subsequent analyses were devoted to clarify the mathematical and physical consequences of pseudo-additivity[25–28] where a deterministic connection between the generalized entropy and the resulting power-law functionals,[29] as well as the duality of nonextensive statistics were recognized.[30] Derived within the context of nonextensive statistics, power-law distributions provided also the missing justification for the use of the hitherto empirically introduced κ-distribution family favored in space plasma modeling from fundamental physics.[2,3,6,7] The corresponding entropic index κ characterizing the degree of nonextensivity of the system, i.e. the degree of long-range interactions or correlations, is not restricted to positive values and thus manifests the duality of nonextensive statistics.

Assuming that particles move independently from each other, i.e. there are no correlations present in the system considered, the BGS statistics is based on the extensive entropy measure $S_B = -k_B \sum p_i \ln p_i$, where p_i is the probability of the ith microstate, k_B is Boltzmann's constant and S_B is extremized for equiprobability. This entropy implies isotropy of the velocity directions and yields straightforward the standard Maxwellian distribution function. Accounting for long-range interactions requires to introduce correlation within the system, which is performed fundamentally in the context of entropy generalization leading to scale-free power-law PDFs. Considering, as example, two sub-systems A and B one can illuminate nonextensivity by the property of pseudo-additivity of the entropies such that

$$S_\kappa(A+B) = S_\kappa(A) + S_\kappa(B) + \frac{k_B}{\kappa} S_\kappa(A) S_\kappa(B) \tag{1}$$

where the entropic index κ quantifies the degree of nonextensivity in the system. For $\kappa = \infty$ the last term on the right hand side cancels leaving the additive terms of the standard BGS statistics. Hence, nonlocality is introduced by the nonlinear term accounting for correlations between the subsystems. In general, the pseudo-additive, κ-weighted term may assume positive or negative definite values indicating a nonextensive entropy bifurcation. Obviously, nonextensive systems are subject to a dual nature since positive κ-values imply the tendency to less organized states where the entropy increases, whereas negative κ-values provide states of a higher level of organization and decreased entropy, as compared to the BGS state, see e.g.[6,18]

The general nonextensive entropy is consistent with the example in Eq. (1) and reads[6,24]

$$S_\kappa = \kappa k_B \left(\sum p_i^{1-1/\kappa} - 1\right) \tag{2}$$

where $\kappa = \infty$ represents the extensive limit of statistical independence. In this case the interaction term in Eq. (1) cancels recovering with respect to Eq. (2) the classical BGS entropy measure. Once the entropy is known the corresponding PDFs are available.

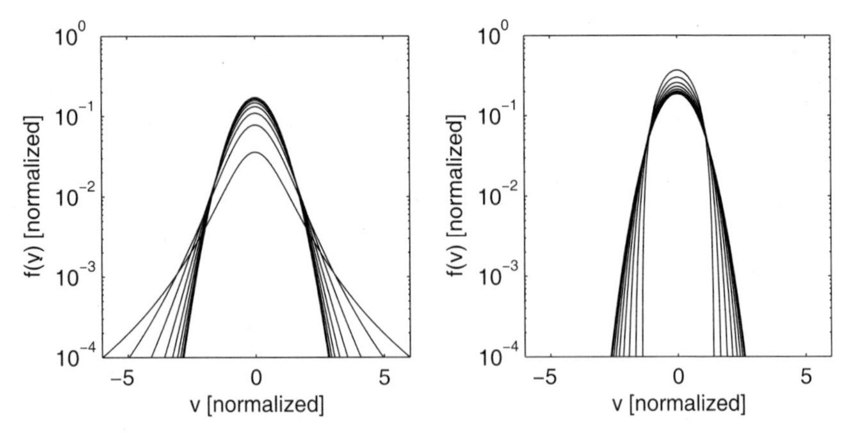

Fig. 1. Schematic plot of normalized velocity distributions according to Eq.(3) for positive (left, halo) and negative (right, core) κ−values. Both sets of curves merge for $\kappa = \infty$ into the same dark curve, representing as BGS solution a Maxwellian, the halo set from outside and the core set from inside.

Extremizing the entropy (2) under conservation of mass and energy the resulting distribution function in velocity space reads

$$f^{\pm} = A^{\pm} \left[1 + \frac{1}{\kappa} \frac{v^2}{v_t^2} \right]^{-\kappa} \tag{3}$$

where v_t corresponds to the mean energy or thermal speed. Hence, the exponential probability function of the Maxwellian gas of an uncorrelated ensemble of particles is replaced by the characteristics of a scale invariant power-law where the sign of κ, indicated by superscripts, governs the corresponding entropy bifurcation. We note that the distribution (3) can be derived by means of Lagrangian multiplyers without introducing any specific form for long-range interactions. Incorporating the sign of κ into Eq. (3) and performing the normalization separately for positive and negative κ-values generates a dual solution subject also to two different, κ-dependent normalizing factors $A^{\pm}(\kappa)$.[6]

The entropy bifurcation appears also in higher order moments yielding for instance κ-dependent generalized pressures and the negative solutions are subject to a cut off in the distribution at $v_{max} = v_t\sqrt{\kappa}$, for details see.[6] Both functions, f^+ and f^- in Eqs. (3) approach one and the same Maxwellian as $\kappa \to \infty$. Fig. 1 demonstrates schematically the non-Maxwellian behavior of both, the suprathermal halo component and the less pronounced core distribution, subject to finite support in velocity space, where the case $\kappa = \infty$ recovers the Maxwellian equilibrium distribution.

The duality of equilibria in nonextensive statistics is manifest in two families, the nonextensive thermodynamic equilibria and the kinetic equilibria, where both families are related via the nonextensive parameter by $\kappa' = -\kappa$.[18,30] κ' and κ denote the corresponding entropic index of the particular family where the transformation $q = 1 - 1/\kappa$ for the transition between the Tsallis q-notation and the symmetric κ-formalism, used here, is applied.[5,7] Positive κ-values are related to the stationary states of thermodynamics and negative κ-values to kinetic stationary states. The limiting BGS state for $\kappa = \infty$ is therefore characterized by self-duality. The nonextensive parameter κ finds also a physical interpretation in terms of the heat capacity C of a medium.[28] A system with $\kappa > 0$ represents an environment with finite positive heat capacity and vice versa, for $\kappa < 0$ the heat capacity is negative. Negative heat capacity is a typical property of self-interacting systems, see e.g.[31] Moreover, contrary to thermodynamic systems where the tendency to dis-organization is accompanied by increasing entropy, self-interaction tends to result in structures of a higher level of organization and decreased entropy. Consistently, "core" refers to negative definite κ and "halo" to positive definite κ-values and the corresponding distribution families merge for $\kappa \to \infty$ into the extensive, selfdual state.

3. Dark Matter and Hot Gas Density Distributions

To date only a few attempts provide physically motivated models for density profiles of astrophysical clusters. The early analytical analysis[32] for the collapse of density perturbations was subsequently further studied[33] and based on infall models.[34,35]

In practice, DM and hot plasma density profiles, as observed in galaxies and clusters or generated in simulations, are widely modeled by empirical fitting functions. The phenomenological $\beta-$model[14] provides a reasonable representation of the hot gas density distribution of clustered structures, further improved by the double β-model with the aim of resolving the $\beta-$discrepancy.[36] Similarly, the radial density profiles of DM halos are analyzed primarily with the aid of phenomenological fitting functions, thus lacking physical support as well.[15,37,38]

Since any astrophysical system is subject to long-range gravitational and/or electromagnetic interactions, this situation motivates again to introduce nonextensive statistics as physical background for the analysis of DM and hot plasma density profiles. In this context the entropy of the standard BGS statistics is generalized, as outlined in section 1, by the pseudo-additive κ-weighted term to mimic the degree of long-range gravitational interactions and correlations within the system.

Extremizing the generalized entropy with regard to conservation of mass and energy in a gravitational potential Ψ yields the energy distribution[18]

$$f^{\pm}(v) = C^{\pm}\left[1 + (v^2/2 - \Psi)/(\kappa\sigma^2)\right]^{-\kappa} \tag{4}$$

As previously, the superscripts refer to the positive or negative intervals of the entropic index κ, accounting for less (+) and higher (-) organized states and thus reflecting the accompanying entropy increase or decrease, respectively. σ represents the mean energy of the distribution and $C^{\pm}$ are the corresponding normalization constants. The density evolution of a system subject to long range interactions in a gravitational potential

$$\rho^{\pm} = \rho_0\left[1 - \Psi/(\kappa\sigma^2)\right]^{(3/2-\kappa)} \tag{5}$$

is found after integration over all velocities. Combining with Poisson's equation $\Delta\Psi = -4\pi G\rho^{\pm}$ provides a second order nonlinear differential equation, determining the radial density profiles of both components, plasma and DM in clustered structures as[18]

$$\frac{d^2\rho}{dr^2} + \frac{2}{r}\frac{d\rho}{dr} - (1-\frac{1}{n})\frac{1}{\rho}(\frac{d\rho}{dr})^2 - \frac{4\pi Gn}{(\frac{3}{2}-n)}\frac{\rho^2}{\sigma^2}(\frac{\rho}{\rho_0})^{-\frac{1}{n}} = 0 \tag{6}$$

where the signs are omitted and $n = 3/2 - \kappa$ is introduced, corresponding to the polytropic index of stellar dynamical systems.[40] As natural consequence of nonextensive entropy generalization the standard isothermal sphere profile[40] bifurcates into two distribution families controlled by the sign and value of the correlation parameter κ.

Physically, we regard the DM halo as an ensemble of self-gravitating, weakly interacting particles in dynamical equilibrium[31,41] and the hot gas component as an electromagnetically interacting high temperature plasma in thermodynamic equilibrium. Hence, astrophysical clusters experience long-range gravitational and/or electromagnetic interactions leading to correlations, such that the standard BGS statistics does not apply again. As discussed previously, the duality of equilibria in nonextensive statistics appears in the nonextensive stationary states of thermodynamics subject to finite positive heat capacity and in the kinetic stationary states with negative heat capacity, a typical property of self-gravitating systems,[31] where both are related only via the sign of the coupling parameter κ. Consequently we have to assign negative κ-values, describing the lower entropy state due to gravitational interaction, to the DM component and the second branch

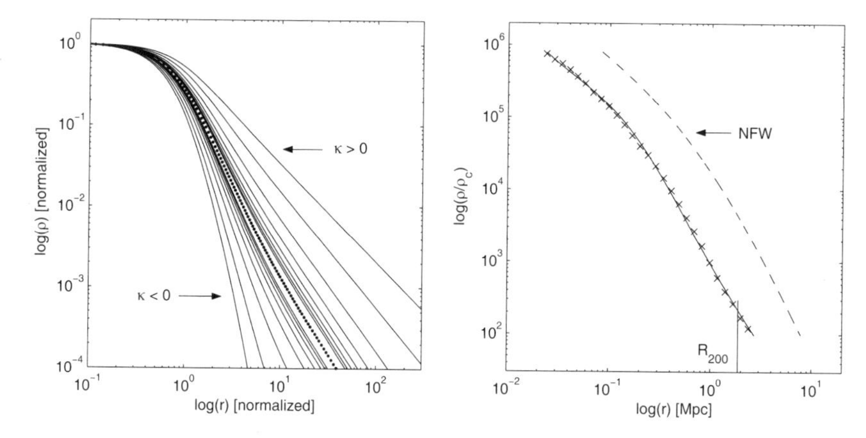

Fig. 2. Left: Normalized nonextensive family of density profiles for $\kappa = 3...10$. For increasing κ both sets of curves converge to the central, limiting isothermal sphere solution ($\kappa = \infty$, dots), ρ^- from inside and ρ^+ from outside. Right: Normalized radial dark matter density profile obtained from simulations (crosses) and a fit of the nonextensive theory (solid) ($\kappa = -15$, $\sigma = 0.12$. The best fitting NFW profile is shifted to the right for comparison (dashed); R_{200} indicates the virial radius..

of positive κ-values and higher entropy, as compared to the BGS self-dual state, to the hot plasma component.

The left panel in Fig. 2 illuminates schematically the radial density profile characteristics for some values of κ for both, DM below and the plasma distributions above the standard exponential BGS solution. Increasing κ values correspond to a decoupling within the system and both branches merge simultaneously in the isothermal sphere profile for $\kappa = \infty$, representing the extensive limit of statistical independence, in analogy to the Maxwellian limit for systems where gravitational interaction is neglected. In Fig. 2, right panel, the result of N-body DM simulations are compared with the nonextensive theoretical approach for $\kappa = -15$ and $\sigma = 0.12$ indicating perfect agreement. For comparison also the best fitting NFW profile[17] is provided.

4. Dark Energy Domain

Currently the universe is known to follow an accelerated expansion where 73% of the total matter content must be assumed to exist as some state of dark energy (DE), 23% as dark matter and 4% as ordinary matter. Favored dark energy models presently under debate can be summarized as follows: (1) association of DE with the vacuum energy of a self-interacting scalar field whose potential energy generates a cosmological constant, (2) quintessence models with slowly evolving scalar fields obeying a nontrivial equation of state, (3) string theoretical candidates of scalar fields and related consequences and (4) some unknown exotic form of matter.

The theoretical context of description is generally based on three components (j) in a homogeneous, isotropic universe under adiabatic expansion where the scale factor evolves as

$$\frac{\ddot{R}}{R} = -\frac{4\pi G}{3}(\rho_{DE} + 3p_{DE} + \rho_m + \rho_r + 3p_r) \tag{7}$$

With the Hubble parameter $H^2 = (\dot{R}/R)^2$, the energy conservation $d\rho_j/dt = 3H(\rho_j + p_j)$ and the equation of state $w_j = p_j/\rho_j$ we find

$$\rho_j \sim R^{-3(1+w_j)} \tag{8}$$

where p_j is the pressure and ρ_j the energy density. Eq(8) provides for ordinary matter (m) $w_m = 0$ and for radiation (r) $w_r = 1/3$, respectively, the conditions $\rho_m \sim R^{-3}$, $\rho_r \sim R^{-4}$ and according to the classical vacuum energy with $w_\Lambda = -1$ the fundamental constraint

$$\frac{3p_{DE}}{\rho_{DE} + \rho_m} < -1 \tag{9}$$

In order to comply with this condition chaotic nonextensive scalar fields were proposed[22] where the chaotic behavior of the strongly self-interacting fields are associated with the vacuum fluctuations. The probability distribution of the fluctuating fields obeys a natural invariant $p(\Phi) = 2/\pi\sqrt(1-\Phi^2)$, equivalent to the generalized nonextensive distribution

$$p(E) \sim (1 + \frac{1}{\kappa}\beta E)^{-\kappa} \tag{10}$$

for $\kappa = -1/2$. Here $E = m\Phi^2/2$ and $\beta^{-1} = m$, i.e. the thermal energy of the nonextensive gas coincides with the scalar field mass.[22]

As consequence of the particular invariant governing the strongly interacting chaotic field fluctuations the dark energy domain is restricted in this context to one specific value of the nonextensive index $\kappa = -1/2$. It can be shown that negative κ-values correspond to positive pressure controlling the DM domain instead, see section 3. However, the nonextensive context provides naturally a dark energy landscape within the interval $1/2 < \kappa < 3/2$, where $C > 0$ and $p_{DE} < 0$. This domain is found from the second moment of the nonextensive distribution $p_{DE} \sim \rho_{DE} \int v^2 f(v) dv$ as

$$\frac{p_{DE}}{\rho_{DE}} = \frac{\kappa}{\kappa - 3/2} \tag{11}$$

Table 1. The nonextensive landscape

domain	pressure	heat capacity	interactions	physical state
$\kappa = \infty$	$p > 0$	$C > 0$	no correlations	ideal gas
$3/2 < \kappa < \infty$	$p > 0$	$C > 0$	correlations	plasma
$1/2 < \kappa < 3/2$	$p < 0$	$C > 0$	high correlations	dark energy
$-\infty < \kappa < 0$	$p > 0$	$C < 0$	all correlations	dark matter

The constraint on the DE equation of state $w_\Lambda = p_{DE}/\rho_{DE} \lesssim -1$ restricts the nonextensive parameter to $\kappa \gtrsim 3/4$. Consistency with the observational constraint on the time variation $w(t)$ by a factor less than 2 yields for the permitted κ-interval the range $3/4 < \kappa < 1$. Moreover, recent observations provide a precise restriction on the DE equation of state by $w_{DE} = -1.14 \pm 0.21$,[42] yielding an accurate value for the dark energy nonextensive correlation index of $\kappa = 0.8$ within the theoretically permitted range $1/2 < \kappa < 3/2$. The nonextensive analysis demonstrates that dark energy behaves like an ordinary gas with positive heat capacity but subject to negative pressure and highest degree of correlations.

5. Summary and Discussion

Nature appears nonlocal and nonlinear on all observable scales requiring a generalization of the standard BGS entropy, applicable exclusively for uncorrelated systems. Nonextensive theory accounts for long-range interactions and correlations and is naturally subject to entropy bifurcation. Manifest in positive or negative definite values of the underlying entropic index κ a tandem character of structures is generated, where higher order states of reduced entropy reside besides lower order states of increased entropy.

The specific physical domains regarding the entropic index are summarized in Table 1 where for $\kappa = \infty$ the standard BGS state appears as special case, representing uncorrelated systems within the entire nonextensive environment applicable for systems subject to long-range interactions and correlations. In particular, astrophysical energy distributions and probability distributions of plasma turbulence are available for positive definite κ-values within the interval $3/2 < \kappa < \infty$, providing accurate representations of observed astrophysical velcity space distributions and probability density structures of intermittent plasma turbulence, as well as the radial profiles of plasma density distributions in gravitationally coupled large-scale structures, as galaxies and galaxy clusters. Nonexistence of the integral excludes the interval ($0 \le \kappa \le 1/2$) for mathematical reasons.[6] Vice versa, subject to negative heat capacity the entire negative valued κ domain represents the corresponding radial dark matter density distributions of large-scale astrophysical structures.

Finally, it is demonstrated that the remaining domain ($1/2 < \kappa < 3/2$) provides, as consequence of the underlying negative equation of state, the required properties of a repulsive dark energy landscape, consistent with the observational constraint for the permitted time variation of the equation of state. Moreover, most recent observational restrictions on the DE equation of state yield the value $\kappa = 0.8$ for the nonextensive DE entropic index, implying strongest degree of correlations. Consistent with previous findings from chaotic scalar field analysis dark energy should behave like a strongly interacting, highly correlated ordinary gas, subject to positive heat capacity and negative pressure.

Acknowledgments

This work was supported by the Austrian Wissenschaftsfonds, grant number P20131-N16.

References

1. D. A. Mendis and M. Rosenberg, *Ann. Rev. Astron. Astrophys.* **32**, 419 (1994).
2. M. P. Leubner, *Planet. Space Sci.* **48**, 133 (2000).
3. M. P. Leubner and N. Schupfer, *J. Geophys. Res.* **106**, 12993 (2001).
4. R. A. Treumann, *Physica Scripta* **59**, 19, (1999).
5. M. P. Leubner, *Astrophys. Space Sci.* **282**, 573 (2002).
6. M. P. Leubner, *Phys. Plasmas* **11**, 1308 (2004).
7. M. P. Leubner, *Astrophys. J.* **404**, 469 (2004).
8. L. Sorriso-Valvo, V. Carbone and P. Veltri, *Geophys. Res. Lett.* **26**, 1801 (1999).
9. M. P. Leubner and Z. Voros, *Astrophys. J.* **618**, 547 (2005).
10. Z. Vörös, M. P. Leubner and W. Baumjohann, *J. Geophys. Res.* **111**, A02102 (2006).
11. P. Bak, C. Tang and K. Wiesenfeld, *Phys. Rev. A* **38**, 364 (1988).
12. P. Bak., *How nature works*, Copernicus, New York (1996).
13. S. C. Chapman and N. W. Watkins, *Space Sci. Rev.* **95**, 293 (2001).
14. A. Cavaliere and R. Fusco-Femiano, *Astron. Astrophys.* **49**, 137 (1976).
15. A. Burkert, *Astrophys. J.* **447**, L25 (1995).
16. H. Zhao, *MNRAS* **278**, 488 (1996).
17. J. F. Navarro, C. S. Frenk and S. D. M. White, *Astrophys. J.* **490**, 493 (1997).
18. M. P. Leubner, *Astrophys. J.* **632**, L1 (2005).

19. M. P. Leubner, in *Dark Matter in Astroparticle and Particle Physics*, H. V. Klapdor-Kleingrothaus and G.F. Lewis, eds., **World Scientific** (2007).
20. E. Pointecouteau, M. Arnaud and G. W. Pratt, *Astron. Astrophys.* **435**, 1 (2005).
21. T. Kronberger, M. P. Leubner and E. van Kampen, *Astron. Astrophys.*, **453**, 21 (2006).
22. C. Beck, *Physica A* **340**, 459 (2004).
23. A. Renyi, *Acta Math. Hungaria* **6**, 285 (1955).
24. C. Tsallis, *J. Stat. Phys.* **52**, 479 (1988).
25. A. R. Plastino, A. Plastino and C. Tsallis, *J. Phys. A: Math. Gen.* **27**, 5707 (1994).
26. C. Tsallis, *Physica A* **221**, 277 (1995).
27. R. Silva, A. R. Plastino and J. A. S. Lima, *Phys. Lett. A* **249**, 401 (1998).
28. M. P. Almeida, *Physica A* **300**, 424 (2001).
29. J. S. Andrade, M. P. Almeida, A. A. Moreira and G. A. Farias, *Phys. Rev. E* **65**, 036121 (2002).
30. I. V. Karlin, M. Grmela and A. N. Gorban, *Phys. Rev. E* **65**, 036128 (2002).
31. C. Firmani, E. D. Onghia, V. Avila-Reese, G. Chincarini and X. Hernandez, *MNRAS* **315**, L29 (2000).
32. J. E. Gunn and J. R. I. Gott, *Astrophys. J.* **176**, 1 (1972).
33. Y. Hoffman, *Astrophys. J.* **328**, 489 (1988).
34. L. L. R. Williams, A. Babul, J. J. Dalcanton, *Astrophys. J.* **604**, 18 (2004).
35. Y. Ascasibar, G. Yepes, S. Gottlöber, V. Müller, *MNRAS* **352**, 1109 (2004).
36. N. A. Bahcall and L. M. Lubin, *Astrophys. J.* **426**, 513 (1994).
37. J. F. Navarro, C. S. Frenk and S. D. M. White, *Astrophys. J.* **462**, 563 (1996).
38. B. Moore, F. Governato, T. Quinn, J. Stadel and G. Lake, *Astrophys. J.* **499**, L5 (1998).
39. B. Moore, T. Quinn, F. Governato, J. Stadel and G. Lake, *MNRAS* **310**, 1147 (1999).
40. J. Binney and S. Tremaine, *Galactic Dynamics*, Princeton Univ. Press., Princeton, 1994.
41. D. N. Spergel and P. J. Steinhard, *Phys. Rev. Lett.* **84**, 3760 (2000).
42. A.Vikhlinin et al., astro-ph/08122720, *Astrophys. J.*, in press (2009).

PART IV

Cosmic Ray Positron Excess

PAMELA AND ATIC ANOMALIES IN DECAYING GRAVITINO DARK MATTER SCENARIO

[a]KOJI ISHIWATA, [b]SHIGEKI MATSUMOTO and [a]TAKEO MOROI

[a]*Department of Physics, Tohoku University, Sendai 980-8578, Japan*

[b]*Department of Physics, University of Toyama, Toyama 930-8555, Japan*

Motivated by the recent results from the PAMELA and ATIC, we study the cosmic-ray electron and positron produced by the decay of gravitino dark matter. We calculate the cosmic-ray electron and positron fluxes and discuss implications to the PAMELA and ATIC data. In this paper, we will show that the observed anomalous fluxes by the PAMELA and ATIC can be explained in such a scenario. We will also discuss the synchrotron radiation flux from the Galactic center in such a scenario.

1. Introduction

In astrophysics, the existence of dark matter (DM) is almost conclusive. According to the recent survey of WMAP,[1] it accounts for 23 % of the total energy density in the universe. In the standard model of particle physics, however, there does not exist candidate for DM, which is one of the reasons to call for beyond the standard model. Supersymmetry (SUSY) is a promising model which can give an answer to the question; in the framework of SUSY, lightest superparticle (LSP) is a viable candidate for DM.

The fluxes of high energy cosmic rays give information about the properties of DM. In the recent years, accuracy of the measurements of the fluxes have been significantly improved. In particular, anomalous signals are reported by PAMELA[2] and ATIC[3] in the observations of cosmic-ray $e^{\pm}$. The PAMELA and ATIC results have attracted many attentions because the anomalies may indicate an unconventional nature of DM. In fact, there have been a sizable number of DM models are proposed to explain the anomalies after the announcements of the PAMELA and ATIC results. Among them, especially in decaying DM scenarios, the observed anomalies can be well explained with the appropriate choice of the lifetime of DM especially in leptonically decaying scenarios. (For works calculating cosmic-ray $e^{\pm}$, see[4,5] and references therein.)

In usual supersymmetric scenario, R-parity conservation is assumed, which protects LSP from decaying into standard model particles and makes it a viable candidate of DM. If we consider the case that R-parity is violated, LSP is no loner stable; however, if R-parity violation (RPV) is weak enough, the lifetime of the LSP can be much longer than the present age of the universe and LSP can play the role of DM.[6] In addition, when the order of the RPV is properly chosen to give the lifetime of $O(10^{26}$ sec), produced cosmic-ray positron gives excellent agreement with PAMELA data.[4]

On the other hand, synchrotron radiation from the decay of DM may give constrains directly to scenarios explaining the PAMELA and ATIC anomalies. Since DM decays into energetic $e^{\pm}$ under the magnetic fields in our galaxy, synchrotron radiation is inevitably induced. Importantly, the WMAP collaboration has observed the radiation in the whole sky, so that the observation gives constraints on the scenarios of the $e^{\pm}$ production due to the decay of DM in the Galactic halo.

In this paper, we consider gravitino (donated as ψ_μ) LSP in RPV. In the scenario, we calculate cosmic-ray $e^{\pm}$ and synchrotron radiation flux induced by them, paying particular attentions to PAMELA and ATIC anomalies. We will see that the PAMELA and ATIC anomalies are simultaneously explained if the lifetime of the gravitino DM is $O(10^{26}$ sec) and the mass is $\sim 1-2$ TeV.[4,5] In addition, synchrotron radiation from the Galactic center is comparable with or smaller than the observation.[7]

2. The Scenario and Model Framework

In this section, we briefly explain the cosmological aspects and the model framework of ψ_μ-DM scenario in RPV. With RPV, the ψ_μ LSP becomes unstable and energetic positron can be produced by the decay. Even if the ψ_μ is unstable, it can be DM if the RPV is weak enough so that the lifetime of the gravitino $\tau_{3/2}$ is much longer than the present age of the universe.[6,8] In fact, such a scenario has several advantages. In the ψ_μ-LSP scenario with RPV, the thermal leptogenesis[9] becomes possible without conflicting the big-bang nucleosynthesis constraints. In addition, the fluxes of the positron and γ-ray can be as large as the observed values, and the anomalies in those fluxes observed by the HEAT[10] and the EGRET[11] experiments, respectively, can be simultaneously explained in such a scenario if $\tau_{3/2} \sim O(10^{26}$ sec).[12,13]

Here, let us consider the bi-linear RPV interactions. Using the bases where the mixing terms between the up-type Higgs and the lepton doublets are eliminated from the superpotential, the relevant RPV interactions are given by

$$\mathcal{L}_{\mathrm{RPV}} = B_i \tilde{L}_i H_u + m^2_{\tilde{L}_i H_d} \tilde{L}_i H_d^* + \mathrm{h.c.}, \tag{1}$$

where $\tilde{L}_i$ is left-handed slepton doublet in i-th generation, while H_u and H_d are up- and down-type Higgs boson doublets, respectively. Then, the ψ_μ decays as $\psi_\mu \to l_i^\pm W^\mp$, $\nu_i Z$, $\nu_i h$, and $\nu_i \gamma$, where $l_i^\pm$ and ν_i are the charged lepton and the neutrino in i-th generation, respectively. Taking account of all the relevant Feynman diagrams, we calculate the branching ratios of these processes.[12] When the gravitino mass $m_{3/2}$ is larger than m_W, the dominant decay mode is $\psi_\mu \to l_i^\pm W^\mp$. In such a case, we see $\tau_{3/2} \simeq 6 \times 10^{25}$ sec $\times (\kappa_i/10^{-10})^{-2}(m_{3/2}/1\ \mathrm{TeV})^{-3}$, where $\kappa_i = (B_i \sin\beta + m^2_{\tilde{L}_i H_d} \cos\beta)/m^2_{\tilde{\nu}_i}$ is the ratio of the vacuum expectation value of the sneutrino field to that of the Higgs boson, with $\tan\beta = \langle H_u^0 \rangle / \langle H_d^0 \rangle$, and $m_{\tilde{\nu}_i}$ being the sneutrino mass. Thus, $\tau_{3/2}$ is a free parameter and can be much longer than the present age of the universe if the RPV parameters B_i and $m^2_{\tilde{L}_i H_d}$ are small enough.

3. Electron and Positron Fluxes

Let us first summarize our procedure to calculate the $e^\pm$ fluxes $\Phi_{e^\pm}$. (For detail, see.[4,5,12]) We solve the diffusion equation to take account of the effects of the propagation of $e^\pm$. The energy spectrum of the $e^\pm$ from DM $f_{e^\pm}(E, \vec{r})$ evolves as[14]

$$\frac{\partial f_{e^\pm}}{\partial t} = K(E) \nabla^2 f_{e^\pm} + \frac{\partial}{\partial E} \left[b(E) f_{e^\pm} \right] + Q. \tag{2}$$

The function K is expressed as $K = K_0 E^\delta_{\mathrm{GeV}}$,[15] where E_{GeV} is the energy in units of GeV, while $b = 1.0 \times 10^{-16} \times E^2_{\mathrm{GeV}}$ GeV/sec. In our numerical calculation, we use the following three sets of the model parameters, called MED, M1, and M2 models, which are defined as $(\delta, K_0[\mathrm{kpc}^2/\mathrm{Myr}], L[\mathrm{kpc}]) = (0.70, 0.0112, 4)$ (MED), $(0.46, 0.0765, 15)$ (M1), and $(0.55, 0.00595, 1)$ (M2), with $R = 20$ kpc for all models. Here, L and R are the half-height and the radius of the diffusion zone, respectively. The MED model is the best-fit to the boron-to-carbon ratio analysis, while the maximal and minimal positron fractions for $E \gtrsim 10$ GeV are expected to be estimated with M1 and M2 models, respectively. We found that the MED and M1 models give similar positron fraction, so only the results with the MED and M2 models are shown in the following. The source term is given as,

$$Q_{\rm dec} = \frac{1}{\tau_{\rm DM}} \frac{\rho_{\rm DM}(\vec{x})}{m_{\rm DM}} \left[\frac{dN_{e^\pm}}{dE}\right]_{\rm dec}, \tag{3}$$

where $\tau_{\rm DM}$ is the lifetime of DM. In the above expressions, $[dN_{e^+}/dE]_{\rm dec}$ is the energy distributions of the $e^\pm$ from single decay processes, respectively, and are calculated by using PYTHIA package[16] for each DM candidate. In addition, $\rho_{\rm DM}$ is the DM mass density for which we adopt the Navarro-Frank-White (NFW) mass density profile:[17] $\rho_{\rm NFW}(\vec{x}) = \rho_\odot r_\odot (r_c + r_\odot)^2 / r(r_c + r)^2$, where $\rho_\odot \simeq 0.30$ GeV/cm^3 is the local halo density around the solar system, $r_c \simeq 20$ kpc is the core radius of the DM profile, $r_\odot \simeq 8.5$ kpc is the distance between the Galactic center and the solar system, and r is the distance from the Galactic center.

Once $f_{e^\pm}$ are given by solving the above equation, the fluxes can be obtained as $[\Phi_{e^\pm}(E)]_{\rm DM} = \frac{c}{4\pi} f_{e^\pm}(E, \vec{x}_\odot)$, where $\vec{x}_\odot$ is the location of the solar system, and c is the speed of light. In order to calculate the total fluxes of $e^\pm$, we also have to estimate the background fluxes. In our study, we adopt the following fluxes for cosmic-ray $e^\pm$ produced by collisions between primary protons and interstellar medium in our galaxy:[14,18] $[\Phi_{e^-}]_{\rm BG} = 0.16 E_{\rm GeV}^{-1.1}/(1 + 11 E_{\rm GeV}^{0.9} + 3.2 E_{\rm GeV}^{2.15}) + 0.70 E_{\rm GeV}^{0.7}/(1 + 110 E_{\rm GeV}^{1.5} + 600 E_{\rm GeV}^{2.9} + 580 E_{\rm GeV}^{4.2})$ GeV^{-1} cm^{-2} sec^{-1} str^{-1} for the electron, and $[\Phi_{e^+}]_{\rm BG} = 4.5 E_{\rm GeV}^{0.7}/(1 + 650 E_{\rm GeV}^{2.3} + 1500 E_{\rm GeV}^{4.2})$ GeV^{-1} cm^{-2} sec^{-1} str^{-1} for the positron.

4. Synchrotron Radiation: Formalism and the Observation

In this section, we first show the formalism for calculation of synchrotron radiation flux. (The detail is in.[7]) Then, we address the implication of the present observation of synchrotron radiation from the Galactic center region.

Synchrotron radiation energy density per unit time and unit frequency is expressed as

$$L_\nu(\vec{x}) = \int dE\ \mathcal{P}(\nu, E) f_{e^\pm}(E, \vec{x}). \tag{4}$$

Here, $\mathcal{P}(\nu, E)$ is synchrotron radiation energy per unit time and unit frequency from single $e^\pm$ with energy E. Adopting the Galactic magnetic flux density of $B \sim 3\ \mu$G, we can see that the synchrotron radiation in the observed frequency band of the WMAP (i.e, $22 - 93$ GHz) is from the $e^\pm$

with the energy of $E \sim 10 - 100$ GeV. For the $e^{\pm}$ in such an energy range, $f_{e^{\pm}}$ can be well approximated by

$$f_{e^{\pm}}^{(\mathrm{local})}(E, \vec{x}) = \frac{1}{\tau_{\mathrm{DM}}} \frac{\rho_{\mathrm{DM}}(\vec{x})}{m_{\mathrm{DM}}} \frac{Y_{e^{\pm}}(>E)}{b(E, \vec{x})}, \tag{5}$$

where $Y_{e^{\pm}}(>E) \equiv \int_E^{\infty} dE' [dN_{e^{\pm}}/dE']_{\mathrm{dec}}$. Thus, we use $f_{e^{\pm}}^{(\mathrm{local})}$ in the calculation of synchrotron radiation flux. Here, we note in this formula that we take into account the effects of both synchrotron radiation and inverse Compton scattering for energy loss rate as $b(E, \vec{x}) = P_{\mathrm{synch}} + P_{\mathrm{IC}}$. This is because, in the Galactic center region, the inverse Compton scattering in the infrared γ-ray from stars becomes the dominant energy-loss process, thus it can not be neglected.

In order to calculate the observed radiation energy flux, we integrate $L_{\nu}(\vec{x})$ along the line of sight (l.o.s.), whose direction is parametrized by the parameters θ and ϕ, where θ is the angle between the direction to the Galactic center and that of the line of sight, and ϕ is the rotating angle around the direction to the Galactic center. (The Galactic plane corresponds to $\phi = 0$ and π.) Then, the synchrotron radiation flux is given by

$$J_{\nu}(\theta, \phi) = \frac{1}{4\pi} \int_{\mathrm{l.o.s.}} d\vec{l} L_{\nu}(\vec{l}). \tag{6}$$

Notice that, adopting the approximation of the constant magnetic flux in the Galaxy, the line of sight and energy integrals factorize.

Radiation flux from Galactic center region has been observed by the WMAP for frequency bands of 22, 33, 41, 61, and 93 GHz.[19,20] Since then, intensive analysis has been performed to understand the origins of the radiation flux. (For recent studies, see.[19–21]) Most of the radiation flux is expected to be from astrophysical origins, such as thermal dust, spinning dust, ionized gas, and synchrotron radiation, which have been studied by the use of other survey data.[22] With the three-year data, the WMAP collaboration claimed that the flux intensity can be explained by the known astrophysical origins.[23] On the contrary, Refs.[19,21] also studied the WMAP three-year data, and claimed that there exists a remnant flux from unknown origin which might be non-astrophysical; the remnant flux is called the "WMAP Haze". However, no clear indication of the WMAP Haze from unknown source was reported by the WMAP collaboration after five-year data.[20]

The existence of the WMAP Haze seems still controversial, and the detailed studies of the WMAP Haze using the data is beyond the scope of

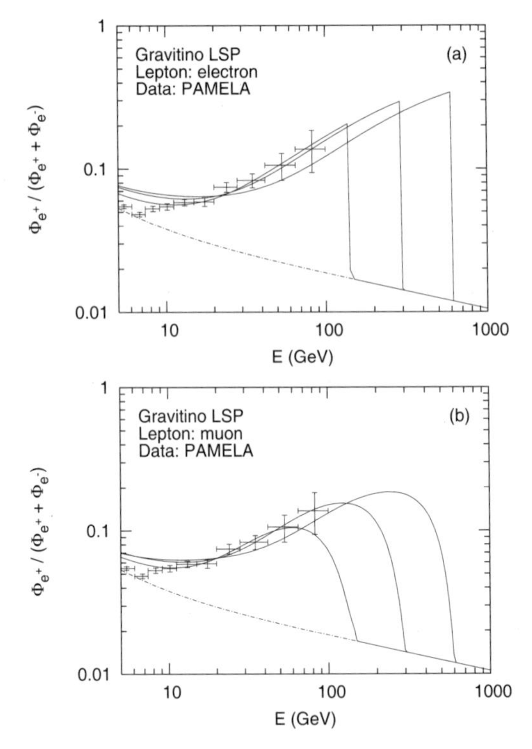

Fig. 1. Positron fractions for the case that ψ_μ dominantly decays to (a) the first-generation lepton in MED model and (b) the second-generation lepton in M2 model. Here, we take $m_{3/2} = 300$ GeV, 600 GeV, and 1.2 TeV (from left to right), with $\tau_{3/2} = 2.0 \times 10^{26}$ sec, 1.1×10^{26} sec, and 8.6×10^{25} sec (9.3×10^{25} sec, 5.8×10^{25} sec, and 5.0×10^{25} sec) in (a) ((b)), respectively. Dot-dashed line is the fraction calculated only by the background fluxes.

our study. Here, we adopt the flux of the WMAP Haze suggested in[21] (i.e. $O(1$ kJy/str$)$) as a reference value.

5. Numerical Results

First, we show the numerical results of the positron fraction. For simplicity, assuming a hierarchy among the RPV coupling constants, we consider the case where the ψ_μ decays selectively into the lepton in one of three generations (plus $W^\pm$, Z, or h). In Fig. 1, we show the positron fraction for the case that the ψ_μ decays only into first- (second-) generation lepton. Here, we use MED (M2) model for first- (second-) generation case and take $m_{3/2} = 300$ GeV, 600 GeV, and 1.2 TeV, with $\tau_{3/2} = 2.0 \times 10^{26}$ sec,

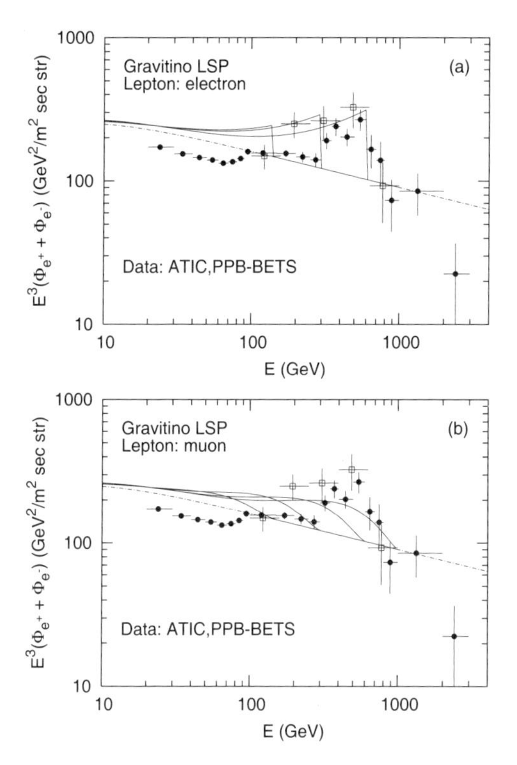

Fig. 2. Total flux: $\Phi_{e^+} + \Phi_{e^-}$ with MED (M2) model for the case that the ψ_μ dominantly decays to the first- (second-) generation lepton in (a) ((b)). Dot-dashed line is the background flux. Here, we use the same parameters in (a) and (b) of Fig. 1, respectively. In (b), we also plot the flux with $m_{3/2} = 2$ TeV and $\tau_{3/2} = 4.6 \times 10^{25}$ sec and PPB-BETS data[24] .

1.1×10^{26} sec, and 8.6×10^{25} sec (9.3×10^{25} sec, 5.8×10^{25} sec, and 5.0×10^{25} sec), which are the best-fit lifetime with PAMELA data, respectively. Here, in order to determine the best-fit lifetime, we calculate χ^2 by the use of PAMELA data. (In our χ^2 analysis, since the positron fraction in the low energy region is sensitive to the background fluxes, we only use the data points with $E \geq 15$ GeV.) From the figure, we see that the positron fraction well agrees with the PAMELA data for $m_{3/2} \gtrsim 100$ GeV irrespective of the gravitino mass if $\tau_{3/2}$ is properly chosen. (Simultaneously, the energetic γ-ray flux is also enhanced, which can be an explanation of the γ-ray excess observed by the EGRET.[11])

Next, we move on to the total flux: $\Phi_{e^+} + \Phi_{e^-}$. The numerical results are shown in Fig.2. Here, we use the best-fit lifetime with the PAMELA data

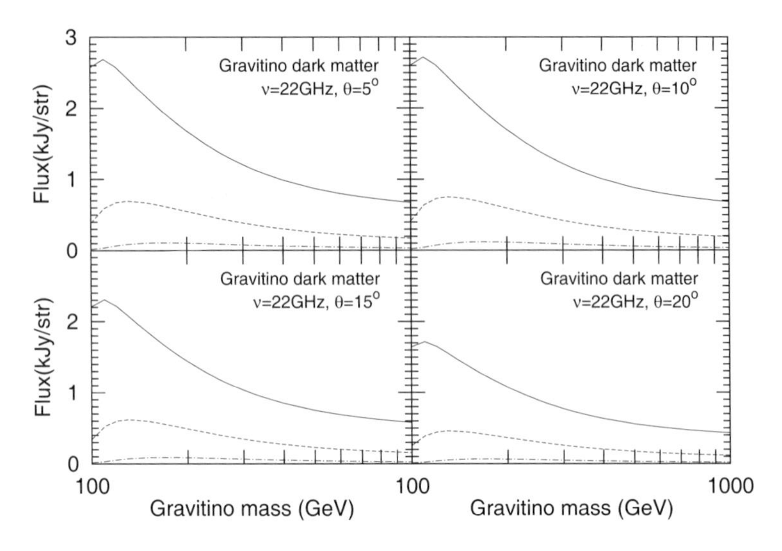

Fig. 3. Synchrotron radiation fluxes at $\nu = 22$ GHz as functions of gravitino mass for angle $\theta = 5°$, $10°$, $15°$, and $20°$. The final-state lepton in the ψ_μ decay is in the first generation. Here, we take $\tau_{3/2} = 5 \times 10^{26}$ sec, and show the cases of B =1, 3, 10 μG (from the bottom to the top) for each figure.

for each $m_{3/2}$, namely the same value in (a) and (b) of Fig. 1, respectively. From the figure, we see that the observed anomalous structure is well reproduced in the both cases. Especially, the result is a good agreement with the observation when $m_{3/2} = 1.2$ TeV (2 TeV) for the case that the final state lepton is the first- (second-) generation. We also note that, in the total flux, the numerical results does not change drastically by the choice of the background. This is because the signal from the gravitino is larger than (or at least comparable to) the background.

Finally, let us discuss the synchrotron radiation flux. The numerical results are shown in Fig.3. In this figure, we consider the case that the ψ_μ mainly decays to first generation lepton and plot for $\nu = 22$ GHz as the function of $m_{3/2}$, taking $\tau_{3/2} = 5 \times 10^{26}$ sec. The angle is set as $\phi = \frac{\pi}{2}$, and $\theta = 5°,\ 10°,\ 15°$, and $20°$, and we take $B = 1,\ 3$, and $10\ \mu$G. For the ψ_μ-DM case, it can be seen that the synchrotron radiation flux is of the order of ~ 1 kJy/str or smaller. As we mentioned, since the the existence of the exotic radiation flux of this size is controversial, it is difficult to confirm or exclude the present scenario without better understandings of the sources of Galactic foreground emission.

6. Conclusions

In this paper, we have studied the cosmic-ray fluxes from the ψ_μ-DM decay in RPV, motivated by the recent observations by PAMELA and ATIC. Assuming that the ψ_μ is the dominant component of DM, we calculate the cosmic-ray $e^\pm$, and found that the both anomalies can be well explained when $\tau_{3/2} \sim O(10^{26}$ sec). In particular, we saw that the ATIC anomaly indicates $m_{3/2} \sim 1 - 2$ TeV in this scenario. We also calculate the synchrotron radiation induced by the cosmic-ray $e^\pm$ from the Galactic center region with the lifetime to explain PAMELA and ATIC anomalies. Then, we obtained the result that the synchrotron radiation flux is $O(1$ kJy/str) or smaller, which does not exclude our scenario by the observation of the Galactic foreground emission.

References

1. G. Hinshaw *et al.* [WMAP Collaboration], Astrophys. J. Suppl. **180**, 225 (2009).
2. O. Adriani *et al.*, arXiv:0810.4995 [astro-ph].
3. J. Chang *et al.*, Nature **456**, 362 (2008).
4. K. Ishiwata, S. Matsumoto and T. Moroi, arXiv:0811.0250 [hep-ph].
5. K. Ishiwata, S. Matsumoto and T. Moroi, arXiv:0903.0242 [hep-ph].
6. F. Takayama and M. Yamaguchi, Phys. Lett. B **485**, 388 (2000).
7. K. Ishiwata, S. Matsumoto and T. Moroi, arXiv:0811.4492 [astro-ph].
8. W. Buchmuller *et al.*, JHEP **0703**, 037 (2007).
9. M. Fukugita and T. Yanagida, Phys. Lett. B **174**, 45 (1986).
10. S. W. Barwick *et al.*,Astrophys. J. **482**, L191 (1997).
11. P. Sreekumar *et al.*,Astrophys. J. **494**, 523 (1998).
12. K. Ishiwata, S. Matsumoto and T. Moroi, Phys. Rev. D **78**, 063505 (2008).
13. A. Ibarra and D. Tran, JCAP **0807**, 002 (2008).
14. E. A. Baltz and J. Edsjo, Phys. Rev. D **59**, 023511 (1999).
15. T. Delahaye *et al.*, Phys. Rev. D **77**, 063527 (2008).
16. T. Sjostrand, S. Mrenna and P. Skands, JHEP **0605**, 026 (2006).
17. J. F. Navarro, C. S. Frenk and S. D. M. White, Astrophys. J. **490**, 493 (1997).
18. I. V. Moskalenko and A. W. Strong, Astrophys. J. **493**, 694 (1998).
19. G. Dobler and D. P. Finkbeiner, Astrophys. J. **680**, 1222 (2008).
20. B. Gold *et al.* [WMAP Collaboration], arXiv:0803.0715 [astro-ph].
21. D. Hooper, D. P. Finkbeiner and G. Dobler, Phys. Rev. D **76**, 083012 (2007).
22. D. P. Finkbeiner, Astrophys. J. **614**, 186 (2004).
23. G. Hinshaw *et al.* [WMAP Collaboration], Astrophys. J. Suppl. **170**, 288 (2007).
24. S. Torii *et al.*, arXiv:0809.0760 [astro-ph].

INDIRECT DARK MATTER SEARCHES VERSUS COSMIC RAY TRANSPORT MODEL UNCERTAINTIES

I. GEBAUER*

Institut für Experimentelle Kernphysik, Universität Karlsruhe (TH),
Postfach 6980, 76128 Karlsruhe, Germany
**E-mail: gebauer@ekp.uni-karlsruhe.de*

The assumption of isotropic diffusion has led to successful models for CR transport, capable of explaining the locally observed CR spectra, as well as the diffuse Galactic gamma rays up to 1 GeV. These models currently form the basis for many indirect DM searches. Galactic winds with speeds of more than 100 km/s have been observed by ROSAT. Such wind speeds are incompatible with isotropic diffusion.

Here, a transport model for Galactic CRs compatible with the wind velocities observed by ROSAT is presented. In such a model the contribution of antiprotons and positrons from Dark Matter annihilation to the local fluxes of CRs is reduced by a factor of $\mathcal{O}(10)$. We compare the model to the INTEGRAL observations of a large bulge/disk ratio, the WMAP haze and the EGRET excess, all of which have been interpreted in the context of Dark Matter annihilation (DMA) and comment on the DMA interpretation of the PAMELA and ATIC/PPB-BETS results.

1. Introduction

Galactic CRs form the main background for indirect DM searches in charged CRs and diffuse Galactic gamma rays. CR transport has been successfully modelled by an isotropic diffusion model with negligible Galactic wind velocities, as implemented in the GALPROP code[a]. The ROSAT Galactic wind observations confirm that our Galaxy launches supernova (SN) driven winds with speeds of about 150 km/s in the Galactic plane. Galactic winds of this strength are incompatible with isotropic models for CR transport, which only allow for wind speeds well below 10km/s in the plane.

[a]The GALPROP code is available from http://galprop.stanford.edu

In order to reproduce our local CRs in the presence of Galactic winds, the anisotropy in diffusion can no longer be neglected, particularly the diffusion in the disk and in the halo has to be different. In addition small scale phenomena such as trapping by molecular cloud complexes and the structure of our local environment (the local bubble or the local fluff) might influence the secondary CR production rate and our local CR density gradients.

We introduce an anisotropic convection driven transport model which is consistent with the Galactic wind observations by ROSAT. The model predicts a large bulge/disk ratio in positron annihilation as observed by INTEGRAL and is able to explain the absence of a positron annihilation signal from molecular clouds (MCs) as observed by INTEGRAL by virtue of a mechanism which confines and isotropizes CRs between MCs. The model reproduces the soft spatial gradient of diffuse gamma ray emission[1,2] while using a flat X_{CO} scaling factor and the supernova remnant (SNR) distribution as the source distribution.

We compare the model to several observations which have been interpreted in the context of DMA. These are the EGRET excess of diffuse gamma rays, the INTEGRAL 511 keV line, the WMAP haze and comment on the DMA interpretations of the PAMELA and ATIC/PPB-BETS results.

2. Anisotropic Convection Driven Transport Models

SNR are assumed to be the main source of CRs. Primary CRs interact with the gas where they produce secondaries such as boron. From the relative amount of secondaries (as given by the ratio of e.g. boron to carbon) and the time it takes CRs to reach the detector (as given by the ratio of radioactive instable isotopes to radioactive stable isotopes, e.g. $^{10}Be/^{9}Be$) one can infer that CRs on average spend most of their time in the low density regions in the halo, before they arrive at Earth.

CRs are thought to scatter resonantly on magnetic turbulences and most models for CR transport are based on a diffusion-convection equation. The spectrum and the origin of the turbulence are unknown. For energies smaller than $10^2 - 10^3$ GeV the turbulence can be generated by the CR plasma itself. For higher energies the growth rates of the resonant modes become smaller than the damping rates and self-confiment is no longer possible.[3] For these energies another mechanism for CR confinement has to be found. A possible solution has been suggested by Ref. 4, where confinement and isotropization by molecular cloud magnetic mirrors have been discussed. This idea receives support from the INTEGRAL observations of the 511 keV positron-electron annihilation line.[5] The width of this line indicates that practically all annihilations occur in the warm or hot neutral hydrogen gas, while the signal from molecular hydrogen is compatible with zero.[6] In principle one can argue that the filling factor is too low for MCs to be found by positrons. However, the high magnetic field in the MCs seems to correlate with the interstellar magnetic field.[7] In this case the positrons are magnetically reflected by the high field regions in MCs. The large average distance between MCs allows to trap particles up to TeV energies, leading to efficient reacceleration.[8] As shown by Ref. 4 magnetic reflection and trapping in MCs could provide a mechanism to confine and isotropize CRs at energies above $10^2 - 10^3 GeV$, where scattering is weak. If trapping by MCs is indeed efficient the CR density inside MCs is small, thus leading to a very weak positron annihilation signal from this phase of the gas.

Although diffusion in our Galaxy is expected to be anisotropic[9,10] and different in the thin halo and the gaseous disk, isotropic transport models succesfully describe the locally observed CRs and the diffuse Galactic gamma rays up to 1 GeV. However, in this case only negligible wind velocities are allowed.[11,12] A recent analysis of the ROSAT X-ray data has confirmed that our Galaxy launches winds with speeds of 150 km/s in the plane and up to 800 km/s in the halo.[13] Although this is a comparably low speed (starburst Galaxies launch winds with speeds up to 3000 km/s), the impact upon CR transport is significant, because CRs drift faster into the halo. As expected from self-consistent Galactic wind calculations,[9] the wind velocities appear to be roughly proportional to the SNR distribution. Transport models with isotropic diffusion can allow for convection velocity gradients of only 10 km/s/kpc (with zero wind speed at z=0 kpc, i.e. 40 km/s in a distance of 4 kpc from the plane), since for larger velocities the constraints from radioactive isotopes and secondary particle production cannot be met.[12] To account for the observed wind velocities anisotropic transport modes have to be taken into account.

INTEGRAL and ROSAT form strong constraints for any transport model. We have implemented the transport picture described above into the publicly available GALPROP code. The main features of our implementation are: i) fast propagation perpendicular to the disk by turbulent diffusion parallel to the magnetic field and convection in agreement with the ROSAT X-ray data ii) slow diffusion in the disk and a possible trapping mechanism by molecular clouds which would increase the relative CR interaction rate in the atomic component of the gas. In addition we allow for small local variations in all transport parameters in order to account for our local Galactic environment. Due to slow diffusion in the Galactic plane the impact of local transport phenomena is larger compared to isotropic models.

Convection is chosen according to a fit to the ROSAT data.[14] The diffusion coefficient in the disk (in R-direction) and the halo (in z-direction) are adjusted independently to best reproduce the local age and secondary production rate and the CR injection spectra are chosen to best reproduce the locally measured proton and electron spectra. Figure 1 shows the local B/C ratio and the beryllium fraction for such an anisotropic convection driven model (aCDM).

Compared to the CR source distribution the spatial gradient of the observed diffuse gamma ray emission is rather flat.[1,2] Isotropic models have to assume either a somewhat flattened source distribution[15] or a flattened gas distribution by virtue of an increasing X_{CO} saling factor[16] in order to reproduce this soft gradient. Galactic winds in agreement with ROSAT naturally explain the observed soft spatial gradient in diffuse gamma-rays[9] without invoking a flattened source distribution or an increase in the X_{CO}-scaling factor:[16] the peak in the gas distribution coincides with the peak in the SNR distribution. Consequently the CR interaction rate in this region, which leads to a relative overproduction of diffuse γ-rays in the isotropic models, is reduced in an aCDM.

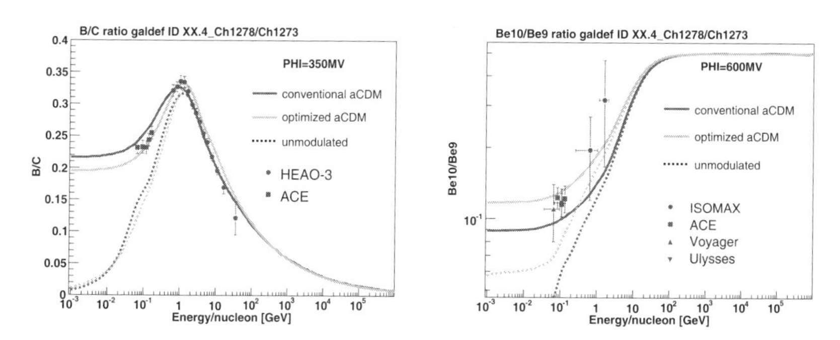

Fig. 1. Local B/C ratio (*left*) and local $^{10}Be/^{9}Be$-fraction (*right*) in an aCDM. Dashed lines represent the local interstellar spectrum (LIS), full lines represent for the LIS modulated to 350MeV (for B/C) and 600 MeV (for $^{10}Be/^{9}Be$). The *conventional* aCDM is tuned for local CR species, leading to the known excess of diffuse gamma-rays above 1 GeV, the *optimized* aCDM is tuned for diffuse gamma-rays and requires additional gradients in local diffusion coefficients, gas densities and Alfvén velocity.

3. The DMA Interpretation of the INTEGRAL Positron Annihilation Signal

INTEGRAL revealed a surprisingly high intensity of the positron annihilation signal from the region of the Galactic bulge, corresponding to a positron injection rate of approximately $1.5 \cdot 10^{43}$ e^+/s in the inner Galaxy[5] and a rather low signal from the disk. The opposite is expected,[17] because MeV positrons largely originate from the decay of radioactive nuclei (mainly ^{56}Co) in the core of dying stars, especially SNIa. In the case of SNIa the core makes up a large fraction of the mass, thus enabling the positrons to escape from the relatively thin layer of the ejecta. The light curves from SNIa suggest that only a few percent of the positrons escape from the ejecta and can annihilate outside after thermalization. These light curves are sustained by gamma rays in the shock waves and later on by the escaping electrons and positrons. The positron annihilation signal from inside the ejecta will undergo additional interactions with the shock waves and is not visible as a single 511 keV line.

If positron transport is neglected, i.e. if one assumes that positrons annihilate close to their sources, an additional source population, confined to the bulge, has to be considered in order to explain the large bulge/disk ratio. As a possible candidate positrons from annihilating[18] or decaying[19] light Dark Matter ($m <$ 100MeV) have been suggested.

However, the observed positron annihilation signal from the disk seems to be saturated already by the positrons from ^{26}Al and ^{44}Ti, leaving no room for additional positrons from SNIa explosions in the disk.[20] Any additional source cannot explain why there is no annihilation signal from SNIa in the disk.

Prantzos[17] pointed out that bulge/disk (B/D) ratios as small as 0.5 are compatible with the INTEGRAL data if the disk positrons diffuse sufficiently away from their sources. He estimated that about 50% of the $\sim MeV$ positrons have to leave the confinement region before slowing down.

The expected B/D ratio, and especially the positron escape fraction from the disk, highly depends on the transport model assumed. In propagation models without convection these low energy positrons annihilate near their source, since diffusion, which is proportional to the energy to some power, is practically absent for MeV particles. However, convection is independent of energy and these particles can be convected away easily to the halo, where they find no electrons to annihilate. The transport to the halo is additionally facilitated by the turbulent SN explosions, which generate high pressure bubbles blowing material out of the Galactic disk into chimneys reaching far into the halo.[21]

In a conventional aCDM About 40% of the 1 MeV positrons from ^{56}Co β^{+}-decays escape from the diffusion zone, for smaller energies the escape fraction rises to 57% at 0.1 MeV due to the smaller diffusion coefficient. The escape fraction is very sensitive to changes in the diffusion coefficient and the strength of the convection velocity, because the ratio of these two transport parameters determines the maximum height from which positrons can still return to the disk. At 0.1 MeV the escape fraction can be as large as 92% for an aCDM with small diffusion in z direction and as small as 14% for a model with convection velocity $10\frac{km}{s}$ at a height of 300pc above the plane.

In the bulge, convection is negligible due to the absence of sources and CR pressure. Consequently, diffusion is the dominant transport mode even for the MeV positrons ejected from SNIa, thus leaving them time to thermalize and annihilate in the large bulge (1-2kpc) before reaching the halo. It should be noted that due to the smaller source strength the number of positrons produced in the bulge is not sufficient to entirely explain the signal from the bulge.[17] Even in an aCDM an additional process or an additional source population in the bulge is required to explain the observed emission from the bulge. This might be a smaller diffusion coefficient for low energies in the bulge which would then increase the positron density at low energies in this region.

Recently it has been found that an additional low mass x-ray binary (LMXRB) population seems to reside in the bulge region showing even the morphological features of the observed annihilation signal.[22] This also cannot explain why there is almost no annihilation signal from positrons from ^{56}Co from the disk, but added to an aCDM this additional LMXRB population might nicely explain the B/D ratio.

The problem of the large B/D ratio for positron annihilation is thus intimately related to the propagation of MeV positrons. Given the fact that the INTEGRAL observations can be entirely understood in the framework of CR transport, we find no need to invoke an additional hypothetical contribution such as light Dark Matter.

4. The DMA Interpretation of the EGRET Excess and Charged CRs from DMA

An excess of diffuse gamma rays has been observed by the EGRET telescope on board of NASA's CGRO (Compton Gamma Ray Observatory).[23] Below

1 GeV the cosmic ray (CR) interactions describe the data well, but above 1 GeV the data are up to a factor two above the expected background. The excess shows all the features of DMA for a WIMP mass between 50 and 70 GeV, as shown on the left hand side of Fig. 2. The DMA interpretation of the EGRET excess is consistent with the expectations from Supersymmetry.[24]

The analysis of the EGRET data was performed with a data-driven background calibration. This is particularly suitable if the shape of the dominant background is well known. For diffuse Galactic gamma rays this is π^0 production in inelastic collisions of CR protons on the hydrogen gas of the disk. The shape of the resulting gamma ray spectrum is known from fixed target accelerator experiments and the shape of the DMA signal is known from $e+e-$ annihilation. Since the signal has a significantly harder spectrum than the background the two shapes can be fitted to the experimental data with a free normalization for each contribution. This way, the contribution from signal and background is obtained for all sky directions in a rather model independent way. Uncertainties in the interstellar background shape arise from solar modulation and in addition from the uncertainties from electron CRs generating gamma-rays by inverse Compton scattering and Bremsstrahlung. However, since the electron flux of CRs is two orders of magnitude below the proton flux, this effect can be included with sufficient accuracy.

DMA signal and background shape have been fitted simultaneously to 180 independent sky directions. The average χ^2 per degree of freedom summed over all ca. 1400 data points is around 1, indicating that the errors are correctly estimated. The good χ^2 implies that the main conditions for a signal of DMA are fulfilled, namely i) the shape of the excess corresponds to the fragmentation of mono-energetic quarks with the same energy in all sky directions and ii) it was found that the intensity distribution of the excess agrees with the mass distribution, as deduced from the rotation curve[25] iii) the background distribution agrees within errors with the expectation from GALPROP, as can be seen from Fig. 3 in Ref. 25.

The DM halo profile derived from this fit shows some unexpected substructure: outside the disk it corresponds to a cored halo profile, but inside the disk it reveals two additional ring-like structures at distances of about 4 and 13 kpc from the Galactic center. Structures are expected from the tidal disruption of dwarf galaxies captured in the gravitational field of our Galaxy. The ring of stars or Monocerus stream (with about $10^8 - 10^9$ solar masses in visible matter) could be the tidal streams of the Canis Major dwarf galaxy (see e.g. Ref. 26,27 and references therein). If so, the tidal streams predicted from N-body simulations are consistent with the ring at 13 kpc.[27] The strong gravitational potential well in this stream can also be observed in reduced gas flaring at the position of the ring.[28] The half-width-half-maximum of the gas layer in the disk is shown on the right-hand panel of Fig. 2. The reduction of gas flaring corresponds to more than 10^{10} solar masses, in agreement with what one would expect from the EGRET data. It should be noted that the peculiar shape of the gas flaring was only understood after the astronomers heard about the EGRET ring. The effect is so large that visible matter cannot explain this peculiar shape, simply because there is not much visible matter above 10 kpc. The change in the slope of the rotation curve at a Galactic radius around 10 kpc can also be explained by this ringlike structure.[25] A similar ring in the outer disk has been discovered in a nearby galaxy, indicating that such infalls may shape the disk and its warps.[29] The ring at 4 kpc might also originate from the disruption of a smaller dwarf galaxy, but here the density of stars is too high to find evidence for tidal streams. However, direct evidence of a stronger gravitational potential well in this region comes from the ring of dust at this location. Since this ring is slightly tilted with respect to the plane its presence and orientation can be explained by the presence of a ringlike structure of DM.

It should be noted that such substructures are most easily discovered by a data-driven approach, which does not rely on propagation and DM halo models.

The local fluxes of electrons, positrons, protons and antiprotons, however, do depend strongly on the transport model. Isotropic transport models usually predict local antiproton fluxes from CRs about 40% too low.[12] However, if antiprotons from DMA, consistent with the EGRET excess are invoked, the predicted fluxes are too large by an order of magnitude.[31]

For a given DM halo profile the local contribution of charged products from DMA depends on the distance CRs travel in the plane before they escape to the halo and their return probability once they are in the halo. If convection plays a major rôle for CR transport, as indicated by the ROSAT observations, the contribution from DMA in local CRs is small, since most of the local positrons and antiprotons from DMA are produced in the halo and the inner ring, thus drifting away by convection.

Figure 3 shows the local antiproton flux for an aCDM. As in the isotropic models the antiproton flux from nuclear interactions tends to be on the low side compared to the data from PAMELA,[32] CAPRICE[33] and BESS.[34] Assuming a 50-70 GeV WIMP as inferred from the EGRET analysis discussed above, this can be remedied by the antiproton flux from DMA, which has a similar shape as the background. The relative production of antiprotons and gamma rays is known from the fragmentation of quarks, as measured in e^+e^- annihilation. No additional fine-tuning is required to describe the data on local antiprotons: convection speeds are fixed by the ROSAT observations and transport parameters are fixed by local CR spectra and constraints from synchrotron radiation. It should be reminded that propagation models without the DMA contribution to antiprotons are challenged by the rather low prediction for local antiprotons.[35] A higher antiproton flux can be obtained, if one resorts to so-called optimized models, in which the global CR spectra are different from the local ones. Such optimized models have been devised to explain the EGRET excess without DMA.[36]

It should be mentioned at this point that the Fermi/LAT observations of the Vela pulsar show a somewhat softer spectrum than previously measured by EGRET.[37] This supports previous hypotheses of a modified spectrum due to detector effects[38] and suggests that the excess in diffuse gamma rays

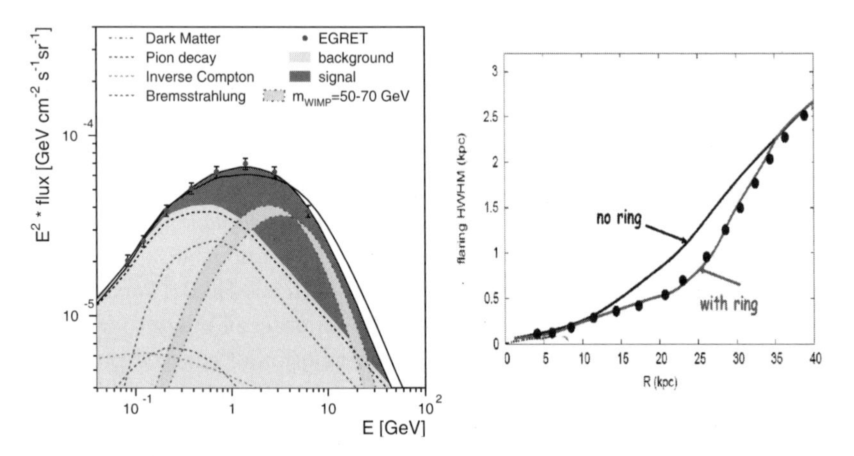

Fig. 2. Left: Fit of the shapes of background and DMA signal to the EGRET data in the direction of the Galactic center. The light shaded (yellow) area indicates the background using the shapes known from accelerator experiments, while the dark shaded (red) area corresponds to the signal contribution from DMA for a 60 GeV WIMP mass, where the small intermediate (blue) shaded area corresponds to a variation of the WIMP mass between 50 and 70 GeV. Taken from Ref. 25. Right: The half-width-half-maximum (HWHM) of the gas layer of the Galactic disk as function of the distance from the Galactic center. Clearly, the fit including a ring of dark matter above 10 kpc describes the data much better. Adapted from data in Ref. 28.

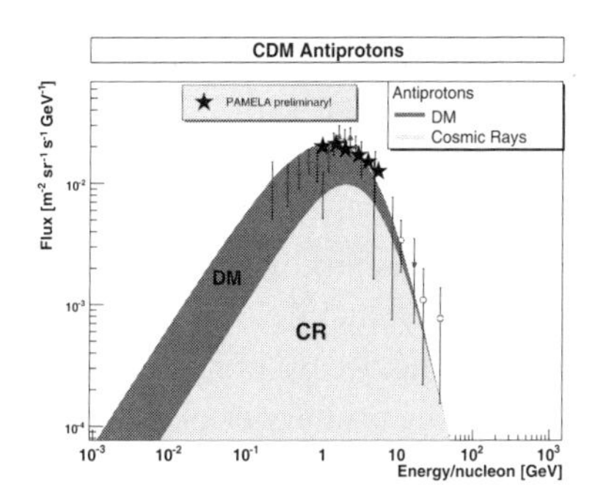

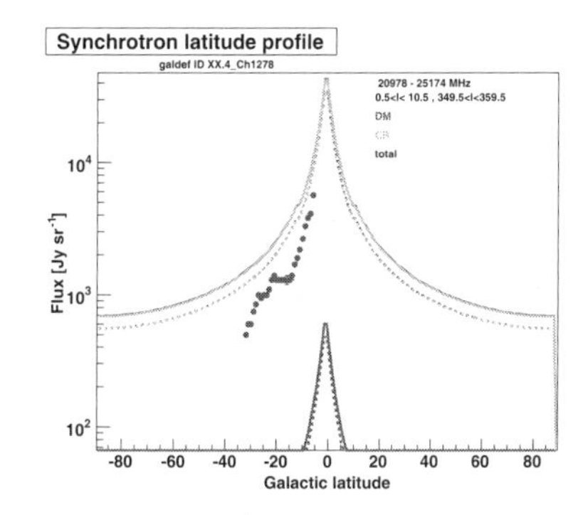

Fig. 3. Left: Local antiproton flux in an aCDM. The contribution from CRs is very similar to the predictions of the isotropic models, while the contribution from DMA is reduced by virtue of the smaller collection volume. Right: Latitude distribution of the synchrotron radiation in the small longitude range were the haze has been measured. The top solid curve is the total flux, the red dots represent the haze (adapted from Ref. 30) and the contribution from the DMA interpretation of the EGRET excess.

will be less significant in the Fermi data. If the EGRET excess is indeed a signature of DMA, then the same spatial features should be present in the Fermi data and a very similar halo profile should improve a fit to the Fermi data.

The right side of Fig. 3 shows the latitude profile of the synchrotron radiation in the so-called WMAP haze region for 22GHz. The WMAP haze consists of an excess of microwave emission from a small region close to the Galactic center. It has been suggested that this signal could be synchrotron emission from relativistic electrons and positrons, possibly originating from DMA in a cuspy halo.[30] However, even in a cored profile the synchrotron radiation from the disk shows a steep increase as shown in Fig. 3, while the intensity of synchrotron radiation from DMA (as required to explain the EGRET excess) is much lower.

5. The DMA Interpretation of the "Anomalous" PAMELA and ATIC/PPB-BETS Results

The positron fraction is defined as the ratio of fluxes of positrons and the sum of electrons and positrons, i.e. $e^+/(e^+ + e^-)$. The PAMELA data on the positron fraction[39] indicate a positron flux much harder than the predictions of the isotropic transport models above ~7 GeV. At the same time the electron spectrum as measured by ATIC[40] and PPB-BETS[41] indicates a "bump" at around 500 GeV. These observations suggest an additional hard positron and electron component which dominates the local spectra for higher energies. Both results have been interpreted as a possible signal from DMA by many authors (see e.g. Ref. 42 for an extensive compilation of references).

Usually the predictions of the conventional GALPROP model[43] are used to constrain the CR contribution in electrons and positrons. CR positrons are produced by the decay of positively charged pions produced by inelastic collisions of CRs with the gas in the disk. Electrons mainly originate from SNRs and the fraction of electrons produced by the decays of the negative pions from CR interactions is small compared to the primary electrons from SNRs. In the absence of additional sources, the local positron spectrum is therefor related to the local spectrum of the nuclei and the energy losses, the local electron spectrum is determined by the electron injection index and the electron energy losses.

Figure 4 shows the local electron and proton spectra. The spectra measured by different experiments differ significantly. Especially the more precise magnetic spectrometer experiments tend to show a power law spectrum above a certain energy with a constant index, while the calorimetric experiments tend to show an excess above the extrapolation for this constant index. E.g. the HEAT and AMS-01 data on electrons are consistent with an electron injection index close to -2.9 in an aCDM. If one includes the calorimetric experiments at higher energies, a harder spectrum with an injection index well above -2.7 is preferred. Local pulsars are expected to dominate the spectrum at higher energies, but without charge measurements of the particles the calorimetric experiments are prone to background from low energetic heavy nuclei.

In Fig. 5 the ratio of protons/electrons for different electron injection indices is compared to the best fitted electron and proton data above 5 GeV. Independent of the electron injection index the p/e ratio shows an increase towards higher energy. Since positrons are produced from protons and nuclei, the positron spectrum is determined by the proton spectrum and the energy losses. Given the increase in p/e the corresponding increase in the positron fraction is not surprising. However, the strength of the increase depends on the transport model (the energy losses for local CRs). The right hand side of Fig. 5 shows the positron fraction for different electron injection indices. The index of -2.9, as preferred from the magnetic spectrometer experiments, also leads to a rising positron fraction, however, the increase observed by PAMELA would require an even softer electron spectrum corresponding to an injection index of -3.17. A positron contribution from local sources in this energy range is a likely possibility.

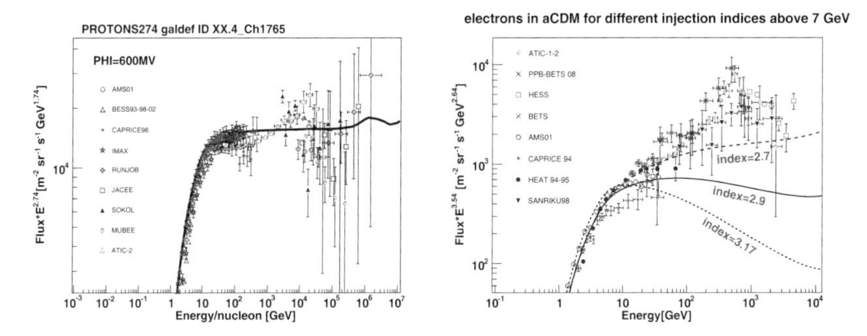

Fig. 4. The spectra of protons, electrons and positrons multiplied with a power of the energy, as indicated on the vertical axis. The data have been taken from a database by Strong[44] were the original references can be found.

It is interesting to note that ATIC measures an increase above the index from spectrometer experiments for both the electron and proton/nuclei spectra.[45] Just at the edge of acceptance the spectra then fall sharply. The increase in protons and nuclei, practically eliminates the DMA interpretation and strongly points to a local source, as suggested e.g. by Ref. 42,46–49, if it is not background from heavy nuclei increasing towards higher energies. In order to conclusively determine the $e^+/(e^+ + e^-)$-fraction expected from CRs one would need well measured electron and positron spectra, instead of just ratios.

Positrons from DMA are produced mainly by the decays of positively charged pions produced after the hadronization of the quarks. The contribution from DMA in the $e^+/(e^+ + e^-)$-fraction is highly uncertain and does depend on the transport model. In an aCDM the contribution as preferred by EGRET is small, as shown by the lower curve in the right hand panel of Fig. 5. This is expected, since most of the local positrons from DMA are produced in the halo and the inner ring, thus drifting away by convection, while the background is produced mainly in the disk.

Alternative DM candidates have been discussed in the literature (see e.g. the list of references in Ref. 42). However, even if a certain candidate matches the constraints from local antiprotons and diffuse gamma rays (which would require a rather fine-tuned DM model), the energetic e^+e^- pairs should be visible in synchrotron radiation, especially in the Galactic center. All this strongly points to a local point source or a group of local point sources, such as pulsars, as has been suggested by many authors.[42,46–49]

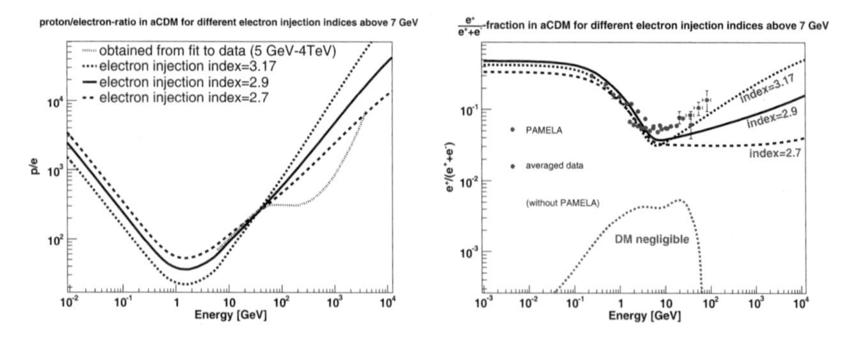

Fig. 5. Left: the ratio of proton and electron fluxes as function of energy compared to a fit to the data. Right: the positron fraction. The averaged HEAT and AMS-01 data are taken from Ref. 50 and the new more precise PAMELA data are from Ref. 39. The different curves correspond to different injection indices of electrons, which are not well constrained (see Fig. 4).

6. Conclusion

Isotropic transport models are inconsistent with wind speeds observed by ROSAT. To be compatible with convective transport modes the assumption of isotropic diffusion has to be dropped.

In this paper an anisotropic convection driven transport model which is compatible with the ROSAT wind observations is described. In such a model the large B/D ratio in positrons, which has been interpreted as DMA, is explained by CR transport. A possible DMA contribution in local CRs is largely reduced by virtue of the smaller collection volume, resulting from the observed Galactic winds. Due to the large uncertainties in local CR fluxes from DMA (factor of $\mathcal{O}(10)$), any DMA interpretation of features in the spectra of charged CRs therefor requires a detailed knowledge of the transport parameters in our local environment.

The increase in electrons at high energies as observed by ATIC and the rise in the positron fraction as observed by PAMELA may be related, but the fact that ATIC also sees an increase in protons and heavier nuclei (see Fig. 4 and Ref. 45) makes a DMA interpretation unlikely. To fully understand the rise in the positron fraction would require much better data for the spectra of electrons, protons and positrons separately instead of only ratios.

Acknowledgments

This work has been done in collaboration with Wim de Boer and Markus Weber.

References

1. A. W. Strong and J. R. Mattox, *Astron. Astrophys.* **308**, L21 (1996).
2. S. W. Digel, I. A. Grenier, A. Heithausen, S. D. Hunter and P. Thaddeus, *Astrophys. J* **463**, 609 (1996).
3. C. J. Cesarsky, *Ann. Rev. Astron. Astrophys.* **18**, 289 (1980), and references therein.
4. B. D. G. Chandran, *Astrophys. J.* **529**, p. 513 (2000).
5. J. Knodlseder *et al.*, *Astron. Astrophys.* **441**, 513 (2005).
6. P. Jean *et al.*, *Astron. Astrophys.* **445**, 579 (2006).
7. J.-L. Han and J. S. Zhang (2006), astro-ph/0611213.
8. V. N. Zirakashvili, Cosmic Ray Acceleration by Magnetic Traps , in *Proc. 26th ICRC 4,439*, (Salt Lake City, USA, 1999).
9. D. Breitschwerdt, V. A. Dogiel and H. J. Völk, *Astron. Astrophys.* **385**, 216 (2002).
10. D. De Marco, P. Blasi and T. Stanev, *JCAP* **0706**, p. 027 (2007).
11. A. W. Strong and I. V. Moskalenko, *Astrophys. J.* **509**, 212 (1998).
12. A. W. Strong, I. V. Moskalenko and V. S. Ptuskin, *Ann. Rev. Nucl. Part. Sci.* **57**, 285 (2007).
13. D. Breitschwerdt, *Nature* **452**, 826 (2008), and references therein.
14. J. E. Everett *et al.*, *Astrophys. J.* **674**, 258 (2008).
15. A. W. Strong, I. V. Moskalenko and O. Reimer, *Astrophys. J.* **537**, 763 (2000).
16. A. W. Strong, I. V. Moskalenko, O. Reimer, S. Digel and R. Diehl, *Astron. Astrophys.* **422**, L47 (2004).
17. N. Prantzos, *Astron. Astrophys.* **449**, 869 (2006).
18. C. Boehm, D. Hooper, J. Silk, M. Casse and J. Paul, *Phys. Rev. Lett.* **92**, p. 101301 (2004).
19. D. Hooper and L.-T. Wang, *Phys. Rev.* **D70**, p. 063506 (2004).
20. N. Prantzos, *New Astron. Rev.* **52**, 457 (2008).
21. M. A. de Avillez and D. Breitschwerdt, *Astrophys. J. Lett.* **665**, L35(August 2007).
22. G. Weidenspointner *et al.*, *Nature* **451**, 159 (2008).
23. S. D. Hunter *et al.*, *Astrophys. J.* **481**, 205(May 1997).
24. W. de Boer *et al.*, *Phys. Lett.* **B636**, 13 (2006).
25. W. de Boer *et al.*, *Astron. Astrophys.* **444**, 51 (2005).
26. N. F. Martin *et al.*, *Mon. Not. Roy. Astron. Soc.* **348**, p. 12 (2004).
27. J. Penarrubia *et al.*, *Astrophys. J.* **626**, 128 (2005).
28. P. M. W. Kalberla, L. Dedes, J. Kerp and U. Haud, *Astron. Astrophys.* **469**, 511(July 2007).
29. D. Martinez-Delgado *et al.*, *Astrophys. J.* **692**, 955 (2009).
30. D. Hooper, D. P. Finkbeiner and G. Dobler, *Phys. Rev.* **D76**, p. 083012 (2007).
31. L. Bergstrom, J. Edsjo, M. Gustafsson and P. Salati, *JCAP* **0605**, 006 (2006).
32. O. Adriani *et al.*, *Phys. Rev. Lett.* **102**, p. 051101 (2009).
33. M. Boezio *et al.*, *Astrophys. J.* **561**, 787 (2001).
34. T. Maeno *et al.*, *Astropart. Phys.* **16**, 121 (2001).

35. I. V. Moskalenko, A. W. Strong, S. G. Mashnik and J. F. Ormes, *Astrophys. J.* **586**, 1050 (2003).
36. A. W. Strong, I. V. Moskalenko and O. Reimer, *Astrophys. J.* **613**, 962 (2004).
37. A. A. Abdo *et al.* (2008), arXiv:astro-ph/0812.2960, to appear in Astrophys. J.
38. F. W. Stecker, S. D. Hunter and D. A. Kniffen, *Astroparticle Physics* **29**, 25(February 2008).
39. O. Adriani *et al.* (2008), arXiv:astro-ph/0810.4995, submitted to Nature.
40. J. Chang *et al.*, *Nature* **456**, p. 362 (2008).
41. S. Torii *et al.* (2008), arXiv:atro-ph/0809.0760, submitted to Astropart. Phys.
42. S. Profumo (2008), arXiv:astro-ph/0812.4457, submitted to Phys.Rev.D.
43. I. V. Moskalenko, A. W. Strong and O. Reimer, *Astron. Astrophys.* **338**, L75 (1998).
44. http://www.mpe.mpg.de/ aws/propagate.html.
45. A. D. Panov *et al.*, The results of ATIC-2 experiment for elemental spectra of cosmic rays, in *Proc. 29th All-Russian Cosmic Ray Conference*, (Moscow, Russia, 2006).
46. P. D. Serpico, *Phys. Rev.* **D79**, p. 021302 (2009).
47. D. Hooper, P. Blasi and P. D. Serpico, *JCAP* **0901**, p. 025 (2009).
48. F. A. Aharonian, A. M. Atoyan and H. J. Voelk, *Astron. Astrophys.* **294**, L41(February 1995).
49. S. Coutu *et al.*, *Astropart. Phys.* **11**, 429 (1999).
50. C. H. Chung *et al.*, The anomaly in the cosmic-ray positron spectrum, in *Proceedings of SUSY 2007*, (Karlsruhe, Germany, 2007).

SIGNATURES OF DARK MATTER ANNIHILATION IN THE LIGHT OF PAMELA/ATIC ANOMALY

KAZUNORI NAKAYAMA

Institute for Cosmic Ray Research, University of Tokyo,
Kashiwa, Chiba 277-8582, Japan
E-mail: nakayama@icrr.u-tokyo.ac.jp

Recent measurements of cosmic-ray electron and positron fluxes by PAMELA and ATIC experiments may indicate the existence of annihilating dark matter with large annihilation cross section. We discuss its possible relation to other astrophysical/cosmological observations : gamma-rays, neutrinos, and big-bang nucleosynthesis. It is shown that they give stringent constraints on some annihilating dark matter models.

Keywords: Dark matter, Cosmic rays, Big-bang nucleosynthesis.

1. Introduction

Recently, the PAMELA satellite experiment[1] reported an excess of the cosmic ray positron flux above the energy $\sim$ 10 GeV, and the ATIC balloon experiment[2] also found an excess of the electron plus positron flux, whose peak energy is around 600 GeV. These results are now drawing lots of attention of particle physicists, since they can be interpreted as signatures of dark matter (DM) annihilation/decay. Although some astrophysical sources are proposed,[3] we focus on the DM annihilation scenario as an explanation of the PAMELA/ATIC anomaly.[4]

It is known that if the positron/electron excess is caused by the DM annihilation, a large annihilation cross section $\langle \sigma v \rangle \sim 10^{-24}$-$10^{-23}$ cm^{3}s^{-1}, which is two or three magnitude of larger than the standard value for reproducing the observed DM abundance in the thermal relic scenario, is needed. Sommerfeld enhancement mechanism may provide a large annihilation rate.[5] Although a DM with large annihilation cross section does not lead to correct DM abundance, DM can also be produced nonthermally, through the decay of long-lived particles, for example.[6] Thus non-thermal DM is a good candidate as a source of the PAMELA/ATIC anomaly. In this study, we show that those DM models are constrained from other observations : gamma-rays, high-energy neutrinos and big-bang nucleosynthesis (BBN).[a] This study is based on the works with J. Hisano, M. Kawasaki, K. Kohri and T. Moroi.[7–9]

2. Cosmic Rays from Dark Matter Annihilation

2.1. *Positrons and Electrons*

In the Galaxy, DM annihilates each other producing high energy particles, including positrons and electrons. High-energy charged particles propagates under an influence by the highly tangled magnetic fields. They also lose their energy during the propagation inside the Galaxy due to the synchrotron emission and inverse Compton scattering processes with CMB photon and diffuse star light. Thus their motion can be regarded as a random walk process with energy loss, and described by the following diffusion loss equation,[10]

$$\frac{\partial f}{\partial t}(E,\vec{x}) = K(E)\nabla^2 f(E,\vec{x}) + \frac{\partial}{\partial E}[b(E)f(E,\vec{x})] + Q(E,\vec{x}), \qquad (1)$$

where $f(E,\vec{x})$ is the positron/electron number density with energy E at the position $\vec{x}$, $K(E)$ is the diffusion constant and $b(E)$ denotes the energy loss due to the synchrotron emission and inverse Compton scattering process off the microwave background photons and diffuse star light. The source term, $Q(E,\vec{x})$, represents the injection from DM annihilation. Then we can solve the diffusion equation in a semi-analytical way.[11] Results are shown in Fig. 1. The top panel shows the ratio of the positron to the sum of electron plus positron flux, and the bottom panel shows the electron plus positron flux multiplied by E^3. We consider three DM models : DM annihilates into e^+e^- with $m = 700$ GeV and $\langle\sigma v\rangle = 5\times10^{-24}$ cm^{3}s^{-1}, into $\mu^+\mu^-$ with $m = 1$ TeV and $\langle\sigma v\rangle = 1.5\times10^{-23}$ cm^{3}s^{-1}, and into $\tau^+\tau^-$ with $m = 1.2$ TeV and $\langle\sigma v\rangle = 2\times10^{-23}$ cm^{3}s^{-1}. Background flux is taken from the simulation by Moskalenko and Strong.[12] Data points of PAMELA,[1] ATIC,[2] BETS[13] and PPB-BETS[14] are also plotted. It is seen that these DM models well reproduce the PAMELA/ATIC anomaly. Notice that no anti-protons are produced in these models.

---

[a]One can avoid these constraints, if possible clumpy structures of the DM halo in the Galaxy are taken into account. However, such small scale structures of the DM halo are not well understood yet. In the following, we assume smooth distribution of DM halo.

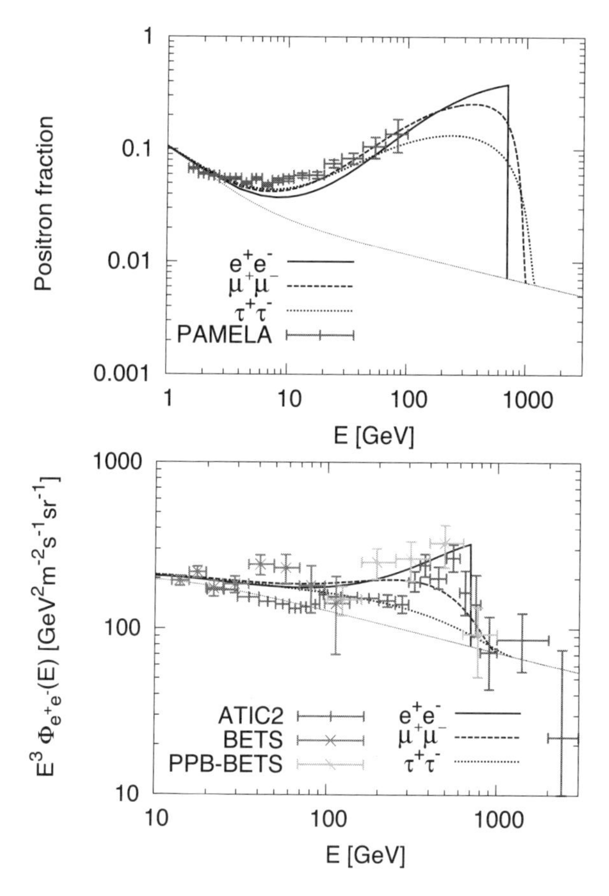

Fig. 1. (Top) The ratio of e^+ to $e^+ + e^-$ flux as a function of its energy, for the case of DM annihilating into e^+e^-, $\mu^+\mu^-$ and $\tau^+\tau^-$. (Bottom) Total $e^+ + e^-$ flux times E^3 for the same DM models as top panel.

2.2. *Gamma-Rays*

We have seen that the DM annihilation model reproducing the PAMELA/ATIC anomaly requires large annihilation cross section, $\langle\sigma v\rangle \sim 10^{-23}$ cm^{3}s^{-1}. Such models would also produce significant gamma-ray flux from the Galactic center. which should be compared with HESS observations.[15] As opposed to the case of electrons/positrons, gamma-rays do not suffer from energy loss and comes to the Earth in a straight way. Thus the evaluation of the gamma-ray flux is rather simple, but it has large uncertainty related to the lack of the understanding of the DM density

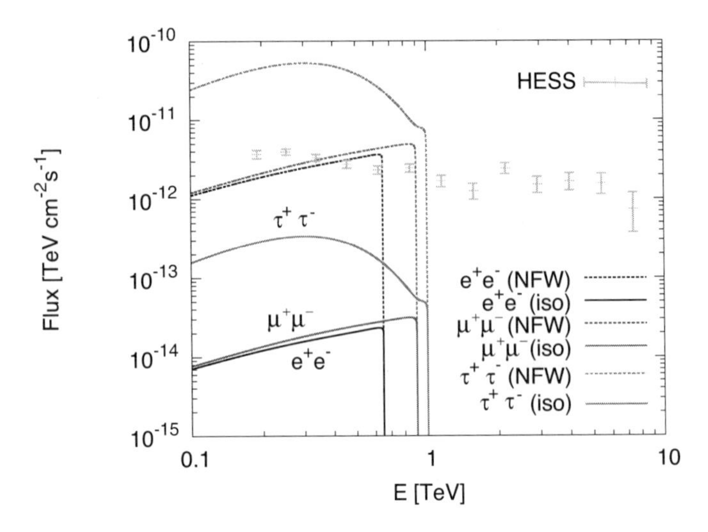

Fig. 2. Gamma-ray flux from the Galactic center for the DM models used in Fig. 1, both for NFW and isothermal density profiles.

profile near the Galactic center, which leads to orders of magnitude uncertainty in the resultant gamma-ray flux. As examples, we adopt two DM halo models : Navarro-Frenk-White (NFW) model[16] where the DM number density scales as $n_{\rm DM}(r) \sim 1/r$ near the Galactic center with distance from the Galactic center r, and isothermal profile which scales as $1/(a^2 + r^2)$ with core radius a. Fig. 2 shows the gamma-ray flux from the region 0.1° around the Galactic center for the DM models used in Fig. 1 with both NFW and isothermal DM density profile. We have included both contributions from internal bremsstrahlung and cascade decay processes. It is seen that the cuspy density profile as NFW profile is not favored, otherwise the gamma-ray flux exceeds the HESS observations. However, if a more moderate density profile is adopted, the gamma-ray constraint is satisfied.

2.3. *Neutrinos*

Similar to the gamma-rays, high-energy neutrinos are also produced by the DM annihilation, which can be observed or constrained by on-going or future neutrino detectors,[17] through the Cherenkov light emission by high-energy muons arising from the interaction of neutrinos with matter inside the Earth. Interestingly, the Super-Kamiokande (SK) results[18] already give constraints on DM models explaining the PAMELA/ATIC anomaly.[9,19] Fig. 3 shows the resulting up-going muon flux for the case of DM annihi-

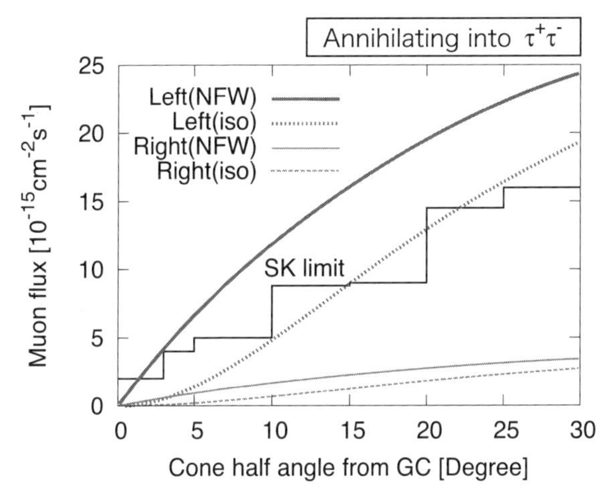

Fig. 3. The up-going muon flux for the DM annihilating into left-handed τ pair ($\tau_L^- \tau_R^+ + \nu_\tau \bar{\nu}_\tau$) and right-handed τ pair ($\tau_R^- \tau_L^+$) for NFW and isothermal profile. Limits from SK are also shown.

lating into $\tau^+\tau^-$ with mass and cross sections are fixed to reproduces the PAMELA/ATIC results.[9] Here we have distinguished two cases : annihilation into left-handed τ's and right-handed τ's, since it is natural to expect that the DM also directly annihilates into neutrino pair if the final state particles are left-handed. Thus we consider two cases where the DM annihilates into left-handed leptons ($\tau_L^- \tau_R^+ + \nu_\tau \bar{\nu}_\tau$) and right-handed leptons ($\tau_R^- \tau_L^+$). It is seen that the former case may conflict with present bound from SK.

3. Constraints from Big-Bang Nucleosynthesis

Finally, we discuss constraint on the DM annihilation cross section from the BBN. If there are some additional energy injection processes during BBN, light elements such as He, D, Li may be destroyed or overproduced, which may be in conflict with observations of primordial light element abundances.[20] Some fraction of the DM still annihilates each other even after the freeze-out epoch where the DM number-to-entropy ratio is fixed, and such effects can have significant effects on BBN if the annihilation cross section is large enough.[21] Effects of DM annihilation on BBN was discussed by Jedamzik,[22] with an emphasis on a possible explanation to the cosmic lithium problem. Here we provide conservative constraints on the DM annihilation rate, in the case of pure radiative annihilation mode.[8]

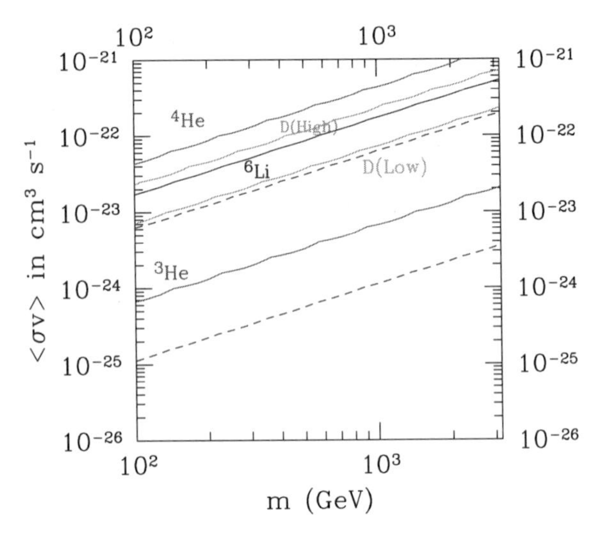

Fig. 4. Constraints on the DM annihilation cross section from observations of various light element abundances. Here pure radiative annihilation is assumed. For reference, the region sandwiched by two dashed lines are allowed from ^{6}Li abundance if no stellar depletion is assumed.

Fig. 4 shows constraints on the annihilation cross section into radiative mode from observation of light element abundances. We adopt the following light element abundances as observed primordial values : $n_{\rm D}/n_{\rm H} = (2.82 \pm 0.26) \times 10^{-5}$ for "Low" value of D and $n_{\rm D}/n_{\rm H} = (3.98^{+0.59}_{-0.67}) \times 10^{-5}$ for "High" value of D,[23] $n_{^3{\rm He}}/n_{\rm D} < 1.10$ for ^{3}He,[24] $Y_{\rm p} = 0.2516 \pm 0.0040$ for ^{4}He,[25,26] $\log_{10}(n_{^7{\rm Li}}/n_{\rm H}) = -9.90 \pm 0.09 + 0.35$ for ^{7}Li,[27,28] $n_{^6{\rm Li}}/n_{^7{\rm Li}} < 0.046 \pm 0.022 + 0.106$ for ^{6}Li[29] taking into account the effect of rotational mixing in the star.[30] Note that the energy fraction going into the radiation ($f_{\rm vis}$) is 35% for the case of μ, and 32% for the case of τ. Thus the vertical axis in Fig. 4 should be regarded as a constraint on $f_{\rm vis}\langle\sigma v\rangle$. We can see that DM models adopted in Fig. 1 are marginally consistent with BBN, or may be disfavored for the case of annihilating into τ's.

4. Conclusions

Motivated by the recent observations of anomalous cosmic-ray electron/positron fluxes by PAMELA and ATIC experiments, we have investigated observational implications of DM annihilation scenario which account for the electron/positron anomalies in a model independent way. Since a large annihilation cross section is required in order to reproduce

PAMELA/ATIC anomaly, it may leave characteristic signatures on other observations: gamma-rays, neutrinos and primordial light element abundances. What we have found is that all these observations give stringent constraints on DM annihilation models accounting for the PAMELA/ATIC anomaly. If the density profile of the DM halo in our Galaxy is cuspy, like NFW profile for example, the HESS results exclude DM annihilation models. This constraint can be avoided if more moderate density profile is adopted. Neutrino constraint may also be so severe for the DM annihilating into left-handed leptons that the up-going muon flux may exceed the SK bound. The BBN constraint is also so stringent that DM annihilation models are only marginally allowed, or some models are already disfavored. All these possible signatures are important for finding or constraining DM annihilation models in connection with cosmic positron/electron anomalies.

Acknowledgments

I would like to thank J. Hisano, M. Kawasaki, K. Kohri and T. Moroi for their collaboration.

References

1. O. Adriani *et al.*, arXiv:0810.4995 [astro-ph].
2. J. Chang *et al.*, Nature **456**, 362 (2008).
3. D. Hooper, P. Blasi and P. D. Serpico, JCAP **0901**, 025 (2009); H. Yuksel, M. D. Kistler and T. Stanev, arXiv:0810.2784 [astro-ph]; S. Profumo, arXiv:0812.4457 [astro-ph]; K. Ioka, arXiv:0812.4851 [astro-ph]; N. J. Shaviv, E. Nakar and T. Piran, arXiv:0902.0376 [astro-ph.HE].
4. L. Bergstrom, T. Bringmann and J. Edsjo, arXiv:0808.3725 [astro-ph]; V. Barger, W. Y. Keung, D. Marfatia and G. Shaughnessy, arXiv:0809.0162 [hep-ph]; M. Cirelli, M. Kadastik, M. Raidal and A. Strumia, arXiv:0809.2409 [hep-ph]; I. Cholis, D. P. Finkbeiner, L. Goodenough and N. Weiner, arXiv:0810.5344 [astro-ph]; Y. Nomura and J. Thaler, arXiv:0810.5397 [hep-ph]; D. Feldman, Z. Liu and P. Nath, arXiv:0810.5762 [hep-ph]; K. Ishiwata, S. Matsumoto and T. Moroi, arXiv:0811.0250 [hep-ph]; Y. Bai and Z. Han, arXiv:0811.0387 [hep-ph]; P. J. Fox and E. Poppitz, arXiv:0811.0399 [hep-ph]; G. Bertone, M. Cirelli, A. Strumia and M. Taoso, arXiv:0811.3744 [astro-ph]; P. Grajek, G. Kane, D. Phalen, A. Pierce and S. Watson, arXiv:0812.4555 [hep-ph].
5. J. Hisano, S. Matsumoto and M. M. Nojiri, Phys. Rev. Lett. **92**, 031303 (2004); J. Hisano, S. Matsumoto, M. M. Nojiri and O. Saito, Phys. Rev. D **71**, 063528 (2005).
6. M. Kawasaki, T. Moroi and T. Yanagida, Phys. Lett. B **370**, 52 (1996); T. Moroi and L. Randall, Nucl. Phys. B **570**, 455 (2000); M. Fujii and

K. Hamaguchi, Phys. Lett. B **525**, 143 (2002); Phys. Rev. D **66**, 083501 (2002); M. Nagai and K. Nakayama, Phys. Rev. D **76**, 123501 (2007); Phys. Rev. D **78**, 063540 (2008); B. S. Acharya, P. Kumar, K. Bobkov, G. Kane, J. Shao and S. Watson, JHEP **0806**, 064 (2008).
7. J. Hisano, M. Kawasaki, K. Kohri and K. Nakayama, Phys. Rev. D **79**, 063514 (2009) [arXiv:0810.1892 [hep-ph]].
8. J. Hisano, M. Kawasaki, K. Kohri, T. Moroi and K. Nakayama, arXiv:0901.3582 [hep-ph].
9. J. Hisano, M. Kawasaki, K. Kohri and K. Nakayama, Phys. Rev. D **79**, 043516 (2009) [arXiv:0812.0219 [hep-ph]].
10. E. A. Baltz and J. Edsjo, Phys. Rev. D **59**, 023511 (1999).
11. J. Hisano, S. Matsumoto, O. Saito and M. Senami, Phys. Rev. D **73**, 055004 (2006).
12. I. V. Moskalenko and A. W. Strong, Astrophys. J. **493**, 694 (1998).
13. S. Torii *et al.*, Astrophys. J. **559**, 973 (2001).
14. S. Torii *et al.*, arXiv:0809.0760 [astro-ph].
15. F. Aharonian *et al.* [The HESS Collaboration], Astron. Astrophys. **425**, L13 (2004); F. Aharonian *et al.* [H.E.S.S. Collaboration], Phys. Rev. Lett. **97**, 221102 (2006) [Erratum-ibid. **97**, 249901 (2006)].
16. J. F. Navarro, C. S. Frenk and S. D. M. White, Astrophys. J. **462**, 563 (1996).
17. J. F. Beacom, N. F. Bell and G. D. Mack, Phys. Rev. Lett. **99**, 231301 (2007); H. Yuksel, S. Horiuchi, J. F. Beacom and S. Ando, Phys. Rev. D **76**, 123506 (2007).
18. S. Desai *et al.* [Super-Kamiokande Collaboration], Phys. Rev. D **70**, 083523 (2004) [Erratum-ibid. D **70**, 109901 (2004)]; Astropart. Phys. **29**, 42 (2008).
19. J. Liu, P. f. Yin and S. h. Zhu, arXiv:0812.0964 [astro-ph].
20. M. Kawasaki, K. Kohri and T. Moroi, Phys. Lett. B **625**, 7 (2005); Phys. Rev. D **71**, 083502 (2005); M. Kawasaki, K. Kohri, T. Moroi and A. Yotsuyanagi, Phys. Rev. D **78**, 065011 (2008); K. Jedamzik, Phys. Rev. D **70**, 063524 (2004); Phys. Rev. D **74**, 103509 (2006).
21. M. H. Reno and D. Seckel, Phys. Rev. D **37**, 3441 (1988); J. A. Frieman, E. W. Kolb and M. S. Turner, Phys. Rev. D **41**, 3080 (1990).
22. K. Jedamzik, Phys. Rev. D **70**, 083510 (2004).
23. J. M. O'Meara *et al.*, Astrophys. J. **649**, L61 (2006).
24. J. Geiss and G. Gloeckler, Space Sience Reviews **106**, 3 (2003).
25. Y. I. Izotov, T. X. Thuan and G. Stasinska, arXiv:astro-ph/0702072.
26. M. Fukugita and M. Kawasaki, Astrophys. J. **646**, 691 (2006).
27. P. Bonifacio *et al.*, arXiv:astro-ph/0610245.
28. S.G. Ryan *et al.*, Astrophys. J. Lett. **530**, L57 (2000).
29. M. Asplund *et al.*, Astrophys. J. **644**, 229 (2006).
30. M. H. Pinsonneault, G. Steigman, T. P. Walker and V. K. Narayanans, Astrophys. J. **574**, 398 (2002).

PART V

X-Rays and Dark Matter, GeV Gamma Rays

FUNDAMENTAL PHYSICS WITH GEV GAMMA RAYS

S. PROFUMO*

Santa Cruz Institute for Particle Physics and
Department of Physics,
University of California, Santa Cruz, CA 95064
**E-mail: profumo@scipp.ucsc.edu*
http://scipp.ucsc.edu/~profumo/

Can we learn about New Physics with astronomical and astro-particle data? Understanding how this is possible is key to unraveling one of the most pressing mysteries at the interface of cosmology and particle physics: the fundamental nature of dark matter. Rapid progress may be within grasp in the context of an approach which combines information from high-energy particle physics with cosmic-ray and traditional astronomical data. I discuss how modifications to the pair annihilation cross section of dark matter with enhanced rates at low relative velocities can lead to a burst of annihilation in the first dark matter halos. I then introduce a novel approach to particle dark matter searches based on the complementarity of astronomical observations across the electromagnetic spectrum, from radio to X-ray and to gamma-ray frequencies.

Keywords: Dark Matter; Cosmology; Supersymmetric models.

1. Introduction

The question of whether or not it is possible to extract information on fundamental, new particle physics properties with astronomical and cosmic ray measurements depends upon our ability to differentiate between standard astrophysical backgrounds and extra-ordinary signals, as well as upon a thorough understanding of the implications of modifications to the cosmological or particle dark matter setup. In the talk I delivered at Dark 2009 I emphasized the message above and illustrated it with four examples. These examples based on:

(1) The role of nearby pulsars in polluting a possible signature in the positron fraction measurement reported by Pamela (from "*Dissecting Pamela (and ATIC) with Occam's Razor: existing, well-known Pulsars naturally account for the 'anomalous*"[1])

(2) Cosmological implications of a $1/v$ term in the dark matter pair annihilation cross section (from "*Early Annihilation and Diffuse Backgrounds in Models of Weakly Interacting Massive Particles in Which the Cross Section for Pair Annihilation Is Enhanced by 1/v*"[2])
(3) Determination of DM particle parameters from gamma-ray measurements with Fermi-LAT (from "*Fitting the Gamma-Ray Spectrum from Dark Matter with DMFIT: GLAST and the Galactic Center Region*"[3])
(4) A multi-wavelength strategy in indirect dark matter detection, applied to the nearby cluster of galaxies Ophiuchus (from "*Non-thermal X-rays from the Ophiuchus cluster and dark matter annihilation*"[4])

In the following, for brevity and lack of space, I will focus on item (2) in the following sec. 2 and item (4) in the concluding sec. 3.

2. Early Annihilation and Diffuse Backgrounds in $1/v$ WIMP Models

In the standard weakly-interacting massive particle (WIMP) scenario, dark matter is composed of particles that have electroweak interactions with ordinary matter. Such particles cease to annihilate to standard-model particles in the primordial plasma to produce a cosmological relic density that is generically in the ballpark of that required to account for the dark matter.

Typical values for the WIMP mass are $m_\chi \sim 10\,\mathrm{GeV}$ – few TeV, and freezeout of annihilations (chemical decoupling) occurs at temperatures $T_f \sim m_\chi/20$. After freezeout, the WIMP temperature remains fixed to the temperature of the primordial plasma via frequent elastic scattering from standard-model particles. When the temperature drops below the kinetic-decoupling temperature $T_{\rm kd}$, which generally falls in WIMP models in the range $T_{\rm kd} \sim 10$ MeV – few GeV,[5] WIMPs kinetically decouple from the plasma, and their temperature subsequently decays with the scale factor R as R^{-2}, rather than R^{-1}. Afterwards, WIMPs are effectively collisionless; they behave in the subsequent Universe like the cold dark matter required to account for detailed measurements of the cosmic microwave background (CMB) and large-scale structure.

There still remains the possibility, though, that once galactic halos form later in the Universe, a tiny fraction of WIMPs could annihilate to produce cosmic gamma rays or cosmic-ray antiprotons or positrons that might be observed. In the standard WIMP scenario, the annihilation cross section (times relative velocity v) can be cast as

$$(\sigma v)_{\rm ann} \simeq a + b(v/c)^2 + \cdots . \tag{1}$$

Roughly speaking, this cross section must evaluate at $v \sim c/2$ to $(\sigma v)_{\rm ann} \simeq 3\times10^{-26}\,{\rm cm}^3\,{\rm s}^{-1}$ to obtain the correct relic density, while annihilation in the Galactic halo occurs at $v \sim 10^{-3}$, where $(\sigma v)_{\rm ann} \simeq a \lesssim 3\times10^{-26}\,{\rm cm}^3\,{\rm s}^{-1}$. The fluxes of cosmic rays from WIMP annihilation are thus generically expected to be very small; detection would require some mechanism to boost the annihilation rate typically by a few orders of magnitude.

One way to boost the annihilation rate is to alter the underlying particle theory so that the WIMP annihilation cross section (times relative velocity) goes as $1/v$ (rather than to a constant, as in the standard scenario) as $v \to 0$. Such a cross section is consistent with s-wave unitarity as long as $(\sigma v)_{\rm ann} \leq 4\pi/(m_\chi^2 v)$ as $v \to 0$. Even larger annihilation cross sections are conceivable if higher partial waves contribute. Specific mechanisms for providing a low-velocity enhancement include the Sommerfeld enhancement and/or the formation of bound particle-antiparticle states in which the particles then annihilate although the functional dependence is not always precisely $1/v$, an issue we elaborate on further below.

Here we parametrize the $1/v$ enhancement by writing $(\sigma v)_{\rm ann} = 3\times10^{-26}\,\sigma_{26}(c/v)\,{\rm cm}^3\,{\rm s}^{-1}$, and we refer to these models as "$1/v$ WIMP models," or just "$1/v$ models." Since the cross section at freezeout required to obtain the relic abundance is roughly $3\times10^{-26}\,{\rm cm}^3\,{\rm s}^{-1}$, we infer that the parameter σ_{26} must be $\sigma_{26} \lesssim 1$ or else the WIMP abundance will be too small. The equality is obtained if there are no other contributions to the annihilation cross section; σ_{26} could be smaller if there are other contributions, such as those in Eq. (1), to the annihilation cross section.

We now describe the bounds to this scenario that arise from upper limits to the diffuse extragalactic gamma-ray background and from CMB constraints to the ionization history. After the Universe becomes matter dominated, perturbations in the cold-dark-matter density are amplified via gravitational infall. The smallest structures undergo gravitational collapse first, and then more massive structures collapse later. The kinetic coupling of WIMPs at temperatures $T_f \gtrsim T \gtrsim T_{\rm kd}$ erases primordial structure on mass scales smaller than[6]

$$M_c \simeq 33\,(T_{\rm kd}/10\,{\rm MeV})^{-3}\,M_\oplus. \tag{2}$$

The first gravitationally-bound structures in the hierarchy therefore have masses M_c. The first objects collapse at a redshift that can be approximated over the range $10^{-6}\,M_\oplus \lesssim M_c \lesssim 100\,M_\oplus$ relevant for WIMPs[5] by $z_c = 140 - \log_{10}(M_c/M_\oplus)$. These protohalos collapse to a virial density ρ that is ≈ 178 times the mean cosmological density at the collapse redshift z_c.

The velocity dispersion in the protohalos can be approximated by $v \sim R/t$, where $t \sim (G\rho)^{-1/2}$ is the dynamical time and $R \sim (M/\rho)^{1/3}$ is the size of the halos. Thus, $v \sim M^{1/3}G^{1/2}\rho^{1/6}$. Numerically, the first halos of mass M_c that collapse at redshift z_c will have velocity dispersions

$$(v/c) \sim 6.0 \times 10^{-9}\,(M_c/M_\oplus)^{1/3}(z_c/200)^{1/2}. \qquad (3)$$

The rate at which WIMPs annihilate in this first generation of halos is $\Gamma = n_\chi(\sigma v)_{\rm ann}$, which evaluates to

$$\Gamma = 2.2 \times 10^{-17}\left(\frac{M_c}{M_\oplus}\right)^{-1/3}\sigma_{26}B_{2.6}\left(\frac{z_c}{200}\right)^{5/2}\left(\frac{m_\chi}{\rm TeV}\right)^{-1}\,{\rm s}^{-1}. \qquad (4)$$

Here we have included a boost factor,

$$B \equiv \frac{\int \rho^2 dV}{V\rho_v^2} = \frac{c_v^3 g(c_v)}{3[f(c_v)]^2}, \qquad (5)$$

for a Navarro-Frenk-White density profile,[7] where c_v is the concentration parameter, $g(c_v) = (1/3)[1-(1+c_v)^{-3}]$, and $f(c_v) = \ln c_v - c_v/(1+c_v)$. This boost factor takes into account the increase in the annihilation rate due to the fact that the dark matter is distributed in these first halos with an NFW density profile $\rho(r)$, rather than uniformly distributed with density ρ. The boost factor varies from $B = 2.6$ for $c_v = 1$ to $B = 50$ for $c_v = 10$. To be conservative, we adopt $B = 2.6$, but include the B dependence through the parameter $B_{2.6} \equiv B/2.6$ in all subsequent expressions.

The first generation of halos survive roughly a Hubble time before they merge into slightly higher-mass (and lower-density) halos. The age of the Universe at redshifts $z \gg 1$ is $t \simeq 2\times10^{14}\,(z_c/200)^{-3/2}$ s. We thus infer that the fraction of dark-matter particles that annihilate in the first generation of halos is

$$f \simeq \Gamma t \simeq 4.4 \times 10^{-3}\left(\frac{M_c}{M_\oplus}\right)^{-1/3}\sigma_{26}B_{2.6}\left(\frac{z_c}{200}\right)\left(\frac{m_\chi}{\rm TeV}\right)^{-1}. \qquad (6)$$

Before considering bounds from diffuse radiation backgrounds, we note that a relatively weak bound to the parameter space can be obtained by noting that measurements of the CMB power spectrum constrain the matter density at recombination to be within 10% of its value today. We thus infer that $f \lesssim 0.1$, independent of any knowledge of the annihilation products.

However, if $1/v$ models are introduced to explain Galactic radiation backgrounds, they require annihilation into e^+e^- pairs and/or gamma rays. As shown in Fig. 2 of Ref. 8, photons injected with energies E_i at redshifts $z_c \sim 100-200$ in the range $100\,{\rm keV} \lesssim E_i \lesssim 300$ GeV propagate freely through the Universe, with energies that decrease with redshift as they propagate. They thus appear to us as a diffuse extragalactic background of gamma rays with energy $E_\gamma = E_i/z_c$. Photons injected at $z \sim 100$ with energies $E_i \lesssim 100$ keV propagate at first through the Universe but then get absorbed at lower redshift by the intergalactic medium (IGM). Photons injected with energies $\gtrsim 300$ GeV get absorbed immediately by the IGM.

Electron-positron pairs injected into the Universe at redshifts $z \sim 100-200$ with energy E_e very rapidly inverse-Compton scatter CMB photons resulting in a gamma ray of energy $E_i \sim (E_e/m_e)^2 T_{\rm CMB}$, where $T_{\rm CMB} \sim 10^{-2}$ eV is the characteristic CMB-photon energy at these redshifts. Thus, electron-positron pairs injected with energies in the range GeV $\lesssim E_e \lesssim$ 2 TeV produce photons in the energy range 100 keV $\lesssim E_i \lesssim$ 300 GeV of the transparency window; these gamma rays then appear to us as a diffuse background. Electrons injected at $z \sim 100 - 200$ with energies $E_e \lesssim$GeV get absorbed by the IGM at low redshifts, and those at $E_e \gtrsim 2$ TeV are absorbed by the IGM immediately at high redshift.

If the photons (or electron-induced photons) are not absorbed by the IGM, then the energy density in photons today from WIMP annihilation in the first halos is simply $\rho_\gamma = f\rho_\chi^0/z_c$; i.e. the fraction of the WIMP energy density that gets converted to radiation through annihilation, scaled by the redshift of the photons. This evaluates numerically to

$$\rho_\gamma = 2.64 \times 10^{-11} \left(\frac{M_c}{M_\oplus}\right)^{-1/3} \sigma_{26} B_{2.6} \left(\frac{m_\chi}{\rm TeV}\right)^{-1} \text{GeV cm}^{-3}. \tag{7}$$

We now compare this with the upper limit $\rho_\gamma \lesssim 5.7 \times 10^{-16} (E_\gamma/\text{GeV})^{-0.1}$ GeV cm^{-3} [8] to obtain the constraint,

$$\sigma_{26} \lesssim 2.2 \times 10^{-5} B_{2.6}^{-1} \left(\frac{M_c}{M_\oplus}\right)^{1/3} \left(\frac{E_\gamma}{\rm GeV}\right)^{-0.1} \left(\frac{m_\chi}{\rm TeV}\right). \tag{8}$$

The EGRET upper bound used here is derived for 30 MeV $\lesssim E_\gamma \lesssim$ 100 GeV. It is a fit to the extragalactic gamma-ray background, much of which comes from unresolved astrophysical sources, and so it should be viewed as a conservative upper limit. Moreover, the recently launched Fermi-GLAST telescope should soon significantly improve the upper limit to this gamma-ray background. Eq. (8) is valid as a conservative upper limit all the way down to energies $\sim$keV. The true upper limits from x-ray- and gamma-ray-background measurements in the energy range 10 keV $\lesssim E_\gamma \lesssim$ 30 MeV are more stringent. Notice that the limit in Eq. (8) does not depend on the collapse redshift z_c.

There will continue to be WIMP annihilation during later, more massive, stages in the structure-formation hierarchy. These will produce higher-energy photons (since they get redshifted less), but the energy density in these higher-energy photons will be far smaller. This can be seen from Eq. (6) by noting that the factor M_c that appears therein will be replaced by $M_*(z)$, the characteristic halo mass at redshift z (interestingly enough, all other redshift dependence in that equation cancels). The mass scale $M_*(z)$ evolves very rapidly with z; e.g., it goes from $10^{-12}\,M_\odot$ to $10^{-4}\,M_\odot$ from $z \simeq 200$ to $z \simeq 100$, and then all the way to $\sim 10^{14}\,M_\odot$ at $z = 0$. The constraint to the model from the diffuse-background flux is thus strongest from the earliest halos.

Let's now consider what happens if the photon (or electron-induced photon) falls outside the transparency window; i.e., if it is injected with energy $E_i \lesssim 100$ keV or $E_i \gtrsim 300$ GeV. In both cases, the photons are absorbed by the IGM before they can reach us, and so there is no constraint from diffuse backgrounds. In both cases, though, very stringent constraints arise from measurements of CMB temperature and polarization.[8,9] The energy deposited in the IGM by the photons ionizes and heats the IGM. The reionized electrons scatter CMB photons thus altering the observed CMB temperature/polarization power spectra. Ref. 9 carried out detailed fits to WMAP3 and large-scale-structure data to constrain the heating/ionization of the IGM. We infer from Fig. 2 in Ref. 9 that no more than a fraction $f \lesssim 10^{-9}$ of the rest-mass energy of the dark matter could have been injected into the IGM at a time $\sim 10^{15}$ sec after the big bang. Although the analysis is detailed, the magnitude of the upper limit can be understood relatively simply: Dark matter outweighs baryons by a factor of 6, and it requires a fraction $(10\,\mathrm{eV/GeV}) \sim 10^{-8}$ of the rest-mass energy of each atom to ionize it.

Thus, if dark matter annihilates to photons with energies $E_i \gtrsim 300$ GeV or to electron-positron pairs with energies $E_e \gtrsim 2$ TeV, then we require

$$\sigma_{26} \lesssim 2.3 \times 10^{-7} \left(\frac{M_c}{M_\oplus}\right)^{1/3} B_{2.6}^{-1} \left(\frac{z_c}{200}\right)^{-1} \left(\frac{m_\chi}{\mathrm{TeV}}\right). \tag{9}$$

The result for photons injected with energies $E_\gamma \lesssim 100$ keV (or electron-positron pairs with $E_e \lesssim$GeV) is similar, but weakened possibly by the redshift of the photon energy that occurs between the time it was injected and the time it was absorbed. The detailed suppression depends on the injected energy and redshift. However, in no case is the suppression stronger than a factor z_c^{-1}. We thus conclude that the bound will be no more than two orders of magnitude weaker than that quoted in Eq. (9).

To summarize: There will be a burst in $1/v$ WIMP models of annihilation in the first gravitationally bound dark-matter halos when they first form at redshifts $z \sim 100 - 200$. There is a weak, albeit final-state–independent, CMB bound that amounts to demanding that no more than $\sim 10\%$ of the dark matter annihilates after recombination. If the WIMP annihilates to photons in the energy range $100\,\mathrm{keV} \lesssim E_i \lesssim 300$ GeV or to electrons in the energy range GeV$\lesssim E_e \lesssim 2$ TeV, then there are constraints, summarized in Eq. (8), to the cross section for annihilation to e^+e^- pairs and photons. If the photons or electrons are injected outside the transparency window, then there is a bound, quoted in Eq. (9), that comes from CMB constraints to ionization of the IGM.

To clarify, the bounds to σ_{26} in Eqs. (8) and (9) are to the cross section for annihilation to photons and e^+e^- pairs, not to the total annihilation cross section. A large hadronic final state branching ratio is however greatly constrained by antiproton data. Note also that (i) $B_{2.6}$ is likely larger than unity, that (ii) z_c is generally larger than 100 for typical WIMP models; and that (iii) M_c is generally smaller than $M_\oplus$. The numerical values in Eqs. (8) and (9) will, for typical WIMP setups, generally be smaller. On the other hand, the numerical value in Eq. (8) may be larger for smaller E_i. For the nominal WIMP values that we have chosen, the bounds are violated by 4-6 orders of magnitude, depending on the gamma-ray energy, and the constraints are still $\sigma_{26} \ll 1$ for almost any combination of the parameters E_γ, M_c, and z_c consistent with typical WIMP values. Specifically, to obtain the enhancements $\mathcal{O}(10^2)$ needed by some dark-matter interpretations of recent cosmic-ray measurements with a Galactic $c/v \sim 10^3$, one requires $\sigma_{26} \gtrsim 0.1$, inconsistent with the limits derived here. Thus, something else must be invoked in a $1/v$ model to evade the gamma-ray and/or ionization-history constraint.

3. Non-Thermal X-Rays from the Ophiuchus Galaxy Cluster and Dark Matter Annihilation

Clusters of galaxies are the largest bound dark matter (DM) structures in the universe. As such, they are natural targets for the search for observational signatures of particle DM.[10] If DM is in the form of weakly interacting massive particles (WIMPs), DM pair annihilations generically produce γ-rays as well as a non-thermal energetic electron-positron ($e^\pm$) population. The latter, in turn, is expected to yield secondary emissions at soft γ-ray, X-ray and radio frequencies via inverse Compton scattering, bremsstrahlung and synchrotron radiation, opening up the possibility of a multi-wavelength approach to particle DM detection.[10,11] A generic feature of the broad-band DM annihilation spectrum is a significant hard X-ray component.[10]

Interestingly, the discovery of a non-thermal hard X-ray emission from the Ophiuchus cluster, detected with relatively robust statistical significance in a 3 Ms observation with the IBIS/ISGRI and JEM-X instruments on board INTEGRAL, was recently reported in Ref. 12. The Ophiuchus cluster is a nearby ($z \simeq 0.028$) rich cluster with a high temperature plasma ($kT \sim 10$ keV), featuring the second brightest emission in the 2-10 keV band. In addition to X-ray observations, the steep-spectrum radio source MSH 17-203 was associated to the Ophiuchus cluster, indicating the presence of relativistic electrons. However, the cluster was not detected at γ-ray frequencies by EGRET.

While ordinary astrophysical mechanisms, including merger shocks, can be invoked to explain the non-thermal electrons presumably responsible for the observed non-thermal activity in galaxy clusters, in the present analysis we propose and investigate a novel scenario where WIMP annihilations produce, or significantly contribute to, said non-thermal population responsible for the hard X-ray detection in the Ophiuchus galaxy cluster. In a model independent approach, we determine the parameters of the particle DM setups that provide the best fits to the INTEGRAL X-ray data, and we compute the resulting multi-wavelength spectra. We then compare these spectra with the radio data and with the γ-ray limits and future prospects for the soon-to-be-launched Gamma-Ray Large Area Telescope (GLAST). The highlights of our analysis are: (1) the DM hypothesis will conclusively be probed with GLAST; (2) the radio emission can in principle also be fitted with the synchrotron emission from DM-annihilation-produced $e^{\pm}$, as long as the average magnetic field in the cluster is of the order of 0.1 μG; (3) GLAST might be able detect the monochromatic γ-rays produced in direct DM pair annihilation into two photons.

The flux of $e^{\pm}$ produced by WIMP pair annihilations depends on the particle DM setup and on the DM density distribution. We define a source function $Q_e(E_e, \vec{x})$, which gives the number of $e^{\pm}$ per unit time, energy and volume element produced locally in space, as

$$Q_e(E_e, \vec{x}) = \langle \sigma v \rangle_0 \sum_f \frac{\mathrm{d}N_e^f}{\mathrm{d}E_e}(E_e)\ B_f\ \mathcal{N}_{\mathrm{pairs}}(\vec{x}). \tag{10}$$

In the equation above, $\langle \sigma v \rangle_0$ is the WIMP annihilation rate at zero temperature, the sum is over all kinematically allowed Standard Model annihilation final states f, each with a branching ratio B_f and an $e^\pm$ distribution $\mathrm{d}N_e^f/\mathrm{d}E_e$, and $\mathcal{N}_{\rm pairs}(\vec{x})$ is the number density of WIMP pairs at a given point $\vec{x}$, i.e. the number of WIMP particle pairs per volume element squared: $\mathcal{N}_{\rm pairs}(\vec{x}) = \rho^2_{\rm DM}(\vec{x})/(2m^2_{\rm DM})$, where $\rho_{\rm DM}$ stands for the DM density. The particle physics framework sets the quantity $\langle\sigma v\rangle_0$, the list of B_f and the mass of the WIMP, $m_{\rm DM}$. The latter also determines the energy scale of the pair annihilation event, and, together with the specific final state f, the $\mathrm{d}N_e^f/\mathrm{d}E_e$ spectral functions, which we numerically compute with the Monte Carlo code Pythia. In addition, $m_{\rm DM}$ enters in the determination of the local number density of WIMP pairs.

Once the source function $Q_e(E_e,\vec{x})$ is determined, the $e^\pm$ spectrum and density are affected by spatial diffusion and energy loss processes, usually described – under the assumptions of negligible convection and re-acceleration effects – by a diffusion-loss equation of the form

$$\frac{\partial}{\partial t}\frac{\mathrm{d}n_e}{\mathrm{d}E_e}(E_e,\vec{x}) = \vec{\nabla}\cdot\left[D(E_e,\vec{x})\vec{\nabla}\frac{\mathrm{d}n_e(E_e,\vec{x})}{\mathrm{d}E_e}\right] + \frac{\partial}{\partial E_e}\left[b(E_e,\vec{x})\frac{\mathrm{d}n_e(E_e,\vec{x})}{\mathrm{d}E_e}\right] + Q_e(E_e,\vec{x}), \tag{11}$$

where $\mathrm{d}n_e/\mathrm{d}E_e$ is the number density of electrons per unit energy, D is the diffusion coefficient, and

$$b(E_e,\vec{x}) = b_{\rm IC} + b_{\rm syn} + b_{\rm Coul} + b_{\rm brem} \tag{12}$$

encodes the various energy loss mechanisms.[10]

Knowledge of the distribution of the DM-induced $e^\pm$ population $\mathrm{d}n_e/\mathrm{d}E_e$, of the magnetic field structure and strength, as well as of the electron, gas and starlight densities allows one to compute the WIMP-induced secondary emissions. Specifically, at radio frequencies the DM-induced emission is dominated by the synchrotron radiation of the relativistic secondary electrons and positrons. IC scattering of the non-thermal $e^\pm$ on target CMB and starlight photons gives rise to a spectrum of photons stretching from below the extreme ultra-violet up to the soft γ-ray band, peaking in the X-ray energy band. Non-thermal bremsstrahlung, *i.e.* the emission of γ-ray photons in the deflection of the charged particles by the electrostatic potential of ionized gas, contributes in the soft γ-ray band. Finally, a hard γ-ray component arises from prompt emission in WIMP pair annihilations, mostly originating from the two photon decay of neutral pions, and, at the high energy end of the spectrum, from internal bremsstrahlung from charged particle final states. The γ-ray spectrum extends up to energies equal to the kinematic limit set by the WIMP mass, $E_\gamma \leq m_{\rm DM}$, and might

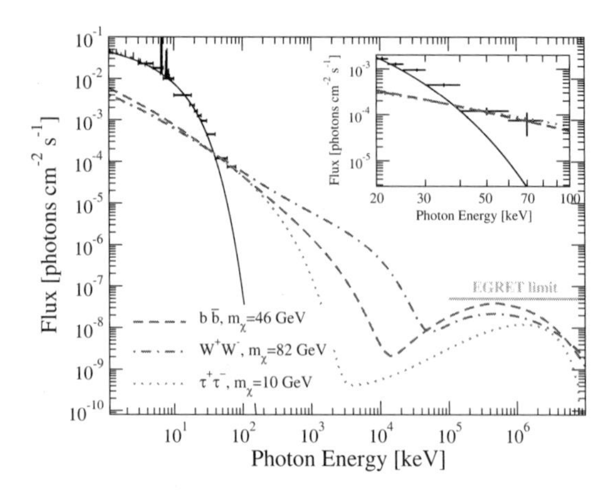

Fig. 1. The hard X-ray and γ-ray spectrum for three DM particle models, plus a single-temperature MEKAL model for the thermal X-ray emission, compared with the INTEGRAL data[12] and with the EGRET upper limit.

feature one or more monochromatic lines associated to two-body annihilation final states where one (or both) of the particles is a photon. We refer the reader to Ref. 10 for details on the computation of the multi-wavelength emission from DM annihilation.

We show in Fig. 1 the photon flux in the hard X-ray and γ-ray bands for three benchmark DM particle model accounting for the INTEGRAL data. We also include a thermal component from the bremsstrahlung emission of the intra-cluster medium, obtained with a single-temperature MEKAL model with the abundance fixed to 0.49 compared to the solar value and $kT = 8.5$ keV.[12] The thermal component was normalized to produce the best fit for the INTEGRAL data below 20 keV. The DM models were instead normalized to obtain the best global fit to the data above 20 keV. We chose DM models with $B_f = 1$ for $f = b\bar{b}$, W^+W^-, $\tau^+\tau^-$, *i.e.* each model pair annihilating into a single Standard Model final state. The three particular final states were selected for two reasons: (1) the resulting $e^\pm$ spectra $\mathrm{d}N_e^f/\mathrm{d}E_e$ range from the softest ($b\bar{b}$) to the hardest ($\tau^+\tau^-$) possible case;[10] (2) the three final states correspond to common well-defined cases found in supersymmetric DM models. For instance, in the minimal supergravity scenario $f \simeq b\bar{b}$ corresponds to the so-called bulk and funnel regions where the neutralino has a relic abundance compatible with the cold DM density, $\tau^+\tau^-$ is found in the coannihilation region (where neutralino pair-annihilation proceeds predominantly through scalar tau exchange) and W^+W^- in the focus point region. This choice of benchmark models fol-

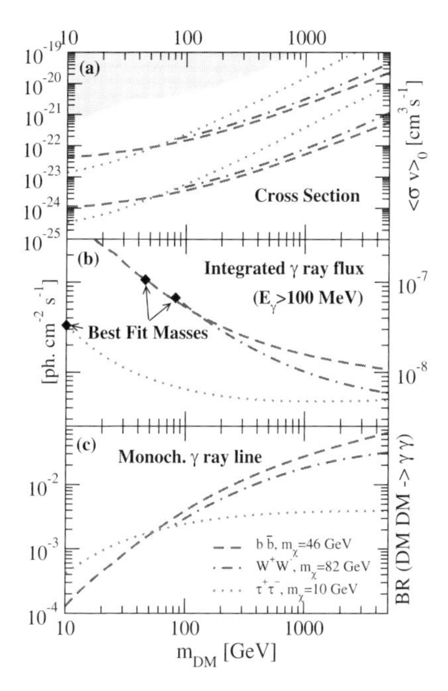

Fig. 2. The preferred pair annihilation cross section (a), the integrated γ-ray flux above 100 MeV (b) and the minimal branching ratio for the detection of the monochromatic γ-ray line at $E_\gamma = m_{\rm DM}$ (c), as a function of the DM particle mass. In panel (a) the upper lines refer to the case of no substructures, the lower lines refer to the substructure setup described in the text, and the gray shaded region is ruled out by EGRET and H.E.S.S. data on the gamma-ray flux from the galactic center region.

lows here and generalizes the approach of[10] – linear combinations of the considered models produce almost any WIMP multi-wavelength emission spectrum.

For each final state, we selected the DM particle mass giving the lowest χ^2 in the fit to the INTEGRAL data: $m_{DM}(f = b\bar{b}) = 46$ GeV, $m_{DM}(f = W^+W^-) = 82$ GeV and $m_{DM}(f = \tau^+\tau^-) = 10$ GeV. What is the DM pair annihilation cross section required to reproduce the spectra shown in Fig. 1? In Fig. 2 (a) we compute the cross section as a function of the DM particle mass, for the three benchmark final states, giving the best fit to the INTEGRAL data. Panel (b) in Fig. 2 shows the integrated γ-ray flux above 0.1 GeV for the best fit models as a function of $m_{\rm DM}$. In all cases we find that the expected γ-ray flux is well above the anticipated GLAST LAT integral flux sensitivity, estimated to be around a few $\times 10^{-10}$ cm^{-2}s^{-1}. Fig. 2 (c) shows the branching ratio $\langle\sigma v\rangle_{\rm tot}/\langle\sigma v\rangle_{\gamma\gamma}$

for the monochromatic DM DM $\rightarrow \gamma\gamma$ channel needed to obtain, for the best fit models, the detection of at least 10 photons with $E_\gamma = m_{\rm DM}$. Notice that the values shown are independent of the assumed DM profile. While $\langle\sigma v\rangle_{\rm tot}/\langle\sigma v\rangle_{\gamma\gamma}$ is entirely model dependent, the range we obtain is generically consistent with what is expected *e.g.* in supersymmetry, and especially in next-to-minimal supersymmetric extensions of the Standard Model. GLAST can therefore easily detect a sizable number of monochromatic energetic γ rays, depending on the specific DM particle model.

In summary, we showed that the origin of the non-thermal particles plausibly responsibly for the recently firmly discovered non-thermal hard X-ray emission from the Ophiuchus cluster might be generated by electrons and positrons produced in WIMP pair annihilations, provided the rate for the latter is large enough. This scenario is compatible with all observational information on the cluster, with particle DM production and searches, and, more importantly, will be thoroughly tested by GLAST. The future γ-ray telescope might even detect the monochromatic two-photon emission provided the particle model has a large enough branching ratio in that channel. Other galaxy clusters exhibiting non-thermal activity will also be outstanding sites for GLAST to look for signatures of WIMP DM pair annihilation. Finally, radio data from the Ophiuchus galaxy cluster can also be accounted for in the DM annihilation scenario, as long as the average magnetic field in the Ophiuchus cluster is below ≈ 0.2 μG.

References

1. S. Profumo (2008).
2. M. Kamionkowski and S. Profumo, *Phys. Rev. Lett.* **101**, p. 261301 (2008).
3. T. Jeltema and S. Profumo, *JCAP* **0811**, p. 003 (2008).
4. S. Profumo, *Phys. Rev.* **D77**, p. 103510 (2008).
5. S. Profumo, K. Sigurdson and M. Kamionkowski, *Phys. Rev. Lett.* **97**, p. 031301 (2006).
6. A. Loeb and M. Zaldarriaga, *Phys. Rev.* **D71**, p. 103520 (2005).
7. J. F. Navarro, C. S. Frenk and S. D. M. White, *Astrophys. J.* **490**, 493 (1997).
8. X.-L. Chen and M. Kamionkowski, *Phys. Rev.* **D70**, p. 043502 (2004).
9. L. Zhang, X. Chen, M. Kamionkowski, Z.-g. Si and Z. Zheng, *Phys. Rev.* **D76**, p. 061301 (2007).
10. S. Colafrancesco, S. Profumo and P. Ullio, *Astron. Astrophys.* **455**, p. 21 (2006).
11. S. Colafrancesco, S. Profumo and P. Ullio, *Phys. Rev.* **D75**, p. 023513 (2007).
12. D. Eckert, N. Produit, S. Paltani, A. Neronov and T. J. L. Courvoisier (2007).

SEARCHING FOR DARK MATTER ANNIHILATION IN X-RAYS AND GAMMA-RAYS

TESLA E JELTEMA* and STEFANO PROFUMO
University of California, Santa Cruz
1156 High St.
Santa Cruz, CA 95060
**E-mail: tesla@ucolick.org*

We discuss the possibilities for the indirect detection of dark matter annihilation with astronomical observations at X-ray and gamma-ray frequencies. In particular, we describe two studies. First, we use recent X-ray observations of local dwarf spheroidal galaxies to constrain the mass and pair annihilation cross section for particle dark matter. Our results indicate that X-ray observations of dwarf galaxies currently constrain dark matter models at the same level or more strongly than gamma-ray observations, although at the expenses of introducing additional assumptions and related uncertainties in the modeling of diffusion and energy loss processes. The limits we find constrain portions of the supersymmetric parameter space, particularly if the effect of dark matter substructures is included. Then, we investigate the possibility for the Fermi Gamma-ray Space Telescope to detect gamma-ray emission from clusters of galaxies, the largest bound dark matter structures. Clusters are expected to emit gamma rays as a result of (1) a population of high-energy primary and re-accelerated secondary cosmic rays fueled by structure formation and merger shocks, active galactic nuclei and supernovae, and (2) particle dark matter annihilation. Using simulated Fermi observations, we study observational handles that might enable us to distinguish the two emission mechanisms, including the gamma-ray spectra, the spatial distribution of the signal and the associated multi-wavelength emissions. Our study indicates that gamma rays from dark matter annihilation with a high particle mass ($m_{\rm WIMP} > 50$ GeV) can be distinguished from a cosmic ray spectrum even for fairly faint sources. Discriminating a cosmic ray spectrum from a light dark matter particle will be instead much more difficult, and will require long observations and/or a bright source.

1. Introduction

A generic feature of weakly interacting massive particle (WIMP) dark matter models is the emission of photons over a broad energy band resulting from the stable yields of dark matter pair annihilation.[1,2] Gamma-ray emission results from prompt production as well as decays, hadronization and radiative processes associated with the annihilation products. In addition to gamma rays, the stable products of WIMP annihilation include energetic electrons and positrons. Electrons (and positrons) diffuse, loose energy and produce secondary radiation through various mechanisms. In the presence of magnetic fields they emit at radio wavelengths via synchrotron radiation. Inverse Compton (IC) scattering off target cosmic microwave background photons and background light at other frequencies gives rise to a broad spectrum of photons, stretching from the extreme ultra-violet up to the gamma-ray band. A further, typically subdominant contribution to secondary photon emission results from non-thermal bremsstrahlung, i.e. the emission of gamma-ray photons in the deflection of the charged particles by the electrostatic potential of intervening gas.

We present the results of two studies with the aim of using the expected multiwavelength emission of dark matter annihilation from astronomical sources to constrain the particle properties of dark matter. First, we use X-ray observations of local dwarf spheroidal galaxies to place constraints on the possible WIMP dark matter particle mass and annihilation cross-section based on the non-detection of diffuse X-ray emission, the first such constraints made using X-ray observations.[3] We then investigate the prospects for the recently launched Fermi Gamma-ray Space Telescope to detect the gamma-ray signal of dark matter annihilation from clusters of galaxies using simulated Fermi observations for a range of models.[4] Clusters are also known to host a variety of high-energy phenomena (merger and accretion shocks, AGN) that could fuel cosmic rays, producing, in turn, gamma rays as a result of collisions with the gas in the intracluster medium (ICM). We use simulated Fermi observations to investigate methods to distinguish the emission from dark matter annihilation from gamma-ray emission due to astrophysical cosmic rays.[4]

2. X-ray Observations of Dwarf Galaxies

Local dwarf spheroidal (dSph) galaxies are an ideal environment for particle dark matter searches with X-rays, being nearby, dark matter dominated systems free of any astrophysical diffuse X-ray background. X-rays are produced as secondary radiation in Inverse Compton scattering off cosmic microwave background photons of electrons and positrons resulting from particle dark matter annihilation. The resulting spectrum is only mildly dependent on the details of the particle dark matter model (the dark matter mass and the dominant final state into which it pair annihilates), and it is, generically, hard (spectral index smaller than ~ 1.5). The normalization of the emission depends on (1) the particle dark matter pair annihilation rate, (2) the diffusion setup, and (3) the dark matter density distribution. For reasonable choices of these three *a priori* unknown inputs of the problem, the X-ray emission is potentially within reach of current X-ray detectors. Interestingly enough, the shape of the spectral energy distribution indicates that for dSph galaxies X-rays have a comparable, if not better, sensitivity to indirect dark matter detection than gamma rays.

We used XMM-Newton archival data on three Local Group dSph galaxies, Ursa Minor, Fornax and Carina, to search for the diffuse X-ray emission expected from dark matter annihilation.[3] We studied the optimal energy and radial range to search for this type of emission, and concluded that for XMM-Newton and for the dSph galaxies under investigation these correspond to an energy band between 0.5 and 8 keV and to a radius of around $6'$. We do not find any significant signal over background, and this, in turn, was turned into constraints on particle dark matter models. The best constraints result from both the Fornax and the Ursa Minor observations, while data from Carina result in bounds that are a factor of a few weaker. Ursa Minor features the largest dark matter density, making it the best candidate target, but has the shortest usable XMM exposure.

In determining the impact on particle dark matter searches of our X-ray constraints, we pointed out the uncertainties resulting from the modeling of cosmic ray diffusion processes, and from the dark matter distribution. In particular, including dark matter substructures can boost our constraints significantly. We phrase the bounds we obtain in terms of the dominant dark matter annihilation final states. For those final states relevant for specific dark matter models, such as supersymmetry, the constraints on the mass versus pair annihilation plane are very similar. We then proceeded to examine how X-ray constraints on particle dark matter annihilation in local dSph galaxies limit the available parameter space of supersymmetric dark matter. The results are shown in Fig. 1. In the most conservative setup with diffusion similar to the Milky Way and no boost from dark matter substructure, only models with rather large annihilation cross sections are excluded. Assuming a smaller diffusion coefficient scaled for the size of a dSph galaxy relative to the Milky Way, or factoring in the effect of dark matter substructures, our constraints fall well within the interesting region where the supersymmetric dark matter can be a thermal relic from the early universe. Also, we were able to set limits on particular supersymmetric dark matter scenarios, such as Wino or Higgsino lightest neutralino dark matter.

An important result of the present analysis is that even assuming a conservative diffusion setup the sensitivity of current X-ray observations and of previous EGRET gamma ray observations to particle dark matter annihilation in dSph galaxies are comparable. For EGRET, we assume a point-source sensitivity of around $10^{-8}\text{cm}^{-2}\text{s}^{-1}$, and an angular acceptance of 1 deg.[5] This fact has two-fold implications: on the one hand, if longer observations of dSph galaxies were carried out with existing telescopes, it is possible that the first astronomical signature of particle dark matter annihilation would come from X-rays. Secondly, should a signature be detected in the future with gamma-ray telescopes like Fermi, it would be extremely important to confirm the nature of the signal via X-ray observations.

In this respect, it is relevant to comment here on how future gamma-ray and X-ray telescopes will improve indirect dark matter searches through observations of nearby dSph galaxies. The LAT instrument on-board the Fermi satellite extends the gamma-ray energy range available to EGRET, with tremendously increased effective area and energy as well as angular resolution. Assuming a diffuse background flux of 1.5×10^{-5} photons cm^{-2} s^{-1} sr^{-1} integrated above 0.1 GeV, and an effective spectral index in the gamma-ray band of 2.1, we find that the Fermi-LAT sensitivity[a] from 5 years of data will improve over the EGRET point-source sensitivity by large factors. In the mass versus pair annihilation cross section plane, and assuming a soft gamma ray spectrum ($b\bar{b}$), Fermi will improve over EGRET by factors ranging between ~ 10 and ~ 100, the first corresponding to a light $m_{\rm DM} \sim 10$ GeV dark matter particle, and the latter to a heavy one ($m_{\rm DM} \sim 1000$ GeV). Assuming a harder gamma-ray spectrum, as appropriate for other dark matter models (e.g. universal extra dimensions), the Fermi performance will be factors between 30 and 300 better than EGRET. A signal of dark matter pair annihilations in gamma-rays appears therefore very promising with Fermi. If detected, such a source would need to be confirmed in its nature, and using a multi-wavelength approach is one of the most promising strategies.

Future X-ray telescopes, like the International X-ray Observatory (IXO), will also have greatly increased effective areas with respect to current instruments. Using projections for the effective area and background, we estimate that the X-ray limits (0.5-8 keV band) placed by a 100 ksec observation of Ursa Minor with IXO would improve over the limits presented here by a factor of roughly 50. Thus even for a conservative diffusion model,

[a]http://www-glast.slac.stanford.edu/software/IS/glast_lat_performance.htm

the future generation of X-ray telescopes will place similar constraints on dark matter annihilation from dwarfs as Fermi, stronger constraints at particle masses below a few hundred GeV. In addition, a signal from Fermi could be confirmed with X-ray observations.

Observations at other wavelengths will also be of great relevance to identify particle dark matter and its properties. In particular, a diffuse radio signal should also be part of the multiwavelength yield of particle dark matter annihilation. However, the level of the radio emission is crucially dependent upon the average magnetic field, which adds further uncertainties both in setting constraints and in understanding the nature of particle dark matter, should a signal be detected. Observations in the hard X-ray band would also be useful; however, as opposed to cluster of galaxies, where the effect of diffusion on the dark matter multi-wavelength SED is typically mild,[2,6] in dSph galaxies high energy electrons and positrons escape more efficiently from the diffusive region, suppressing the hard X-ray emission.

In short, we showed that X-rays can play an important role in exploring the nature of particle dark matter and in pinpointing its properties. This role is complementary, but not subsidiary, to searches with gamma rays, and we believe very exciting results at both frequencies might be just around the corner.

3. Gamma-rays from Clusters

Diffuse gamma-ray emission has not yet been detected from clusters of galaxies, but under reasonable assumptions, the recently launched Fermi Gamma-ray Space Telescope could detect clusters for the first time at gamma-ray wavelengths.[4,7,8] Being the largest bound dark matter structures, it is reasonable to expect that clusters feature a significant gamma-ray emission from the annihilation of dark matter particles. Clusters are also expected to emit gamma rays as a result of non-thermal activity, such as cluster mergers, AGN and supernovae, fueling a population of high-energy cosmic rays.[8,9] Gamma-ray emission results from collisions of cosmic ray protons with non-relativistic protons in the intracluster medium producing neutral pions which decay to yield gamma-rays. Gamma-ray emission from either or both of these processes may be detectable with Fermi. We present a theoretical study using simulated Fermi observations of the expected gamma-ray signal from clusters and specifically investigate how to tell apart these two potential mechanisms of gamma-ray production.[4]

By making simple analytical assumptions on the cosmic ray spectra and source distribution and on the dark matter particle properties and density

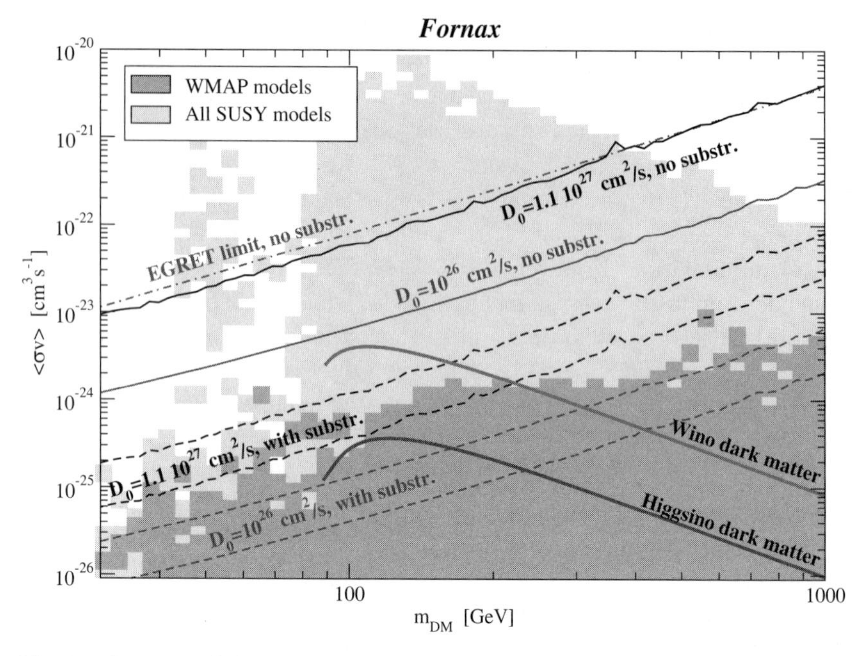

Fig. 1. Overview of the constraints from X-ray observations of Fornax on supersymmetric WIMP dark matter models.[3] We show the limits obtained using our conservative reference diffusion setup and a diffusion coefficient $D_0 = 10^{26}$ cm^2/s. The solid lines correspond to the case of no substructures, while the dashed lines indicate the range where one could expect the limit to be set when substructures are included. The yellow area corresponds to values of the dark matter mass and annihilation cross section found for neutralino models within the minimal supersymmetric extension of the Standard Model. The orange area indicates those supersymmetric models that also produce a thermal relic neutralino abundance in agreement with the inferred cold dark matter density. The solid green and indigo lines locate the predictions for "vanilla" Wino and Higgsino dark matter models.

distribution, we simulated a set of benchmark models for both gamma-ray production mechanisms under investigation. We believe the set of models we considered is representative of the variety of possibilities one can realistically expect to encounter in clusters of galaxies. Full details of the models can be found in Jeltema, Kehayias, & Profumo (2008). We simulate the expected 1 and 5 years gamma-ray signals with the Fermi Science Tools, and analyzed our results. Examples, of the simulated 5 year Fermi observations of a cluster with the mass and distance of the Coma cluster are shown in Fig. 2 and Fig. 3 for cosmic rays and dark matter annihilation, respectively.

The main results of this theoretical study are summarized below.

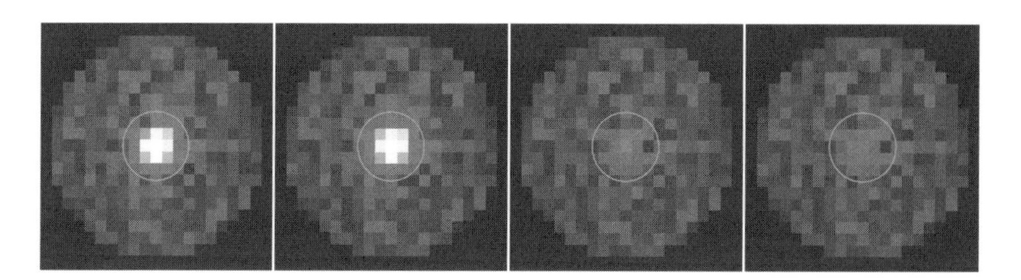

Fig. 2. Simulated gamma-ray emission from cluster cosmic rays.[4] Shown are simulations of five year Fermi observations of a cluster with the distance and mass of Coma and a cosmic ray to thermal energy density ratio of 0.1. Simulations include the EGRET extragalactic background,[10] and the images are binned to have 1 degree pixels. The panels show different simulations reflecting a reasonable range in the possible spatial and spectral cosmic ray distributions. All of these models are easily detected, and even for lower cosmic ray densities ($\sim 2\%$), in five years of data Fermi is expected to detect a few clusters.

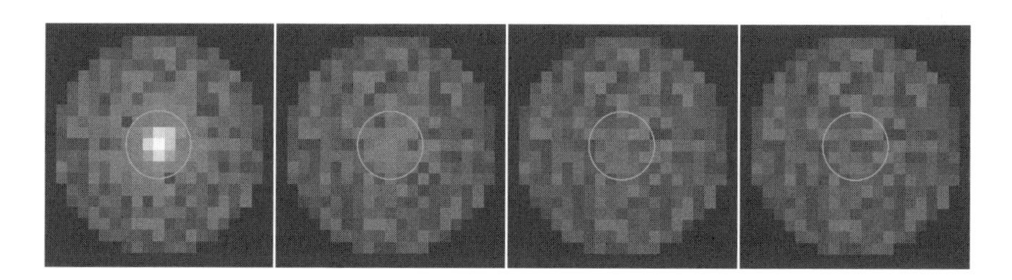

Fig. 3. Simulated gamma-ray emission from dark matter annihilation for five year Fermi observations of a cluster with the distance and mass of Coma.[4] Simulations include the EGRET extragalactic background,[10] and the images are binned to have 1 degree pixels. The panels show different simulations reflecting variations in the WIMP mass (40 and 110 GeV) and the contribution of substructure to the density distribution. The color bar matches the color bar in Fig. 2.

- The spectral analysis of the simulated signal for our benchmark models shows that gamma ray emission from dark matter annihilation with a relatively large dark matter particle mass ($m_{\rm WIMP} > 50$ GeV) can be distinguished from a cosmic ray spectrum even for fairly faint sources. Distinguishing a cosmic ray spectrum from a low dark matter particle mass appears to be more challenging, and would require deep data and/or a bright source.
- The level of the gamma-ray emission from clusters produced by cosmic rays can be comparable to that from dark matter annihilation. In the case of a mix of the two emissions, we find that for bright enough sources, if

the dark matter contribution to the cluster flux is significant and the particle mass is not very low ($m_{WIMP} > 40$ GeV) the presence of a dark matter component can be seen even in the presence of a significant gamma-ray flux from cosmic rays. However, tight constraints on the model parameters (m_{WIMP} and α_p) may be problematic.

- Our cluster gamma-ray simulated emissions appear, after data reduction, as extended rather than point sources, with extensions which depend on the spatial models and on the emission mechanism. However, determining the spatial distribution with Fermi will be a challenging task requiring an optimal control of the backgrounds.
- We showed that the ratio of the integrated gamma-ray to hard X-ray flux in galaxy clusters can be used as a diagnostics for the discrimination of the origin of non-thermal phenomena. The generic expectation is that a dark matter induced signal would produce a brighter hard X-ray emission as opposed to cosmic rays, for a given gamma-ray flux.
- We presented X-ray data-driven predictions for the gamma-ray flux from 130 clusters and groups of galaxies in the HIGFLUCS[11] and GEMS[12] catalogs. We found that the clusters with the brightest gamma-ray emission from cosmic rays include the Perseus, Coma, Ophiuchus, Abell 3627 and Abell 3526 clusters; the most luminous clusters in dark matter emission are predicted to be the Fornax group, Ophiuchus, Coma, Abell 3526 and Abell 3627.
- We discovered that the objects with the largest dark matter to cosmic ray gamma-ray luminosity in our sample are groups and poor clusters. In particular, the highest ratios are associated to the groups NGC 5846, 5813 and 499, M49 and HCG 22. Of these, M49 ranks overall 6th/130 in terms of predicted dark matter induced gamma-ray emission, NGC 5846 ranks 12th and NGC 5813 15th; Fornax also has a relatively high dark matter to cosmic ray gamma-ray flux, and is the brightest object in dark matter emission. All these objects are very promising candidates for a search for gamma-ray emission with the Fermi telescope that could potentially be related to particle dark matter.

Our analysis indicates that given a detection of gamma rays from clusters of galaxies with Fermi, distinguishing between a cosmic ray and a dark matter origin will be a challenging task. We outlined in detail the three best handles we will have to accomplish that task: a spectral analysis, the study of the source spatial extent, and multiwavelength studies with deep hard X-ray observations.

References

1. E. A. Baltz and L. Wai, *Phys. Rev. D* **70**, 023512 (2004).
2. S. Colafrancesco, S. Profumo and P. Ullio, *Astron. Astrophys.* **455**, 21 (2006).
3. T. E. Jeltema and S. Profumo, *Astrophys. J.* **686**, 1045 (2008).
4. T. E. Jeltema, J. Kehayias and S. Profumo, arXiv:0812.0597 [astro-ph].
5. R. C. Hartman *et al.* [EGRET Collaboration], *Astrophys. J. Suppl.* **123**, 79 (1999).
6. S. Profumo, *Phys. Rev. D* **77**, 103510 (2008).
7. S. Ando and D. Nagai, *Mon. Not. Roy. Astron. Soc.* **385**, 2243 (2008).
8. C. Pfrommer, *Mon. Not. Roy. Astron. Soc.* **385**, 1242 (2008).
9. P. Blasi, S. Gabici and G. Brunetti, *Int. J. Mod. Phys. A* **22**, 681 (2007).
10. P. Sreekumar *et al.* [EGRET Collaboration], *Astrophys. J.* **494** 523 (1998).
11. T. H. Reiprich and H. Boehringer, *Astrophys. J.* **567**, 716 (2002).
12. J. P. F. Osmond and T. J. Ponman, *Mon. Not. Roy. Astron. Soc.* **350** 1511 (2004).

WEIGHING SUPER-MASSIVE BLACK HOLES WITH X-RAY–EMITTING GAS

D. A. BUOTE* and P. J. HUMPHREY

Department of Physics and Astronomy,
University of California, Irvine,
Irvine, CA 92697-4575, USA
**E-mail: buote@uci.edu*

F. BRIGHENTI

Dipartmento di Astronomia, Università di Bologna,
Via Ranzani 1, Bologna 40127, Italy

K. GEBHARDT

Astronomy Department, University of Texas,
Austen, TX 78712, USA

W. G. MATHEWS

UCO/Lick Observatory,
University of California, Santa Cruz,
Santa Cruz, CA 95064, USA

We present a new approach to measure the mass of a super-massive black hole (SMBH) located at the center of a giant elliptical galaxy. This method applies the well-known technique of using the hot, X-ray emitting plasma as a tracer of the large-scale gravitational potential of a giant galaxy (or galaxy cluster) and extends it far down into the central region of a galaxy using high-resolution X-ray data from the Chandra X-ray Observatory. We report the first detection of a SMBH using this method in the Virgo elliptical galaxy, NGC 4649, and present results of preliminary detections in 3 other systems. In addition to providing interesting constraints on the black-hole masses, we show that the stellar mass-to-light ratios of the galaxies computed from this approach agree very well with the prediction from stellar population synthesis models, thus providing strong support for the underlying assumptions of the method (e.g., hydrostatic equilibrium).

Keywords: X-rays; galaxies – galaxies; elliptical and lenticular; cD – galaxies; ISM – black hole physics.

1. Introduction

The masses of super-massive black holes (SMBHs) are of special interest because they correlate with properties of the host galaxy. In particular, the correlation of SMBH mass and stellar velocity dispersion provides a crucial link between the formation of both the SMBH and the galaxy as a whole.[1,2] To understand this connection, it is necessary to measure the $M_{\rm BH}-\sigma$ relation as accurately as possible. The largest number of accurate SMBH measurements are derived from two dynamical methods – stellar dynamics and ionized gas disks.[3,4] However, stellar dynamics measurements must make assumptions to resolve the velocity dispersion anisotropy. Studies of ionized gas disks typically assume Keplerian motion in circular orbits. But these disks are often disturbed,[5] suggesting that turbulence and non-circular motion lead to additional systematic uncertainty. In order to determine the mean $M_{\rm BH}-\sigma$ relation and the scatter about it, the systematic errors in the different measurement methods needs to be understood.

Comparisons between $M_{\rm BH}$ measured using the ionized disk and stellar dynamics methods can be used to probe the systematic errors inherent in each approach, but such comparisons are rare and have yielded mixed results to date.[6–8] To better understand the systematics in the $M_{\rm BH}-\sigma$ relation it is therefore desirable to consider measurements from another method with a different set of assumptions. One such method was proposed about 10 years ago by Brighenti and Mathews,[9] who suggested using the density and temperature profile of the hot, X-ray–emitting gas in an elliptical galaxy to probe the mass of a quiescent SMBH. Their study, which assumed simple cooling flow models, showed that the gas temperature should increase substantially within the central 100 pc of an elliptical galaxy with a SMBH. In almost 10 years of Chandra observations of elliptical galaxies, this effect was not observed.

It is not too surprising that this effect had not been reported. Its discovery requires precise temperature measurements on scales of a few 100 pc, which pushes the limits of the capability of the *Chandra* X-ray Observatory. First, since enough counts are required to extract temperature and density information from the spectrum, only the brightest X-ray elliptical galaxies are suitable. Second, the galaxy must be very nearby to resolve as close as possible to the SMBH's sphere of influence considering the 1-arcsecond spatial resolution of *Chandra*. Finally, the hot gas, especially near the SMBH, cannot be too disturbed so as to make the approximation of hydrostatic equilibrium too poor to be useful.

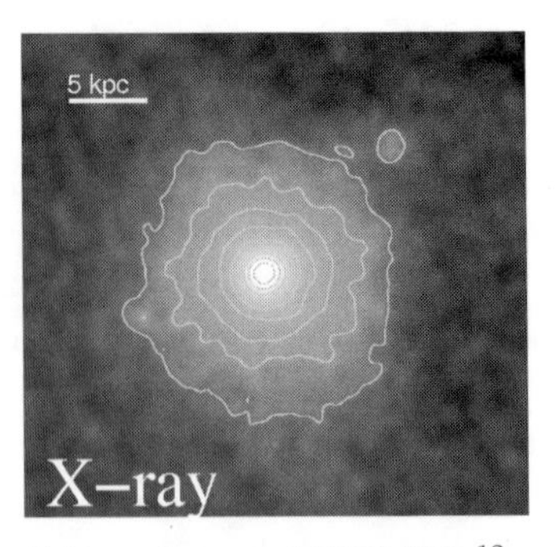

Fig. 1. (*Left Panel*) IR and (*Right Panel*) X-ray images of NGC 4649.[12]

This final point should not be understated. Observations with *Chandra* reveal prominent disturbances in the hot gas in several elliptical galaxies, with M84 (NGC 4374) and NGC 4636 being particularly noteworthy examples;[10,11] e.g., in M84 the cavities traced out by the X-rays are effectively filled by the observed radio emission emanating from the low-level active nucleus. In such systems it is likely that the hydrostatic equilibrium approximation is not very accurate.

2. NGC 4649

With these considerations, we are led to NGC 4649, a well-studied giant elliptical galaxy in the Virgo cluster. NGC 4649 has the desirable characteristics – it is indeed nearby (15.6 Mpc), very bright and has very regular optical and X-ray morphologies. The system also has a well-constrained SMBH mass from stellar dynamics,[13] which allows us, by comparison, to assess how reasonable are the assumptions, in particular hydrostatic equilibrium, required for the X-ray method.

We obtained a deep exposure (80 ks) of this galaxy with *Chandra* in AO-8 to study its large-scale mass distribution, which confirms the very relaxed X-ray morphology. We note that there have been occasional reports of significant asymmetrical features, possibly cavities, at the center, using previous lower quality X-ray data.[14,15] We performed several tests of the image morphology, including constructing a χ^2 residual image of the *Chandra* data obtained by subtracting a model of elliptical isophotes within the central regions. We do not find any statistically significant features with these higher quality data. At any rate, by comparing our SMBH mass to that determined from the stellar dynamics we obtain a quantitative assessment of the utility of the hydrostatic equilibrium approximation.

Briefly, the X-ray observations are translated into a mass profile in the following way. One partitions the *Chandra* X-ray image into concentric circular annuli, and the spectrum is extracted from each annulus. Each spectrum contains primarily emission from hot plasma containing thermal bremsstrahlung radiation and optically thin line emission. Therefore, focusing on the thermal bremsstrahlung component, the normalization of the spectrum essentially provides the gas density while the spectral shape gives the temperature. From these measurements of density and temperature, one computes the gas pressure. Then by assuming hydrostatic equilibrium the balance of the forces from gas pressure and gravity allows the mass profile to be computed.

In practice, one can specify other combinations of thermodynamic quantities to go along with the mass in the equation of hydrostatic equilibrium. For our default analysis, we input models of the entropy (S) and mass profiles into the equation of hydrostatic equilibrium and then compute the appropriately emission-weighted and instrument-convolved temperature (T) and entropy profiles to match to the observations. So the S and T data are used to constrain the input S and mass (M) models. We have the following free parameters in our models – 1 for the overall normalization of the pressure profile, 5 for the entropy (broken power law plus a constant), 4 for the mass profile (1 for the SMBH, 2 for the NFW DM halo, 1 for the stellar M/L). In all we have 10 free parameters constrained by 30 entropy and temperature data points.

We paid special care to make an accurate determination of the stellar light (and therefore stellar mass) profile of the galaxy using both archival *HST* and *2MASS* data. While we found a simple Sersic profile ($n = 2.6$) provided a decent fit to the light profile, there were significant residuals throughout the entire radial range fitted. To obtain a more accurate result, we performed a non-parametric deprojection using the same spherical "onion peeling" approach that we used for the X-ray data analysis. This "onion peel" method is a purely geometrical calculation and involves no fitting.[16,17]

The results for the temperature profile are displayed in Figure 2. Outside of ~ 2 kpc, the T profile rises in the usual way seen for more massive cool core groups and clusters. However, below 1 kpc, the T profile reverses direction and eventually peaks at the center with a value near 1.1 keV. We have included a hard spectral component in our fits to account for unresolved point sources, including possible AGN activity – the temperature peak seen here arises from a thermal gas component.

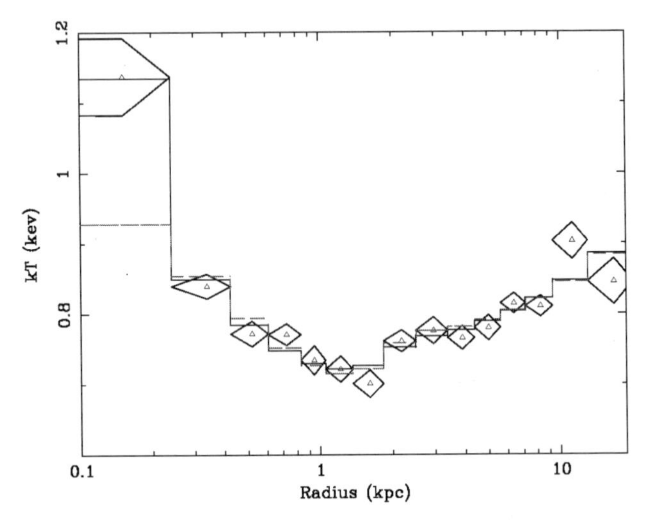

Fig. 2. Radial temperature profile of NGC 4649.[12] The (red) dashed line is the model without a SMBH, and the (black) solid line is the model including the SMBH. The improvement in the fit when including the SMBH is about 4σ.

If we do not include a SMBH in our mass models, we obtain a poor fit. It is noteworthy that the centrally concentrated stellar light is able to produce an inverted T profile, but it is not concentrated enough to explain the crucial central radial bin. If we include a SMBH, the fit is improved substantially, with most of the improvement (4σ) coming from the better description of the central bin.

(We note that the entropy profile (not shown) is consistent with that seen in more massive systems, the groups and clusters. The broken power-law is a good fit to the data, and the slope outside $\sim$ 2 kpc is consistent with that expected from cosmological gravitational infall ($S \sim r^{1.1}$).)

We obtain a mass of $3.35^{+0.67}_{-0.95} \times 10^9\, M_\odot$ (90% conf.) for the SMBH, which make it one of the most massive SMBHs known. (See Figure 3 for the derived mass profile.) Our X-ray measurement, therefore, provides a critical data point at the upper mass end of the $M_{\rm BH} - \sigma$ relation, and the good agreement with the mass estimate from stellar dynamics[13] ($(2 \pm 0.5) \times 10^9\, M_\odot$) and the precision of the measurement, are very competitive with other measurements of SMBHs. The agreement between the two methods provides an important consistency check and validation of both methods, in particular the accuracy of the approximation of hydrostatic equilibrium in the X-ray analysis.

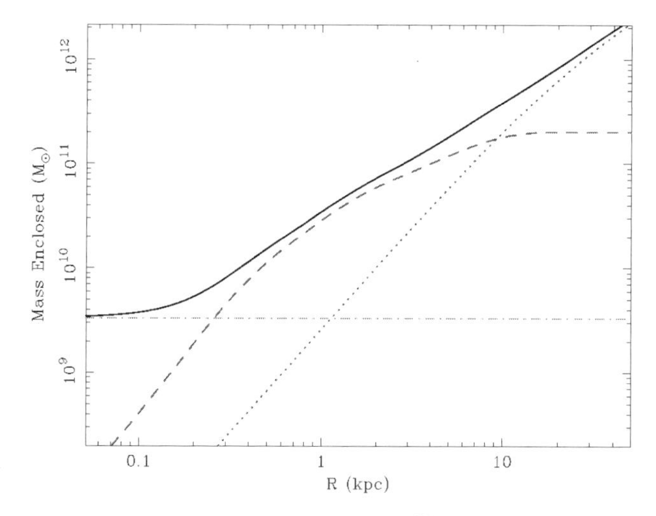

Fig. 3. Best-fitting radial mass profile of NGC 4649.[12] We show the total enclosed mass (solid: black), the stars (dashed: red), the dark matter (dotted: blue), and the SMBH (dash-dot: orange). The small contribution from the hot gas is not shown.

These results for NGC 4649, representing the first detection of a SMBH by its gravitational influence on hydrostatic X-ray emitting gas, have been published in Humphrey et al. (2008).[12]

3. Preliminary Results for 3 Systems

As noted above, this type of measurement pushes *Chandra* to the limits of its capability. But for this method to be really useful, it must be applicable to more targets. We have recently applied the X-ray method to additional systems that are not so optimal as NGC 4649. We selected our targets from the LEDA galaxy database in the following way. First, we considered only galaxies within a distance of 30 Mpc, because we need to resolve close to the SMBH sphere of influence. We used only galaxies with relatively high central velocity dispersions, to ensure a large SMBH mass according to the $M_{\rm BH} - \sigma$ relation. Also we need the galaxies to be bright in X-rays and have an existing *Chandra* observation. Finally, we wanted to avoid systems with strong AGN.

These selection criteria yielded 6 galaxies, 1 of which is NGC 4649. We excluded NGC 1399 and NGC 4374 because they exhibit obvious large-scale disturbances, leaving us with 3 galaxies: NGC 1332, NGC 4261, and NGC 4472. In fact, 4261 does possess fairly prominent X-ray emission from

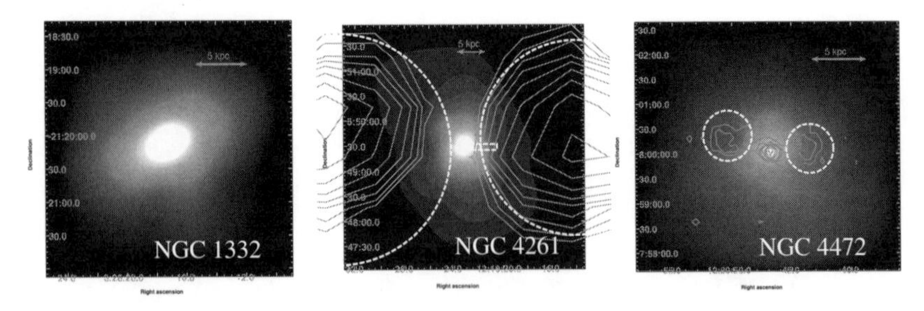

Fig. 4. Smoothed, point-source-subtracted *Chandra* X-ray images of three early-type galaxies from Humphrey et al. (2009, in preparation). Overlaid are radio contours from the NVSS survey for NGC 4261 and the VLA FIRST for NGC 4472.

an AGN, but we include it because we are able to separate the spectral signature of the AGN from that of the hot gas.

In Figure 4 we display the *Chandra* images of the 3 galaxies, which we have lightly smoothed with an adaptive Gaussian filter. Unlike NGC 4649, the X-ray isophotes are moderately flattened. For these systems we over-plot the radio contours showing clear radio lobes in two cases, but we emphasize the X-ray images do not show prominent disturbances associated with the lobes. Nevertheless, in our analysis we have excluded the parts of the lobes contained within the dashed white lines shown in the figure. By studying galaxies such as these, we can examine whether the mere presence of such radio lobes seriously compromises the reliability of the X-ray measurement of the SMBH mass.

The *Chandra* spectra of these galaxies, particularly within the central radial bin (corresponding to $\sim$ 200 pc in each case), consist of multiple components. First, there is the contribution from unresolved Low-Mass X-Ray Binaries (LMXBs), which are most noticeable at higher energies (above $\sim$ 2 keV). There is also a small contribution from unresolved stellar sources and cataclysmic variables, which is truly negligible. In NGC 4261 there is clear emission from an AGN, but it is highly absorbed and only contributes significantly at higher energies (above $\sim$ 2 keV). Fortunately, the hot gas dominates the spectra of all of the galaxies around 1 keV, which allows good constraints on the gas properties, and thus the SMBH mass, to be obtained.

As with NGC 4649, we constructed hydrostatic mass models and fitted them to the X-ray data, again assuming a simple broken power-law form for the entropy profile. We have improved our analysis method in a few

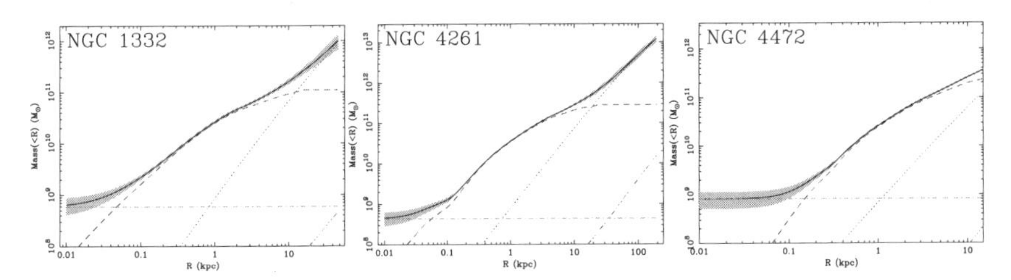

Fig. 5. Radial mass profiles for 3 early-type galaxies derived from fits to the *Chandra* X-ray data from Humphrey et al. (2009, in preparation). For each galaxy the solid (black) line is the total enclosed mass (associated grey region is 1σ error), the dashed (red) line is the stellar mass, the dotted (blue) line is the dark matter, the dash-dot (orange) line is the SMBH, and the dash-dot-dot-dot line is the hot gas.

respects, most notably (1) we deproject the optical light profile using a Lucy-Richardson algorithm that allows for a more general axisymmetric light distribution, and (2) we employ a Bayesian procedure to estimate the model parameters using a Monte Carlo method (MultiNest code[18]).

We find that the fits of the hydrostatic models to the temperature and density profiles of the hot gas are good in all 3 cases. Unlike NGC 4649, for these systems there is no pronounced central T peak. For NGC 1332 the temperature does rise at the center, but we find this rise can be fully explained by the increasing stellar density at the center. The lack of a central T spike caused by the central SMBH is also reflected in the entropy profiles, which do not flatten toward the center as in NGC 4649. In fact, for both NGC 1332 and NGC 4641, models without a central SMBH fit nearly as well as models that include a SMBH. Only for NGC 4472 do we see an interesting improvement in the fit when adding the SMBH. However, while adding the SMBH does increase the temperature slightly in NGC 4472, the key effect is to allow for a higher central density, which agrees better with the data. In sum, while all 3 galaxies are consistent with a SMBH, only for NGC 4472 does adding the SMBH clearly improve the fit – it is significant at the 2.4σ level.

It is not too surprising that the signature of the SMBH is less significant in these systems compared to NGC 4649 – see Figure 5. We find that $M_{\rm BH} \lesssim 10^9\,M_\odot$ in the 3 systems compared to $M_{\rm BH} \sim 3 \times 10^9\,M_\odot$ for NGC 4649. Moreover, the SMBH mass in NGC 1332 and NGC 4261 is much less than the stellar mass in the central radial bin (~ 200 pc), making detection difficult. In contrast, the SMBH mass is comparable to the stellar mass in the central radial bin for NGC 4472, where we do have a marginally significant detection.

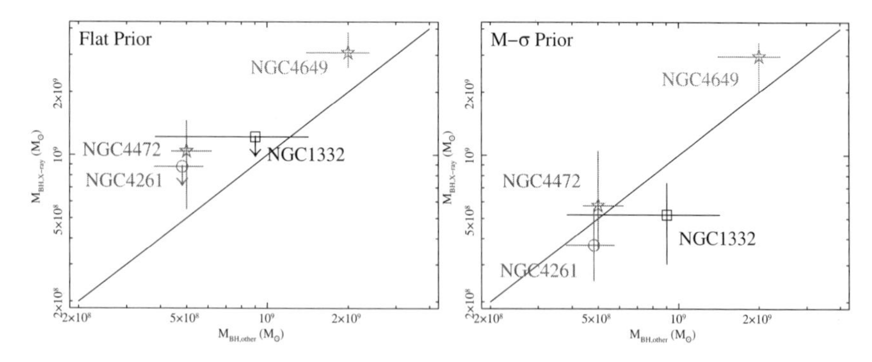

Fig. 6. (*Left Panel*) The derived $M_{\rm BH}$ from the X-ray data, assuming a flat prior for $M_{\rm BH}$, shown versus $M_{\rm BH}$ determinations from stellar dynamics (NGC 4472 and NGC 4649), gas disc dynamics (NGC 4261), or the $M_{\rm BH} - \sigma$ relation (NGC 1332). For NGC 1332 and NGC 4261 we show the 3σ upper limit for the X-ray–determined mass. The dotted line indicates $y = x$. (*Right Panel*) The same, but using the $M_{\rm BH} - \sigma$ relation as a prior on $M_{\rm BH}$. We find excellent agreement between $M_{\rm BH}$ measured with the difference techniques (Humphrey et al. 2009, in preparation).

In Figure 6 we display the SMBH masses we obtained from the X-ray data and compare them to to measurements from other methods. First, we examined using the $M_{\rm BH} - \sigma$ relation as the Bayesian prior for the SMBH mass. The primary consequence of using the $M_{\rm BH} - \sigma$ prior is to exclude unphysically low SMBH masses allowed by the data for NGC 4261 and NGC 1332. We find good agreement between the X-ray method and the other methods in this case. While the $M_{\rm BH} - \sigma$ prior may be the physical choice, it is important to compare to results using a flat prior on the SMBH mass, which essentially means assuming any SMBH mass is equally likely. When using the flat prior we find that the SMBH masses do shift a little, but the good agreement with other methods is preserved. However, the constraints for NGC 1332 and NGC 4261 are now reduced to upper limits, as expected. It will take higher quality X-ray data to detect the SMBHs in NGC 1332 and NGC 4261 in a manner that is less sensitive to the choice of the prior.

Another test of the viability of the hydrostatic method is to compare the stellar M/L ratio measured from the X-rays to the predictions from single-burst stellar population synthesis models, which represent non-dynamical mass estimates. As shown in Figure 7 we find overall good agreement between the measured and predicted stellar M/L values (particularly for models assuming a Kroupa/Chabrier IMF), which provides additional support for the assumptions underlying the X-ray method.

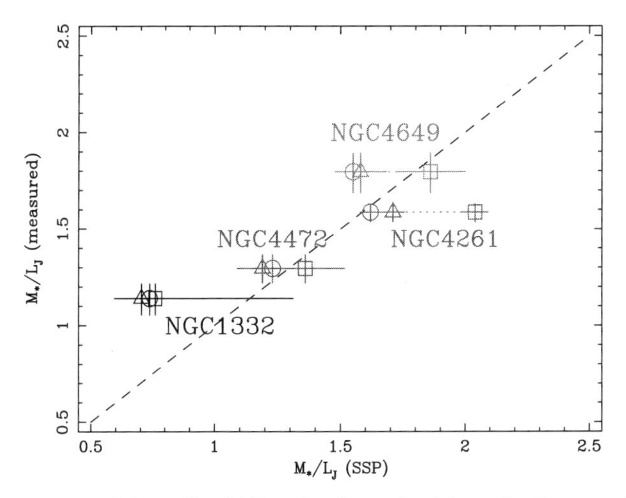

Fig. 7. Comparison of the stellar M/L ratios determined from the X-ray analysis to the predictions of three different single-burst stellar population synthesis models. The key is square: Maraston[19] (Salpeter IMF), circle: PEGASE[20] (Salpeter IMF), triangle: Bruzual & Charlot[21] (Kroupa/Chabrier IMF) (see Humphrey et al. 2009, in preparation).

The results in this section will be presented in detail in Humphrey et al. (2009, in preparation).

4. Conclusions

We conclude that hydrostatic X-ray plasma offers a practical, competitive means to measure SMBH masses in early-type galaxies. The good agreement we find between $M_{\rm BH}$ measured from X-rays and other dynamical techniques provides key support for the assumptions and approximations of the various methods and indicates that the dynamical methods are accurate.

While we expect that *Chandra* will eventually allow for interesting measurements of $M_{\rm BH}$ for a small number of systems, maybe as many as 10, next-generation telescopes with better resolution and collecting area, in particular the proposed *Generation-X* mission,[22] would likely increase the number of SMBH measurements by over an order of magnitude.

References

1. K. Gebhardt, R. Bender, G. Bower, A. Dressler, S. M. Faber, A. V. Filippenko, R. Green, C. Grillmair, L. C. Ho, J. Kormendy, T. R. Lauer, J. Magorrian, J. Pinkney, D. Richstone and S. Tremaine, *ApJ* **539**, L13(August 2000).

2. L. Ferrarese and D. Merritt, *ApJ* **539**, L9(August 2000).
3. L. Ferrarese and H. Ford, *Space Science Reviews* **116**, 523(February 2005).
4. A. W. Graham, *Publications of the Astronomical Society of Australia* **25**, 167(November 2008).
5. M. Cappellari, E. K. Verolme, R. P. van der Marel, G. A. V. Kleijn, G. D. Illingworth, M. Franx, C. M. Carollo and P. T. de Zeeuw, *ApJ* **578**, 787(October 2002).
6. K. L. Shapiro, M. Cappellari, T. de Zeeuw, R. M. McDermid, K. Gebhardt, R. C. E. van den Bosch and T. S. Statler, *MNRAS* **370**, 559(August 2006).
7. J. D. Silge, K. Gebhardt, M. Bergmann and D. Richstone, *AJ* **130**, 406(August 2005).
8. N. Neumayer, M. Cappellari, J. Reunanen, H.-W. Rix, P. P. van der Werf, P. T. de Zeeuw and R. I. Davies, *ApJ* **671**, 1329(December 2007).
9. F. Brighenti and W. G. Mathews, *ApJ* **527**, L89(December 1999).
10. C. Jones, W. Forman, A. Vikhlinin, M. Markevitch, L. David, A. Warmflash, S. Murray and P. E. J. Nulsen, *ApJ* **567**, L115(March 2002).
11. A. Finoguenov, M. Ruszkowski, C. Jones, M. Brüggen, A. Vikhlinin and E. Mandel, *ApJ* **686**, 911(Octtober 2008).
12. P. J. Humphrey, D. A. Buote, F. Brighenti, K. Gebhardt and W. G. Mathews, *ApJ* **683**, 161(August 2008).
13. K. Gebhardt, D. Richstone, S. Tremaine, T. R. Lauer, R. Bender, G. Bower, A. Dressler, S. M. Faber, A. V. Filippenko, R. Green, C. Grillmair, L. C. Ho, J. Kormendy, J. Magorrian and J. Pinkney, *ApJ* **583**, 92(January 2003).
14. S. W. Randall, C. L. Sarazin and J. A. Irwin, *ApJ* **636**, 200(January 2006).
15. K. Shurkin, R. J. H. Dunn, G. Gentile, G. B. Taylor and S. W. Allen, *MNRAS* **383**, 923(January 2008).
16. A. C. Fabian, E. M. Hu, L. L. Cowie and J. Grindlay, *ApJ* **248**, 47(August 1981).
17. G. A. Kriss, D. F. Cioffi and C. R. Canizares, *ApJ* **272**, 439(September 1983).
18. F. Feroz and M. P. Hobson, *MNRAS* **384**, 449(February 2008).
19. C. Maraston, *MNRAS* **362**, 799(September 2005).
20. M. Fioc and B. Rocca-Volmerange, *A&A* **326**, 950(Octtober 1997).
21. G. Bruzual and S. Charlot, *MNRAS* **344**, 1000(Octtober 2003).
22. R. A. Windhorst, R. A. Cameron, R. J. Brissenden, M. S. Elvis, G. Fabbiano, P. Gorenstein, P. B. Reid, D. A. Schwartz, M. W. Bautz, E. Figueroa-Feliciano, R. Petre, N. E. White and W. W. Zhang, *New Astronomy Review* **50**, 121(March 2006).

PART VI

Dark Energy, Dark Matter and Dark Radiation

DARK MATTER AND DARK RADIATION

LOTTY ACKERMAN, MATTHEW R. BUCKLEY, SEAN M. CARROLL and MARC KAMIONKOWKSI

California Institute of Technology, Pasadena, CA 91125, USA

We explore the feasibility and astrophysical consequences of a new long-range $U(1)$ gauge field ("dark electromagnetism") that couples only to dark matter, not to the Standard Model. The dark matter consists of an equal number of positive and negative charges under the new force, but annihilations are suppressed if the dark matter mass is sufficiently high and the dark fine-structure constant $\hat{\alpha}$ is sufficiently small. The correct relic abundance can be obtained if the dark matter also couples to the conventional weak interactions, and we verify that this is consistent with particle-physics constraints. The primary limit on $\hat{\alpha}$ comes from the demand that the dark matter be effectively collisionless in galactic dynamics, which implies $\hat{\alpha} \lesssim 10^{-3}$ for TeV-scale dark matter. These values are easily compatible with constraints from structure formation and primordial nucleosynthesis. We raise the prospect of interesting new plasma effects in dark matter dynamics, which remain to be explored. This proceedings is based on the work presented originally in.[1]

1. Introduction

A wide variety of cosmological observations seem to point to a two-component dark sector, in which approximately 73% of the energy density of the universe is in dark energy and 23% is in non-baryonic dark matter (DM). Ordinary matter constitutes the remaining 4%.[2] The physics of the dark matter sector is plausibly quite minimal: an excellent fit to the data is obtained by assuming that dark matter is a cold, collisionless relic, with only the relic abundance as a free parameter. The well-known "WIMP miracle"[3] is the fact that a stable, neutral particle with weak-scale mass and coupling naturally provides a reasonable energy density in DM. Such particles are found in many well motivated extensions of the Standard Model (SM) that solve the hierarchy problem. In the contemporary universe, they would be collisionless as far as any conceivable dynamical effects are concerned.

Nevertheless, it is also possible to imagine a rich phenomenology within the dark sector. In this proceedings, we explore the possibility of a long-range *gauge* force coupled to DM, in the form of a new unbroken abelian field, dubbed the $U(1)_D$ "dark photon." We imagine that this new gauge boson $\hat{\gamma}$ couples to a DM fermion χ, but not directly to any SM fields. Our model is effectively parameterized by only two numbers: m_χ, the mass of the DM, and $\hat{\alpha}$, the dark fine-structure constant. For appropriate values, the DM freezes out in the early universe. In the present day, DM is a plasma, which could conceivably lead to interesting collective effects in the halo dynamics.

Remarkably, the allowed values of m_χ and $\hat{\alpha}$ seem quite reasonable. We find that the most relevant constraint comes from demanding that accumulated soft scatterings do not appreciably perturb the motion of DM particles in a galaxy over the lifetime of the universe, which can be satisfied by $\hat{\alpha} \sim 10^{-3}$ and $m_\chi \sim$ TeV. For values near these bounds, the alterations in DM halo shapes may in fact lead to closer agreement with observation.[4] If, in addition, χ possesses $SU(2)_L$ quantum numbers, we obtain the correct relic abundance.

We consider a number of other possible observational limits on dark electromagnetism, and show that they do not appreciably constrain the parameter space. Since the DM halo is overall neutral under $U(1)_D$, there is no net long-range force that violates the equivalence principle. Although there are new light degrees of freedom, their temperature is naturally lower than that of the SM plasma, thereby avoiding constraints from Big-Bang Nucleosynthesis (BBN). On the other hand, we find that there are plasma instabilities (*e.g.* the Weibel instability) that can potentially play an important role in the assembly of galactic halos; however, a detailed analysis of these effects is beyond the scope of this work. For additional references on dark matter and new gauge forces, see Refs[5–16].

2. Dark Radiation and the Early Universe

We postulate a new "dark" abelian gauge group $U(1)_D$ with gauge coupling constant $\hat{g}$ and dark fine-structure constant $\hat{\alpha} \equiv \hat{g}^2/4\pi$. In the simplest case, the dark matter sector consists of a single fermion $\chi/\bar{\chi}$ with $U(1)_D$ charge of $+1$ and mass m_χ As the limits on new long range forces on SM fields are very stringent, we assume that all the SM fields are neutral under $U(1)_D$. In this Section, we will derive constraints on the mass m_χ and coupling $\hat{\alpha}$ from the evolution of dark matter in the early universe. Two considerations drive these constraints: the dark matter must provide the right relic abundance at thermal freeze-out, and the dark radiation from the $U(1)_D$ cannot contribute too greatly to relativistic degrees of freedom at BBN.

The degrees of freedom in the dark sector are thus the heavy DM fermions χ and massless dark photons $\hat{\gamma}$. We assume that the mixing term $c\hat{F}_{\mu\nu}F^{\mu\nu}$ is set to zero at some high scale. This is a self-consistent choice, since if there is no mixing between the dark and visible sectors, $c = 0$ is preserved by the RGE. We may therefore ignore paraphoton constraints.[7,18]

We now follow the thermal history of the dark sector. Our analysis follows that of Ref. [13]. After inflation, the visible and dark sectors could conceivably reheat to different temperatures. Even if the temperatures are initially equal, after they decouple, entropy deposited from frozen-out degrees of freedom in one sector will generally prevent the dark temperature $\hat{T}$ from tracking the visible sector temperature T. The ratio $\xi = \hat{T}/T$ will depend on the spectrum of both sectors, and is itself a function of T.

As the temperature drops below a particle's mass, the temperature of that sector declines more slowly, as the associated degrees of freedom freeze out and dump entropy into their respective sectors (dark or visible). As the entropy density s of the visible sector and $\hat{s}$ of the dark sector are individually conserved after decoupling, we must separately count the degrees of freedom in these two sectors. There are two definitions of degrees of freedom of interest to us: g_* and g_{*S}. The former goes as the 4th power of T, while the latter goes as T^3. If we restrict the visible sector to the SM, then the term $g_{*\text{vis}}$ is 106.75 above the top mass, $g_{*\text{vis}} = 10.75$ between 100 MeV $\gtrsim T \gtrsim$ 1 MeV, and drops again to 3.36 in the present day. (See *e.g.* Ref.[17] for more details.)

We may split the dark g_{*S} into heavy and light degrees of freedom: g_{heavy} and g_{light}, where the heavy degrees of freedom are non-relativistic at BBN. We are interested in the number of degrees of freedom at BBN ($T \sim 1$ MeV), where we find that[13]

$$\frac{g_{\text{light}}\xi(T_{\text{BBN}})^3}{(g_{\text{heavy}} + g_{\text{light}})\xi(T_{\text{RH}})^3} = \frac{g_{*\text{vis}}(T_{\text{BBN}})}{g_{*\text{vis}}(T_{\text{RH}})}. \tag{1}$$

In order for the dark sector to satisfy the BBN bound,[19]

$$g_{\text{light}}\left[\frac{g_{\text{heavy}} + g_{\text{light}}}{g_{\text{light}}}\frac{10.75}{g_{*\text{vis}}(T_{\text{RH}})}\right]^{4/3}\xi(T_{\text{RH}})^4 \leq 2.52 \quad \text{(95\% confidence)}. \tag{2}$$

Since the high energy completion of the visible sector must at minimum include the SM fields, $g_{*\text{vis}}(T_{\text{RH}}) \geq 106.75$; a bound on the dark sector g_{light} and g_{heavy} can be derived for a fixed value of $\xi(T_{\text{RH}})$ (see Fig. 1). A similar bound on relativistic degrees of freedom can be derived from the cosmic microwave background, but provides a weaker 2σ exclusion limit.[13,20]

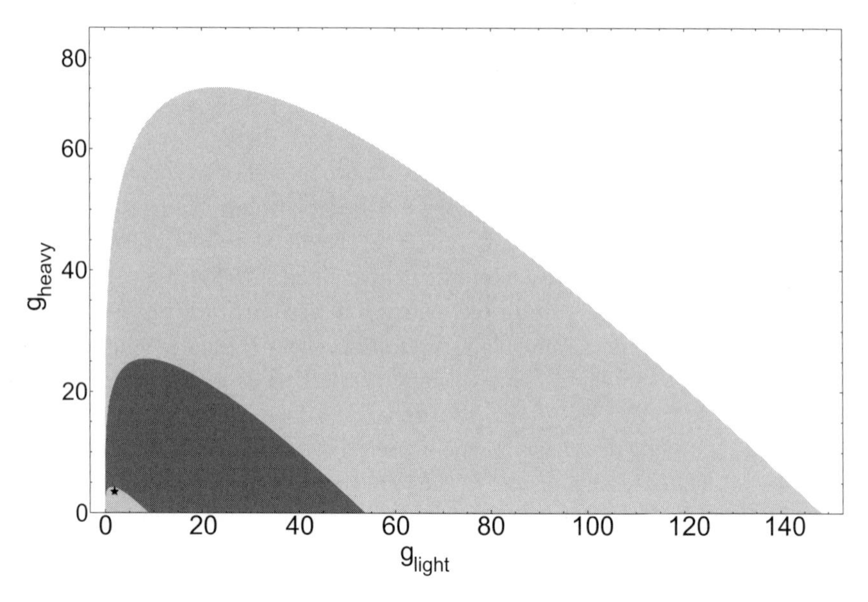

Fig. 1. The allowed values of dark g_{light} (those degrees of freedom relativistic at T_{BBN}) and g_{heavy} (the remaining dark degrees of freedom) arising from BBN constraints Eq. (2). The allowed regions correspond to 95% confidence levels for $\xi(T_{\text{RH}}) = 1$ and a visible sector $g_{*\text{vis}} = 106.75$ (red), $\xi(T_{\text{RH}}) = 1$ and $g_{*\text{vis}} = 228.75$ (corresponding to MSSM particle content, in blue), and $\xi(T_{\text{RH}}) = 1.4(1.7)$ and $g_{*\text{vis}} = 106.75(228.75)$ (in yellow). The minimal dark sector model of this paper is noted by a black star at $g_{\text{light}} = 2$ and $g_{\text{heavy}} = 3.5$.

We now turn to bounds on the coupling $\hat{\alpha}$ and dark matter mass m_χ coming from the dark matter abundance. At high $\hat{T}$ χ particles are kept in thermal equilibrium with the $\hat{\gamma}$ via t-channel pair annihilation/creation. To leading order, $\langle\sigma v\rangle$ is independent of v: $\langle\sigma v\rangle \approx \pi\hat{\alpha}^2/2m_\chi^2$. Using this, the relic density of the χ particles may be easily calculated.[17] With the measured value[2] $\Omega_{DM}h^2 \doteq 0.106 \pm 0.08$, and using the standard formulae for thermal freeze-out we may solve for the allowed values of $\hat{\alpha}$ as a function of m_χ. The resulting band is shown in Fig. 2.

We now consider how changing our assumptions on g_* and ξ can change our conclusions on the allowed parameter space. The parameter $\xi(T_f)$ does not enter explicitly into the calculation for $\Omega_{\text{DM}}h^2$, however it does affect the number of active degrees of freedom at freeze-out directly through g_* and $g_{*\text{S}}$, and indirectly by allowing the temperature T to differ from $\hat{T}$. If $\xi < 1$, $\hat{T} < T$ and there could be many more heavy visible degrees of freedom still active when χ freezes out. $\xi > 1$ would reduce the visible degrees of freedom. As we have seen in Eq. (2), it is difficult to construct a

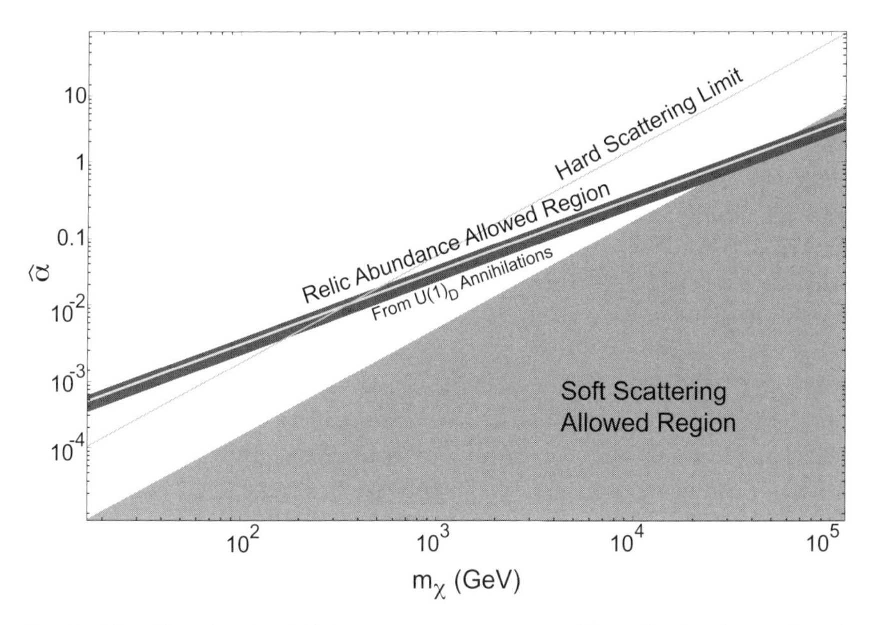

Fig. 2. The allowed regions of $\hat{\alpha}$ vs. m_χ parameter space. The relic abundance allowed region applies to models in which $U(1)_D$ is the only force coupled to the dark matter; in models where the DM is also weakly interacting, this provides only an upper limit on $\hat{\alpha}$. The thin yellow line is the allowed region from correct relic abundance assuming $\Omega_{\rm DM}h^2 = 0.106 \pm 0.08$, $\xi(T_{\rm RH}) = 1$, $g_{*\rm vis} \approx 100$, and $g_{\rm heavy} + g_{\rm light} = 5.5$ while the surrounding blue region is $g_{*\rm vis} = 228.75(60)$, $\xi(T_{\rm RH}) = 1(0.1)$, and $g_{\rm heavy} + g_{\rm light} = 100(5.5)$ at the lower(upper) edge. The diagonal green line is the upper limit on $\hat{\alpha}$ from effects of hard scattering on galactic dynamics; in the red region, even soft scatterings do not appreciably affect the DM dynamics. We consider this to be the allowed region of parameter space.

scenario with large ξ, short of a massive increase in $g_{*\rm vis}$ and small values of $g_{*\rm heavy} + g_{*\rm light}$. We include in Fig. 2 the bounds from both a large and small value of g_*. The large limit is $g_{*\rm vis}(T_f) = 228.75$, (*i.e.* equivalent to the MSSM degrees of freedom), $\xi(T_{\rm RH}) = 1$, and $g_{\rm heavy} + g_{\rm light} = 100$, while the small value is given by $g_{*\rm vis}(T_f) = 60$, (*i.e.* equivalent to the SM degrees of freedom at $\Lambda_{\rm QCD}$), $\xi(T_{\rm RH}) = 0.1$, and $g_{\rm heavy} + g_{\rm light} = 5.5$.

3. Galactic Dynamics

Although freezeout in our scenario is similar to that in the standard WIMP scenario, the long-range DM-DM interactions may lead to considerably different DM phenomenology in the current Universe, and in particular in galactic halos. In this scenario, DM halos are composed of an equal mixture of χ and $\bar{\chi}$. The overall halo will be $U(1)_D$ neutral, eliminating long-range forces that are incompatible with experiment.

However, nearest-neighbor interactions between χ particles remain, and these interactions can be constrained by observations that suggest that dark matter is effectively collisionless.[4,21–23] Roughly speaking a conservative upper bound is less than one DM-DM interaction per age of the universe or $\sigma/m_\chi \lesssim 0.1\ \text{cm}^2/\text{g}$. A separate bound of $\sigma/m_\chi < 1.25$ can be derived from the Bullet Cluster,[24,25] but as this is less restrictive we ignore it here.

In our model, the leading constraint is soft scattering on the allowed values of $\hat{\alpha}$ and m_χ, where a hard scatter is a single interaction that exchanges $\mathcal{O}(\text{K.E.})$. Here we consider the approach of one χ particle towards another $\chi(\bar{\chi})$ at impact parameter b. By calculating the δv^2 for a single collision and integrating over b, we find the number τ/τ_{dyn} of orbits it will take for the dark-matter particle to have $\Delta v^2/v^2 \sim \mathcal{O}(1)$ is

$$\frac{\tau_{\text{soft}}}{\tau_{\text{dyn}}} = \frac{G^2 m_\chi^4 N}{8\hat{\alpha}^2} \ln^{-1}\left(\frac{GNm_\chi^2}{2\hat{\alpha}}\right) \gtrsim 50. \tag{3}$$

Here $N \sim 10^{64}(m_\chi/\text{TeV})^{-1}$ is the number of DM particles in the Galaxy. There is a logarithmic suppression in soft scattering relative to hard due to the long-range Coulomb force generated by the $U(1)_D$. As can be seen in Fig. 2, the allowed region from these considerations of Galactic dynamics completely exclude the $\hat{\alpha}/m_\chi$ band that gives the correct relic abundance up to $m_\chi \sim 30$ TeV. We surmise that a bound even stronger than that estimated here can be obtained from the dwarf galaxies that exhibit the highest observed dark-matter phase-space densities (and thus tighter bounds on σ/m_χ).[27]

However, if we expand our model so that the χ particles possess $SU(2)_L$ quantum numbers in addition to a $U(1)_D$ charge, then the cross section for freeze-out in the early universe is dominated by the weak interaction $\sigma \sim \alpha^2/m_\chi^2$, and the $U(1)_D$ contribution is negligible for the small values of $\hat{\alpha}$ under consideration. At late times the situation is reversed. The weak cross section remains small, as it is the result of a short-range force. However the long range cross section for soft scattering increases as the dark matter cools and slows.

In outlining our original model in Section 2, we set the coefficient of the mixing term $F_{\mu\nu}\hat{F}^{\mu\nu}$ to zero at the high scale. Clearly loops involving χ would generate a non-zero mixing if the χ field possesses non-zero hypercharge Y. In order to avoid this complication, we set $Y = 0$. This requires χ to sit in an n-plet of $SU(2)_L$ where n is odd; we take $n = 3$, so the χ triplet contains the neutral χ^0 and EM charged $\chi^\pm$, all with $U(1)_D$ charges of +1. Due to $SU(2)_L$ loops, the $\chi^\pm$ are 166 MeV heavier than the χ^0, and decay before BBN. If the dark matter mass is $m_\chi = 2.4$ TeV, the correct dark matter abundance (including production and then decay of $\chi^\pm$) results from thermal freeze out.[28] We note that our model does have the nice feature of automatically suppressing unwanted decays of χ into SM particles, as by assumption χ is the lightest particle charged under $U(1)_D$.

This model is anomaly free, satisfies all BBN bounds, and $\hat{\gamma}$ does not couple to SM fermions or γ at all orders. The lowest order coupling of SM fermions requires two dark photons and thus occurs at $\alpha^2\hat{\alpha}$, thus the $\hat{\gamma}$s are unlikely to be directly detected. The presence of a new unbroken $U(1)_D$ in the dark sector could only be probed via its effect on galactic dynamics. Values of $\hat{\alpha}$ near the maximum allowed from soft-scattering should have a measurable effect on the halo structure, as in this regime the dark matter is no longer completely collisionless. A full study of this effect requires simulations beyond the scope of this paper.

4. Plasma Instabilities

The existence of a dark matter 'plasma' may have additional effects that could significantly affect structure formation. We focus here on the Weibel instability in galactic halos. This may have significant and visible effects in the halo, but requires simulation beyond the scope of this paper.

As a simple example we consider the Weibel instability,[30] an exponential magnetic-field amplification that arises if the plasma particles have an anisotropic velocity distribution. Such anisotropies could arise, for example, during hierarchical structure formation as subhalos merge to form more massive halos. Similar instabilities in the baryonic gas have been postulated to account for the magnetic fields in galaxy clusters.[31] In our Galaxy, the growth rate Γ of the field is $\Gamma \sim 10^{-2}\text{s}^{-1} \times \hat{\alpha}^{1/2}/(m_\chi/\text{TeV})$.

To be relevant for galactic-halo formation, the timescale Γ^{-1} for magnetic-field amplification should be shorter than the dynamical timescale τ of the merging subhalos. The instability will be therefore be of interest when $(m_\chi/\text{TeV}) \lesssim 10^{11}\hat{\alpha}^{1/2}\left(\tau/10^6 \text{ yrs}\right)$. This range of $\hat{\alpha}$ and m_χ encompasses the entire parameter space of interest for any reasonable value of τ. Therefore, we suspect that galactic structure will be affected by plasma effects in the dark matter due to the $U(1)_D$ even when $\hat{\alpha}$ is not near the boundary of allowed values from soft scattering. A more detailed study will be required to assess these effects.

5. Conclusions

Given how little direct information we have about the nature of dark matter, it is of crucial importance to explore models in which the DM sector has an interesting phenomenology of its own. In many ways, an unbroken $U(1)$ gauge field coupled to dark matter is a natural way to obtain a long-range interaction between DM particles. In contrast to the case of hypothetical long-range scalar fields, the masslessness of the gauge field is protected by a symmetry, and the absence of long-range violations of the equivalence principle is naturally explained by the overall charge neutrality of the dark plasma. New unbroken $U(1)$'s can appear naturally in unified models.

While a dark $U(1)$ may be realized as a broken symmetry with massive vector bosons, it has been pointed out that there are few constraints on the massless, unbroken case from the early universe. We have verified that the minimal model, with just a single massive Dirac fermion for the dark matter and a massless dark photon, is consistent with limits obtained from the number of relativistic degrees of freedom at BBN, with relatively mild assumptions on the reheating temperature of the dark sector. More complicated models are also allowed.

We found that one cannot build a dark matter model charged under a hidden unbroken $U(1)_D$ in which this new gauge group is responsible for thermal freeze out. As can be seen in Fig. 2, the required values of $\hat{\alpha}$ and m_χ required for the χ particles to form a thermal relic would violate bounds coming from limits on hard and soft scattering of dark matter in the Galactic halo. As an important consequence of this argument, models in which dark matter couples to an exact copy of ordinary electromagnetism (in particular, with $\hat{\alpha} = \alpha$) are ruled out unless $m_\chi >$ a few TeV. This constrains the parameter space of models[13] with hidden copies of the SM or the MSSM in which the dark matter is electrically charged where the stau was suggested as a dark matter candidate.

By adding additional interactions to increase the annihilation cross-section, it is possible to build a scenario with an unbroken dark $U(1)$ and the correct relic abundance. Introducing another short-range force coupling to the χ, for example the familiar $SU(2)_L$, can provide an appropriately large cross section for $\chi/\bar{\chi}$ annihilation. The new coupling $\hat{\alpha}$ must then be relatively small (compared to the $SU(2)_L$ α) in order to evade Galactic dynamics bounds.

The simplest model which realizes this situation is a Dirac fermion in a triplet of $SU(2)_L$ (in order to avoid $U(1)_Y/U(1)_D$ mixing). Bounds from the early universe then force m_χ to be on the order of a few TeV, which implies $\hat{\alpha} \lesssim 10^{-2}$. Looking for the effects on Galactic dynamics arising from a soft scattering mediated by a long-range force would be the only search strategy. Clearly, as $\hat{\alpha}$ goes to zero, the model becomes indistinguishable from minimal weakly coupled dark matter. However, if the coupling is near the limit from soft scattering, one would expect detectable deviations from the assumptions of collisionless dark matter currently used in simulations.

Additionally, since the $U(1)_D$ effectively makes the dark halo a plasma (albeit a very cold, tenuous one), there may be other effects on structure formation that constrain this model. We have estimated that the timescale for the Weibel instability in our model is short compared to relevant timescales for galactic dynamics. If this instability has a dramatic effect when subhalos collide during the assembly of a galactic halo, our $U(1)_D$ could be excluded for the entire range of interesting parameters. Further work is required to before we reliably understand the quantitative effects of such instabilities on galactic dynamics.

M.B. would like to thank the organizers of the DARK 2009 conference and the Physics Department in the University of Canterbury. This work was supported by DoE DE-FG03-92-ER40701 and the Gordon and Betty Moore Foundation.

References

1. L. Ackerman, M. R. Buckley, S. M. Carroll and M. Kamionkowski, Phys. Rev. D **79**, 023519 (2009) [arXiv:0810.5126 [hep-ph]].
2. C. Amsler *et al.* [Particle Data Group], Phys. Lett. B **667**, 1 (2008).
3. G. Jungman, M. Kamionkowski and K. Griest, Phys. Rept. **267**, 195 (1996) [arXiv:hep-ph/9506380].
4. D. N. Spergel and P. J. Steinhardt, Phys. Rev. Lett. **84**, 3760 (2000) [arXiv:astro-ph/9909386]; M. Kaplinghat, L. Knox and M. S. Turner, Phys. Rev. Lett. **85**, 3335 (2000) [arXiv:astro-ph/0005210]; A. Tasitsiomi, Int. J. Mod. Phys. D **12**, 1157 (2003) [arXiv:astro-ph/0205464].
5. D. Hooper and K. M. Zurek, Phys. Rev. D **77**, 087302 (2008) [arXiv:0801.3686 [hep-ph]].
6. A. De Rujula, S. L. Glashow and U. Sarid, Nucl. Phys. B **333**, 173 (1990).
7. B. Holdom, Phys. Lett. B **166**, 196 (1986).
8. S. Davidson, S. Hannestad and G. Raffelt, JHEP **0005**, 003 (2000) [arXiv:hep-ph/0001179].
9. S. Dimopoulos, D. Eichler, R. Esmailzadeh and G. D. Starkman, Phys. Rev. D **41**, 2388 (1990).
10. L. Chuzhoy and E. W. Kolb, arXiv:0809.0436 [astro-ph].
11. S. S. Gubser and P. J. E. Peebles, Phys. Rev. D **70**, 123510 (2004) [arXiv:hep-th/0402225].
12. J. L. Feng and J. Kumar, arXiv:0803.4196 [hep-ph].

13. J. L. Feng, H. Tu and H. B. Yu, arXiv:0808.2318 [hep-ph].
14. B. A. Dobrescu, Phys. Rev. Lett. **94**, 151802 (2005) [arXiv:hep-ph/0411004].
15. M. Pospelov, A. Ritz and M. B. Voloshin, Phys. Lett. B **662**, 53 (2008) [arXiv:0711.4866 [hep-ph]].
16. D. V. Ahluwalia, C. Y. Lee, D. Schritt and T. F. Watson, arXiv:0712.4190 [hep-ph].
17. E. Kolb and M.S. Turner, "The Early Universe," Addison-Wesley Publishing Company, Redwood City, (1990)
18. L. B. Okun, Sov. Phys. JETP **56**, 502 (1982) [Zh. Eksp. Teor. Fiz. **83**, 892 (1982)].
19. R. H. Cyburt, B. D. Fields, K. A. Olive and E. Skillman, Astropart. Phys. **23**, 313 (2005) [arXiv:astro-ph/0408033].
20. T. L. Smith, E. Pierpaoli and M. Kamionkowski, Phys. Rev. Lett. **97**, 021301 (2006) [arXiv:astro-ph/0603144].
21. J. Miralda-Escude, arXiv:astro-ph/0002050.
22. R. Dave, D. N. Spergel, P. J. Steinhardt and B. D. Wandelt, Astrophys. J. **547**, 574 (2001) [arXiv:astro-ph/0006218].
23. N. Yoshida, V. Springel, S. D. M. White and G. Tormen, arXiv:astro-ph/0006134.
24. D. Clowe, M. Bradac, A. H. Gonzalez, M. Markevitch, S. W. Randall, C. Jones and D. Zaritsky, Astrophys. J. **648**, L109 (2006) [arXiv:astro-ph/0608407].
25. S. W. Randall, M. Markevitch, D. Clowe, A. H. Gonzalez and M. Bradac, arXiv:0704.0261 [astro-ph].
26. O. Y. Gnedin and J. P. Ostriker, arXiv:astro-ph/0010436.
27. J. J. Dalcanton and C. J. Hogan, Astrophys. J. **561**, 35 (2001) [arXiv:astro-ph/0004381]; C. J. Hogan and J. J. Dalcanton, Phys. Rev. D **62**, 063511 (2000) [arXiv:astro-ph/0002330].
28. M. Cirelli, N. Fornengo and A. Strumia, Nucl. Phys. B **753**, 178 (2006) [arXiv:hep-ph/0512090].
29. P. J. E. Peebles, "Principles of physical cosmology," *Princeton, USA: Univ. Pr. (1993) 718 p*
30. E. S. Weibel, Phys. Rev. Lett. **2**, 83 (1959); B. D. Fried, Phys. Fluids **2**, 337 (1959).
31. M. V. Medvedev, L. O. Silva and M. Kamionkowski, Astrophys. J. **642**, L1 (2006) [arXiv:astro-ph/0512079].
32. J. March-Russell, S. M. West, D. Cumberbatch and D. Hooper, JHEP **0807**, 058 (2008) [arXiv:0801.3440 [hep-ph]].

COSMOGRAPHIC ANALYSIS OF DARK ENERGY

MATT VISSER and CÉLINE CATTOËN

School of Mathematics, Statistics, and Operations Research
Victoria University of Wellington
New Zealand
E-mail: {matt.visser, celine.cattoen}@msor.vuw.ac.nz
http://msor.victoria.ac.nz/

The Hubble relation between distance and redshift is a purely cosmographic relation that depends only on the symmetries of a FLRW spacetime, but does not intrinsically make any dynamical assumptions. This suggests that it should be possible to estimate the parameters defining the Hubble relation without making any dynamical assumptions. To test this idea, we perform a number of inter-related cosmographic fits to the legacy05 and gold06 supernova datasets, paying careful attention to the systematic uncertainties. Based on this supernova data, the "preponderance of evidence" certainly suggests an accelerating universe. However we would argue that (unless one uses additional *dynamical* and *observational* information, and makes additional *theoretical assumptions*) this conclusion is not currently supported "beyond reasonable doubt". As part of the analysis we develop two particularly transparent graphical representations of the redshift-distance relation — representations in which acceleration versus deceleration reduces to the question of whether the relevant graph slopes up or down.

Keywords: Cosmography, Hubble parameter, deceleration parameter, jerk.

1. Introduction

When analyzing the case for "dark energy", it is critically important to realize that the standard luminosity distance versus redshift relation,[1,2]

$$d_L(z) = \frac{c\,z}{H_0}\left\{1 + \frac{[1-q_0]}{2} z + O(z^2)\right\}, \tag{1}$$

and its higher-order extension,[3–6]

$$d_L(z) = \frac{c\,z}{H_0}\left\{1 + \frac{[1-q_0]}{2} z - \frac{1}{6}\left[1 - q_0 - 3q_0^2 + j_0 + \frac{kc^2}{H_0^2\,a_0^2}\right] z^2 + O(z^3)\right\}, \tag{2}$$

are purely cosmographic results applicable to *any* FLRW universe, regardless of the assumed dynamics. Following the spirit of Hubble's original proposal,[7] one could in principle fit such a relation directly to the supernova data,[8–13] thereby estimating cosmological parameters (such as H_0, q_0, and the jerk j_0) without making any dynamical assumptions — but we shall see that there are ways of pre-processing the Hubble relation to make the result (and potential problems) stand out in greater clarity.[14–18]

For instance, it is sometimes observationally more convenient to count photons rather than energy, and consider the "photon flux distance"[14–16]

$$d_F = \frac{d_L}{(1+z)^{1/2}}, \tag{3}$$

for which, defining

$$d_H = \frac{c}{H_0}; \qquad \text{and} \qquad \Omega_0 = 1 + \frac{kc^2}{H_0^2 a_0^2} = 1 + \frac{k\, d_H^2}{a_0^2}; \tag{4}$$

one derives

$$d_F(z) = d_H z \left\{ 1 - \frac{q_0 z}{2} + \frac{[3 + 10q_0 + 12q_0^2 - 4(j_0 + \Omega_0)]}{24} z^2 + O(z^3) \right\}. \tag{5}$$

Furthermore, using the "distance modulus" in terms of which the supernova data is actually reported[8–13]

$$\mu_D = 5\ \log_{10}[d_L/(10\ \text{pc})] = 5\ \log_{10}[d_L/(1\ \text{Mpc})] + 25, \tag{6}$$

one has the simple relation

$$\ln[d_F/(z\ \text{Mpc})] = \frac{\ln 10}{5}[\mu_D - 25] - \ln z - \frac{1}{2}\ln(1+z), \tag{7}$$

leading to a particularly useful form of the Hubble relation:

$$\ln\left[\frac{d_F}{z\ \text{Mpc}}\right] = \ln\left[\frac{d_H}{\text{Mpc}}\right] - \frac{q_0 z}{2} + \frac{[3 + 10q_0 + 9q_0^2 - 4(j_0 + \Omega_0)]}{24} z^2 + O(z^3). \tag{8}$$

Note that the question of whether or not the universe is accelerating or decelerating now reduces to the simple question of whether or not the curve obtained by plotting $\ln[d_F/(z\ \text{Mpc})]$ versus z slopes up or down.

For the data plots presented in this article we have used data from the supernova legacy survey (legacy05)[8,9] and the Riess *et. al.* "gold" dataset of 2006 (gold06).[11] (The gold06 dataset is a larger dataset that contains most but not all of the legacy05 supernovae.) Note that figures 1 and 2 are not plots of "statistical residuals" obtained after curve fitting — rather they can be interpreted as plots of "theoretical residuals", obtained by first splitting off the linear part of the Hubble law (which is now encoded in the

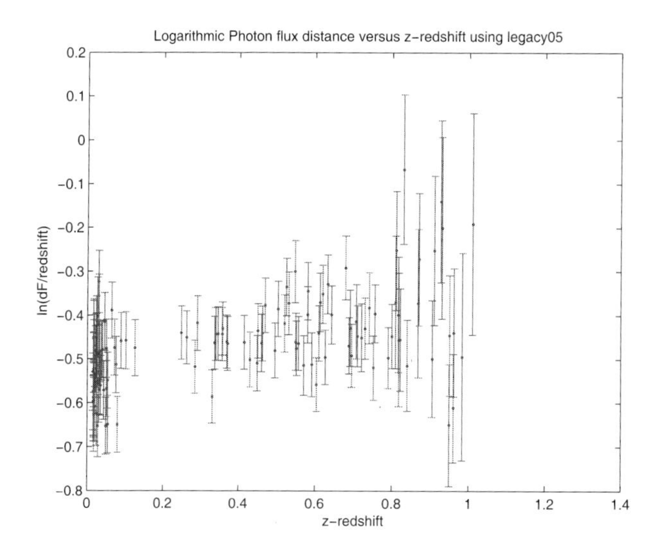

Fig. 1. The normalized logarithm of the photon flux distance, $\ln(d_F/[z \text{ Mpc}])$, as a function of the z-redshift using the legacy05 dataset.[8,9] As is traditional in the field, the plotted error bars do not include estimates of the systematic errors.

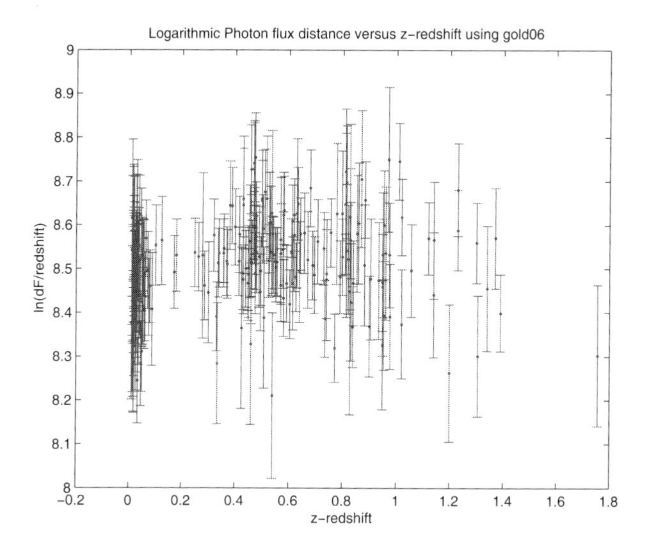

Fig. 2. The normalized logarithm of the photon flux distance, $\ln(d_F/[z \text{ Mpc}])$, as a function of the z-redshift using the gold06 dataset.[10,11] As is traditional in the field, the plotted error bars do not include estimates of the systematic errors.

intercept with the vertical axis), and secondly choosing the quantity to be plotted so as to make the slope of the curve at redshift zero particularly easy to interpret in terms of the deceleration parameter.

The plots presented in figures 1 and 2 are considerably more ambiguous than we had initially expected. In generating these plots and performing the statistical analysis to be described below (considerably more detail can be found at[14–16]) we had initially hoped to verify the robustness of the Hubble relation, and to possibly obtain improved estimates of cosmological parameters such as the deceleration and jerk parameters, thereby complementing other recent cosmographic and cosmokinetic analyses such as,[19–23] as well as other analyses that take a sometimes skeptical view of the totality of the observational data.[24–28]

In view of the rather disturbing visual impact of figures 1 and 2 we resolved to see if they could be improved by further transformations of the data. For instance, we looked at the possibility of transforming the redshift variable, we looked at the possibility of adopting a number of other distance surrogates, and we performed a detailed statistical analysis of the data paying careful attention to the question of estimating the systematic uncertainties.[14] While the "preponderance of evidence" certainly suggests an accelerating universe, we would argue that (unless one uses additional *dynamical* and *observational* information, and makes additional *theoretical* assumptions) this conclusion is not currently supported "beyond reasonable doubt". The supernova data (considered in isolation) certainly *suggests* an accelerating universe, but it is not sufficient to allow us to reliably conclude that the universe *is* accelerating.[a]

2. New Redshift Variable: $y = z/(1+z)$

Because much of the recent supernova data is being acquired at large redshift ($z \gtrsim 1$), there are a number of theoretical reasons why it might be more appropriate to adopt the modified redshift variable[14–16]

$$y = \frac{z}{1+z}; \qquad z = \frac{y}{1-y}. \tag{9}$$

In the past (of an expanding universe)

$$z \in (0, \infty); \qquad y \in (0, 1); \tag{10}$$

[a]From recent discussions with a broad cross section of the community, it appears that this result now seems to have become part of the standard "folklore". Typically, statistically strong arguments for cosmic acceleration rely on working within a particular dynamical framework (such as ΛCDM), and on extra observational data (such as independent constraints on Ω_0 coming from CMB observations).

while in the future

$$z \in (-1, 0); \qquad y \in (-\infty, 0). \tag{11}$$

Thus the variable y is both easy to compute, and when extrapolating back to the Big Bang has a nice finite range $(0, 1)$. Furthermore, Taylor series in terms of the y variable have improved convergence properties at high redshift.[14–16] We will refer to this variable as the *y-redshift.* [b] In terms of the variable y:

$$d_L(y) = d_H\, y \left\{ 1 - \frac{[-3 + q_0]}{2} y + \frac{[12 - 5q_0 + 3q_0^2 - (j_0 + \Omega_0)]}{6} y^2 + O(y^3) \right\}. \tag{12}$$

It is now useful to define a quantity

$$d_Q = \frac{d_L}{(1+z)^{3/2}} = \frac{d_F}{1+z} = (1-y)\, d_F, \tag{13}$$

which we shall refer to as the "deceleration distance". This quantity has the nice feature that

$$\ln[d_Q/(y \text{ Mpc})] = \frac{\ln 10}{5}[\mu_D - 25] - \ln y + \frac{3}{2}\ln(1-y), \tag{14}$$

whence

$$\ln\left[\frac{d_Q}{y \text{ Mpc}}\right] = \ln\left[\frac{d_H}{\text{Mpc}}\right] - \frac{q_0\, y}{2} + \frac{[3 - 2q_0 + 9q_0^2 - 4(j_0 + \Omega_0)]}{24} y^2 + O(y^3). \tag{15}$$

Thus plotting $\ln[d_Q/(y \text{ Mpc})]$ versus y results in a curve whose slope at redshift $y = 0$ is directly proportional to the deceleration parameter: The question of whether or not the universe is accelerating or decelerating now reduces to the simple question of whether or not the curve obtained by plotting $\ln[d_Q/(y \text{ Mpc})]$ versus y slopes up or down.

Visually, the plots presented in figures 3 and 4 are again considerably more ambiguous than we had initially expected. Note that up to this point we have not performed any statistical analyses, we have "merely" found a dramatic way of visually presenting the observational data.

---

[b]Similar expansion variables have certainly been considered before. See, for example, Chevalier and Polarski,[29] who effectively worked with the dimensionless quantity $b = a(t)/a_0$, so that $y = 1 - b$. Similar ideas have also appeared in several related works.[30–33] Note that these authors have typically been interested in parameterizing the so-called w-parameter, rather than specifically addressing the Hubble relation.

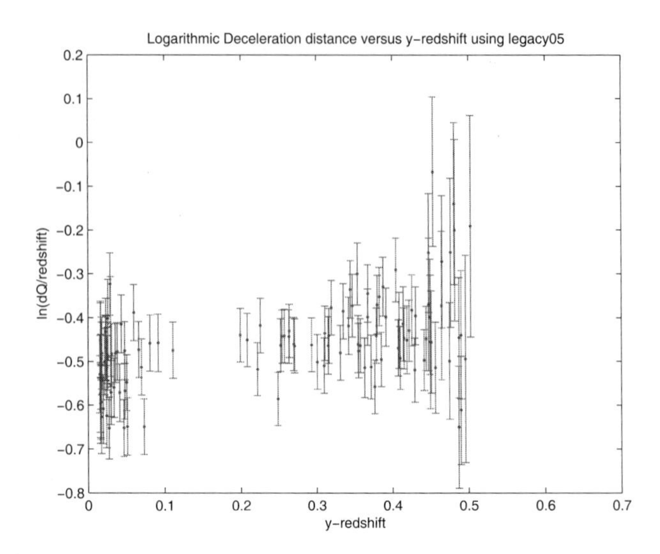

Fig. 3. The normalized logarithm of the deceleration distance, $\ln(d_Q/[y \text{ Mpc}])$, as a function of the y-redshift using the legacy05 dataset.[8,9] As is traditional in the field, the plotted error bars do not include estimates of the systematic errors.

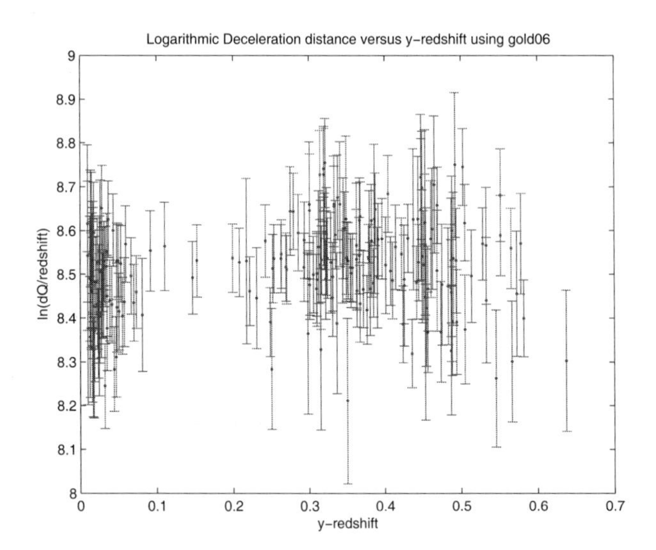

Fig. 4. The normalized logarithm of the deceleration distance, $\ln(d_Q/[y \text{ Mpc}])$, as a function of the y-redshift using the gold06 dataset.[10,11] As is traditional in the field, the plotted error bars do not include estimates of the systematic errors.

3. Data Fitting: Statistical Tests

In view of the somewhat ambiguous and possibly alarming nature of the plots presented in figures 1–4, we then performed a number of statistical tests and analyses to check the extent to which robust and reliable results could be obtained from the supernova data considered in isolation. (For extensive technical details see,[14,15] and related theoretical discussion in[16]). In performing the statistical analyses reported below we compared and contrasted results using several notions of cosmological distance, and two different versions of redshift. The distance surrogates we used were:[14–16]

- The "luminosity distance" d_L.
- The "photon flux distance": $d_F = d_L\ (1+z)^{-1/2}$.
- The "photon count distance": $d_P = d_L\ (1+z)^{-1}$.
- The "deceleration distance": $d_Q = d_L\ (1+z)^{-3/2}$.
- The "angular diameter distance": $d_A = d_L\ (1+z)^{-2}$.

The z-based versions of the Hubble law we used were:[14,16]

$$\ln\left[\frac{d_L}{z\ \mathrm{Mpc}}\right] = \frac{\ln 10}{5}[\mu_D - 25] - \ln z \tag{16}$$

$$= \ln\left[\frac{d_H}{\mathrm{Mpc}}\right] - \frac{[-1+q_0]}{2} z + \frac{\left[-3 + 10q_0 + 9q_0^2 - 4(j_0 + \Omega_0)\right]}{24} z^2 + O(z^3).$$

$$\ln\left[\frac{d_F}{z\ \mathrm{Mpc}}\right] = \frac{\ln 10}{5}[\mu_D - 25] - \ln z - \frac{1}{2}\ln(1+z) \tag{17}$$

$$= \ln\left[\frac{d_H}{\mathrm{Mpc}}\right] - \frac{q_0 z}{2} + \frac{\left[3 + 10q_0 + 9q_0^2 - 4(j_0 + \Omega_0)\right]}{24} z^2 + O(z^3).$$

$$\ln\left[\frac{d_P}{z\ \mathrm{Mpc}}\right] = \frac{\ln 10}{5}[\mu_D - 25] - \ln z - \ln(1+z) \tag{18}$$

$$= \ln\left[\frac{d_H}{\mathrm{Mpc}}\right] - \frac{[1+q_0]}{2} z + \frac{\left[9 + 10q_0 + 9q_0^2 - 4(j_0 + \Omega_0)\right]}{24} z^2 + O(z^3).$$

$$\ln\left[\frac{d_Q}{z\ \mathrm{Mpc}}\right] = \frac{\ln 10}{5}[\mu_D - 25] - \ln z - \frac{3}{2}\ln(1+z) \tag{19}$$

$$= \ln\left[\frac{d_H}{\mathrm{Mpc}}\right] - \frac{[2+q_0]}{2} z + \frac{\left[15 + 10q_0 + 9q_0^2 - 4(j_0 + \Omega_0)\right] 1}{24} z^2 + O(z^3).$$

$$\ln\left[\frac{d_A}{z\ \mathrm{Mpc}}\right] = \frac{\ln 10}{5}[\mu_D - 25] - \ln z - 2\ln(1+z) \tag{20}$$

$$= \ln\left[\frac{d_H}{\mathrm{Mpc}}\right] - \frac{[3+q_0]}{2} z + \frac{\left[21 + 10q_0 + 9q_0^2 - 4(j_0 + \Omega_0)\right]}{24} z^2 + O(z^3).$$

Similarly, the y-based versions of the Hubble law we used were:[14,16]

$$\ln\left[\frac{d_L}{z\ \mathrm{Mpc}}\right] = \frac{\ln 10}{5}[\mu_D - 25] - \ln y \tag{21}$$
$$= \ln\left[\frac{d_H}{\mathrm{Mpc}}\right] - \frac{[-3+q_0]}{2}y + \frac{[21 - 2q_0 + 9q_0^2 - 4(j_0+\Omega_0)]}{24}y^2 + O(y^3).$$

$$\ln\left[\frac{d_F}{z\ \mathrm{Mpc}}\right] = \frac{\ln 10}{5}[\mu_D - 25] - \ln y + \frac{1}{2}\ln(1-y) \tag{22}$$
$$= \ln\left[\frac{d_H}{\mathrm{Mpc}}\right] - \frac{[-2+q_0]}{2}y + \frac{[15 - 2q_0 + 9q_0^2 - 4(j_0+\Omega_0)]}{24}y^2 + O(y^3).$$

$$\ln\left[\frac{d_P}{z\ \mathrm{Mpc}}\right] = \frac{\ln 10}{5}[\mu_D - 25] - \ln y + \ln(1-y) \tag{23}$$
$$= \ln\left[\frac{d_H}{\mathrm{Mpc}}\right] - \frac{[-1+q_0]}{2}y + \frac{[9 - 2q_0 + 9q_0^2 - 4(j_0+\Omega_0)]}{24}y^2 + O(y^3).$$

$$\ln\left[\frac{d_Q}{z\ \mathrm{Mpc}}\right] = \frac{\ln 10}{5}[\mu_D - 25] - \ln y + \frac{3}{2}\ln(1-y) \tag{24}$$
$$= \ln\left[\frac{d_H}{\mathrm{Mpc}}\right] - \frac{q_0}{2}y + \frac{[3 - 2q_0 + 9q_0^2 - 4(j_0+\Omega_0)]}{24}y^2 + O(y^3).$$

$$\ln\left[\frac{d_A}{z\ \mathrm{Mpc}}\right] = \frac{\ln 10}{5}[\mu_D - 25] - \ln y + 2\ln(1-y) \tag{25}$$
$$= \ln\left[\frac{d_H}{\mathrm{Mpc}}\right] - \frac{[1+q_0]}{2}y + \frac{[-3 - 2q_0 + 9q_0^2 - 4(j_0+\Omega_0)]}{24}y^2 + O(y^3).$$

Fits were carried out for all five distance surrogates, and for both definitions of redshift, using polynomial approximants to the Hubble relation up to 7th-order. [c] The F-test was then used to discard statistically meaningless terms, and it was seen that quadratic fits were the best that could meaningfully be adopted. [d] The results are presented in tables 1–4.

[c] Note that because the uncertainty in the redshift is encoded in the uncertainty of the distance modulus, the uncertainty in logarithmic distance is just scaled by a factor of $\ln(10)/5$. Therefore, if the uncertainty is gaussian in the distance modulus, it is also gaussian in logarithmic distance, which is crucial for least squares fitting.

[d] Note that in a cosmographic framework, where one is most closely following the spirit of Hubble's original methodology,[7] one does not have a dynamical model to fit the data to, and the use of least-squares fits to a truncated Taylor series is the best one can possibly hope for. Ultimately, the truncated Taylor series method is not really a very radical approach, being firmly based in quite standard statistical techniques.[34–39]

Deceleration and jerk parameters (legacy05 dataset, y-redshift).

distance	q_0	$j_0 + \Omega_0$
d_L	-0.47 ± 0.38	-0.48 ± 3.53
d_F	-0.57 ± 0.38	$+1.04 \pm 3.71$
d_P	-0.66 ± 0.38	$+2.61 \pm 3.88$
d_Q	-0.76 ± 0.38	$+4.22 \pm 4.04$
d_A	-0.85 ± 0.38	$+5.88 \pm 4.20$

With 1-σ statistical uncertainties.

Deceleration and jerk parameters (legacy05 dataset, z-redshift).

distance	q_0	$j_0 + \Omega_0$
d_L	-0.48 ± 0.17	$+0.43 \pm 0.60$
d_F	-0.56 ± 0.17	$+1.16 \pm 0.65$
d_P	-0.62 ± 0.17	$+1.92 \pm 0.69$
d_Q	-0.69 ± 0.17	$+2.69 \pm 0.74$
d_A	-0.75 ± 0.17	$+3.49 \pm 0.79$

With 1-σ statistical uncertainties.

Deceleration and jerk parameters (gold06 dataset, y-redshift).

distance	q_0	$j_0 + \Omega_0$
d_L	-0.62 ± 0.29	$+1.66 \pm 2.60$
d_F	-0.78 ± 0.29	$+3.95 \pm 2.80$
d_P	-0.94 ± 0.29	$+6.35 \pm 3.00$
d_Q	-1.09 ± 0.29	$+8.87 \pm 3.20$
d_A	-1.25 ± 0.29	$+11.5 \pm 3.41$

With 1-σ statistical uncertainties.

Deceleration and jerk parameters (gold06 dataset, z-redshift).

distance	q_0	$j_0 + \Omega_0$
d_L	-0.37 ± 0.11	$+0.26 \pm 0.20$
d_F	-0.48 ± 0.11	$+1.10 \pm 0.24$
d_P	-0.58 ± 0.11	$+1.98 \pm 0.29$
d_Q	-0.68 ± 0.11	$+2.92 \pm 0.37$
d_A	-0.79 ± 0.11	$+3.90 \pm 0.39$

With 1-σ statistical uncertainties.

Even after we have extracted these numerical results there is still a considerable amount of interpretation that has to go into understanding their physical implications.[14,15] In particular note that the differences between the various models, (Which distance do we use? Which version of redshift do we use? Which dataset do we use?), often dwarf the statistical uncertainties within any particular model. If better quality (smaller scatter) data were to become available, then one could hope that the cubic term would survive the F-test. This would have follow-on effects in terms of making the differences between the various estimates of the deceleration parameter smaller,[14,15] which would give us greater confidence in the reliability and robustness of the conclusions.

The fact that there are such large differences between the cosmological parameters deduced from the different models based on physically plausible distance indicators should give one pause for concern. These differences do not arise from any statistical flaw in the analysis, nor do they in any sense represent any "systematic" error, rather they are an intrinsic side-effect of what it means to do a least-squares fit — to a finite-polynomial approximate Taylor series — in a situation where it is physically unclear as to which if any particular measure of "distance" is physically preferable, and which particular notion of "distance" should be fed into the least-squares algorithm. (This "feature" — some may call it a "limitation" — of the least-squares algorithm in the absence of a clear physically motivated dynamical model is an often overlooked confounding factor in data analysis.[34–39])

4. Systematic Uncertainties

In addition to the purely statistical uncertainties discussed above, one needs to make an estimate of the systematic uncertainties, and following NIST guidelines,[40] combine the statistical and systematic uncertainties in quadrature

$$\sigma_{\text{combined}} = \sqrt{\sigma^2_{\text{statistical}} + \sigma^2_{\text{total-systematic}}}. \tag{26}$$

Estimating systematic uncertainties is notoriously difficult. A careful description of our own preferred way of estimating systematic uncertainties is fully discussed in,[14,15] wherein we consider both modelling and historical uncertainties. It should be emphasized that the (to our minds) overly optimistic estimates of systematic uncertainties commonly found in the literature do not greatly change our conclusions below.

5. Expanded Uncertainties

After due allowance is made for estimating the systematic uncertainties, the NIST guidelines[40] recommend defining an "expanded uncertainty" by

$$U_k = k\,\sigma_{\text{combined}}. \tag{27}$$

Here the factor k is chosen for scientific (or legal) reasons to be such that one is "certain" that the true result lies within the stated range. The tradition within the social and medical sciences is to accept $k = 2$ (that is, two-sigma, corresponding approximately to 95% confidence intervals) as being sufficient to draw valid conclusions. Particle physics has traditionally adopted $k = 3$ as the minimum standard for claiming "evidence for" a given hypothesis. (This is the origin of the aphorism: "If it's not three-sigma, it's not physics".) Over the last 20 years or so, particle physics has moved to the more conservative consensus that $k = 5$ is the minimum standard for claiming "discovery" of "new physics". Our best estimates for the combined and expanded uncertainties are presented in tables 5–6.

Deceleration parameter summary: Combined and expanded uncertainties.

dataset	redshift	$q_0 \pm \sigma_{\text{combined}}$	$q_0 \pm U_3$	$q_0 \pm U_5$
legacy05	y	-0.66 ± 0.42	-0.66 ± 1.26	-0.66 ± 2.10
legacy05	z	-0.62 ± 0.23	-0.62 ± 0.70	-0.62 ± 1.15
gold06	y	-0.94 ± 0.39	-0.94 ± 1.16	-0.94 ± 1.95
gold06	z	-0.58 ± 0.23	-0.58 ± 0.68	-0.58 ± 1.15

Jerk parameter summary: Combined and expanded uncertainties.

dataset	redshift	$(j_0 + \Omega_0) \pm \sigma_{\text{combined}}$	$(j_0 + \Omega_0) \pm U_3$	$(j_0 + \Omega_0) \pm U_5$
legacy05	y	$+2.65 \pm 4.63$	$+2.65 \pm 13.9$	$+2.65 \pm 23.2$
legacy05	z	$+1.94 \pm 1.72$	$+1.94 \pm 5.17$	$+1.94 \pm 8.60$
gold06	y	$+6.47 \pm 4.75$	$+6.47 \pm 14.2$	$+6.47 \pm 23.8$
gold06	z	$+2.03 \pm 1.75$	$+2.03 \pm 5.26$	$+2.03 \pm 8.75$

6. Conclusions

What can we conclude from this? While the "preponderance of evidence" is certainly that the universe is currently accelerating, $q_0 < 0$, this is not yet a "gold plated" result, *at least not without bringing in other physical assumptions and observations*; such as a specific dynamical model [*e.g.*, ΛCDM] and/or invoking knowledge of Ω_m or the CMB data, all of which are subject to their own additional theoretical assumptions. It is certainly more likely that the expansion of the universe is accelerating, than that the expansion of the universe is decelerating — but this is not the same as having definite evidence in favour of acceleration.

We wish to emphasize the point that, regardless of one's views on how to combine formal estimates of uncertainty, the very fact that different distance scales yield data-fits with such widely discrepant estimates for the cosmological parameters strongly suggests the need for extreme caution in interpreting the supernova data.

There are a number of other more sophisticated statistical methods that might be applied to the data to possibly improve the statistical situation. For instance, ridge regression, robust regression, and the use of orthogonal polynomials and "loess curves" could all be adopted and adapted to focus more carefully on the region near redshift zero.[34–39] However one should always keep in mind the difference between *accuracy* and *precision*.[41] More sophisticated statistical analyses may permit one to improve the precision of the analysis, but unless one can further constrain the systematic uncertainties such precise results will be no more accurate than the current situation.

However, we are certainly not claiming that all is grim on the cosmological front — and do not wish our views to be misinterpreted in this regard — there are clearly parts of cosmology where there is plenty of high-quality data, and more coming in, constraining and helping refine our models. But regarding some specific cosmological questions the catch cry should still be "Precision cosmology? Not just yet".[42] In closing, we strongly encourage readers to carefully contemplate figures 1–4 as an inoculation against over-interpretation of the supernova data.

Ultimately, it is the fact that figures 1–4 do not *exhibit any overwhelmingly obvious trend that makes it so difficult to make a robust and reliable estimate of the sign of the deceleration parameter.*

Finally we remind the reader that it is the putative acceleration of the expansion of the universe, no matter how derived, that then (via the Freidmann equations), is taken to imply the existence of "dark energy". In the absence of truly compelling model-independent evidence for cosmic acceleration one has to be at least a little cautious regarding the existence of "dark energy".

References

1. S. Weinberg, *Gravitation and cosmology: Principles and applications of the general theory of relativity*, (Wiley, New York, 1972).
2. P. J. E. Peebles, *Principles of physical cosmology*, Princeton University Press, 1993.
3. T. Chiba and T. Nakamura, Prog. Theor. Phys. **100** (1998) 1077 [arXiv:astro-ph/9808022].
4. V. Sahni, T. D. Saini, A. A. Starobinsky and U. Alam, JETP Lett. **77** (2003) 201 [Pisma Zh. Eksp. Teor. Fiz. **77** (2003) 249] [arXiv:astro-ph/0201498].
5. M. Visser, Class. Quant. Grav. **21** (2004) 2603 [arXiv:gr-qc/0309109].
6. M. Visser, Gen. Rel. Grav. **37** (2005) 1541 [arXiv:gr-qc/0411131].
7. E. P. Hubble, Proc. Natl. Acad. Sci. USA **15** (1929)168173.
8. P. Astier *et al.*, Astron. Astrophys. **447** (2006) 31 [arXiv:astro-ph/0510447]. Electronic data available at http://snls.in2p3.fr/conf/papers/cosmo1/
9. SNLS — Super Nova Legacy Survey, http://legacy.astro.utoronto.ca/ http://cfht.hawaii.edu/SNLS/ Electronic data available at http://snls.in2p3.fr/conf/papers/cosmo1/
10. A. G. Riess *et al.*, "Type Ia Supernova Discoveries at $z > 1$ From the Hubble Space Telescope: Evidence for Past Deceleration and Constraints on Dark Energy Evolution", arXiv:astro-ph/0402512. Electronic data available at http://braeburn.pha.jhu.edu/~ariess/R06/
11. A. G. Riess *et al.*, "New Hubble Space Telescope Discoveries of Type Ia Supernovae at $z > 1$: Narrowing Constraints on the Early Behavior of Dark Energy", arXiv:astro-ph/0611572. Electronic data available at http://braeburn.pha.jhu.edu//~ariess/R06/
12. S. Jha, A. G. Riess and R. P. Kirshner, "Improved Distances to Type Ia Supernovae with Multicolor Light Curve Shapes: MLCS2k2", arXiv:astro-ph/0612666. Electronic data available at http://astro.berkeley.edu//~saurabh/mlcs2k2/
13. W. M. Wood-Vasey *et al.*, "Observational Constraints on the Nature of the Dark Energy: First Cosmological Results from the ESSENCE Supernova Survey," arXiv:astro-ph/0701041.
14. C. Cattoën and M. Visser, "Cosmography: Extracting the Hubble series from the supernova data," arXiv:gr-qc/0703122.
15. C. Cattoën and M. Visser, Phys. Rev. D **78** (2008) 063501 [arXiv:0809.0537 [gr-qc]].
16. C. Cattoën and M. Visser, Class. Quant. Grav. **24** (2007) 5985 [arXiv:0710.1887 [gr-qc]].
17. C. Cattoën and M. Visser, Class. Quant. Grav. **25** (2008) 165013 [arXiv:0712.1619 [gr-qc]].
18. C. Cattoën and M. Visser, JCAP **0811** (2008) 024 [arXiv:0806.2186 [gr-qc]].
19. D. Rapetti, S. W. Allen, M. A. Amin and R. D. Blandford, "A kinematical approach to dark energy studies", arXiv:astro-ph/0605683.
20. R. D. Blandford, M. Amin, E. A. Baltz, K. Mandel and P. J. Marshall, "Cosmokinetics", arXiv:astro-ph/0408279.

21. C. Shapiro and M. S. Turner, "What Do We Really Know About Cosmic Acceleration?", arXiv:astro-ph/0512586.
22. R. R. Caldwell and M. Kamionkowski, "Expansion, Geometry, and Gravity", arXiv:astro-ph/0403003.
23. O. Elgaroy and T. Multamaki, JCAP **0609** (2006) 002 [arXiv:astro-ph/0603053].
24. H. K. Jassal, J. S. Bagla and T. Padmanabhan, Phys. Rev. D **72** (2005) 103503 [arXiv:astro-ph/0506748].
25. H. K. Jassal, J. S. Bagla and T. Padmanabhan, "The vanishing phantom menace", arXiv:astro-ph/0601389.
26. T. R. Choudhury and T. Padmanabhan, Astron. Astrophys. **429** (2005) 807 [arXiv:astro-ph/0311622].
27. T. Padmanabhan and T. R. Choudhury, Mon. Not. Roy. Astron. Soc. **344** (2003) 823 [arXiv:astro-ph/0212573].
28. V. Barger, Y. Gao and D. Marfatia, Phys. Lett. B **648** (2007) 127 [arXiv:astro-ph/0611775].
29. M. Chevallier and D. Polarski, Int. J. Mod. Phys. D **10** (2001) 213 [arXiv:gr-qc/0009008].
30. E. V. Linder, "Probing gravitation, dark energy, and acceleration", Phys. Rev. D **70** (2004) 023511 [arXiv:astro-ph/0402503].
31. E. V. Linder, "Biased Cosmology: Pivots, Parameters, and Figures of Merit", arXiv:astro-ph/0604280.
32. B. A. Bassett, P. S. Corasaniti and M. Kunz, Astrophys. J. **617**, L1 (2004) [arXiv:astro-ph/0407364].
33. D. Martin and A. Albrecht, "Talk about pivots", arXiv:astro-ph/0604401.
34. N. R. Draper and H. Smith, *Applied regression analysis*, (Wiley, New York, 1998).
35. D. C. Montgomery, E. A. Peck, and G. G. Vining, *Introduction to linear regression analysis*, (Wiley, New York, 2001).
36. R. D. Cook and S. Weisberg, *Applied regression including computing and graphics*, (Wiley, New York, 1999).
37. W. J. Kennedy, Jr. and J. E. Gentle, *Statistical computing*, (Marcel Dekker, New York, 1980).
38. R. J. Carroll and D. Rupert, *Transformation and weighting in regression*, (Chapman and Hall, London, 1998),
39. G. J. S. Ross, *Nonlinear estimation*, (Spinger–Verlag, New York, 1990).
40. Barry N. Taylor and Chris E. Kuyatt, *Guidelines for Evaluating and Expressing the Uncertainty of NIST Measurement Results*, NIST Technical Note 1297.
Online version available as http://physics.nist.gov/cuu/Uncertainty/index.html
41. P. R. Bevington, *Data reduction and analysis in the physical sciences*, (McGraw–Hill, New York, 1969).
42. S. L. Bridle, O. Lahav, J. P. Ostriker and P. J. Steinhardt, Science **299** (2003) 1532 [arXiv:astro-ph/0303180].

PART VII

Dark Mass at the Center of the Galaxy

LUMINOUS ACCRETION ONTO THE DARK MASS AT THE CENTER OF THE MILKY WAY

A. ECKART[1,2,*], M. GARCÍA-MARÍN[1], S. KÖNIG[1], D. KUNNERIATH[1,2], K. MUŽIĆ[1], C. STRAUBMEIER[1], G. WITZEL[1] and M. ZAMANINASAB[1,2]

[1] *I.Physikalisches Institut, Universität zu Köln, Zülpicher Str.77, 50937 Köln, Germany*

[2] *Max-Planck-Institut für Radioastronomie, Auf dem Hügel 69, 53121 Bonn, Germany*

**E-mail: eckart@ph1.uni-koeln.de*

We report on the results of two successful, simultaneous observations of Sagittarius A* at the center of the Milky Way. The observations were carried out in 2004 and 2008 using telescopes operating from the mm-radio domain to the X-ray domain, and detected strong flux density variations in all wavelength bands. Modeling suggests that a combination of a synchrotron self Compton process and an adiabatic expansion of source components are at work. The luminous flare emission of Sagittarius A* also supports the presence of an accreting super massive black hole at that position. We also discuss the potential of NIR interferometry for further detailed investigations of the accretion process in SgrA*.

Keywords: Black hole physics; infrared; general; accretion; accretion disks; Galaxy; center; nucleus; Black Holes; individual; SgrA*.

1. Introduction

As the closest example of a galactic nucleus, at a distance of only about 8 kpc, the center of the Milky Way presents an exquisite laboratory for learning about the accretion process onto supermassive black holes (SMBH). Here stellar orbits have convincingly proven the existence of a SMBH of mass $\sim 4\times 10^6 M_\odot$ at the position of the compact radio, infrared, and X-ray source Sagittarius A* (Sgr A*; Eckart & Genzel 1996, 1997, Eckart et al. 2002, Schödel et al. 2002, Eisenhauer et al. 2003, Ghez et al. 2000, 2005a). Additional strong evidence of a SMBH at the position of Sgr A* comes from the observation of rapid flare activity both in the X-ray and NIR wavelength domains (Baganoff et al. 2001; Genzel et al. 2003; Ghez et al. 2004, Eckart et al. 2006abc). Due to its proximity, Sgr A* provides us with an unique opportunity to understand the physics and possibly the evolution of SMBHs at the nuclei of galaxies. Variability at radio through sub-millimeter wavelengths has been studied extensively. It shows that variations occur on timescales from hours to years (e.g. Mauerhan et al. 2005, Eckart et al. 2006a, Yusef-Zadeh et al. 2008, Marrone et al. 2008). Several flares have provided evidence of decaying millimeter and sub-millimeter emission following simultaneous NIR/X-ray flares. Although much effort has been invested in monitoring the galactic center variability at these wavelengths, only a handful of simultaneous observations aimed at analyzing their correlation have been carried out up to date (see for instance Eckart et al. 2006a, 2008ab; Yusef-Zadeh et al. 2006ab, 2007, 2008, Marrone et al. 2008).

Here we report on the first measurements of Sgr A* that successfully detected simultaneous flare emission in the near-infrared and sub-millimeter domain using the ESO VLT and the APEX sub-mm telescopes[a]. The paper is organized as follows: In the second section we describe the observations and data reduction followed by an outline of the flare modeling in section 3. In section 4 we discuss the potential of NIR interferometry in studying details of the accretion physics in SgrA*. We conclude with a discussion and summary in sections 5.

For optically thin synchrotron emission we refer throughout this paper to photon spectral indices (α) using the convention $S_\nu \propto \nu^{-\alpha}$ and to spectral indices (p) of electron power-law distributions using $N(E) \propto E^{-p}$ with $p = (1 + 2\alpha)$. The assumed distance to Sgr A* is 8 kpc (Reid 1993), consistent with recent results (e.g. Lu et al. 2008, Ghez et al. 2005a, 2008, Eisenhauer et al. 2003).

2. Observations

The sub-mm regime is of special interest for simultaneous flare measurements, as it provides important constraints on the modeling. In this general wavelength range synchrotron source components that also radiate in the infrared domain become optically thick, and represent the dominant reservoir of photons that are then scattered to the X-ray domain through the inverse Compton process. On 3 June 2008, substantial progress was made

[a]Based on observations with the ESO telescopes at the Paranal Observatory under programs IDs:077.B-0028, 79.B-0084, and 81.B-0648. The sub-mm data are based on observations with the Atacama Pathfinder Experiment (APEX). APEX is a collaboration between the Max-Planck-Institut für Radioastronomie, the European Southern Observatory, and the Onsala Space Observatory

in this matter: as part of a larger campaign, Sgr A* was simultaneously observed in the sub-millimeter and NIR wavelength domains using the European Southern Observatory facilities. These observations result in the first simultaneous successful detection of strongly variable emission in both wavelength domains using the VLT and the APEX telescope. In the following subsections we describe the data acquisition and reduction for the individual telescopes.

The simultaneous observations data for the flare observed in 2004 were obtained using the following telescopes. In the infrared we used the ESO NACO adaptive optics (AO) instrument[b], the ACIS-I instrument on board the *Chandra X-ray Observatory* as well as the Submillimeter Array SMA[c] on Mauna Kea, Hawaii, and the Very Large Array[d] in New Mexico.

For the flare described by Eckart et al. (2008c) the observations were carried out in May/June 2008. Near-infrared (NIR) observations of the Galactic center (GC) were carried out with the NIR camera CONICA and the adaptive optics (AO) module NAOS (briefly "NACO") at the ESO VLT unit telescope 4 (YEPUN) on Paranal, Chile, during the night between 2-3 June 2008. The Galactic center 870 μm data were taken with the newly commissioned LABOCA bolometer array, located on the Atacama Pathfinder EXperiment (APEX) telescope at the Llano Chajnantor (Fig. 1).

3. Flare Analysis

Prominent flux density excursions of SgrA* can be explained using known physical processes like adiabatic expansion and the synchrotron and synchrotron self Compton mechanism. In the following we summaries some basics we have used in our modeling.

3.1. *Adiabatically expanding source components*

Our basic assumption to model the sub-mm light curves is the presence of an expanding uniform blob of relativistic electrons with an energy spectrum

[b]Based on observations at the Very Large Telescope (VLT) of the European Southern Observatory (ESO) on Paranal in Chile; Program: 271.B-5019(A).

[c]The Submillimeter Array is a joint project between the Smithsonian Astrophysical Observatory and the Academia Sinica Institute of Astronomy and Astrophysics, and is funded by the Smithsonian Institution and the Academia Sinica.

[d]The VLA is operated by the National Radio Astronomy Observatory which is a facility of the National Science Foundation operated under cooperative agreement by Associated Universities, Inc.

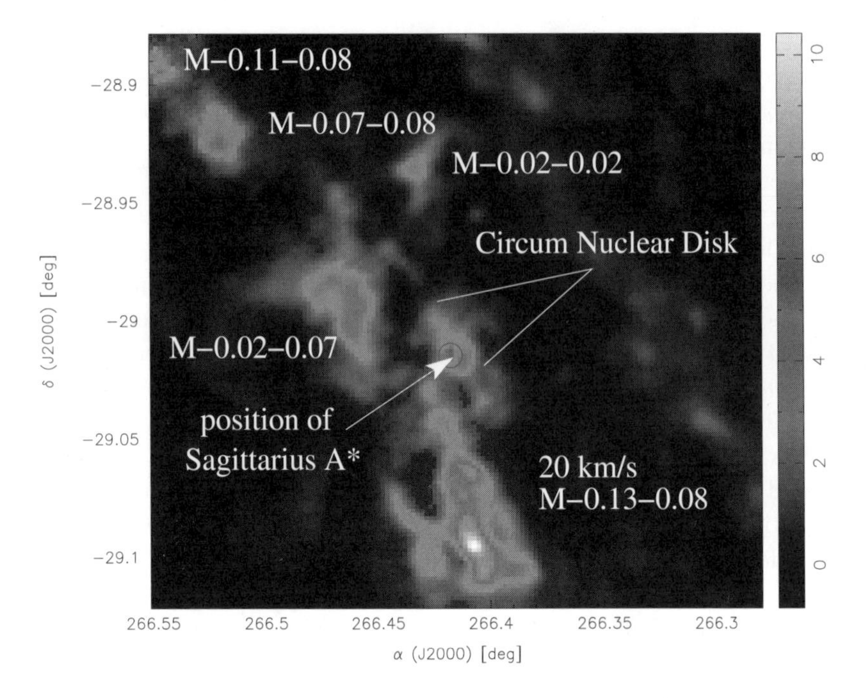

Fig. 1. Map of the 345 MHz sub-mm continuum emission from the Galactic Center region obtained with the LABOCA bolometer on APEX (Eckart et al. 2008c, Garcia-Marin et al. 2009). Outlined are some of the cloud complexes close to the Circum Nuclear Disk (CND) that are responsible for the bulk of the line of sight absorption for cold gas.

$n(E) \propto E^{-p}$ threaded by a magnetic field. As a consequence of the blob expansion, the magnetic field inside the blob declines as R^{-2}, the energy of relativistic particles as R^{-1} and the density of particles as R^{-3} (van der Laan 1966). The synchrotron optical depth at frequency ν then scales as

$$\tau_\nu = \tau_0 \left(\frac{\nu}{\nu_0}\right)^{-(p+4)/2} \left(\frac{R}{R_0}\right)^{-(2p+3)} \tag{1}$$

and the flux density scales as

$$S_\nu = S_0 \left(\frac{\nu}{\nu_0}\right)^{5/2} \left(\frac{R}{R_0}\right)^{3} \frac{1-\exp(-\tau_\nu)}{1-\exp(-\tau_0)} \tag{2}$$

Here R_0, S_0, and τ_0 are the size, flux density, and optical depth at the peak frequency of the synchrotron spectrum ν_0. The goal of the present model is to combine the description of an adiabatically expanding cloud with a synchrotron self-Compton formalism, as this is the most likely physical scenario to explain the delay between the sub-mm and the simultaneous near-IR and

X-ray peaks. Thus, we use the definition of τ_0 as the optical depth corresponding to the frequency at which the flux density is a maximum (van der Laan 1966), rather than the definition of τ_0 as the optical depth at which the flux density for any particular frequency peaks (Yusef-Zadeh et al. 2006b). This implies that τ_0 depends only on p through the condition

$$e^{\tau_0} - \tau_0(p+4)/5 - 1 = 0 \tag{3}$$

and ranges from, e.g., 0 to 0.65 as p ranges from 1 to 3. Therefore, given the particle energy spectral index p and the peak flux S_0 in the light curve at some frequency ν_0, this model predicts the variation in flux density at any other frequency as a function of the expansion factor (R/R_0).

Finally, a model for $R(t)$ is required to convert the dependence on radius to time: we adopt a simple linear expansion at constant expansion speed v_{exp}, so that $R - R_0 = v_{exp}\,(t - t_0)$. For $t \leq t_0$ we have made the assumption that the source has an optical depth that equals its frequency dependent initial value τ_ν at $R = R_0$. So in the optically thin part of the source spectrum the flux initially increases with the source size at a constant τ_ν and then decreases due to the decreasing optical depth as a consequence of the expansion. For the $\sim 4\times10^6 M_\odot$ super-massive black hole at the position of Sgr A*, one Schwarzschild radius is $R_s = 2GM/c^2 \sim 10^{10}$ m and the velocity of light corresponds to about 100 R_s per hour. For $t > t_0$ the decaying flank of the curve can be shifted towards later times first, by increasing the turnover frequency ν_0 or the initial source size R_0, and second, by lowering the spectral index α_{synch} or the peak flux density S_0. Increasing the adiabatic expansion velocity v_{exp} shifts the peak of the light curve to earlier times. Adiabatic expansion will also result in a slower decay rate and a longer flare timescale at lower frequencies.

3.2. *Description and properties of the SSC model*

We have employed a simple SSC model to describe the observed radio to X-ray properties of SgrA* using the nomenclature given by Gould (1979) and Marscher (1983). Inverse Compton scattering models provide an explanation for both the compact NIR and X-ray emission by up-scattering sub-mm-wavelength photons into these spectral domains. Such models are considered as a possibility in most of the recent modeling approaches and may provide important insights into some fundamental model requirements. The models do not explain the entire low frequency radio spectrum and IQ (intemediate quiescent phase) state X-ray emission. Here IQ stands for

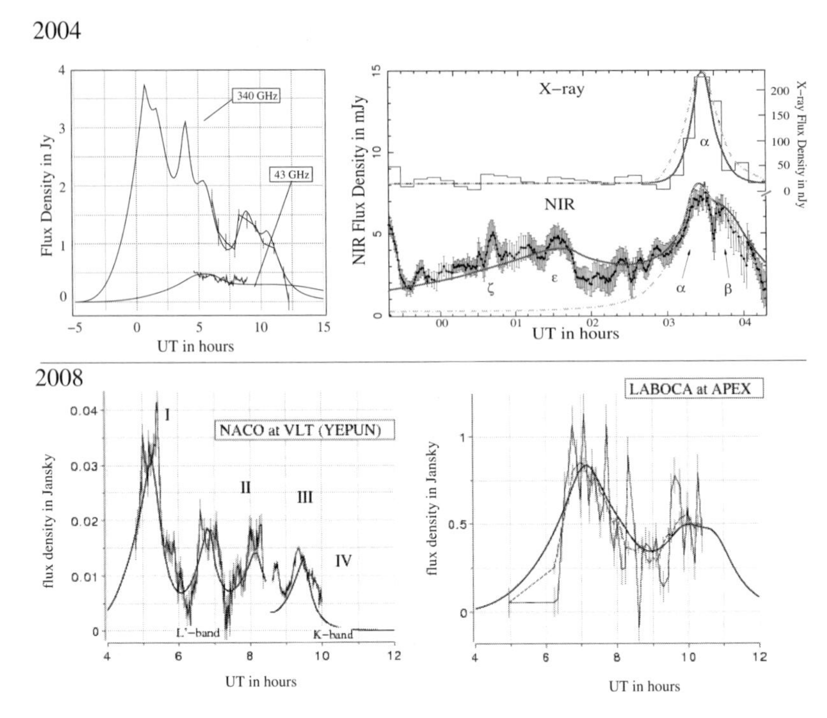

Fig. 2. **Top left:** The variable part of the observed 43 GHz and 340 GHz light curves shown in comparison to model the favoured model by Eckart et al. (2009). **Top right:** A comparison between the X-ray and NIR 2.2 μm light curves and the modeling results for the A, B (both in green), and C (red) models. The contributions of individual model components as listed by Eckart et al. (2009). **Bottom:** Our favoured model compared to the infrared (left) and sub-mm (right) light curve of Sagittarius A*. The data are represented by vertical red bars ($\pm 1\sigma$) and a black connection line between them For both panels see details in Eckart et al. 2009, 2008c.

intemediate quiescent phase i.e. the phase between large flare excursions. However, they give a description of the compact IQ and flare emission originated from the immediate vicinity of the central black hole. A more detailed explanation is also given by Eckart et al. (2004).

We assume a synchrotron source of angular extent θ. The source size is of the order of a few Schwarzschild radii R_s=2GM/c^2 with $R_s \sim 10^{10}$ m for a $\sim 4\times10^6 M_\odot$ black hole. One R_s then corresponds to an angular diameter of $\sim$8 μas considering a distance to the Galactic Center of 8 kpc (Reid 1993, Eisenhauer et al. 2003, Ghez et al. 2005). The emitting source becomes optically thick at a frequency ν_m with a flux density S_m, and has an optically thin spectral index α following the law $S_\nu \propto \nu^{-\alpha}$. This allows us to calculate the magnetic field strength B and the inverse Compton scattered flux density S_{SSC} as a function of the X-ray photon energy E_{keV}. The synchrotron self-Compton spectrum has the same spectral index as the synchrotron spectrum that is up-scattered i.e. $S_{SSC} \propto E_{keV}{}^{-\alpha}$, and is valid within the limits E_{min} and E_{max} corresponding to the wavelengths λ_{max} and λ_{min} (see Marscher et al. 1983 for further details). We find that Lorentz factors γ_e for the emitting electrons of the order of typically 10^3 are required to produce a sufficient SSC flux in the observed X-ray domain. A possible relativistic bulk motion of the emitting source results in a Doppler boosting factor $\delta = \Gamma^{-1}(1\text{-}\beta\cos\phi)^{-1}$. Here ϕ is the angle of the velocity vector to the line of sight, β the velocity v in units of the speed of light c, and Lorentz factor $\Gamma = (1\text{-}\beta^2)^{-1/2}$ for the bulk motion. Relativistic bulk motion is not a necessity to produce sufficient SSC flux density, but we have used modest values for Γ=1.2-2 and δ ranging between 1.3 and 2.0 (i.e. angles ϕ between about 10° and 45°) since they will occur in cases of relativistically orbiting gas as well as relativistic outflows - both of which are likely to be relevant to SgrA*.

3.3. *Modeling results for the 2004 flare*

We report on new modeling results based on the mm- to X-ray emission of the SgrA* counterpart associated with the massive $\sim 4 \times 10^6 M_\odot$ black hole at the Galactic Center. We investigate the physical processes responsible for the variable emission from SgrA*. The observations revealed several flare events in all wavelength domains. Here we show that the flare emission can be described with a combination of a synchrotron self-Compton (SSC) model followed by an adiabatic expansion of the source components. The SSC emission at NIR and X-ray wavelengths involves up-scattered submillimeter photons from a compact source component. At the start of the flare, spectra of these components peak at frequencies ranging between several 100 GHz and 2 THz. The adiabatic expansion then accounts for the variable emission observed at sub-mm/mm wavelengths. The derived physical quantities that describe the flare emission give a blob expansion speed of $v_{exp} \sim 0.005c$, magnetic field of B around 60 G or less and spectral indices of α=0.8 to 1.4, corresponding to particle spectral index p$\sim$2.6 to 3.8. A combined SSC and adiabatic expansion model can fully account for the observed flare flux densities and delay times covering the spectral range from the X-ray to the mm-radio domain. The derived model parameters suggest that the adiabatic expansion takes place in source components that have a bulk motion larger than v_{exp} or the expanding material contributes to a corona or disk, confined to the immediate surroundings of SgrA*. The data and the prefered model are shown in the top panels of Fig. 2.

3.4. *Modeling results for the 2008 flare*

The adiabatic expansion results in a time difference between the peaks in the VLT and APEX light curves of about 1.5±0.5 hours, and compare well with the values obtained in a global, multi-wavelength observing campaign by our team in 2007. Back then, two bright NIR flares were traced by CARMA (Combined Array for Research in mm-wave Astronomy; 100 GHz) in the US, ATCA (Australia Telescope Compact Array; 86 GHz) in Australia, and the MAMBO bolometer at the IRAM 30m in Spain (230 GHz; first results given by Kunneriath et al. 2008, Eckart et al. 2008b). This light curve complements our parallel 13, 7, and 3 mm VLBA run (Lu et al. 2008). Other recent simultaneous multi-wavelength observations also indicate the presence of adiabatically expanding source components with a delay between the X-ray and sub-mm flares of about 100 minutes (Eckart et al. 2006a, Yusef-Zadeh et al. 2008, Marrone et al. 2008). As pointed out by Eckart et al. (2008ab), a combination of a temporary accretion disk with a short jet can explain most of the properties associated with infrared/X-ray Sgr A* light curves (Eckart et al. 2008ab). The data and the prefered model are shown in the bottom panels of Fig. 2.

In the following we highlight a few aspects of our modeling.
The adiabatic expansion model: The May 2007 polarimetric NIR measurements (Eckart et al. 2008a) showed a flare event with the highest sub-flare contrast observed until now. In the relativistic disk model these data provide evidence of a spot expansion and its shearing due to differential rotation. An expansion by only 30% will lower the Synchrotron-Self-Compton (SSC) X-ray flux significantly. In the framework of the spot model this flare event provides additional support for expansion of individual source components.

The expansion speed: The rapid decay of the observed NIR/X-ray flares (e.g. Baganoff et al. 2001; Genzel et al. 2003; Ghez et al. 2004, Eckart et al. 2006abc), as well as the current results from coordinated observing campaigns including sub-mm monitoring (Eckart et al. 2006b; Yusef-Zadeh et al. 2006b; Marrone et al. 2008 and this work) suggest that non-radiative cooling processes, such as adiabatic expansion, are essential, although the adiabatic cooling model results in very low expansion speeds. From modeling the mm-radio flares Yusef-Zadeh et al. (2008) invoke expansion velocities in the range from v_{exp}=0.003-0.1c. This compares well with the expansion velocity of the order of 0.0045±0.0020 c that we obtain in our case with the June 2008 data. These velocities are low compared to the expected relativistic sound speed in orbital velocity in the vicinity of the SMBH. The low expansion velocities suggest that the expanding gas can not escape from Sgr A* or must have a large bulk motion (see discussions in Marrone et al. 2008 and Yusef-Zadeh et al. 2008). Therefore the adiabatically expanding source components either have a bulk motion larger than v_{exp} or the expanding material contributes to a corona or disk, confined to the immediate surroundings of Sgr A*. An expansion of source components through shearing due to differential rotation within the accretion disk may explain the low expansion velocities. The recent theoretical approach of hot spot evolution due to shearing is highlighted in Eckart et al. (2008a) and Zamaninasab et al. (2008, 2009; see also Pecháček et al., 2008).

Structure of the light curve: The flux density variations of Sgr A* can be explained in a disk or jet model (see e.g. discussion in Eckart et al. 2006ab, 2008a), or they could be seen as a consequence of an underlying physical process that can mathematically be described as red-noise (Do et al. 2008, Meyer et al. 2008). In our case we find a light curve structure that consists of maxima separated by about 70 to 110 minutes with additional fluctuations of smaller amplitude. Assuming the presence of a disk and by simultaneous fitting of the previously obtained light curve fluctuations and the time-variable polarization angle, we have shown that the data can be successfully modeled with a simple relativistic hot spot/ring model (Meyer et al. 2006ab, 2007, Eckart et al. 2006ab, 2008ab, see also Zamaninasab 2009). In this model the broad near-infrared flares ($\sim$100 minutes duration) of Sgr A* are due to a sound wave that travels around the SMBH once. The sub-flares, superimposed on the broad flare, are thought to be due evolving (i.e. expanding) hot-spots that may be relativistically orbiting the central SMBH. The spot emission would then be due to transiently heated and accelerated electrons which can be modeled as a plasma component (scenarios in which spiral wave structures contribute to the observed variability are also under discussion, e.g. Karas et al. 2007).

In the case of a jet (see e.g. Markoff, Bower & Falcke 2007, Markoff, Nowak & Wilms 2005), the observed flux density variations may more likely be a result of the variations in the accretion process (or jet instabilities) - possibly followed by an adiabatic expansion of the jet components - rather than being a result of a modulation from an orbiting spot. In this case one may expect that red-noise variations on these short times scales are a natural extension of the variability found for longer periods.

4. The Potential for Interferometry in the NIR

Important aspects and details of the SgrA* accretion physics and the way in which matter has been transformed into stars or provided for the accretion process can be studied using interferometry in the infrared.

High resolution studies of the Galactic Center with the Large Binocular Telescope LBT, specifically with the imaging interferometric beam combiner LINC-NIRVANA, will allow us to clearly determine the structures of these young star candidates in the K-band, were hot dust close to the stars can be detected (e.g. Herbst et al. 2008, Eckart et al. 2008c, Bertram et al. 2008, Rost et al. 2008a, Straubmeier et al. 2006). Goal of these studies will be to consolidate the identification with young stars that are still surrounded by relics of their dust disks, which will certainly suffer from

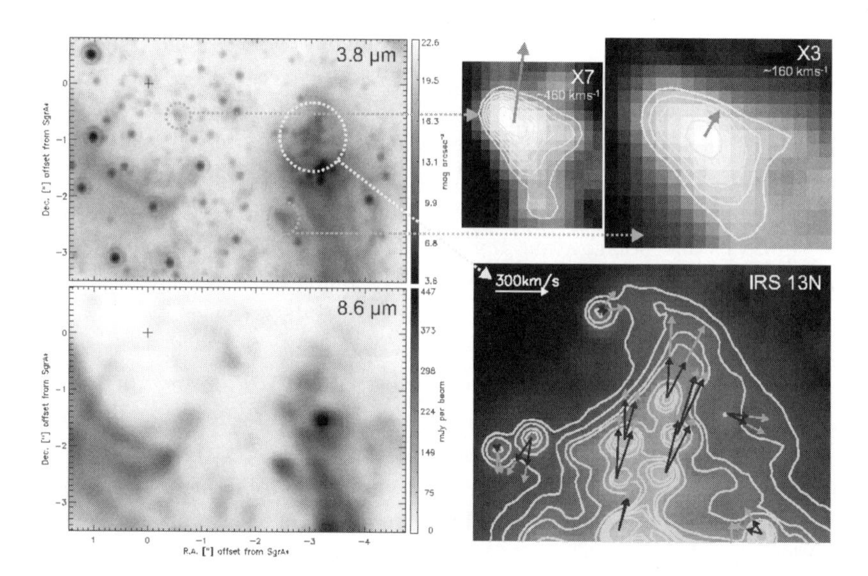

Fig. 3. **Left top and bottom:** A $4{\times}6arcsec^2$ section of the central stellar cluster at 3.8μm and 8.68μm wavelength including the position of SgrA* (marked by the black cross) and IRS13N just north of the cluster of bright stars of IRS 13 proper. **Top right:** The sources IRS13N, X3 and X7 in the Galactic Center stellar cluster (Muzic et al. 2007). **Bottom right:** Proper motions of the dust embedded IRS13N sources (Muzic et al. 2008). The small cluster represents a comoving group of stars showing that this particular group of stars is not only young with respect of their spectroscopic properties, but that it is also a dynamically young stellar association.

traveling through the dense stellar cluster. High angular resolution studies that are sensitive to both point like stars and extended emission will allow us to identify and study dust embedded candidates for young stars. A detailed compilation of stand-off distances to the sections dominated by hot dust and proper motion velocities will help to determine the stellar winds that are involved and allow us to narrow down the stellar types even in dust enshrouded systems. Comparative studies in the Galactic Center starburst clusters Arches and Quintuplet will help to understand the star formation in the center of the Milky Way. This will substantially contribute to the understanding of the complex disk evolution scenario in dense clusters.

In the MIR and long-NIR high angular resolution is required to distinguish between the feeble emission from SgrA* and the more extended dust emission of the mini-spiral. The spectral index of SgrA* in the NIR/MIR is not well known. Here synchrotron losses may dominate the spectral shape. Variability and polarization studies in the MIR can only be done with LINC-NIRVANA (sensitivity and resolution). Especially in the 3-4 micron region the situation will be complicated, since polarization through dust scattering and emission will contribute to the contamination of the signal from SgrA*.

On of the key scientific projects is the investigation of the stellar population at the GC with an unprecedented statistical quality through multicolor imaging and supported by imaging spectroscopy (e.g with Lucifer at the LBT). One further goal will be the investigation in the immediate vicinity of Sgr A*. How much do the individual stellar populations (e.g. He-stars at distances of 1 to 3 arcseconds or B-star at distances of 1 arcsecond) contribute to the accretion flow onto SgrA* and how much do they contribute to the wind from the central few arcseconds? Some evidences for such a wind from the SgrA* are shown in Fig. 3.

For monitoring the variability of SgrA* in coordinated campaigns (radio/mm/MIR/X-ray) and in polarized radio/NIR emission, the LBT will be best suited to separate SgrA* especially during faint phases from the surrounding high velocity stars. Similarly, mm-interferometers can separate Sgr A* from the thermal emission of the CND and the mini-spiral. Therefore the combination of the LBT with mm-interferometers like CARMA, ATCA and ALMA is well suited to study the evolution of evolving source components.

Another interesting goal will be to investigate the sizes, shape, kinematics, and excitation of the thin, elongated filaments in the ISM and of the dust shells associated with luminous stars. The results are essential for the understanding of the physics of the stars and the GC interstellar medium (ISM) as well as the interaction between both. Is there a one-to-one identification between the narrow dust features and radio features? Do they have the same proper motions? Is the central wind dominated by the hot stars or (in certain directions?) by a possible wind from SgrA* (see Fig. 3)?

We aim at investigating of a larger number of individual stars and their shells as well as stellar aggregates and their relative proper motions (like IRS 13 and IRS 13N) in the MIR/NIR. How do these stars interact with the mini-spiral material - both with the few 100 K warm gas and dust as well as the 10^4 K hot thermal plasma?

Deeper insight into the accretion physics of SgrA* will be obtained with new instrumentation that is being built for the ESO VLTI. The GRAVITY experiment is specifically designed to observe highly relativistic motions of matter close to the event horizon of Sgr A* (e.g. Eisenhauer et al. 2009, Straubmeier et al. 2008, Haubois et al. 2006, Gillessen et al. 2006). It includes an integrated optics, 4-telescope, dual feed beam combiner operated in a cryogenic vessel; near infrared wavefront sensing adaptive optics; fringe tracking on secondary sources within the field of view of the VLTI and a novel metrology concept. Simulations show that the planned design matches the scientific needs; in particular that 10μas astrometry is feasible for a source with a magnitude of K=15 like Sgr A*, given the availability of suitable phase reference sources.

5. Summary and Discussion

We have presented new simultaneous measurements of the near-IR sub-mm flare emission of the Sgr A* counterpart associated with the SMBH at the Galactic center. The data were obtained in global campaigns carried out in 2004 using the NACO at VLT, the VLA, and SMA, and in 2008, using NACO and LABOCA at APEX.

The highly variable near-IR emission observed in 2004 presents four major events (I - IV), whose peaks are separated by about $\sim$80 min. The peak of the sub-mm light curve is delayed with respect to that of the near-IR data. We argue that this delay is due to the adiabatic expansion of synchrotron source components that become optically thin. The expansion velocity is about 1500 km/s (0.5% of the speed of light). We model the four flare events with 4 source components. In this case the sub-mm light curves of the individual components are blended, with a time delay for the peaks of about 1.5$\pm$0.5 hours with respect to the near-IR data.

The 2008 light curve structure is consistent with the previously found variability in 2004 and on other occasions (e.g. Eckart et al. 2004, Yusef-Zadeh et al. 2008, Marrone et al. 2008). It could be interpreted as emission from material in relativistic orbits around the SMBH. However, variable emission from a jet or an explanation as a dominant flux density contribution from a red-noise process cannot be excluded as well.

These data show that the VLT/APEX combination is especially well suited for very long simultaneous light curves between the NIR and the sub-mm domain. Further simultaneous radio/sub-mm data, NIR K- and L-band measurements in combination with X-ray observations should lead to a set of light curves that will allow us to prove the proposed model and to discriminate between the individual higher and lower energy flare events. Here Chandra's high angular resolution is ideally suited to separate for weak flares the thermal non-variable bremsstrahlung and the non-thermal variable part of the Sgr A* X-ray flux density.

Explaining SgrA* with alternative solutions for a SMBH is difficult. A universal Fermion ball solution for compact galactic nuclei is made unatractive by the results from stellar orbits near SgrA* (see references in Eckart et al. 2008abc). Alternative explanation of the central mass as a massive Boson star (Torres, Capozziello & Lambiase 2000, Lu & Torres 2003 and references therein) are severely challenged by the general agreement between the measured polarized flare structure and the theoretical predictions (e.g. Eckart et al. 2006b) as well as the indication of a quasi-periodicity in the data. If the indicated quasi-periodicity is due to orbital motion then also a stationary Boson star can be excluded (see also result for the nucleus of MCG-6-30-15 by Lu & Torres 2003) as an alternative solution for SgrA*. In this case one expects the orbital periods to be larger. Especially in the light of the reported violent and luminous accretion phenomena and the mounting evidence of very recent and possibly ongoing star formation in the central cluster indicate the requirements to keep Fermion or Boson balls stable are most probably not meet. Future insight into the accretion process will certainly be deliverd by interferometry in the (sub-)millimeter and infrared wavelength domain.

Acknowledgments

We are grateful to all the ESO PARANAL and Sequitor staff, and especially to the members of the NAOS/CONICA, VLTI, and APEX team. The observations were made possible through a special effort by the APEX/ONSALA staff to have the LABOCA bolometer ready for triggering. Macarena

García-Marín is supported by the German federal department for education and research (BMBF) under the project numbers: 50OS0502 & 50OS0801. M. Zamaninasab and D. Kunneriath are members of the International Max Planck Research School (IMPRS) for Astronomy and Astrophysics at the MPIfR and the Universities of Bonn and Cologne.

References

1. Baganoff, F. K., Bautz, M.W., Brandt, W.N., Chartas, G., Feigelson, E.D., Garmire, G.P., Maeda, Y., Morris, M., Ricker, G.R., Townsley, L.K., & Walter, F. 2001, Nature 413, 45
2. Bertram, Thomas; Eckart, Andreas; Lindhorst, Bettina; Rost, Steffen; Straubmeier, Christian; Tremou, Evangelia; Wang, Yeping; Wank, Imke; Witzel, Gunther; Beckmann, Udo; and 3 coauthors; The LINC-NIRVANA fringe and flexure tracking system; 2008, SPIE 7013, 66
3. Blum, R.D., Sellgren, K., Depoy, D.L., 1996, ApJ 470, 864
4. Diolaiti, E., Bendinelli, O., Bonaccini, D., et al. 2000, A&A Suppl. 147, 335
5. Do, T.; Ghez, A.M.; Morris, M.R.; Yelda, S.; Meyer, L.; Lu, J.R.; Hornstein, S.D.; Matthews, K. 2008, arXiv0810.0446D, in press
6. Eckart, A., Genzel, R., 1996, Nature 383, 415
7. Eckart, A., Genzel, R, 1997 , MNRAS 284, 576
8. Eckart, A., Genzel, R., Ott, T. and Schödel, R., 2002 , MNRAS 331, 917
9. Eckart, A., Baganoff, F. K., Morris, M., Bautz, M. W., Brandt, W. N., Garmire, G. P., Genzel, R., Ott, T., Ricker, G. R., Straubmeier, C., Viehmann, T., Schödel, R., Bower, G. C., Goldston, J. E., 2004, A&A 427, 1
10. Eckart, A., Baganoff, F. K., Schödel, R., Morris, M., Genzel, R., Bower, G. C., Marrone, D., et al., 2006a , A&A 450, 535
11. Eckart, A., Schödel, R., Meyer, L., Trippe, S., Ott, T., Genzel, R., 2006b, A&A 455, 1
12. Eckart, A., Schödel, R., et al., 2006c, ESO Messenger 125, 2
13. Eckart, A., Baganoff, F. K., Zamaninasab, M., Morris, M. R., Schödel, R., Meyer, L., Muzic, K., Bautz, M. W., Brandt, W. N., Garmire, G. P., and 11 coauthors, 2008a, A&A 479, 625
14. Eckart, A., Schödel, R., Baganoff, F.K., Morris, M., et al. 2008b, JPhCS, in press
15. Eckart, A.; Schödel, R.; Garcia-Marin, M.; Witzel, G.; Weiss, A.; Baganoff, F. K.; Morris, M. R.; Bertram, T.; Dovciak, M.; Duschl, W. J.; and 21 coauthors, 2008c, A&A 492, 337
16. Eckart, A., Baganoff, F. K., Morris, M. R., Kunneriath, D., Zamaninasab, M., Witzel, G., Schödel, R., García-Marí, M., Meyer, L., Bower, G.C., Marrone, D., Bautz, M.W., Brandt, W.N., Garmire, G.P., Ricker, G.R., Straubmeier, C., Roberts, D.A., Muzic, K., Mauerhan, J., Zensus, A., 2009, accepted by A&A.
17. Eckart, Andreas; Witzel, Gunther; Kunneriath, Devaky; Koenig, Sabine; Straubmeier, Christian; Bertram, Thomas; Zamaninasab, Mohammad; Schoedel, Rainer; Muzic, Koraljka; Tremou, Evangelia; and 6 coauthors;

Prospects for observing the Galactic Center: combining LBT LINC-NIRVANA observations in the near-infrared with observations in the mm/sub-mm wavelength domain; 2008, SPIE 7013, 144
18. Eisenhauer, F., Schödel, R., Genzel, R., Ott, T., Tecza, M., Abuter, R., Eckart, A., Alexander, T., 2003, ApJ 597, L121
19. Eisenhauer, F., Genzel, R., Alexander, T., Abuter, R., Paumard, T., Ott, T., Gilbert, A., Gillessen, S., 2005, ApJ 628, 246
20. Eisenhauer, F.; Perrin, G.; Brandner, W.; Straubmeier, C.; Boehm A.; Baumeister, H.; Cassaing, F.; Clét, Y.; Dodds-Eden, K.; Eckart, A.; and 35 coauthors; GRAVITY: Microarcsecond Astrometry and Deep Interferometric Imaging with the VLT; Science with the VLT in the ELT Era, Astrophysics and Space Science Proceedings, Volume. ISBN 978-1-4020-9189-6. Springer Netherlands, 2009, p. 361
21. García-Marín, M., 2008, in prep.
22. Genzel, R., Schödel, R., Ott, T., et al., 2003, Nature 425, 934
23. Ghez, A., Morris, M., Becklin, E.E., Tanner, A. & Kremenek, T., 2000, Nature 407, 349
24. Ghez, A.M., Wright, S.A., Matthews, K., et al., 2004, ApJ 601, 159
25. Ghez, A.M., Salim, S., Hornstein, S. D., Tanner, A., Lu, J. R., Morris, M., Becklin, E. E., Duchêne, G., 2005a, ApJ 620, 744
26. Ghez, A.M., Hornstein, S.D., Lu, J.R., Bouchez, A., Le Mignant, D., van Dam, M.A., Wizinowich, P., Matthews, K., Morris, M., Becklin, E. E., et al. 2005b, ApJ 635, 1087
27. Gierlinski, M., Middleton, M., Ward, M., Done, C., 2008, Nature 455, 369
28. Gillessen, S., et al., 2006, JPhCS 54, 411
29. Gillessen, S.; Perrin, G.; Brandner, W.; Straubmeier, C.; Eisenhauer, F.; Rabien, S.; Eckart, A.; Lena, P.; Genzel, R.; Paumard, T.; Hippler, S. ; GRAVITY: the adaptive-optics-assisted two-object beam combiner instrument for the VLTI; 2006, SPIE 6268, 33
30. Gould, R.J., 1979, A&A 76, 306
31. Karas, V., Dovciak, M., Eckart, A., Meyer, L., Proc. of the Workshop on the Black Holes and Neutron Stars, eds. S. Hledik and Z. Stuchlik, 19-21 September 2007 (Silesian University, Opava), astro-ph0709.3836
32. Haubois, X.; Eisenhauer, F.; Perrin, G.; Rabien, S.; Eckart, A.; Lena, P.; Genzel, R.; Abuter, R.; Paumard, T.; Brandner, W. ; GRAVITY, Probing Space-time And Faint Objects In The Infrared.; Astronomical Facilities of the Next Decade, 26th meeting of the IAU, Special Session 1, 16-17 August, 2006 in Prague, Czech Republic, SPS1, No.4
33. Herbst, T. M.; Ragazzoni, R.; Eckart, A.; Weigelt, G.; LINC-NIRVANA: achieving 10 mas imagery on the Large Binocular Telescope; 2008, SPIE 7014, 43
34. Hornstein, S. D., Matthews, K., Ghez, A. M., Lu, J. R., Morris, M., Becklin, E. E., Rafelski, M., Baganoff, F. K., 2007, ApJ 667, 900
35. Hornstein, S. D., Matthews, K., Ghez, A. M., Lu, J. R., Morris, M., Becklin, E. E., Baganoff, F. K., Rafelski, M. 2006, JPhCS 54, 399
36. Krabbe, A., Iserlohe, C., Larkin, J. E., et al., 2006, ApJ 642, L145

37. Kunneriath, D., et al. 2008, JPhCS, in press
38. Lu, J. R., Ghez, A. M., Hornstein, S. D., Morris, M. R., Becklin, E. E., Matthews, K., accepted to ApJ, 2008arXiv0808.3818L
39. Lu, R.-S., et al., 2008, JPhCS, in press
40. Lu, Youjun; Torres, Diego F., 2003, IJMPD 12, 63
41. Markoff, S., Bower, G.C., Falcke, H., 2007, MNRAS 379, 1519
42. Markoff, S., Nowak, M.A., Wilms, J., 2005, ApJ 635, 1203
43. Marrone, D. P., Baganoff, F. K., Morris, M., Moran, J. M., Ghez, A. M., Hornstein, S. D., Dowell, C. D., Munoz, D. J., Bautz, M. W., Ricker, G. R., and 7 coauthors, 2008, ApJ 682, 373
44. Marscher, A.P. 1983, ApJ, 264, 296
45. Mauerhan, J.C., Morris, M., Walter, F., Baganoff, F.K., 2005, ApJ 623, L25
46. Meyer, L., Eckart, A., Schödel, R., Duschl, W. J., Muciz, K., Dovciak, M., Karas, V., 2006a, A&A 460, 15
47. Meyer, L., Schödel, R., Eckart, A., Karas, V., Dovciak, M., Duschl, W. J., 2006b, A&A 458, L25
48. Meyer, L., Schödel, R., Eckart, A., Duschl, W. J., Karas, V., Dovciak, M., 2007, A&A 473, 707
49. Meyer, L., Do, T., Ghez, A., Morris, M.R., Witzel, G., Eckart, A., Bèlanger, G., Schödel, R., 2008, ApJ, submitted.
50. Pecháček, T., Karas, V., Czerny, B., 2008, A&A 487, 815
51. Reid, M.J., 1993, ARA&A 31, 345
52. Schödel, R., Ott, T., Genzel, R., Hofmann, R., Lehnert, M., Eckart, A., Mouawad, N., Alexander, T., 2002, Nature 419, 694
53. Rost, Steffen; Bertram, Thomas; Lindhorst, Bettina; Straubmeier, Christian; Tremou, Evangelia; Wang, Yeping; Witzel, Gunther; Eckart, Andreas ; The LINC-NIRVANA Fringe and Flexure Tracker: testing piston control performance; 2008, SPIE 7013, 99
54. Siringo, G., Weiss, A.el, Kreysa, E., et al., 2007, ESO Messenger 129, 2
55. Straubmeier, Christian; Eisenhauer, Frank; Perrin, Guy; Brandner, Wolfgang; Eckart, Andreas; Opto-mechanical design of the spectrometers of GRAVITY: the 6-Baseline K-Band Interferometer for the VLTI; 2008, SPIE 7013, 93
56. Straubmeier, Christian; Bertram, Thomas; Eckart, Andreas; Rost, Steffen; Wang, Yeping; Herbst, Tom; Ragazzoni, Roberto; Weigelt, Gerd ; The imaging fringe and flexure tracker of LINC-NIRVANA: basic opto-mechanical design and principle of operation; 2006, SPIE 6268, 48
57. Torres, D.F., Capoziello, S., Lambiase, G., 2000, Phys. Rev. D 62, 104012
58. van der Laan, H., 1966, Nature 211, 1131
59. Weiss, A., Kovacs, A., Guesten, R., Menten, K. M., Schuller, F., Siringo, G., Kreysa, E., 2008, accepted by A&A, 2008arXiv0808.3358W
60. Yusef-Zadeh, F., Wardle, M., Cotton, W. D., Heinke, C. O., Roberts, D. A. 2007 ApJ 668, L47
61. Yusef-Zadeh, F., Bushouse, H., Dowell, C. D., Wardle, M., Roberts, D., Heinke, C., Bower, G. C., Vila-Vilaró, B., Shapiro, S., Goldwurm, A., Bélanger, , 2006a, ApJ 644, 198

62. Yusef-Zadeh, F., Roberts, D., Wardle, M., Heinke, C.O., Bower, G.C., 2006b, ApJ 650, 189
63. Yusef-Zadeh, F., Wardle, M., Heinke, C., Dowell, C. D., Roberts, D., Baganoff, F. K., Cotton, W., 2008 ApJ 682, 361
64. Zamaninasab et al. (2009), submitted to A&A
65. Zamaninasab, M., Eckart, A., Meyer, L., Schödel, R., Dovciak, M., Karas, V., Kunneriath, D., Witzel, G., Giessuebel, R., König, S., Straubmeier, C., Zensus, A., 2008, Proc. of a conference on 'Astrophysics at High Angular Resolution (AHAR 08)' held 21-25 April 2008 in Bad Honnef, Germany, 2008arXiv0810.0138Z

THE DARKSTARS CODE: A PUBLICLY AVAILABLE DARK STELLAR EVOLUTION PACKAGE

PAT SCOTT* and JOAKIM EDJSÖ

Oskar Klein Centre for Cosmoparticle Physics & Department of Physics, Stockholm University, AlbaNova University Centre, SE-106 91 Stockholm, Sweden
**pat@fysik.su.se*

MALCOLM FAIRBAIRN

PH-TH, CERN, Geneva, Switzerland & King's College London, WC2R 2LS, UK

We announce the public release of the 'dark' stellar evolution code DarkStars. The code simultaneously solves the equations of WIMP capture and annihilation in a star with those of stellar evolution assuming approximate hydrostatic equilibrium. DarkStars includes the most extensive WIMP microphysics of any dark evolution code to date. The code employs detailed treatments of the capture process from a range of WIMP velocity distributions, as well as composite WIMP distribution and conductive energy transport schemes based on the WIMP mean-free path in the star. We give a brief description of the input physics and practical usage of the code, as well as examples of its application to dark stars at the Galactic centre.

Keywords: Cosmology; stellar evolution; dark matter; WIMPs; galactic centre; dark stars.

The last two years have seen strong interest in the impacts of dark matter upon stellar structure and evolution. The predominant focus has been on self-annihilating WIMP (weakly-interacting massive particle) dark matter, because it has the ability to affect stellar structure by annihilating in stellar cores[1–5] and collapsing protostellar halos.[6–9] Interest has been driven by the prospect of providing constraints upon the nature of dark matter,[2,5] by intrinsic curiosity in the resultant 'dark stars' themselves,[4,9,10] and by their possible impacts upon early-universe processes like reionisation.[11,12]

We have previously discussed the possibility that main-sequence dark stars could exist at the centre of our own Galaxy.[3,5,13,14] In those papers we utilised a form of the standard stellar evolution code STARS[15–17] modified to include the effects of dark matter capture and annihilation. This modified code is DarkStars, and in these proceedings we announce its public release. DarkStars is written in Fortran95, and can be freely downloaded from `http://www.fysik.su.se/~pat/darkstars`. Below we give outlines of the code's input physics and practical usage, along with some simple examples of stars evolved with it.

DarkStars includes gravitational capture of WIMPs from the galactic halo via the full equations of Gould,[18] including both spin-dependent and spin-independent scattering on the 22 most important atomic nuclei. The capture routines are adapted from the solar capture code in DarkSUSY.[19] Capture can be performed semi-analytically from either a standard isothermal WIMP halo or an isothermal halo where the WIMP velocity distribution has been truncated at the local escape velocity. Alternatively, numerical capture calculations can be performed on a velocity distribution derived[20] from the Via Lactea[21] simulation of a Milky Way-type galaxy, or any other arbitrary, user-supplied velocity distribution.

The distribution of WIMPs with height in a star is obtained by interpolating between two limiting distributions according to the value of the WIMP mean-free path in the star: one corresponding to WIMPs with very long mean-free paths, the other to WIMPs with very short mean-free paths. Conductive energy transport by weak-scattering events between atomic nuclei and WIMPs is taken into account in a manner consistent with this distribution: the conductive luminosity at each height is approximated by rescaling the known expression for the conductive luminosity at short mean-free paths, according to the actual value of the mean free path in the star. The annihilation luminosity at each height in the star is simply calculated as the product of the annihilation cross-section and square of the local WIMP density, and fed along with the conductive luminosity into the luminosity equation in the stellar solver. Full technical details of the input physics for DarkStars can be found in Ref. 5.

DarkStars operates with a simple text-file input, containing a series of switches and physical parameters with which to perform a particular evolutionary run. Switches allow choices between analytical and numerical capture, different halo velocity distributions, the inclusion or exclusion of annihilation and conductive energy transport effects, and the option to run in a special 'reconvergence mode' where the solution obtained at each timestep is converged twice (see Ref. 5 for details). Runs can be saved and restarted at will, and the input format provides the ability to make periodic saves during the course of a single evolutionary run.

The user can specify the WIMP mass, spin-dependent, spin-independent and annihilation cross-sections, as well as the stellar mass and metallicity, the initial population of WIMPs in the star and the ultimate percentage of energy lost to neutrinos in each annihilation. One may also opt to specify a constant stellar velocity through a WIMP halo with some particular local density and velocity dispersion, located at a position with a single well-defined Galactic escape velocity. Alternatively, runs can be performed along user-specified orbits, where these four parameters become dynamic quantities given in an additional text file. Orbits can also be looped if desired.

The code presently allows metallicities of $Z = 0.03$–0.0001 with full evolutionary functionality, and a $Z = 0$ mode valid only for protostellar evolution. The latter includes opacities taken from Ref. 22, but does not yet contain an implementation of the full opacity tables required to treat the case where a Pop III star passes from the $Z = 0$ regime into the non-metal-free one by nuclear burning. DarkStars comes with ZAMS starting models for all non-zero metallicities; protostellar models must be supplied by the user.

In Fig. 1 we give some example evolutionary tracks computed with DarkStars, for a $Z = 0.01$, $1\,\mathrm{M}_\odot$ star. The three different paths result from immersing the star in different ambient halo densities of dark matter: one essentially without dark matter, another in a halo with a moderate dark matter density, and a third in a very dense dark matter environment. In this case we have chosen to use a simple, non-truncated, isothermal halo velocity distribution with dispersion $\sigma = 270\,\mathrm{km\,s}^{-1}$, and set the star moving through it at $220\,\mathrm{km\,s}^{-1}$. The runs in Fig. 1 have been halted when the star either leaves the main sequence, or ceases to move any further in the HR diagram. In the most extreme case, where the ambient density is highest, the energy provided by WIMP annihilation pushes the star partially back up the Hayashi track and continues to provide enough energy to keep it there well beyond the age of the Universe; the star has become a 'dark star', powered entirely by dark matter annihilation.

The interior changes producing these different evolutionary histories are summarised in Fig. 2. Here we show the changing contributions of nuclear burning and WIMP annihilation over time in the three stars from Fig. 1, as well as the evolution of their central temperatures, densities and pressures.

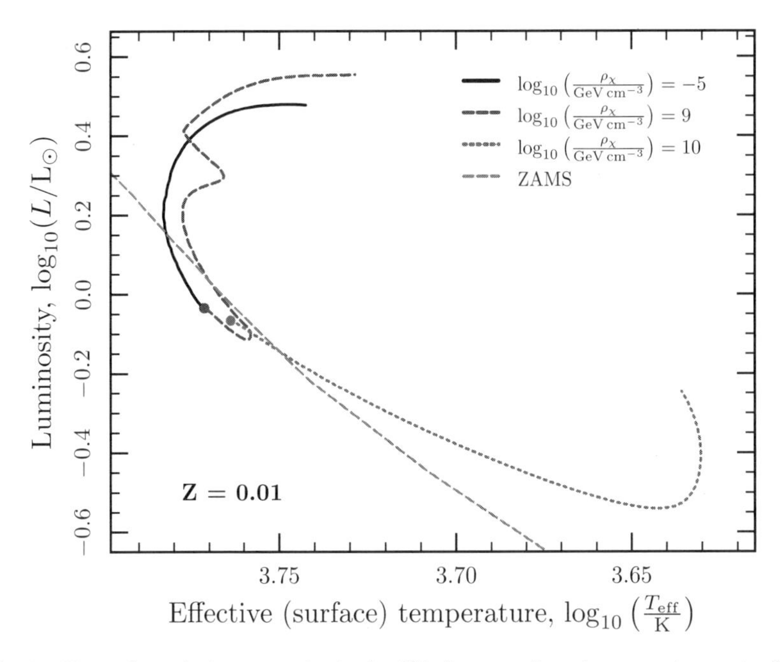

Fig. 1. Example evolutionary paths in the HR-diagram of a solar-mass star evolved in three different dark matter environments. Filled circles indicate the approximate starting point of each track; the starting point of the star evolved in the densest halo is offset from the others only because of the large adjustment to its structure required to accommodate the effects of dark matter in converging the initial model. In each case, stars were given a proper velocity of $220\,\mathrm{km\,s^{-1}}$ through their dark matter halo, which was modelled as an isothermal sphere with velocity dispersion $\sigma = 270\,\mathrm{km\,s^{-1}}$. Runs were halted when the star either left the main sequence (as for the two upper curves), or ceased to move any further (the case for the lower curve). In the case of the lower curve, WIMP annihilation provides enough energy to push the star back up the Hayashi track and hold it there, turning it into a dark star with a lifetime well beyond the current age of the Universe.

In the case of the dark star, we see that as annihilation energy dominates over fusion power, the star's core cools and expands, driving it back up the protostellar cooling curve and effectively turning nuclear burning off. The star evolved in the intermediately-dense dark matter halo also exhibits clear differences to the normal star, spending about 50% longer on the main sequence. Despite the steadily diminishing role of dark matter in this star relative to hydrogen burning over time, it also experiences a much more sudden exhaustion of its nuclear fuel than the normal star just before moving away from the main sequence, as seen in the very steep increase in its core properties just beyond 12 Gyr.

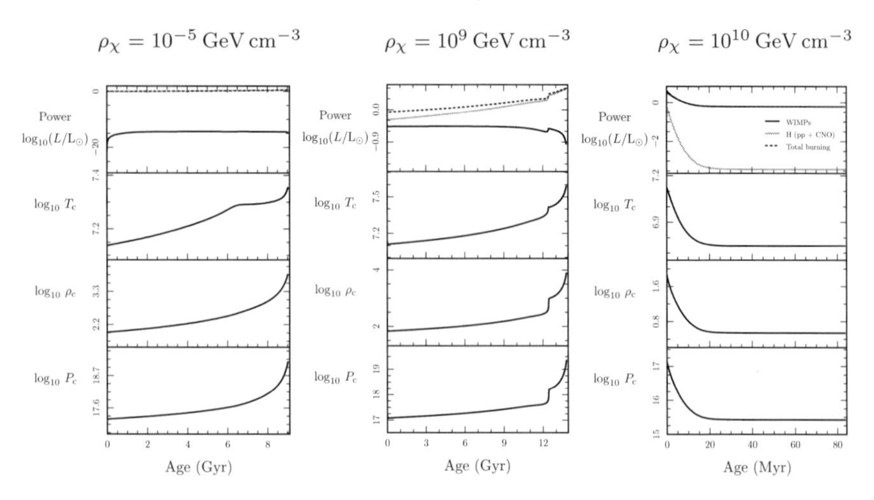

Fig. 2. Evolution of the partial luminosities provided by WIMP annihilation and fusion, as well as the central temperatures, densities and pressures of the example stars shown in Fig. 1. Note the marked drop in central temperature, density and pressure in the dark star of the rightmost panel as it regresses up the Hayashi track, drastically reducing the power provided by nuclear fusion. Note also the extended main sequence lifetime of the middle star as compared to the normal star on the left, and its very sudden exhaustion of core hydrogen, as indicated by very steep changes in its core properties at an age of ~12 Gyr. This sudden and violent exit from the main sequence might result in dynamical instability and/or mass ejection.

As a taste of the sorts of more detailed studies which can be performed with DarkStars, in Fig. 3 we give an example from Ref. 5 of the sorts of WIMP-to-nuclear burning ratios that dark stars might achieve on orbits of different eccentricities, close the the centre of of the Milky Way. All orbits in this plot had periods of 10 years. Typically, stars in which WIMP annihilation produces more energy than nuclear fusion (i.e. above 0 on the y-axis of Fig. 3) are considered good candidates for detection as dark stars. This plot includes curves corresponding to two different dark matter density profiles near the Galactic centre: an adiabatically contracted halo with a gravitationally-induced spike around the black hole which has been allowed to diffuse away over time ('AD+spike'), and an NFW profile with a similar spike ('NFW+spike'). Here we see that in the adiabatic contraction scenario, WIMP annihilation can provide up to 100 times the energy of fusion for stars on realistic orbits in our own Galaxy. The curves shown here correspond to capture from the Via Lactea-derived velocity distribution mentioned earlier, though it should be noted that the applicability of results from large-scale N-body simulations to velocity distributions near

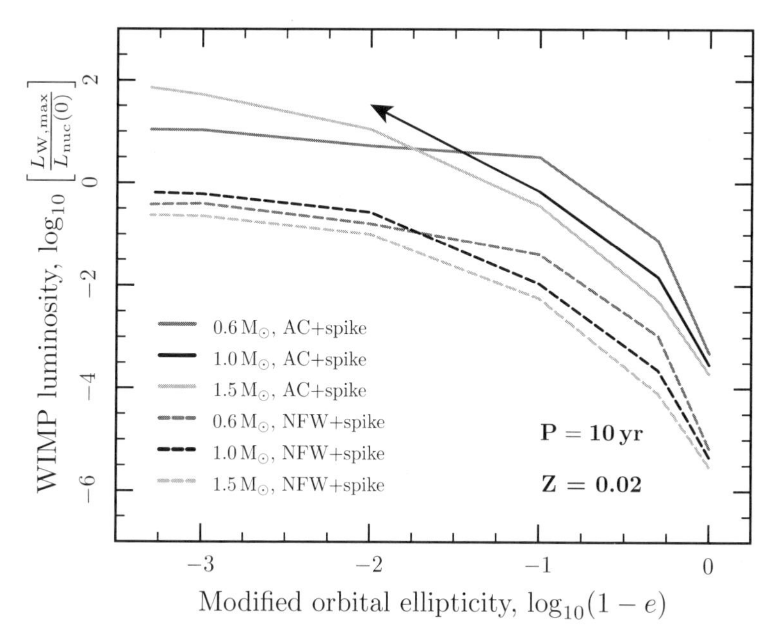

Fig. 3. WIMP-to-nuclear luminosity ratios achieved by stars on orbits with 10-year periods around the Galactic centre. Dark matter annihilation can produce up to 100 times as much power as nuclear fusion in stars on realistic orbits in our own Galaxy. If the Galactic halo has been adiabatically contracted (AC+spike), annihilation can equal nuclear fusion in stars on orbits with eccentricities greater than $e = 0.9$, for masses less than about $1.5\,M_\odot$. If not (NFW+spike), stars of a solar mass or less require $e \gtrsim 0.99$ to approach break-even between annihilation and fusion. The arrow indicates that the $1\,M_\odot$, AC+spike curve is expected to probably continue in this direction, but converging stellar models becomes rather difficult for such high WIMP luminosities. From Ref. 5.

the central supermassive black hole is debatable; the expected distribution of WIMP velocities at the centre of the Galaxy is still quite uncertain.

DarkStars uses the EZ flavour[23] of STARS, making it relatively accessible to those who wish to extend or modify the code to perform a particular task. The current version is prepared for future modifications to the WIMP velocity distributions and nuclear form factors, as well as the inclusion of WIMP evaporative effects and/or full metal-free evolution. A legacy mode for computing capture rates by the Sun, and an (experimental) mode for computing evolution in the case where WIMPs take a non-negligible time to thermalise inside a star, are also included.

Input files for the three example evolutionary runs discussed above are included in the DarkStars release, along with the expected outputs from the code. Ruby plotting scripts using Tioga[a] are also included for generating all the plots shown in these proceedings and Ref. 5, along with the original plotting scripts included in EZ. An option (which draws partially on the Tioga scripts originally shipped with EZ) is also provided to generate movies of a star's evolution.

Acknowledgments

PS is grateful to the G & E Kobbs and Helge Axelsson Johnsons Foundations for enabling his attendance at Dark2009 through their generous financial support, to Ross Church and Richard Stancliffe for helpful discussions about STARS, and to Bill Paxton for permission to include large parts of EZ in the DarkStars release. JE thanks the Swedish Research Council for funding support.

References

1. I. V. Moskalenko and L. L. Wai, *ApJ* **659**, L29 (2007).
2. G. Bertone and M. Fairbairn, *Phys. Rev. D* **77**, 043515 (2008).
3. M. Fairbairn, P. Scott and J. Edsjö, *Phys. Rev. D* **77**, 047301 (2008).
4. F. Iocco, *ApJ* **677**, L1 (2008).
5. P. Scott, M. Fairbairn and J. Edsjö, *MNRAS* **394**, 82 (2009).
6. D. Spolyar, K. Freese and P. Gondolo, *Phys. Rev. Lett.* **100**, 051101 (2008).
7. K. Freese, P. Bodenheimer, D. Spolyar and P. Gondolo, *ApJ* **685**, L101 (2008).
8. F. Iocco, A. Bressan, E. Ripamonti, R. Schneider, A. Ferrara and P. Marigo, *MNRAS* **390**, 1655 (2008).
9. D. Spolyar, P. Bodenheimer, K. Freese and P. Gondolo, (2009). arXiv:0903.3070
10. K. Freese, D. Spolyar and A. Aguirre, *JCAP* **11**, 14 (2008).
11. A. Natarajan and D. J. Schwarz, *Phys. Rev. D* **78**, 103524 (2008).
12. D. R. G. Schleicher, R. Banerjee and R. S. Klessen, (2008). arXiv:0809.1519
13. P. Scott, J. Edsjö and M. Fairbairn, in *Dark Matter in Astroparticle and Particle Physics: Dark 2007.*, eds. H. K. Klapdor-Kleingrothaus and G. F. Lewis (World Scientific, Singapore, 2008). arXiv:0711.0991
14. P. Scott, M. Fairbairn and J. Edsjö, in *Identification of dark matter 2008, Proceedings of Science*, PoS(idm2008)073 (SISSA, Trieste, Italy, 2008). arXiv:0810.5560
15. P. P. Eggleton, *MNRAS* **151**, 351 (1971).
16. P. P. Eggleton, *MNRAS* **156**, 361 (1972).
17. O. R. Pols, C. A. Tout, P. P. Eggleton and Z. Han, *MNRAS* **274**, 964 (1995).
18. A. Gould, *ApJ* **321**, 571 (1987).

[a]`http://tioga.rubyforge.org/`

19. P. Gondolo, J. Edsjö, P. Ullio, L. Bergström, M. Schelke and E. A. Baltz, *JCAP* **7**, 8 (2004).
20. M. Fairbairn and T. Schwetz, */jcap* **1**, 37 (2009).
21. J. Diemand, M. Kuhlen and P. Madau, *ApJ* **657**, 262 (2007).
22. J. J. Eldridge and C. A. Tout, *MNRAS* **348**, 201 (2004).
23. B. Paxton, *PASP* **116**, 699 (2004).

RELATIVISTIC SIGNATURES AT THE GALACTIC CENTER

ECATERINA MARION HOWARD

Department of Physics and Engineering,
Macquarie University, Sydney, NSW, Australia
**E-mail: khoward@ics.mq.edu.au*
www.mq.edu.au

Studies of the inner few parsecs at the Galactic Centre provide evidence of a $4\times10^6 M_\odot$ supermassive black hole, associated with the unusual, variable radio and infrared source Sgr A*.

Our major aim is the study and analysis of the physical processes responsible for the variable emission from the compact radio source Sgr A*. In order to understand the physics behind the observed variability, we model the time evolution of the flare emitting region by studying light curves and spectra of emission originating at the surface of the accretion disk, close to the event horizon, near the marginally stable orbit of a rotating black hole.

Here we discuss the methods used in the analysis of the time-variable spectral features and subsequently present preliminary modeling results.

Keywords: Sgr A*; light curve; hot spot; black hole physics; Galactic Center; accretion disc; relativity effects.

1. Introduction

Recent simultaneous observations from radio to X-rays provide information about irregular outbursts of emission from Sgr A*, lasting between a few minutes and a few hours.[1]

The flare activity originates from within a few milli-arcseconds of the radio position of Sgr A*. The temporal correlation between the variability at radio, near-infrared and X-ray wavelengths suggests a common physical origin of the phenomena.

Due to its relative proximity, SgrA* provides favorable circumstances for a better understanding of the processes responsible for the observed time-dependent emission. Apparently, the emission from Sgr A* originates in an accretion flow in keplerian motion, with a peak occuring within several Schwarzschild radii ($r_S \equiv 2GM/c^2$) of the centre.

Motivated by the evidence of variability at different wavelengths, we investigate the viability of the hot-spot model for the SgrA* flaring activity.[2]

Theoretical models are trying to address the question of what the variability may tell us about the black hole properties and analyze the characteristics of the accretion disc close to the event horizon.[3] We take into account emission from within the last stable orbit of a rotating black hole or from the region near above the marginally stable orbit.

Using fully relativistic ray-tracing methods, we aim to analyze the constraining parameters and study the major imprints of both special and general relativistic effects on the time-resolved emission. We include all relativistic effects (energy shift, aberration, light bending, lensing and relative time delay) near a Kerr black hole. General and special relativistic effects play a crucial role in the time-dependent behaviour, particularly for a maximally rotating Kerr black hole.

2. Basic Assumptions and Objectives

By integrating the photon geodesic paths between a position inside a spot located within the accretion disk and an observer positioned at infinity, we obtain time dependent spectra for an orbiting or infalling spot co-moving with the accretion disk, in a deep gravitational potential. A detailed time-resolved analysis makes possible the study of the evolution of the emitting region, close to the event horizon as well as the diagnosis of the light curves of the variable emission region, for various spectral emissivity profiles, different viewing directions of the distant observer and different locations of the spot relatively to the local observer, the event horizon and the center of the disk.

We consider a complete system comprising a black hole, an accretion disc and a co-rotating spot within the cold accretion disk. The gravitational field is described in terms of Kerr metric, for a rotating black hole and particularly for a non-rotating black hole. Therefore, both static Schwarzschild and rotating Kerr black hole cases are considered. The orbiting accretion disc is geometrically thin and optically thick, therefore we take into account only photons coming from the equatorial plane directly to the observer.[4] The spot is orbiting within the disk near the rotating black hole. We also assume that the matter in the accretion disc is cold and neutral.

A fully relativistic ray tracing code is used to account for relativistic effects, mapping the intrinsic disk emission in a co-moving (local) frame, to a distant observer located at infinity. Relativistic effects alter the shape of the spectra, particularly at large inclination angles and play an important

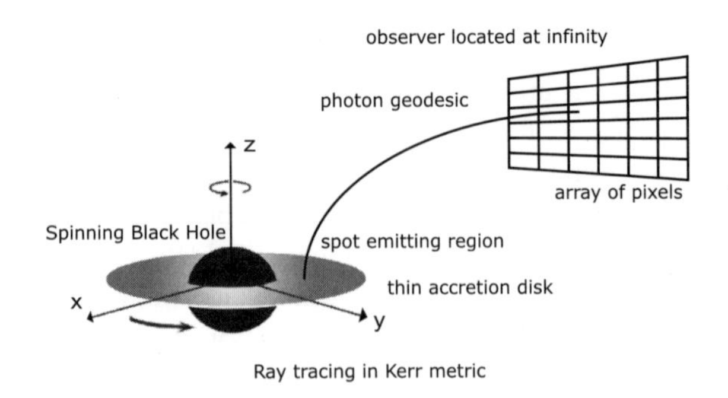

Fig. 1. Ray tracing in Kerr geometry.

role in the time-dependent emission processes, especially in the case of a maximally rotating Kerr black hole.

3. Methods, Modeling and Discussion

We try to address various questions concerning the validity of the simple spot model and the role of the relativistic effects in the observed phenomena. We calculate time-dependent spectra of both cases of an orbiting or a free-falling spot.

The code integrates the local emission in polar coordinates on the disc and, as a consequence, we handle emissions originated in a non-axisymmetric area of integration within the accretion flow.[5]

We are able to study free-falling photons from the marginally stable orbit down towards the black hole horizon. This provides us with information about the plunge region, the area between the Event Horizon and the last stable orbit. Emission starts within a localized spot on the accretion disc. We consider two cases, whether the spot moves with a Keplerian velocity along a stable circular orbit or, if close to the marginally stable orbit, it plunges into the black hole. The intrinsic emissivity at each point depends on the energy shift of the photons and it is time dependent. Photons emerging from the spot area would be affected by all relativistic imprints. The KYSPOT code by Dovciak et al. takes all special and general relativistic effects into account by using pre-calculated sets of transfer functions that map various properties of the emission region in the accretion disk onto the sky plane (Cunningham 1975, 1976).[6] The transfer function, obtained by integration of the geodesic equation, maps the flux in the local frame

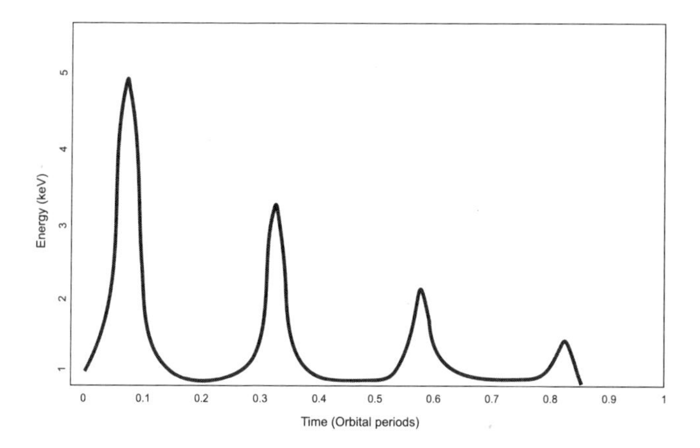

Fig. 2. Apparent spectrum of an in-falling spot emitting monoenergetic photons close to the horizon of an extreme Kerr black hole. The spin parameter is a=0.998 and the rest energy is 6.4 keV. Energy is on the ordinate, time on the abscissa.

comoving with the disk, to the flux as seen by an observer located at infinity. Spectral profiles are obviously affected by relativistic smearing but due to the power-law character of the relativistic imprints, the main shape of the primary continuum profile remains intact. Only the central gravity potential influences the trajectories of the photons towards the observer. The functions are calculated for different values of observer inclination angle and black hole horizons. The binary extensions contain information for different radii and different values of g-factor, defined as the ratio between the photon energy observed at infinity and the local photon energy as emitted from the disc.[7]

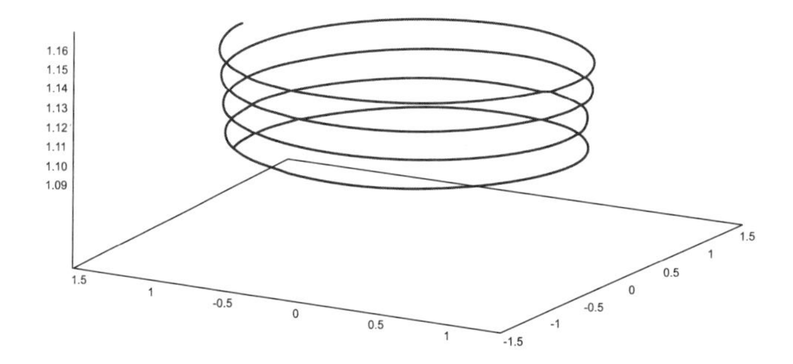

Fig. 3. Plunging trajectory of a photon into the event horizon of an extreme Black Hole.

We obtain time-dependent spectra for various viewing directions of a distant observer based on different emissivity profiles and various angular momenta of the black hole. The relativistic corrections to the local emission are parameterized by the black hole spin and the observer's inclination angle.[8]

The intrinsic emissivity is specified in the frame co-moving with the disc medium and it is defined as a function of r, φ and t in the equatorial plane. When the geodesic integration is ended, after the transfer of photons to the distant observer is performed, Boyer-Lindquist coordinates will replace the initial Kerr ingoing coordinate system. The observed spectra depend on the position of the spot with respect to the disk normal.

We obtain light curves of the variable emission region, for different positions and angles (azimuthal and polar) of the spot relative to the distant observer, the event horizon and the centre of the disk. We study light curves for different inclination angles of the observer relative to the disk axis (θ_o) and analyze the dependence of the variability on the inclination. We analyze the general and special relativity effects that influence the photons paths along null geodesics towards an observer located at infinity.

The figures show the spectrum and plunging trajectory of a spot near the Event Horizon of an extreme black hole.

We also consider different spin parameter values, various viewing angles and different sizes of the emitting spot.

References

1. Yusef-Zadeh, F., Roberts, D., Wardle, M., Heinke, C.O., Bower, G.C., (2006) *ApJ*, 650, 189.
2. Yusef-Zadeh, F., Bushouse, H., Dowell, C. D., Wardle, M., Roberts, D., Heinke, C., Bower, G. C., Vila Vilaro, B., Shapiro, S., Goldwurm, A., and Belanger, G. (2006) *ApJ*, 644, 198-213.
3. Schnittman J. D., & Bertschinger E. (2004) *ApJ*, 606, 1098.
4. Fanton C., Calvani M., de Felice F., & Čadež A. (1997) *PASJ*, 49, 159.
5. Dovčiak M. *PhD Thesis*, (2004) (Charles University, Prague).
6. Cunningham, C. T., (1975) *ApJ*, 202, 788.
7. Dovčiak M., Karas V., Martocchia A., Matt G. and Yaqoob T. (2004) *Proceedings of RAGtime, 4/5 Workshops on black holes and neutron stars, Opava 2003*, 14-16/13-15.
8. Dovčiak M., Karas V., & Yaqoob T. (2004) *ApJS*, 153, 205.

PART VIII

Dark and Baryonic Matter in Galaxies

NON-STANDARD BARYON-DARK MATTER INTERACTIONS

B. FAMAEY[1,*], J.-P. BRUNETON[2]

[1]*IAA, Université Libre de Bruxelles, Bvd du Triomphe, 1050 Bruxelles, Belgium*
[2]*SISSA/ISAS, Via Beirut 2–4, 34014, Trieste and INFN Sezione di Trieste, Via Valerio, 2, 34127 Trieste, Italy*
[*]*E-mail: bfamaey@ulb.ac.be*

After summarizing the respective merits of the Cold Dark Matter (CDM) and Modified Newtonian Dynamics (MOND) paradigms in various stellar systems, we investigate the possibility that a non-standard interaction between baryonic matter and dark matter could reproduce the successes of CDM at extragalactic scales while making baryonic matter effectively obey the MOND field equation in spiral galaxies.

Keywords: Gravitation; dark matter.

1. Introduction

Data on large scale structures point towards a Universe dominated by dark matter and dark energy. Discovering the nature of these mysterious components of the Universe is, without a doubt, the major challenge of modern astrophysics. Nowadays, the dominant paradigm is that dark matter is made of non-baryonic weakly interacting massive particles, the so-called *cold dark matter* (CDM), and that dark energy is well represented by a cosmological constant (Λ) in Einstein equations. The ΛCDM cosmological model has known a remarkable success in explaining and predicting diverse data sets corresponding to the Universe at its largest scales. Nevertheless, an ever growing number of observations at galactic scales are in conflict with the predictions of the ΛCDM cosmological model. One often assumes that these problems are due to the fact that baryons are playing an important role in the dynamics of galaxies, and that their physics is not incorporated rigorously enough in cosmological simulations down to galactic scales.[1] However, a standard feedback from the baryons in the form of supernova explosions and stellar winds would require an extreme fine-tuning to explain the conspiracy between the baryonic and DM distributions in spiral galaxies, which is actually *independent* of the particular history of each stellar system. Signatures of this conspiracy are (i) the baryonic Tully-Fisher relation,[2] (ii) the fact that the mass discrepancy in galaxies always appears at the same gravitational acceleration $a_0 \sim 10^{-10}\mathrm{m/s}^2$, or (iii) the fact that galaxy rotation curves display obvious features (bumps or wiggles) that are also clearly visible in the stellar or gas distribution. In short, a one-to-one correspondence appears to exist between the integrated baryonic mass and the integrated dark mass at all radii in rotating galaxies, and not only in the outskirts. This relation is known as *Milgrom's law*,[3,4] and was originally interpreted as a modification of Newtonian dynamics (MOND).

Although MOND is very successful at galactic scales, this phenomenological law encounters some serious difficulties at larger scales. It is now well established that interpreting the data using MOND's law does not completely remove the need for DM in clusters of galaxies. A large amount of unseen matter is still necessary to fully explain both lensing and dynamics of these objects. Another problem related to MOND's phenomenology is that it still lacks a well-defined relativistic formulation,[5] that would be both natural (free of fine-tunings) and consistent from a field theory point of view. Related to this is the fact that the physical origin of a new acceleration scale, and of the modification of the physical laws that it implies, are rather unclear at present. Some very interesting numerical coincidences ($a_0 \sim H_0 \sim \Lambda^{1/2}$) may pave the way to some unified picture, but to date, no major workable ideas have been proposed in this direction.

Having therefore noticed (i) the difficulty to encapsulate the MOND phenomenology in a fully consistent theory, (ii) the abscence of clear physical origin to it, and (iii) the fact that the phenomenology anyway fails at extragalactic scales, we are naturally lead to the idea that MOND could be only an effective picture, only valid at galactic scales, of a more general theory. We will therefore explore the possibility that the modification of gravity assumed in MOND may originate from a new interaction between DM and baryons, that could explain Milgrom's law at galactic scales while also reproducing the successes of CDM at larger scales.

Section 2 deals more precisely with the successes and problems of the CDM and MOND paradigms at different scales. This will help us to discuss in Section 3 some model-independent considerations about the possibilty of a unification of CDM and MOND. We will notably show that some new dimensionful constants (besides a_0) may be needed to unify properly these two paradigms in a single framework. We will finally present a class of models that encapsulate most of the desired phenomenology, although this should be viewed only as a very first step, as the resulting model does suffer from naturalness problems, and is still far from explaining the physical origin of a_0.

2. CDM vs. MOND

In this Section, we compare the respective merits of CDM and MOND in various stellar systems and summarize this in a table, + and − signs being attributed to each paradigm, with the inherent subjectivity that this kind of classification encompasses. Obviously, one could argue that the concordance ΛCDM model is a fully defined cosmological model, and that putting it on equal footings with a simple phenomenological law is unfair. On the other hand this phenomenological law is fully predictive in stationary systems, and its lack of present-day physical basis can be considered as less harmful than the fine-tuning problems that the CDM paradigm encounters. Let us also note that the actual situation is often much more complex than what can be summarized with a straightforward + or − sign. We take here the most pessimistic attitude, which is to put a − sign whenever the paradigm encounters a single problem: this results in a globally very pessimistic overview of both paradigms, but it does not necessarily mean that these problems cannot be solved within each paradigm. With all these important caveats in mind, we proceed hereafter to the comparison.

In high surface brightness (HSB) spiral galaxies such as the Milky Way, MOND predictions are extremely successful[4] while the CDM paradigm suffers from obvious problems such as the cusp and the fine-tuning problem resulting from MOND's success. What is more, the transfer of angular momentum from the disk to the halo prevents from forming large enough baryonic disks in the process of galaxy formation (the angular momentum problem). In low surface brightness galaxies (LSB), the situation is even worse for CDM, where the bumps and wiggles of the rotation curves following the light distribution are hard to understand. On the other hand, the high mass-discrepancy in these systems was predicted by MOND even before the first LSB rotation curves were measured. For tidal dwarf galaxies (TDG), formed in the tidal tails resulting from the collision of two spiral galaxies, CDM simulations predict that the objects should be devoid of collisionless DM since they are formed out of the material that was in the rotating disks of the progenitors. However, rotation curves of three TDGs in the NGC 5291 system indicate the presence of large amounts of DM.[6] The solution for CDM would then be to assume large amounts of baryonic

DM in the disks of the progenitors. However the mass-discrepancy in these TDGs perfectly obeys the MOND law,[7] which would thus require a really *ad hoc* distribution of baryonic DM, similar to the fine-tuned distribution of CDM in other rotationally supported galaxies.

On the other hand, while the velocity dispersion profiles of dwarf spheroidal (dSph) satellites of the Milky Way are generally in accordance with MOND, Sextans and Draco would require unrealistically high stellar mass-to-light ratios,[8] thus meaning that these galaxies may be problematic for MOND. They are also problematic for the CDM paradigm, since they also suffer from the cusp problem, and their planar distribution around the Milky Way could argue as them actually being old TDG relics rather than CDM building blocks.[9] In some large elliptical galaxies, the dearth of DM in the central parts can be problematic for the CDM paradigm,[10] but on the other hand some ellipticals (essentially at the center of clusters) require unrealistically high mass-to-light ratio in the MOND context.[11]

At subgalactic scales, the velocity dispersion of stars in the Pal 14 globular cluster are consistent with Newtonian dynamics and no DM, which is in accordance with the CDM paradigm, but might be problematic for MOND:[12] however, more studies of this kind are required since the conclusion was reached based on very few stars for one gaussian fit to the total velocity distribution (while a Newtonian core and MONDian outskirts might be expected in MOND), no anisotropy was considered (which could severely affect the conclusions depending on the position of the tracer stars in the cluster), and the MOND prediction was based on a purely circular orbit for Pal 14.

At extragalactic scales, the CDM paradigm is extremely succesful, while MOND still needs large amounts of unseen mass in galaxy clusters, essentially in the central parts, leading to a total discrepancy of a factor 2-3 in the outskirts. For large-scale structure (LSS), no simulations comparable to what is done in CDM have been performed, notably because of the absence of a fundamental theory underpinning the MOND paradigm, and because of the unknown nature of the missing mass in galaxy clusters. Finally, present-day Lorentz-covariant extensions of MOND[13–15] have not been able to produce a high third-peak in the angular power spectrum of the CMB without producing too much power at large angular scales, or without adding a substantial non-baryonic hot dark matter component[16] (which could also solve the galaxy cluster problem).

This overview thus leads us to the following conclusions: while CDM is successful at extragalactic and subgalactic scales, it fails at galactic scales, where it requires an extreme and unrealistic fine-tuning to explain the observations; MOND, on the other hand, is an extremely powerful and predictive phenomenological law in rotationally supported systems, but it fails in *some* pressure-supported systems.

	CDM	**MOND**
HSB	−	+
LSB	−−	++
TDG	−−	++
dSph	−	−
Ellipticals	−	−
Globular Clusters	+	−
Galaxy Clusters	+	−−
LSS	++	−
CMB	++	−−

3. Baryon-DM Interactions to Reproduce the Successes of Both MOND and CDM?

In view of the detailed comparison of the successes of MOND and CDM in the above section, we are naturally led to the idea that both of them could be only effective descriptions valid in different regimes. The first step in such a direction could be a theory which would reduce to General Relativity (GR) plus cold and weakly interacting Dark Matter at extragalactic scales, but that would produce a MONDian behavior for baryons through non-trivial dynamics of the dark sector at galactic scales. However, in view of the above comparison, it should be clear that such a theory would only be a first rough characterization of the desired phenomenology. It is nevertheless already sufficient in order to discuss some non trivial and model-independent facts. We indeed expect the generalized Dark Matter picture to involve Milgrom's acceleration a_0 as a parameter, be it a fundamental or effective constant, and also a mass scale, of the order of the TeV, to describe CDM in the appropriate regime. These two basic parameters of the theory may however not be enough to disentangle between galaxy clusters and (spiral) galaxies. As a matter of fact, the internal accelerations within rich clusters and within spirals are almost the same, so that without any other order parameter in the theory, we will have to consider that the same physics applies for both these types objects, in contradiction with the desired phenomenology.

The point is that, in the unified picture we are looking for, there is no reason why a_0 should play both the role of a characteristic acceleration at which dynamics departs from Newton, and the role of an order parameter that separates the MONDian regime from the CDM one. In the case where a_0 plays this double role, i.e. when the unified picture only involves a_0 and the mass m as parameters, the theory generically implements a transition from MOND to Newtonian (or GR) dynamics plus Dark Matter, instead of the usual transition from MOND to Newtonian dynamics without Dark Matter. For instance, it has been suggested that DM could carry a space-like four-vector in analogy with dipoles:[17] in this formalism, the Lagrangian in the matter action for DM has an ordinary mass term for a pressureless perfect fluid, but also has terms depending on a vector field (representing the dipole four-vector moment carried by the DM fluid) and on its covariant time derivative. This implies that the contribution of the DM fluid to the energy density involves a monopolar and a dipolar term: the monopolar term can then play the role of CDM in the early Universe, and the dipolar term takes over in galaxies and creates the MOND phenomenology through what can be called "gravitational polarization". The success of this model relies on a "weak-clustering" hypothesis, namely that, in galaxies, the DM fluid does not cluster much and is at rest because the internal force of the fluid precisely balances the gravitational force, in such a way that the polarization field is precisely aligned with the gravitational one in order to create the MOND effect.

More generally, in the absence of a weak clustering hypothesis, these kinds of models where a_0 plays a double-role generically predict the presence of Dark Matter (Cold if the mass m is large) within every collapsed region where internal accelerations are high with respect to a_0. This is the case for central regions of both HSB and rich galaxy clusters. As it is however well-known, there is no need for (and thus not much room for) Dark Matter in the central regions of HSB (until radii of order several kpc). Hence such a mini-halo of Dark Matter predicted by this kind of theory should always be, for some reason, of negligible mass (weak clustering hypothesis). On the contrary, we have seen that a large amount of Dark Matter is still needed in MOND at cluster scales. This missing mass could then perfectly well originate from the corresponding mini-halo in the central parts of galaxy clusters (where internal accelerations are several times a_0). It is however generally unclear why the mini-halo should have a negligible mass within galaxies but a large one within clusters. Whether this is indeed the case or not can only be checked through simulations of structure formation in the unified ΛCDM - MOND models.

There are then two possible ways out. One is to consider that the mass of the Dark Matter field is actually small, hence being a Hot Dark Matter (HDM) candidate. It is then natural that the HDM halo at the center of HSB weighs negligibly, whereas it can cluster significantly at larger scales. Cosmology of course becomes GR+HDM, which is known to have some problems, although some of them may be avoided by MONDian effects at late times (enhanced gravity during the formation of structures).

The other way around rests on the initial program, i.e. designing a framework in which clusters must be CDM dominated (no MONDian effects at all), whereas spirals are MONDian. This way, the mass discrepancy within clusters is explained as in the CDM paradigm, and the weak clustering property is only needed for HSB. Such theories, as stressed above, require the introduction of new scales in order to separate clusters from the central parts of HSB (as the internal acceleration cannot play this role). The size of these collapsed systems may be a relevant quantity, as they differ by orders of magnitude. However, apart from using a mass term (already discussed hereabove), the size of collapsed objects is a non-local quantity, which is difficult to obtain as a combination of local variables (fields) within a local theory. Their total mass may also be a relevant parameter. This is already more natural than size, as there is at least one example of a theory (which, however, is unstable and thus ill-defined) in which MONDian or Newtonian behavior do not only arise at some critical acceleration, but also for some critical mass (Phase Coupling Gravitation[18]). Another possibility, in view of our comparison of the respective merits of MOND and CDM in various systems in Section 2, would be to simply distinguish MONDian rotationally supported systems from CDM pressure-supported systems, but this distinction is also impossible to express in a local field-theory. So, in the end, the simplest and most natural candidate to separate MONDian systems from CDM ones seems to be the local baryonic medium density (or gas density): the typical medium density in galactic disks is $\rho \sim 5\times10^{-22}\text{kg/m}^3$, whereas the typical density in galaxy clusters is $\rho \sim 10^{-25}\text{kg/m}^3$. Hence we may introduce a new parameter of density, ρ_0, of order 10^{-24}kg/m^3, i.e. of the order the MONDian density a_0^2/Gc^2 up to a pure number, in order to separate galaxy clusters from spiral galaxies. A theory with such a new order parameter would reduce to GR+CDM everywhere except in low acceleration *and* high medium density environments (low density in the sense of the trace of the stress-energy tensor for baryons $T_{\rm mat}$). With such a new order parameter, HSB, LSB and TDG would be MONDian, but fluffy globular clusters in the outskirts of galaxies may be either MONDian or Newtonian depending on their gas density, while gas poor ellipticals and dwarf spheroidal satellites of the Milky Way would also behave according to CDM rather than to MOND. Let us also note that, owing to the hierarchical scenario, clusters collapse after galaxies, and therefore in a very low density environment. This ensures that they are indeed CDM dominated through their whole history.

We defined[19] a whole class of models that are able to implement the basic ideas we have discussed so far. The models all consist of standard GR including a cosmological constant, minimally coupled standard matter fields ψ, a simple massive scalar field χ acting as Dark Matter, and a new term in the action describing the interaction between standard and dark matter fields. Hence the action reads

$$S = S_{\rm EH}[g_{\mu\nu}] + S_{\rm Mat}[\psi, g_{\mu\nu}] + S_{\rm DM}[\chi, g_{\mu\nu}] + S_{\rm Int}[\chi, \psi, g_{\mu\nu}]. \qquad (1)$$

The interaction term is defined as

$$S_{\rm Int}[\chi, \psi, g_{\mu\nu}] = \frac{1}{2c}\int \sqrt{-g} d^4x \, T^{\mu\nu}_{\rm Mat} h_{\mu\nu}, \qquad (2)$$

where $h_{\mu\nu}(\chi, \psi, g)$ is some rank-two symmetric tensor that depends on χ and its derivatives.[19] It is then clear that the model reduces exactly to ΛCDM in vacuum (for baryons), and approximately when the interaction is negligible. The tensor h can moreover be defined in such a way that, close to any distribution of baryonic matter the Dark Matter fields couples to standard matter and gives it a MONDian dynamics, whenever its acceleration is small with respect to a_0. These models thus describe the phenomenology we were discussing above, where no additional order parameters are present besides a_0 and m. Varying m then describe either a "MONDian HDM" or a "MONDian CDM". Introducing the density as a new parameter is straightforward. It suffices to multiply the above tensor h by some function of the density, $F(|T_{\rm Mat}|/\rho_0)$, with the relevant asymptotic behavior: F should vanish for small density but saturate to 1 at large density.

This class of simple models should be viewed only as a very first step, as they suffer from naturalness problems, and are still far from explaining the physical origin of a_0 or any possible link with dark energy. Many things should still be investigated within such a class of models: (i) the high medium density in the disks of spiral galaxies should produce a MONDian metric, but in the halo, nothing a priori prevents the dynamics to behave according to CDM: the solution to this problem should lie within the boundary conditions in a static situation, and within numerical simulations of galaxy formation; (ii) in the high-acceleration regime close to the center of HSB, nothing a priori prevents the formation of a mini-halo of dark matter discussed above: solutions to this problem may rest on numerical simulations of galaxy formation or by making the mass of DM also vary as a function of the medium density; (iii) when deriving the equations of motions for a simple toy-model in which the baryons are represented by a standard scalar field, one notices some mixing between the baryon and DM fields: it will be of high interest to investigate the consequences of such mixings on a possible creation of baryonic matter from DM (and conversely) when the interaction term becomes non-negligible in the process of galaxy formation.

Finally, let us note that the fact that the medium density (or any other parameter) is indeed a relevant parameter controlling the transition from MOND to CDM in low acceleration regimes is in principle a testable hypothesis: falsifying it would require to find a low acceleration and high medium density stellar system not strictly obeying Milgrom's law.

References

1. S. Mashchenko et al., 2008, Science **319** 174
2. S. McGaugh, 2005, ApJ **632** 859
3. M. Milgrom, 1983, ApJ **270** 365
4. R.H. Sanders, S. McGaugh, 2002, ARA&A **40** 263
5. J.-P. Bruneton, G. Esposito-Farèse, 2007, PRD **76** 124012
6. F. Bournaud, et al., 2007, Science **316** 1166
7. G. Gentile, et al., 2007, A&A **472** L25
8. G. Angus, 2008, MNRAS **387** 1481
9. M. Metz, P. Kroupa, 2007, MNRAS **376** 387
10. O. Tiret, et al., 2007, A&A **476** L1
11. T. Richtler, et al., 2008, A&A **478** L23
12. K. Jordi, et al., 2009, arXiv:0903.4448
13. J. Bekenstein, 2004, PRD **70** 083509
14. H.S. Zhao, B. Li, 2008, arXiv:0804.1588
15. C. Skordis, et al., 2006, PRL **96** 011301
16. G. Angus, 2009, MNRAS **394** 527
17. L. Blanchet, A. Le Tiec, 2009, arXiv:0901.3114
18. R.H. Sanders, 1988, MNRAS **235** 105
19. J.-P. Bruneton, et al., 2009, JCAP **03** 021

ANNIHILATION IN THE MILKY WAY: SIMULATED DISTRIBUTION OF DARK MATTER ON SMALL SCALES

F. S. STOEHR

ST-ECF, ESO
Garching, 85748, Germany
E-mail: fstoehr@eso.org
www.eso.org

We use a high-resolution simulation of a galaxy-sized dark matter halo, published simulated data as well as four cluster-sized haloes[1] to study the inner halo structure in a Lambda cold dark matter cosmology. We find that the circular velocity curves are substantially better described by SWTS[2,3] profiles than by NFW[4] or Moore[5] profiles. Our findings confirm that no asymptotic slope is reached and that the corresponding extrapolated density profiles reach a finite maximum density. We analyse the impact of our findings on the detectability of the gamma-ray signal from the central regions of the Milky Way (MW) and from dark matter substructures in its halo, if the dark matter in the Universe is made of weakly self-interacting particles which self-annihilate. We discuss detection strategies for current gamma-ray detectors. We in addition review recent work by other authors and comment on possible boosts due to effects of structure on very small scales.

Keywords: Galaxies; Dark Matter; Simulations; gamma-rays; annihilation.

1. Introduction

Especially the past decade has witnessed overwhelming progress in the understanding of the distribution of matter in the Universe and the arrival of the era of precision cosmology has been claimed by many authors. A combination of observations of the cosmic microwave background radiation, deep galaxy surveys and Super-Novae (SN) distance-measurements indicate that the Universe is flat with 4% of baryonic content, 26% of non-baryonic cold dark matter (DM) and 72% of dark energy. The nature of the latter two components remains yet unknown. Still, this ΛCDM model has been remarkably successfull in describing the large scale structure of the Universe (e.g. Ref. 6). As far as could be tested, the model also provides quite good descriptions of the DM distribution on galactic scales but if it holds down to the smallest scales (a kpc or less) is still controversial and under heavy investigation.

Knowing the distribution of the DM within the halo of the MW is essential, however, for e.g. the estimate of possible gamma-ray flux from self-annihilation radiation which might be expected if the dark matter particle is a Weakly Interacting Massive Particle (WIMP). Such a particle arises naturally in a number of supersymmetric extensions of the standard model of particle physics.

2. Halo Profiles

We ran a high-resolution simulation of a galaxy-sized ΛCDM halo with GADGET-2[7] and termed it GA3new (for details see Ref. 8). There are 11 562 566 particles within r_{200} and the simulation was until recently the best-resolved simulation of its kind. Fig. 1 shows the circular velocity curve of GA3new (thick solid, red) together with the two most-used analytical fitting functions, the NFW[4] profile (dotted) and the Moore[5] profile (dashed) as a function of radius from the centre of the DM halo. The analytical

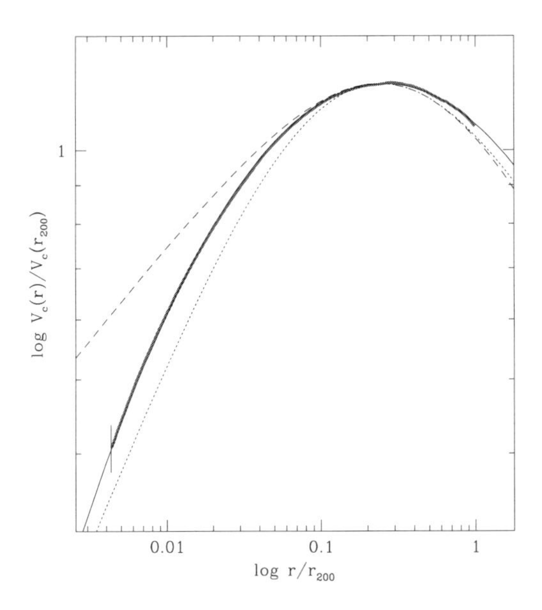

Fig. 1. Circular velocity curve of GA3new from the converged radius r_{conv} to the virial raius r_{200} with SWTS (thin solid), NFW (dotted) and Moore et al. (dashed) fits.

profiles were (arbitrarily) required to pass through the maximum of the circular velocity curve which fixes their free parameters. They are double power-laws and have asymptotic inner slopes, i.e. the slopes of the corresponding density profiles rise like r^{-1} and $r^{-1.5}$ in the NFW and Moore case, respectively.

The actual data however does not show such behaviour. It indeed curves all the way from the virial radius down to r_{conv} the radius to which the profiles have converged. The simplest analytical profile with this behaviour, a parabola that curves at a constant rate in a log-log plot, was proposed by us earlier[2] for subhalo profiles. This SWTS profile is shown in Fig. 1 as thing solid line and does fit the GA3new profile extremely well. Fits of DM simulations carried out by other authors can be found in Ref. 8.

It is important to realise that especially for the NFW profile no simulation had been reached into the region where the slope of the density profile would become constant. The fitting only ever was done in the region where the analytic NFW profile changes *continuously* from the inner slope of -1 to the outer slope of -3. This can be seen clearly in Fig. 2 that shows the slope of the density profile of GA3new as a function of radius. The vertical lines indicate the converged radius and the virial radius of the

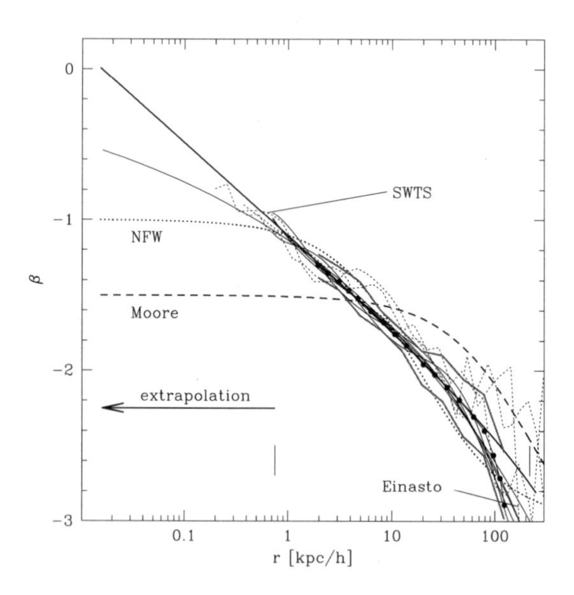

Fig. 2. Simulated density profile slopes β compared to those of the SWTS, Einasto, NFW and Moore et al. analytical fitting functions.

simulation. The thick short solid lines show the bounds of the slop measurements from Ref. 9, the thin dotted lines are two recent very high-resolution simulations[10,11] scaled to GA3new.

The thin solid line is a profile that was found by Einasto[12] and lately was used to fit simulated DM haloes. The SWTS profile and the Einasto profile, which have both one free parameter more than the NFW or Moore profile, are nearly indistinguishable within the converged region of the simulations.

3. Annihilation

One of the most likely candidates for the DM particle is the WIMP, for example a neutralino which is the lightest stable partice of supersymmetric extensions of the standard model. If it can self-annihilate, then the rate of produced gamma-rays is given by

$$\Gamma = \frac{\sigma v}{m_{DM}^2} \int_V \rho_{DM}^2(r)\, dV = \frac{\sigma v}{m_{DM}^2} \int_V \rho_{DM}^2(r)\, r^2\, dr \qquad (1)$$

As can be seen from the spherically symmetric version (middle expression), for density profiles with slopes of $r^{-1.5}$ or steeper, the annihilation flux would formally diverge. As we have shown, even when taking the uncertainties into account, at least when baryonic physics is neglected, such steep profiles seem to be ruled out. In contrast to our findings,[3] it seems that simulations with ever greater particle numbers reveal that the cumulated flux of the substructure haloes does outshine the contribution of the main halo for a distant observer.[11] However, although this is the case, the centre of the DM halo, i.e. the Galactic Centre (GC), is still the brightest source on the sky: At the sun's position, tidal forces are strong enough to disrupt substructure haloes efficiently and no subhalo is expected to have survived very nearby that could outshine the galactic centre.[3,13] The unresolved smooth flux within the MW from substructure, as seen from an observer close to the GC, appears to be nearly isotropic and indiscernible from the extragalactic gamma-ray background.[13] The GC itself, is source of a not-well determined gamma-ray background (produced by cosmic-rays hitting the gas in the bulge and astrophysical sources) so that observing the main galactic halo excluding the bulge and disk probably yields the highest detectability. For a good number of supersymmetric models FERMI might detect that component.[3,14]

It can be seen from Eq. 1 that *any* deviation from a completely smooth DM distribution will increase the annihilation flux. Such deviations could very well be present but not simulated, as the mass resolution of today's simulations is still about 10 to 20 orders of magnitude above the theoretical clumping limit (cut-off mass) of WIMPS which is expected to be between 10^{-6} and 10^{-12} $M_{\odot}$. We mention some of the possible small-scale effects:

Although *caustics* can arise in the DM distribution, several authors have shown that the increase in annihilation flux is marginal.[15,16] As mentioned before, *substructure in the MW halo* are unlikely to be detected although they would in principle be an ideal testbed as they are supposed to be DM dominated.[3,11,17,18] The same holds true for the *smooth unresolved contribution of substructure haloes.*[11,17,18] Substructure in *extragalactic sources* certainly would produce a strong gamma-ray flux but the distance to even the closest object, M31, is too large to outshine the GC.[11,17,18] Most authors agree that when gasphysics is included into the study, then probably the DM density at the GC would be larger than that estimated from DM only simulations (*adiabatic compression*) (e.g. Ref. 19 and references therein). No quantitative consens has been yet reached, but it is clear that the gamma-ray background from astrophysical sources at the galactic centre is difficult to account for. Finally, the *Black Hole* at the GC could increase the DM density in its surroundings (e.g. Ref. 20 and references therein). Again it is unclear what annihilation flux could be expected and the background problem is the same as for the possible enhancement due to adiabatic compression.

4. Conclusions

We have re-investigated the velocity and density profiles of DM haloes using high-resolution simulations and have discussed possible indirect DM detection though gamma-rays. We find that DM halo profiles do not show asymptotic slopes and that therefore double-powerlaw analytical fits like NFW or Moore profiles are not the best possible descriptions. Our proposed SWTS profile or the Einasto profile provide much better fits in the converged regions of the simulations. As they differ from each other outside of the converged regions of the simulations, however, extrapolations have to be done with great care.

The brightest source for gamma-rays from possible self-annihilation for an observer at the position of the sun is the galactic centre but a high gamma-ray background make it likely, that the best detectable source is the main galactic halo when bulge and disk are excluded.

Large uncertainties in the detectability estimation remain, of course beginning with the possibility that DM is not made of WIMPS at all or that they do not self-annihilate, but also including the uncertain gamma-ray background estimation and the limited knowledge on the exact distribution of the DM on very small scales. Still, with a good measure of luck FERMI might detect DM annihilation in the very near future.

References

1. T. Fukushige, A. Kawai and J. Makino, *ApJ* **606**, 625(May 2004).
2. F. Stoehr, S. D. M. White, G. Tormen and V. Springel, *MNRAS* **335**, L84(Octtober 2002).
3. F. Stoehr, S. D. M. White, V. Springel, G. Tormen and N. Yoshida, *MNRAS* **345**, 1313(November 2003).
4. J. Navarro, C. Frenk and S. White, *ApJ* **490**, 493(December 1997).
5. B. Moore, F. Governato, T. Quinn, J. Stadel and G. Lake, *ApJL* **499**, L5+(May 1998).
6. V. Springel, C. S. Frenk and S. D. M. White, *Nature* **440**, 1137(April 2006).
7. V. Springel, *MNRAS* **364**, 1105(December 2005).
8. F. Stoehr, *MNRAS* **365**, 147(January 2006).
9. D. Reed, F. Governato, L. Verde, J. Gardner, T. Quinn, J. Stadel, D. Merritt and G. Lake, *MNRAS* **357**, 82(February 2005).
10. J. Stadel, D. Potter, B. Moore, J. Diemand, P. Madau, M. Zemp, M. Kuhlen and V. Quilis, *ArXiv e-prints* (August 2008).
11. V. Springel, S. D. M. White, C. S. Frenk, J. F. Navarro, A. Jenkins, M. Vogelsberger, J. Wang, A. Ludlow and A. Helmi, *Nature* **456**, 73(November 2008).
12. J. Einasto, *Trudy Inst. Astrofiz. Alma-Ata* **5**, p. 87 (1965).
13. V. Springel, J. Wang, M. Vogelsberger, A. Ludlow, A. Jenkins, A. Helmi, J. Navarro, C. Frenk and S. White, *MNRAS* **391**, 1685(December 2008).
14. E. A. Baltz, B. Berenji, G. Bertone, L. Bergström, E. Bloom, T. Bringmann, J. Chiang, J. Cohen-Tanugi, J. Conrad, Y. Edmonds, J. Edsjö, G. Godfrey, R. E. Hughes, R. P. Johnson, A. Lionetto, A. A. Moiseev, A. Morselli, I. V. Moskalenko, E. Nuss, J. F. Ormes, R. Rando, A. J. Sander, A. Sellerholm, P. D. Smith, A. W. Strong, L. Wai, P. Wang and B. L. Winer, *Journal of Cosmology and Astro-Particle Physics* **7**, 13(July 2008).
15. C. J. Hogan, *Phys. Rev. D* **64**, 063515(September 2001).
16. M. Vogelsberger, S. D. M. White, A. Helmi and V. Springel, *MNRAS* **385**, 236(March 2008).
17. J. Diemand, M. Kuhlen and P. Madau, *ApJ* **657**, 262(March 2007).
18. M. Kuhlen, J. Diemand, P. Madau and M. Zemp, *Journal of Physics Conference Series* **125**, 012008(July 2008).
19. D. H. Weinberg, S. Colombi, R. Davé and N. Katz, *ApJ* **678**, 6(May 2008).
20. O. Y. Gnedin and J. R. Primack, *Physical Review Letters* **93**, 061302(August 2004).

PRIMORDIAL MOLECULES AND FIRST STRUCTURES

D. PUY*

University of Sciences Montpellier II, Astrophysics Group,
GRAAL/CNRS UMR 5024, CC72, Montpellier 34090, France
**E-mail: Denis.Puy@graal.univ-montp2.fr*

The primordial molecules, which appear during the phase of cosmological recombination, play an important role in the mechanisms of formation of the first gravitational structures. Their thermal influence on the gravitational dynamics of collapse can generate, under certain conditions, an instability which leads to the fragmentation of the initial collapsing structure. In this framework it is crucial to establish the initial conditions of the mechanism of gravitational collapse, in particular the abundances of molecules susceptible to have a thermal influence as the primordial molecules.

In a Universe made of baryonic and non-baryonic dark matter, we need to discuss how each component couples with the other, and how this coupling may modify the primordial chemistry during the gravitational growth of structures. Since dark matter is assumed to be affected only by gravity and is collisionless, there is no effective pressure term in its equation of evolution. The linearized continuity equation in Fourier modes describes the dark matter and baryon fluids by two second-order differential equations which couple the baryon chemistry and gas dynamics to dark-matter by gravity.

In this contribution besides to remind the chemical processes in the early Universe, we shall analyze the influence that these last ones can have on the formation of the first baryonic structures gravitational. Thus, we present calculations in the linear approximation of density fluctuations, but in the full non-linear regime of chemical abundances about the primordial molecule formation in a uniform medium perturbed by small density inhomogeneities at various spatial wavelengths. We analyze the differential abundances of the primordial molecules H_2, HD and LiH. As the Universe expands, the baryonic fluctuations increase and induce strong contrasts on the primordial molecular abundances.

The main result is that the chemical abundances at the transition between the linear and non-linear regimes of density fluctuations (such as in proto-collapsing structures) are already very inhomogeneous and scale dependent. These results indicate that pronounced inhomogeneous chemical abundances are present already before and during the dark age. This must have a direct consequence on the mass spectrum of the first bound objects since gas cooling depends then mainly on the particular abundances of H_2 and HD.

1. Introduction

The question of the formation of baryonic structures of the Universe, which exists in very big variety (see Fukugita & Peebles[1]), remains a question still opened this day. The formation of these gravitational structures depends essentially on the quantity of dark matter (non baryonic component), which is going to influence the Jeans mechanism of gravitational instability, and the nature of the baryons. Baryons can indeed have a thermal influence on the dynamic processes. Molecules perform perfectly this role because they can, according to the dynamic situation where they evolve, to produce a thermal cooling.

The standard Big Bang nucleo-synthesis (SBBN) model for the Universe predicts the nuclei abundances of mainly hydrogen, helium and lithium and their isotopes. Similarly the subsequent chemistry of these light elements and their respective isotopic forms may be called the standard Big Bang chemistry (SBBC). Although the matter dilution acts against molecule formation, the temperature drops is sufficiently fast for allowing the formation of simple molecules, see Lepp & Shull,[3] Puy et al.,[4] Galli & Palla.[5] All these authors clearly pointed out that trace amounts of molecules (the primordial molecules) such as H_2 and HD were formed during the post-recombination period of the Universe. The ongoing physical reactions are numerous after the nuclei recombination, partly due to the cosmic microwave background radiation (CMBR), see Stancil & Dalgarno[6] and Lepp et al.[7]

In this paper we have investigated in the linear regime of density fluctuations the full chemistry in Fourier modes characterized by their wavenumbers. This approach allows us to quantify the differential growth of molecular abundances and provides the initial conditions of perturbation in the non-linear density fluctuation regime. This last point is crucial to reach a better understanding of the conditions prevailing before the period where the first stars formed.

The plan of this contribution is as follow: in Sect. 2 we analyze the evolution of matter fluctuations, in Sect. 3 we perform an analysis of the chemical evolution in pure Fourier modes, still possible and convenient in the linear regime. Finally, in Sect. 4, we discuss some possible implications of this study.

2. Primordial Chemistry and Evolution of Matter Fluctuations

In the hot big-bang picture, after the nucleosynthesis period, it is possible to form atoms by recombination of a primordial nucleus with free electrons, and depends on the degree to which the recombination is inhibited by the presence of cosmic background radiation. The cosmological recombination process is not instantaneous, because the electrons, captured into different atomic levels, could not cascade down to the ground. Atoms reached the ground state either through the cosmological redshifting of the Lyman-α line protons or by the $2s-1s$ two-photon process. Nevertheless the Universe expanded and cooled faster the recombination could be completed, and a small fraction of free electrons and protons remained.

Once neutral He is signicantly abundant, this atom is the first neutral atom which appeared in the Universe), charged transfer with ions is possible, allowing for the formation of other neutral species. Any H_2 formed in the uniform background is dissociated by the radiation, until the density is too low to produce it. Charge transfer from H_2^+ is the most likely alternative to form H_2, as the collisions reactions with H. HD forms in ways similar to H_2, but the formation is very significant when H_2 appears and the mechanism if dissociative collision with H_2 is very efficient.

Discussion of how irregularities grow is greatly simplified by the fact that a limiting approximation of general relativity, Newtonian mechanics, applies in a region small compared to the Hubble radius. Here we analyze the evolution of the molecular abundances in the linear regime of the density fluctuations including baryons and dark matter.

The mathematical description of a lumpy Universe revolves around the dimensionless density perturbation, $\delta(\vec{r})$, which is obtained from the spatially varying baryons $\rho_{\rm b}(\vec{r})$ and dark matter $\rho_{\rm d}(\vec{r})$ mass density, where $\vec{r}$ is the comoving coordinate. The index b is relative to baryons, and d to dark matter.

In a Universe made of non-baryonic dark matter and baryons, we need to discuss both components together, since each component is coupled to the other by gravity. Since cold dark matter is affected only by gravity and is presumably collisionless, in the fluid limit we can ignore the velocity dispersion of the cold dark matter particles. There is no effective pressure term in the equation of evolution for dark matter. Moreover we can neglect in the linear approximation the influence of a peculiar velocity, which introduce second order transport terms. The system of second-order differential equations for the perturbations are coupled with the chemical and baryons evolution equations of temperature and of density.

We start our simulation well before the redshift of hydrogen recombination $z_{\rm s} = z_{\rm rec,H} = 3000$. We pull the initial abundances out of the standard Big Bang chemistry, given by Puy & Pfenniger[8] where we consider a cold

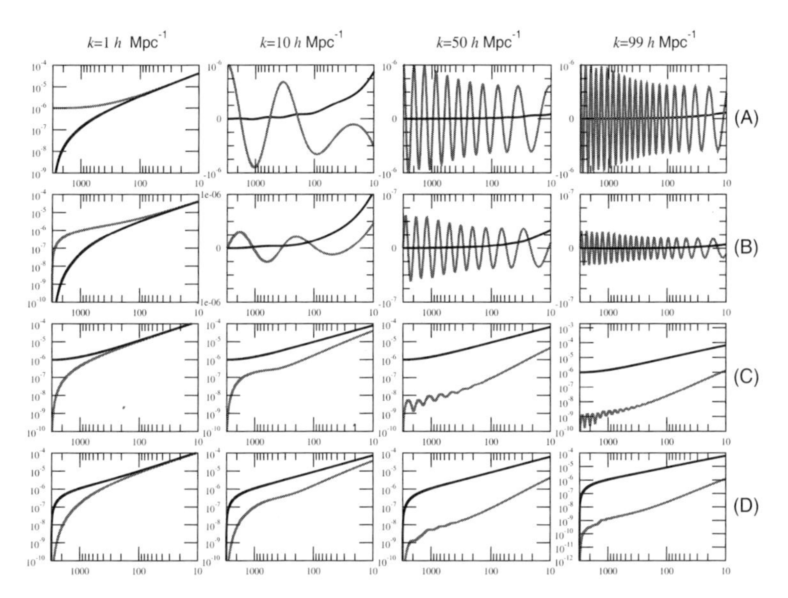

Fig. 1. Evolution of the pure Fourier modes of the density fluctuations, as a function of redshift z for different comoving wave-numbers and elementary initial conditions. The black lines are the dark matter density fluctuations, the red lines are the baryons density fluctuations. Each column defines a particular scale ($k = 1$, $k = 10$, $k = 50$ and $k = 99$ h Mpc^{-1}), each row is relative to a particular initial condition such as (A): $\delta_{\mathrm{b,s}} = 10^{-6}/\sqrt{k}$ and $\delta_{\mathrm{d,s}} = \dot{\delta}_{\mathrm{b,s}} = \dot{\delta}_{\mathrm{d,s}} = 0$, (B): $\dot{\delta}_{\mathrm{b,s}} = 10^{-6}H/\sqrt{k}$ and $\delta_{\mathrm{b,s}} = \delta_{\mathrm{d,s}} = \dot{\delta}_{\mathrm{d,s}} = 0$, (C): $\delta_{\mathrm{d,s}} = 10^{-6}H/\sqrt{k}$ and $\delta_{\mathrm{b,s}} = \dot{\delta}_{\mathrm{b,s}} = \dot{\delta}_{\mathrm{d,s}} = 0$, (D): $\dot{\delta}_{\mathrm{d,s}} = 10^{-6}H/\sqrt{k}$ and $\delta_{\mathrm{b,s}} = \dot{\delta}_{\mathrm{b,s}} = \dot{\delta}_{\mathrm{b,s}} = 0$, curves from Puy & Pfenniger.[8]

dark matter power spectrum. Fig. (1) shows the evolution of density fluctuations (matter and dark matter) for different comoving wavenumber k and different initial conditions.

The solutions for the baryons change from oscillatory to growing modes if k is larger or smaller than the critical Jeans wavenumber k_{J}. Fig. (1) shows the evolution of density fluctuations (matter and dark matter) for different comoving wavenumber k and different initial conditions.

From $k = 1\,h$ Mpc^{-1} to $k = 5\,h$ Mpc^{-1}, i.e., from large to medium scales, the evolution of baryons fluctuation are strongly coupled with the dark matter fluctuations. For scales $k > 10\,h$ Mpc^{-1} the oscillations of baryonic fluctuations are important and are decoupled from the dark matter evolution. At the smallest scales, $k = 99\,h$ Mpc^{-1}, as expected the baryons fluctuate fast, since the pressure term is proportional to k^2. This pressure effect is, of course, absent for collisionless dark matter.

3. Molecular Fluctuations

We solve and analyze the full chemical equations for different Fourier modes and take the difference of the solutions for the abundances of H_2 and HD with the ones of the unperturbed SBBC case. Doing this way we preserve the full non-linearities of the chemical network in the linear density perturbations.

To compare the chemical fluctuations with eventual future observations it is useful to calculate typical differential abundances in physical space. For the sake of simplicity and display convenience, we consider here only a 2D map. A 3D generalization of the following experiment is straightforward, but would be much more computer intensive without bringing much different results.

By Fourier transforming the integration results between $z_{\rm s} = 3000$ and $z_{\rm f} = 10$ we obtain the density fluctuations $\delta_{\rm b}(\vec{x})$, $\delta_{\rm d}(\vec{x})$ as functions of z, as well as the abundance fluctuations $\delta_{\xi}(\vec{x})$ for each species ξ. The series of such 2D maps of density or molecular abundance fluctuations can then be combined to make films of the abundance fluctuation runs. In Puy & Pfenniger[8] we presented different frames of the H_2 fluctuation films in order to illustrate the evolution of molecular abundances in real space (box of length $l = 2\pi\, h^{-1} \sim 8.85$ Mpc) at the successive redshifts $z = 3000$, $z = 1425$, $z = 600$ and $z = 10$. At the initial redshift $z_{\rm s}$ the H_2 distribution is relatively homogeneous. However as the Universe expands, baryonic fluctuations increase and induce strong contrasts of H_2. A very lumpy distribution of H_2 and HD abundances results at $z_{\rm f} = 10$. We analysed some frames in the evolution of the HD abundance in real space (box of length $l = 2\pi\, h^{-1} \sim 8.85$ Mpc) at the successive redshifts $z = 3000$, $z = 1425$, $z = 100$ and $z = 10$. At the initial redshift $z_{\rm s}$, the HD distribution is less homogeneous than the H_2 distribution. Apparently the HD chemistry is more sensitive to the differential growth of the baryonic fluctuations. The contrast becomes more important and leads, at $z_{\rm f} = 10$, to a more lumpy distribution than the H_2 distribution. We notice that some regions reveal HD fluctuations close to the value 15, see Puy & Pfenniger.[8]

4. Discussions

The results indicate that a lumpy distribution of molecules is already present during the dark age period of the Universe (roughly $10 < z < 3000$). The clear conclusion is that well before the first stars form the molecular Universe is already strongly inhomogeneous. Moreover the peak of molecular contrasts appears precisely in the range of redshift $z = 500 - 100$. These redshifts are the typical turn-around redshifts between the linear and non-linear regime of small-scale structures leading to stars. Thus it is crucial to have a precise view of the chemistry at this period, which determines the fragmentation conditions and the properties and mass spectrum of the first collapsing objects.

The picture of a molecular lumpy Universe suggests that in the non-linear regime of perturbations, strong molecular abundances contrasts could be very important. Pfenniger & Combes[9] pointed out the possible fractal distribution of cold molecular gas in the interstellar medium, a similar distribution could be present already during the dark age. Thus our results should be useful for numerous applications.

Acknowledgments

We acknowledge the french program PNCG (Programme National de Cosmologie et Galaxies) for their financial assistance.

References

1. M. Fukugita and P.J.E Peebles *ApJ* **616**, 643, 2004
2. M. Signore and D. Puy *Eur. Phys. J. C* **59**, 117, 2009
3. S. Lepp and M. Shull *ApJ* **280**, 465, 1984
4. D. Puy et al. *AA* **267**, 337, 1993
5. D. Galli and F. Palla *AA* **335**, 403, 1998
6. P. Stancil et al. *ApJ* **509**, 1, 1998
7. S. Lepp et al. *J. Phys. B* **35**, R57, 2002
8. D. Puy D. and D. Pfenniger submitted to *AA* 2009
9. Pfenniger, D., & Combes, F. *AA* **285**, 94, 1994

DARK COMPANION OF BARYONIC MATTER IN SPIRAL GALAXIES

Y. Sobouti

Department of Physics, Institute for Advanced Studies in Basic Sciences-Zanjan
P. O. Box 45195-1159, Zanjan 45195, Iran
**E-mail: sobouti@iasbs.ac.ir*
www.iasbs.ac.ir

Flat or almost flat rotation curves of spiral galaxies can be explained by logarithmic gravitational potentials. The field equations of GR admit of spacetime metrics with such behaviors. The scenario can be interpreted either as an alternative theory of gravitation or, equivalently, as a dark matter paradigm. In the latter interpretation, one is led to assign a dark companion to the baryonic matter who's size and distribution is determined by the mass of the baryons. The formalism also opens up a way to support Milgrom's idea that the acceleration of a test object in a gravitational field is not simply the newtonian gravitational force g_N, but rather an involved function of (g_N/a_0), a_0 MOND's universal acceleration.

Keywords: Dark matter; alternative GR; spiral galaxies; rotation curves.

1. Introduction

The goal of the paper is to understand the idiosyncrasy of the rotation curves of spiral galaxies. The newtonian or the GR gravitation of the observable matter is not sufficient to explain the large asymptotic speeds of test objects in orbits around the galaxies, nor their slow decline with increasing distances. In search of the missing gravity, alternative theories of gravitation and/or of dark matter are proposed. In a recent work[1] we pointed out that no one has reported a case where there is no baryonic matter, but there is a dynamical issue to be settled. We argued that if the dark matter reveals itself only in the presence of the baryonic one, it is logical to assume that the two are twin companions. On the other hand, both dark matter scenarists and (at least some) alternative theorists explain the rotation curves of spirals equally satisfactorily. We argue, if two people give correct answers to the same question, they ought to be saying the same thing, albeit in different languages. And since in an alternative theory one gives a definite rule for the gravity field, there must be rules to govern the mutual companionship of the dark and baryonic matters.

We begin with a GR formalism and show that spacetime metrics with logarithmic behaviors are accommodated by Einstein's field equations and can adequately explain the anomalous features of the dynamics of the spirals. Conclusions are interpretable either in terms of an alternative theory of gravitation, or as a dark matter paradigm. With an advantage, however: the questions, how much dark matter accompanies a given baryonic mass, how it is distributed, and what is its equation of state, are also answered.

2. Model and Formalism

We are concerned with the outer reaches of spiral galaxies (a baryonic vacuum), where the rotation curves display non classical features. Their asymptotic speeds do not have a Keplerian decline and follow the Tully-Fisher relation. We approximate the galaxy by a spherically symmetric distribution of baryonic matter[a]. The spacetime around it will accordingly be spherically symmetric and static:

$$ds^2 = -B(r)dt^2 + A(r)dr^2 + r^2\left(d\theta^2 + \sin^2\theta d\varphi^2\right). \tag{1}$$

We adopt a dark matter language and assume that the galaxy possesses a static dark perfect gas companion of density $\rho_d(r)$, of pressure $p_d(r) << \rho_d(r)$, and of covariant 4-velocities $U_t = -B^{1/2}, U_i = 0,\ i = r, \theta, \varphi$. Einstein's field equations become.

$$R_{\mu\nu} - \frac{1}{2}g_{\mu\nu}R = -T_{\mu\nu} = -[p_d g_{\mu\nu} + (p_d + \rho_d)U_\mu U_\nu], \tag{2}$$

where we have let $8\pi G = c^2 = 1$. To respect the Bianchi identities and the conservation laws of the baryonic matter, one must have $T^{\mu\nu}{}_{;\nu} = 0$. The latter, in turn, leads to the hydrostatic equilibrium for the dark fluid, that is, if one wishes to attribute such notions to a hypothetical entity.

[a]At a distance r, the ratio of the quadrupole to the monopole gravitational field of a flattened galaxy is of the order of $(R_{gyr}/r)^2$, where R_{gyr} is the gyration radius of the galaxy about one of the principle axes of the ellipsoid of inertia of the system. In a galaxy with R_{gyr} of the order of few tenths of $R_{visible}$, and at a distance r of about few times $R_{visible}$, this ratio can easily fall below few parts in thousand.

From Eq. (2) the two combinations $R_{tt}/B + R_{rr}/A + 2R_{\theta\theta}/r^2$ and $R_{tt}/B + R_{rr}/A$ give

$$\frac{1}{r^2}\left[\frac{d}{dr}\left(\frac{r}{A}\right) - 1\right] = -\rho_d, \tag{3}$$

$$\frac{1}{rA}\left(\frac{B'}{B} + \frac{A'}{A}\right) = \rho_d + p_d, \tag{4}$$

respectively. Neglecting p_d in comparison with ρ_d and eliminating ρ_d between the two equations gives

$$\frac{B'}{B} = \frac{1}{r}(A-1). \tag{5}$$

We now assume that $A(r) - 1$ is a well behaved and differentiable function of r and has a series expansion in negative powers of r,

$$A - 1 = \sum_{n=0} \frac{s_n}{r^n}, \quad s_n \text{ constant}. \tag{6}$$

Substituting this expansion in Eq. (5) and integrating the resulting expression gives

$$B = \left(\frac{r}{r_0}\right)^{s_0} \exp\left(-\sum_{n=1} \frac{s_n}{nr^n}\right) \approx \left[1 + s_0 \ln\left(\frac{r}{r_0}\right) - \frac{s_1}{r} - \cdots\right]. \tag{7}$$

We note that s_0 is dimensionless and s_n, $n \geq 1$, has the dimension $(\text{length})^n$. The right hand side expression is the weak field approximation and holds for $s_n/r^n << 1$, $\forall\ n$.

With A and B known, the density ρ_d can be calculated from either of Eqs. (3) or (4). Here, however, we adopt a weak field point of view, $B(r) = 1 + 2\phi_{grav}/c^2$, and calculate ρ_d from Poisson's equation. Thus,

$$4\pi G\rho_d = \frac{1}{2}c^2\nabla^2(B-1) = \frac{c^2}{2r^2}\left[s_0 - \sum_{n=2}(n-1)\frac{s_n}{r^n}\right]. \tag{8}$$

Hereafter, we restore the physical dimensions $8\pi G$ and c^2 for clarity. The pressure of the companion fluid is obtained from $T^{\mu\nu}{}_{;\nu} = 0$,

$$\frac{p_d'}{p_d + \rho_d} \approx \frac{p_d'}{\rho_d} = -\frac{1}{2r}(A-1). \tag{9}$$

Integration is straight forward. The first two terms in the series are

$$p_d = \frac{c^2 s_0}{16\pi G r^2}\left[\frac{1}{2}s_0 + \frac{s_1}{3r}\right]. \tag{10}$$

Upon elimination of r between Eqs. (8) and (10) one obtains the equation of state, $p_d(\rho_d)$. It is barotropic. We conclude this section by writing down the dynamical acceleration of a test object circling the galaxy with the speed v

$$a_{\rm dyn} = \frac{v^2}{r} = \frac{1}{2}c^2 B' = \frac{1}{2}c^2\left[\frac{s_0}{r} + \frac{s_1}{r^2} + \cdots + \frac{s_n}{r^{n+1}} + \cdots\right]. \tag{11}$$

3. What are s_n's

The s_1 term in Eqs. (6)-(11) represents the classic gravitation of the baryonic matter with a force range of r^{-2}. Magnitude-wise, s_1, should be identified with the Schwarzschild radius of the spherical galaxy, $s_1 = 2GM/c^2$. The s_0-term is not a classical term. It has a force range r^{-1} and dominates all other terms at large distances. It is responsible for the large asymptotic speeds and their non Keplerian decline at far reaches of the spirals. In[2] and[1] we resorted to the Tully-Fisher relation (the proportionally of the asymptotic speed, $v_\infty = c(s_0/2)^{1/2}$, to the fourth root of the mass of the host galaxy) and arrived at

$$s_0 = \alpha\left(\frac{M}{M_\odot}\right)^{1/2}, \quad \alpha \text{ constant.} \tag{12}$$

In weak accelerations (less than certain 'universal acceleration' a_0), Milgrom's MOND[3] anticipates a force field $(a_0 g_N)^{1/2}$, instead of the newtonian gravitation, $g_N = GM/r^2$. The far distance limit of Eq. (11) with α given by Eq. (12) is of Milgrom's form. Comparing the two formalisms, one finds $\alpha = 2(a_0 GM_\odot)^{1/2}c^{-2}$. Either from this expression, with $a_0 \approx 1.2 \times 10^{-8}\text{cm/sec}^2$,[4] or from a direct statistical analysis of the asymptotic speeds of spirals[2] one finds

$$\alpha \approx 2.8 \times 10^{-12}, \text{ dimensionless 'universal constant'.} \tag{13}$$

The remaining s_n-terms, $n \geq 2$, in Eqs. (6)-(11) are also nonclassical. The range of their force is $r^{-(n+1)}$ (not to be confused with the multipole fields of extended objects). There is no compelling observational evidence for their existence in regions external to a spherical distribution of matter. Nevertheless, we retain them for a possible formal support they may give to Milgrom's MOND, to be elaborated below.

A conjecture: There is a surprise in Eq. (11). Upon elimination of r in favor of $g_N = GM/r^2$, one may write it as

$$\frac{a_{\rm dyn}}{a_0} = \left(\frac{g_N}{a_0}\right)^{1/2} + \left(\frac{g_N}{a_0}\right) + \cdots + \lambda_n\left(\frac{g_N}{a_0}\right)^{(n+1)/2} + \cdots. \tag{14}$$

where λ_n's can be expressed in terms of s_n's through a term-by-term comparison of Eqs. (11) and (14). One obtains

$$\lambda_n = \frac{c^2 s_n}{2a_0}\left(\frac{a_0}{GM}\right)^{(n+1)/2}, n = 2, 3, \cdots, \tag{15}$$

or

$$s_n = \frac{2a_0}{c^2}\lambda_n\left(\frac{GM}{a_0}\right)^{-(n+1)/2}. \tag{16}$$

All λ_n's are dimensionless. Apparently, Eq. (14) is an expansion of the dynamical acceleration in a power series of $(g_N/a_0)^{1/2}$. The coefficient of the first term is the 'universal constant' 1 because of the 'universal' Tully-Fisher relation. The coefficient of the second term is 1 because of the universal law of newtonian gravitation in the weak field regime. Now the conjecture: If there is any significance attached to the series expansion of Eq. (14) beyond the first two terms, is it possible that in the remaining terms

"All λ_n's are universal constants (not necessarily 1), and independent from the mass of the host baryonic matter centered at the origin"?

The proof or disproof of the conjecture should come from observations. We recall, however, Milgrom's stand that the dynamical acceleration of a test body is not simply proportional to g_N, but it is a involved function of g_N/a_0, and vice versa, g_N is a function of a_{dyn}/a_0. His suggestion for this function is through an, almost arbitrary, interpolating function. If the conjecture above holds, Eq. (14) can be considered as a series expansion of one such function, and a support for Milgrom's idea.

4. Concluding Remarks

That logarithmic potentials are natural solutions of Einstein's field equations is the highlight of the paper. They enable one to arrive at a law of gravitation alternative to that of Newton and/or to those known to GR. Equivalently, one may choose to attribute dark companions to baryonic matters. In the case of a spherically symmetric baryonic mass, the size and distribution of the density and pressure of the companion, outside the baryonic mass, are given by Eqs. (8) and (10).

The spacetime is a baryonic vacuum but not a dark matter one. The consequences are noteworthy. For example:

The spacetime is not flat. Contraction of Eq. (2) gives

$$R = -(3p_d + \rho_d) \approx -\frac{s_0}{r^2} + O\left(r^{-4}\right).$$

The 3-space is not flat. Direct calculation with $g^{(3)}_{ij}, i, j = r, \theta, \varphi$, yields

$$R^{(3)} = -\frac{2}{r^2}\frac{d}{dr}(r\rho_d) \approx -2\frac{s_0}{r^2} + O\left(r^{-4}\right).$$

There is an excess lensing.[7] Contribution from the s_0 term alone is

$$\delta\beta = \frac{1}{2}\pi s_0.$$

Due to the smallness of both s_0 and Sun's mass, effects in the scale of the solar system are immeasurably small.[1]

That the dynamical acceleration of a test object in the external gravitational field of a spherical mass could have a series expansion in (g_N/a_0), in accord with Milgrom's idea, is an intriguing idea. The support for it should come from observations.

A word of caution: The paper relies heavily on observations pertaining to spiral galaxies. Its conclusions may be scale dependent, not applicable to systems with scales larger than galactic scales. In a recent paper Bernal et al[6] analyze weak lensing data from clusters of galaxies on the basis of the metric field of[2] (similar to those of Eqs. 6 and 9). They conclude, in the notation of this paper, $s_0 \propto M^{1/4}$, instead of $M^{1/2}$ of Eq. (12). This finding while raises an alarm against extrapolation to larger systems, clusters of galaxies and beyond, at the same time opens the question that deviations from the newtonian or GR gravitations may have a hierarchical structure depending on the size of the system under study.

Shortcomings of the paper and the open questions it leaves behind should also be mentioned. The theory developed here is for a spherical distribution of mass. Extension to extended objects and to many body systems is not a trivial task. It may require further assumptions not contemplated so far. The difficulty lies in the facts that a) the added s_0- and s_n- terms, $n \geq 2$, are not linear in the mass of the baryonic matter. The nonlinearity is much more complicated than that of GR. b) In the parlance of a dark matter paradigm, the dark companion of a localized baryonic matter is not localized and extends to infinity. As a way out of the dilemma, we are planning to expand an extended object into its localized monopole and higher multipole moments, and see if it is possible to find a dark multipole moment for each baryonic one, more or less in the way done for the monopole moment.

Acknowledgment

The author wishes to acknowledge a discussion with Sergio Mendoza that eventually lead to the expansion of Eq. (14).

References

1. Sobouti, Y., arXiv:0810.2198[gr-gc].
2. Sobouti, Y., arXiv:astro-ph/0603302v4; and A&A **464**, 921, 2007; Saffari, R., & Sobouti, Y.,**472**, 833, 2007.
3. Milgrome, M., ApJ, **270**, 365, 1983.
4. Begeman, K. G., Broeils, A. H., & Sanders, R. H., MNRAS, **249**, 523, 1991; Sanders, R. H., & Verheijen, M. A. W., arXiv:astro-ph/9802240, 1998; Sanders, R. H., & Mc Ghough, S. S., arXiv:astro-ph/0307358, 2002.
5. Tully, R. B., & Fisher, J. R., A&A, **54**, 661, 1977.
6. Bernal, T., Mendoza, S., arXiv:0811.1800v1 [astro-ph], 2008.
7. Mendoza, S., Rosas-Guevara, Y. M., A&A, **472**, 317, 2007.

PART IX

Fifth Family, Dark Energy Perturbations and Dark Matter

OFFERING THE MECHANISM FOR GENERATING FAMILIES: THE APPROACH UNIFYING SPINS AND CHARGES PREDICTS A NEW STABLE FAMILY FORMING DARK MATTER CLUSTERS

N.S. MANKOČ BORŠTNIK

Department of Physics, FMF, University of Ljubljana,
Jadranska 19, 1000 Ljubljana, Slovenia
**E-mail: norma.mankoc@fmf.uni-lj.si*
http://www.fmf.uni-lj.si

The approach[1–4] unifying all the internal degrees of freedom—the family quantum number, the spin and all the charges into only two kinds of the spin—is offering a new way of understanding the properties of quarks and leptons: not only their charges and their connection to the corresponding gauge fields but also the appearance of families and the Yukawa couplings. In this talk I present a simple starting Lagrange density for a spinor—carrying in $d = 1+13$ only two kinds of the spin, no charges, and interacting with the corresponding gauge fields—the vielbeins and the two kinds of spin connection fields. The way of breaking the starting symmetries determines the observed properties of the families of spinors and of the gauge fields, predicting that there are four families at low energies and that a much heavier fifth family with zero Yukawa couplings to the lower four families, might, by forming baryons in the evolution of the universe, contribute a major part to the dark matter. I report on the limitations that the cosmological and the direct experimental evidences might put on this stable family of quarks and leptons.

Keywords: Origin of families; new stable family; dark matter candidate; higher dimensional spaces; unifying theories; dark matter candidates; Kaluza-Klein-like theories.

1. Introduction

The standard model of the electroweak and colour interactions (extended by the right handed neutrinos) fits with around 30 parameters and constraints all the existing experimental data. It leaves, however, unanswered many open questions, among which are also the questions about the origin of charges ($U(1), SU(2), SU(3)$), of families, and correspondingly of the

Yukawa couplings of quarks and leptons and the Higgs mechanism. Answering the question about the origin of families and their masses is the most promising way leading beyond the today knowledge about the elementary fermionic and bosonic fields.

Starting with a simple Lagrange density for spinors, which carry in $d = 1 + 13$ two kinds of spins, represented by the two kinds of the Clifford algebra objects[5] $S^{ab} = \frac{i}{4}(\gamma^a \gamma^b - \gamma^b \gamma^a)$ and $\tilde{S}^{ab} = \frac{i}{4}(\tilde{\gamma}^a \tilde{\gamma}^b - \tilde{\gamma}^b \tilde{\gamma}^a)$, with $\{\gamma^a, \gamma^b\}_+ = 2\eta^{ab} = \{\tilde{\gamma}^a, \tilde{\gamma}^b\}_+, \{\gamma^a, \tilde{\gamma}^b\}_+ = 0$, and no charges, and interact correspondingly only with the vielbeins and the two kinds of spin connection fields, the approach unifying spins and charges[1–4] ends up at observable energies with the observed families of quarks and leptons coupled through the charges to the known gauge fields in the way assumed by the standard model, and carrying masses, determined by a part of a simple starting action [a]. The approach predicts an even number of families, among which is the candidate for forming the dark matter clusters.

The approach confronts several problems (some of them are the problems common to all the Kaluza-Klein-like theories), which we [b] are studying step by step when searching for possible ways of spontaneous breaking of the starting symmetries and conditions, which might lead to the observed properties of families of fermions and of gauge and scalar fields, and looking for predictions the approach might make.

In what follows I briefly present in the first part of the talk the starting action of the approach for fermions and the corresponding gauge fields. The representation of one Weyl spinor of the group $SO(1,13)$ in $d = 1+13$, analyzed with respect to the properties of the subgroups $SO(1,7) \times SU(3) \times U(1)$ of this group and further with respect to $SO(1,3) \times SU(2) \times U(1) \times SU(3)$ manifests the left handed weak charged quarks and leptons and the right handed weak chargeless quarks and leptons.

The way of braking symmetries leads first to eight families at low energy region and then to twice four families. It is a part of a starting Lagrange density for a spinor in $d = 1 + 13$ which manifests as Yukawa couplings in $d = 1 + 3$. The lowest three of the lower four families are the observed families of quarks and leptons, with all the known properties assumed by

[a]This is the only theory in the literature to my knowledge, which does not explain the appearance of families by just postulating their numbers on one or another way, but by offering the mechanism for generating families.

[b]I started the project named the approach unifying spins and charges fifteen years ago, proving alone or together with collaborators step by step, that such a theory has the chance to answer the open questions of the Standard model. The names of the collaborators and students can be found on the cited papers.

the Standard model. Our rough estimations predict that there is the fourth family with possibly low enough masses that it might be seen at LHC.

The fifth family, which decouples in the Yukawa couplings from the lower four families, might have a chance in the evolution of our universe to form baryons and is accordingly the candidate to form the dark matter. In the second part of the talk I present properties of this stable fifth family, as required by the approach and as limited by the cosmological evidences and the direct measurements.

All estimates are very approximate and need serious additional studies. Yet these rough estimations give a good guide to further studies.

2. The Approach Unifying Spin and Charges

The approach[1–4] assumes that in $d \geq (1+13)$-dimensional space a Weyl spinor carries nothing but two kinds of the spin (no charges): The Dirac spin described by γ^a's defines the ordinary spinor representation, the second kind of spin,[5] described by $\tilde{\gamma}^a$'s and anticommuting with the Dirac one, defines the families of spinors [c].

$$\begin{aligned} &\{\gamma^a, \gamma^b\}_+ = 2\eta^{ab} = \{\tilde{\gamma}^a, \tilde{\gamma}^b\}_+, \quad \{\gamma^a, \tilde{\gamma}^b\}_+ = 0, \\ &S^{ab} := (i/4)(\gamma^a \gamma^b - \gamma^b \gamma^a), \quad \tilde{S}^{ab} := (i/4)(\tilde{\gamma}^a \tilde{\gamma}^b - \tilde{\gamma}^b \tilde{\gamma}^a). \end{aligned} \tag{1}$$

Defining the vectors (the nilpotents and projector)[5]

$$\begin{aligned} &\overset{ab}{(\pm i)}: = \frac{1}{2}(\gamma^a \mp \gamma^b), \; \overset{ab}{[\pm i]}:= \frac{1}{2}(1 \pm \gamma^a \gamma^b), \quad \text{for } \eta^{aa}\eta^{bb} = -1, \\ &\overset{ab}{(\pm)}: = \frac{1}{2}(\gamma^a \pm i\gamma^b), \; \overset{ab}{[\pm]}:= \frac{1}{2}(1 \pm i\gamma^a \gamma^b), \quad \text{for } \eta^{aa}\eta^{bb} = 1, \end{aligned} \tag{2}$$

and noticing that the above vectors are eigen vectors of S^{ab} as well as of $\tilde{S}^{ab}$

$$S^{ab} \overset{ab}{(k)} = \frac{k}{2} \overset{ab}{(k)}, \quad S^{ab} \overset{ab}{[k]} = \frac{k}{2} \overset{ab}{[k]}, \tilde{S}^{ab} \overset{ab}{(k)} = \frac{k}{2} \overset{ab}{(k)}, \quad \tilde{S}^{ab} \overset{ab}{[k]} = -\frac{k}{2} \overset{ab}{[k]}, \tag{3}$$

and recognizing that γ^a transform $\overset{ab}{(k)}$ into $\overset{ab}{[-k]}$, while $\tilde{\gamma}^a$ transform $\overset{ab}{(k)}$ into $\overset{ab}{[k]}$

$$\gamma^a \overset{ab}{(k)} = \eta^{aa} \overset{ab}{[-k]}, \; \gamma^b \overset{ab}{(k)} = -ik \overset{ab}{[-k]}, \; \gamma^a \overset{ab}{[k]} = \overset{ab}{(-k)}, \; \gamma^b \overset{ab}{[k]} = -ik\eta^{aa} \overset{ab}{(-k)}, \tag{4}$$

[c] There is no third kind of the Clifford algebra objects: If the Dirac one corresponds to the multiplication of any object (any product of the Dirac γ^a's) from the left hand side, can the second kind of the Clifford objects be understood (up to a factor) as the multiplication of any object from the right hand side.

$$\tilde{\gamma}^a \overset{ab}{(k)} = -i\eta^{aa}\overset{ab}{[k]},\ \tilde{\gamma}^b \overset{ab}{(k)} = -k\overset{ab}{[k]},\ \tilde{\gamma}^a \overset{ab}{[k]} = i\overset{ab}{(k)},\ \tilde{\gamma}^b \overset{ab}{[k]} = -k\eta^{aa}\overset{ab}{(k)}, \tag{5}$$

one sees that $\tilde{S}^{ab}$ form the equivalent representations with respect to S^{ab} and the families of quarks and leptons certainly do (before the break of the electroweak symmetry in the standard model) manifest the equivalent representations.

Let us make a choice of the Cartan subalgebra set of the algebra S^{ab} as follows: $S^{03}, S^{12}, S^{56}, S^{78}, S^{9\,10}, S^{11\,12}, S^{13\,14}$. Then we can write as a starting basic vector of one left handed ($\Gamma^{(1,13)} = -1$) Weyl representation of the group $SO(1,13)$, the quark u_R^{c1}. It is the eigen state of all the members of the Cartan subalgebra and it is the right handed (with respect to $\Gamma^{(1+3)}$), and has the properties: $Y\ u_R^{c1} = 2/3\ u_R^{c1}$, $\tau^{2i}\ u_R^{c1} = 0$ and $(\tau^{33}, \tau^{38})\ u_R^{c1} = (\frac{1}{2}, \frac{1}{2\sqrt{3}})\ u_R^{c1}$. Written in terms of nilpotents and projectors it looks like:

$$\overset{03}{(+i)}\overset{12}{(+)} \mid \overset{56}{(+)}\overset{78}{(+)} \mid\mid \overset{9\,10}{(+)}\overset{11\,12}{(-)}\ \overset{13\,14}{(-)}\ |\psi\rangle = \frac{1}{2^7}$$
$$(\gamma^0-\gamma^3)(\gamma^1+i\gamma^2)|(\gamma^5+i\gamma^6)(\gamma^7+i\gamma^8)||(\gamma^9+i\gamma^{10})(\gamma^{11}-i\gamma^{12})(\gamma^{13}-i\gamma^{14})|\psi\rangle.$$

The eightplet (the representation of $SO(1,7)$ with the fixed colour charge, $\tau^{33} = 1/2$, $\tau^{38} = 1/(2\sqrt{3})$), of one of the eight families (equivalent representations), looks like in TABLE 1.

One can notice (when using Eq.(4)) that $\gamma^0\gamma^7$ and $\gamma^0\gamma^8$ rotate the right handed weak chargeless quark into the left handed weak charged quark of the same colour charge and the same spin.

The generators $\tilde{S}^{ab}$ transform one vector of the representation of S^{ab} into the vector with the same properties with respect to S^{ab}, in particular both vectors bellow describe a right handed u_R-quark of the same colour and the same spin and the same hyper charge

$$2i\tilde{S}^{01}\ \overset{03}{(+i)}\overset{12}{(+)} \mid \overset{56}{(+)}\overset{78}{(+)} \mid\mid \overset{9\,10}{(+)}\overset{11\,12}{(-)}\overset{13\,14}{(-)} = \overset{03}{[+i]}\overset{12}{[+]} \mid \overset{56}{(+)}\overset{78}{(+)} \mid\mid \overset{9\,10}{(+)}\overset{11\,12}{(-)}\overset{13\,14}{(-)}\,. \tag{6}$$

Since the term $-\frac{1}{2}\,\tilde{S}^{ab}\tilde{\omega}_{abc}$ transforms in general one equivalent representation into all the others, we expect that it generates, together with the corresponding gauge fields, the Yukawa couplings. Let us[1–4] now make a choice of a simple action for a spinor which carries in $d = (1+13)$ only two kinds of the spin (no charges)

$$S = \int d^d x\ \mathcal{L},\quad \mathcal{L} = \frac{1}{2}(E\bar{\psi}\gamma^a p_{0a}\psi) + h.c.$$
$$p_{0a} = f^{\alpha}{}_a p_{0\alpha} - \frac{1}{2E}\{p_\alpha, Ef^{\alpha}{}_a\}_-,\quad p_{0\alpha} = p_\alpha - \frac{1}{2}S^{ab}\omega_{ab\alpha} - \frac{1}{2}\tilde{S}^{ab}\tilde{\omega}_{ab\alpha}. \tag{7}$$

Table 1. The 8-plet of quarks - the members of $SO(1,7)$ subgroup, belonging to one Weyl left handed ($\Gamma^{(1,13)} = -1 = \Gamma^{(1,7)} \times \Gamma^{(6)}$) spinor representation of $SO(1,13)$. It contains the left handed weak charged quarks and the right handed weak chargeless quarks of a particular colour $((1/2, 1/(2\sqrt{3})))$. Here $\Gamma^{(1,3)}$ defines the handedness in $(1+3)$ space, S^{12} the ordinary spin (which can also be read directly from the basic vector), τ^{23} the weak charge and Y defines the hyper charge. Let the reader notice (by taking into account the relations $\gamma^a \stackrel{ab}{(k)} = \eta^{aa} \stackrel{ab}{[-k]}$, $\stackrel{ab}{(-k)}\stackrel{ab}{(k)} = \eta^{aa} \stackrel{ab}{[-k]}$) that $\gamma^0 \stackrel{78}{(-)}$ (appearing in $-\mathcal{L}_Y = \psi^\dagger \gamma^0 \{\stackrel{78}{(+)} p_{0+} + \stackrel{78}{(-)} p_{0-}\}\psi$) transforms u_R^{c1} of the 1st row into u_L^{c1} of the 7th row, while $\gamma^0 \stackrel{78}{(+)}$ transforms d_R^{c1} of the 3rd row into d_L^{c1} of the 5th row, doing what the Higgs and γ^0 do in the standard model.

i		$\vert^a\psi_i>$	$\Gamma^{(1,3)}$	S^{12}	τ^{23}	Y
		Octet of quarks				
1	u_R^{c1}	$\stackrel{03}{(+i)}\stackrel{12}{(+)} \vert \stackrel{56}{(+)}\stackrel{78}{(+)} \vert\vert \stackrel{9\,10}{(+)}\;\stackrel{11\,12}{(-)}\;\stackrel{13\,14}{(-)}$	1	$\frac{1}{2}$	0	$\frac{2}{3}$
2	u_R^{c1}	$\stackrel{03}{[-i]}\stackrel{12}{[-]} \vert \stackrel{56}{(+)}\stackrel{78}{(+)} \vert\vert \stackrel{9\,10}{(+)}\;\stackrel{11\,12}{(-)}\;\stackrel{13\,14}{(-)}$	1	$-\frac{1}{2}$	0	$\frac{2}{3}$
3	d_R^{c1}	$\stackrel{03}{(+i)}\stackrel{12}{(+)} \vert \stackrel{56}{[-]}\stackrel{78}{[-]} \vert\vert \stackrel{9\,10}{(+)}\;\stackrel{11\,12}{(-)}\;\stackrel{13\,14}{(-)}$	1	$\frac{1}{2}$	0	$-\frac{1}{3}$
4	d_R^{c1}	$\stackrel{03}{[-i]}\stackrel{12}{[-]} \vert \stackrel{56}{[-]}\stackrel{78}{[-]} \vert\vert \stackrel{9\,10}{(+)}\;\stackrel{11\,12}{(-)}\;\stackrel{13\,14}{(-)}$	1	$-\frac{1}{2}$	0	$-\frac{1}{3}$
5	d_L^{c1}	$\stackrel{03}{[-i]}\stackrel{12}{(+)} \vert \stackrel{56}{[-]}\stackrel{78}{(+)} \vert\vert \stackrel{9\,10}{(+)}\;\stackrel{11\,12}{(-)}\;\stackrel{13\,14}{(-)}$	-1	$\frac{1}{2}$	$-\frac{1}{2}$	$\frac{1}{6}$
6	d_L^{c1}	$\stackrel{03}{(+i)}\stackrel{12}{[-]} \vert \stackrel{56}{[-]}\stackrel{78}{(+)} \vert\vert \stackrel{9\,10}{(+)}\;\stackrel{11\,12}{(-)}\;\stackrel{13\,14}{(-)}$	-1	$-\frac{1}{2}$	$-\frac{1}{2}$	$\frac{1}{6}$
7	u_L^{c1}	$\stackrel{03}{[-i]}\stackrel{12}{(+)} \vert \stackrel{56}{(+)}\stackrel{78}{[-]} \vert\vert \stackrel{9\,10}{(+)}\;\stackrel{11\,12}{(-)}\;\stackrel{13\,14}{(-)}$	-1	$\frac{1}{2}$	$\frac{1}{2}$	$\frac{1}{6}$
8	u_L^{c1}	$\stackrel{03}{(+i)}\stackrel{12}{[-]} \vert \stackrel{56}{(+)}\stackrel{78}{[-]} \vert\vert \stackrel{9\,10}{(+)}\;\stackrel{11\,12}{(-)}\;\stackrel{13\,14}{(-)}$	-1	$-\frac{1}{2}$	$\frac{1}{2}$	$\frac{1}{6}$

The above action can further be rewritten as

$$\mathcal{L} = \bar{\psi}\,\gamma^m (p_m - \sum_{A,i} g^A \tau^{Ai} A^{Ai}_m)\psi + \{\sum_{s=7,8} \bar{\psi}\gamma^s p_{0s}\,\psi\} + \text{the rest}, \tag{8}$$

with the meaning

$$\tau^{Ai} = \sum_{a,b} c^{Ai}{}_{ab}\, S^{ab}, \quad \{\tau^{Ai}, \tau^{Bj}\}_- = i\delta^{AB} f^{Aijk}\tau^{Ak}, \tag{9}$$

where $A = 1$ stays for $U(1)$, $i = \{1\}$, which is the hyper charge Y in the standard model notation, $A = 2$ stays for the $SU(2)$ weak charge, $i = \{1,2,3\}$, $A = 3$ stays for the colour $SU(3)$ charge, $i = \{1, \cdots, 8\}$. All the spinors, which appear in $2^{8/2-1}$ families before the break of the $SO(1,7)$ symmetry, are massless [d], while the term $\sum_{s=7,8} \bar{\psi}\gamma^s p_{0s}\,\psi$ in

[d] In the references[6] we present for the toy model the proof that the break of symmetry can preserve masslessness.

Eq.(8) form what the standard model postulates as the Yukawa couplings. Let us rewrite it, naming it $\mathcal{L}_Y$

$$-\mathcal{L}_Y = \psi^\dagger \gamma^0 \gamma^s p_{0s} \psi \quad = \psi^\dagger \, \gamma^0 \{ \overset{78}{(+)} \; p_{0+} + \overset{78}{(-)} \; p_{0-} \} \psi, \qquad (10)$$

with

$$\begin{aligned} p_{0\pm} &= (p_7 \mp i \, p_8) - \frac{1}{2} S^{ab} \omega_{ab\pm} - \frac{1}{2} \tilde{S}^{ab} \tilde{\omega}_{ab\pm}; \\ \omega_{ab\pm} &= \omega_{ab7} \mp i \, \omega_{ab8}, \quad \tilde{\omega}_{ab\pm} = \tilde{\omega}_{ab7} \mp i \, \tilde{\omega}_{ab8}. \end{aligned} \qquad (11)$$

One can see in ref.[2–4] how does this term behave after particular breaks of symmetries and what predictions for the masses and the mixing matrices does it make.

The action for the gauge fields is the Einstein one:[6] linear in the curvature

$$\begin{aligned} S &= \int \, d^d x \; E \; (R + \tilde{R}), \\ R &= \frac{1}{2} \left[f^{\alpha[a} f^{\beta b]} \; (\omega_{ab\alpha,\beta} - \omega_{ca\alpha} \omega^c{}_{b\beta}) \right] + h.c., \\ \tilde{R} &= \frac{1}{2} \left[f^{\alpha[a} f^{\beta b]} \; (\tilde{\omega}_{ab\alpha,\beta} - \tilde{\omega}_{ca\alpha} \tilde{\omega}^c{}_{b\beta}) \right] + h.c.. \end{aligned}$$

Here [e] $f^{\alpha[a} f^{\beta b]} = f^{\alpha a} f^{\beta b} - f^{\alpha b} f^{\beta a}$. The action 12 manifests after the break of symmetries all the known gauge fields and the Higgs fields [f].

2.1. *The Yukawa couplings, the masses of families and the mixing matrices*

Let us analyze the Yukawa couplings $-\mathcal{L}_Y = \psi \quad = \psi^\dagger \, \gamma^0 \{ \overset{78}{(+)} \; p_{0+} + \overset{78}{(-)} \; p_{0-} \} \psi$ (Eq.10, 11) in order to see what predictions we can make. We do not know the way of breaking the starting symmetry $SO(1,13)$, but it should

[e] $f^\alpha{}_a$ are inverted vielbeins to $e^a{}_\alpha$ with the properties $e^a{}_\alpha f^\alpha{}_b = \delta^a{}_b$, $e^a{}_\alpha f^\beta{}_a = \delta^\beta_\alpha$. Latin indices $a, b, .., m, n, .., s, t, ..$ denote a tangent space (a flat index), while Greek indices $\alpha, \beta, .., \mu, \nu, ..\sigma, \tau..$ denote an Einstein index (a curved index). Letters from the beginning of both the alphabets indicate a general index ($a, b, c, ..$ and $\alpha, \beta, \gamma, ..$), from the middle of both the alphabets the observed dimensions $0, 1, 2, 3$ ($m, n, ..$ and $\mu, \nu, ..$), indices from the bottom of the alphabets indicate the compactified dimensions ($s, t, ..$ and $\sigma, \tau, ..$). We assume the signature $\eta^{ab} = diag\{1, -1, -1, \cdots, -1\}$.

[f] I am studying how does the break of symmetries of $SO(1,7) \times SU(3) \times U(1)$ to $SO(1,3) \times U(1) \times U(1) \times SU(3)$ influence the gauge fields, leading to not only all the gauge fields, but also to (since the symmetry breaks twice to two kinds of) scalar (that is Higgs) fields.

be the way, which leads to all the starting assumptions of the standard model. Since the handedness in $d = 1 + 3$, which obviously concerns spin, and the weak charge are related in the Standard model, the breaking must go through $SO(1,7)$, where the spin and the handedness are manifestly correlated as seen in TABLE 1. We assume accordingly[1–4] the following way of breaking: First $SO(1,13) \to SO(1,7) \times SU(3) \times U(1)$, then $SO(1,7) \times SU(3) \times U(1) \to SO(1,3) \times SU(2) \times U(1) \times SU(3)$, and finally $\to SO(1,3) \times U(1) \times U(1) \times SU(3)$, which is just the observed symmetry. These breaking must appear in both sectors: $\omega_{ab\alpha}$ and $\tilde{\omega}_{ab\alpha}$, not necessarily with the same parameters.

The way of the above suggesting breaking leads to the charges and gauge fields as assumed in Eq.(8) in the sector $-\frac{1}{2}\, S^{ab}\, \omega_{ab\alpha}$, while it leads in the sector $-\frac{1}{2}\, \tilde{S}^{ab}\, \tilde{\omega}_{ab\alpha}$ to two times four decoupled families. In references[2–4] this decoupling is analyzed and the predictions made. The way of breaking and correspondingly the symmetries imposed on the fields $\omega_{ab\sigma}$ and $\tilde{\omega}_{ab\sigma}$, $\sigma = \{5,6,7,8\}$ influence our roughly estimated properties of quarks and leptons of the first four families, predicting on the tree level the masses of the fourth family and the mixing matrices of the first four families. This rough estimation[1–4] allows the masses of the fourth family quarks to lie at around 250 GeV or higher, the fourth family neutrino mass at around 80 GeV or higher and the fourth family electron mass at around 200 GeV or higher. We predict the mixing matrices for quarks and leptons. The fourth family quarks have possibly a chance to be seen at LHC.

The lower of the upper four families, which is stable (has zero Yukawa couplings to the lower four families), must have accordingly the masses above 1 TeV. Being stable the neutral (with respect to the weak and colour charge) clusters of the fifth family members are candidates for forming the dark matter.

These rough estimations, although to my understanding a good guide to the properties of families, need much more sophisticated calculations to be really trustful.

3. The Fifth Family as the Candidate for Forming the Dark Matter Clusters

In this section I present limitations,[8] which the dark matter puts on our stable fifth family quarks forming the dark matter clusters. The limitations follow from i) the evolution of our universe, manifesting in the cosmological observations, ii) the gravitational effects, manifesting in (almost) equal velocities of all the stars and ionized hydrogen atoms within galaxies and out of galaxies, in scattering of galaxies and of clusters and in gravitational lensing, and iii) in the direct measurements.[9]

Since the approach unifying spins and charges does predict a stable (fifth) family, it is a great chance that the dark matter is made out of a neutral (with respect to the colour and the electromagnetic charge) clusters of the fifth family, provided that the approach is the right theory to explain physics beyond the standard model.

The break of the $SO(1,7)$ symmetry (occurring in the approach in two steps and predicting two kinds of scalar fields) leads to twice in the Yukawa couplings decoupled four families, all eight families carrying the colour, the electromagnetic and one additional Q' charge after the final break of the electroweak symmetry, just as the lower three families do. Accordingly has the fifth family the quantum numbers of the first three families, distinguishing from them in the family index and consequently in masses. Having zero Yukawa couplings to the lower four families, the fifth family members are stable [g]. The fifth family neutrons, neutrinos, or atoms, formed during the evolution of our universe might accordingly contribute today the main part of the matter ((5-7)times as much as the ordinary matter).

There are several candidates for the massive dark matter constituents in the literature, known as WIMPs (weakly interacting massive particles), the references can be found in.[9,12]

3.1. *Properties of clusters of the fifth family quarks*

Our fifth family members (u_5, d_5, ν_5, e_5), since carrying the same charges as the three known families, form baryons (protons p_5 ($u_5 u_5 d_5$), neutrons n_5 ($u_5 d_5 d_5$), Δ_5^-, Δ_5^{++}, etc.) and atoms. The properties of these baryons can be estimated by using the non relativistic Bohr-like model with the $\frac{1}{r}$ dependence of the potential between a pair of quarks $V = -\frac{3\alpha_c}{r}$, where α_c is in this case the colour (3 for three possible colour charges) coupling constant. Equivalently goes for anti-quarks. This is a meaningful approximation as long as the one gluon exchange is the dominant contribution to the interaction among quarks (the electromagnetic and weak interaction contributions are more than 10^{-3} times smaller) that is as long as excitations of a cluster are not influenced by the linearly rising part of the

[g] Heavy enough the fifth family baryons can have for many orders of magnitude different time scale for forming (if at all) solid matter than the ordinary matter, while their neutrinos do not contribute to the width of the neutral weak bosons. Even the four family members are already heavy enough not to be in contradiction with the experimental evidences.[3,7,15]

potential. Let us tell that a simple bag model evaluation does not contradict such a simple model. Which one of p_5, n_5, or may be Δ_5^- or Δ_5^{++}, is a stable fifth family baryon, depends on the mass difference $m_{u_5} - m_{d_5}$, as well as on the weak and the electromagnetic interactions among quarks. If m_{d_5} is appropriately lighter than m_{u_5}, then n_5 is a colour singlet electromagnetic chargeless stable cluster of quarks, with the weak charge $-1/2$. If m_{d_5} is heavier enough with respect to stronger electromagnetic repulsion among the two u_5 than among the two d_5, then m_{u_5}, the proton p_5, which is a colour singlet stable nucleon with the weak charge $1/2$ needs the electron e_5 or e_1 to form an electromagnetic chargeless cluster. ν_5, carrying only the weak charge, behaves if it is stable ($m_{e_5} > m_{\nu_5}$) similarly as n_5 when the weak force dominates over the "nuclear" force.

I shall (for simplicity) assume in this talk that n_5 (and $\bar{n}_5$) is a stable baryon, leaving all the other possibilities, with ν_5 included [h] for further studies.[13] In the Bohr-like model the binding energy and the average radius are (up to a factor of the order π) equal to $E_{c_5} = -\frac{3}{2}\, m_{q_5} c^2 (3\alpha_c)^2$, $\quad r_{c_5} = \frac{\hbar c}{3\alpha_c m_{q_5} c^2}$ The mass of the cluster is approximately $m_{c_5} c^2 = 3 m_{q_5} c^2 (1 - \frac{1}{2}(3\alpha_c)^2)$. Assuming that the coupling constant of the colour charge α_c runs with the kinetic energy $(-E_{c_5}/3)$ of quarks $(\alpha_c(E^2) = \frac{\alpha_c(M^2)}{1+\frac{\alpha_c(M^2)}{4\pi}(11-\frac{2N_F}{3})\ln(\frac{E^2}{M^2})}$, with $\alpha_c((91\ \mathrm{GeV})^2) = 0.1176(20)$, the number of flavours $N_F = 8$) we estimated the properties of a baryon as presented on TABLE 2. The binding energy is approximately of two orders of magnitude smaller than the mass of the cluster. The baryon n_5 is lighter than the baryon p_5, if $(m_{u_5} - m_{d_5})$ is smaller then $(0.6, 60, 600)$ GeV for

Table 2. Properties of a cluster of the fifth family quarks within the Bohr-like model. m_{q_5} in TeV/c^2 is the assumed fifth family quark mass.

$\frac{m_{q_5} c^2}{\mathrm{TeV}}$	α_c	$\frac{E_{c_5}}{\mathrm{TeV}}$	$\frac{r_{c5}}{10^{-7}\mathrm{fm}}$	$\frac{\pi r_{c5}^2}{(10^{-7}\mathrm{fm})^2}$
10^2	0.09	5.4	150	$6.8 \cdot 10^4$
10^4	0.07	$3 \cdot 10^2$	1.9	12
10^6	0.05	$2 \cdot 10^4$	0.024	$1.9 \cdot 10^{-3}$

[h] One can worry whether the two more species of neutrinos could contradict the experimental data. Our very rough estimations,[7] not yet trustable, show, that the fourth family neutrinos have the mass around 80 GeV or higher and the precise electroweak data analyses[15] show that such neutrinos are not in disagreement with the experimental data. Also heavy enough neutrinos, stable or with small mixing matrices, do not influence the astrophysical data.

the three values of the m_{q_5} on TABLE 2, respectively. The corresponding "nucleon-nucleon" force determines for many orders of magnitude smaller cross section than in the case of the first family nucleons.

If a cluster of the heavy (fifth family) quarks and leptons and of the ordinary (the lightest) family is made, then, since ordinary family dictates the radius and the excitation energies of a cluster, its properties are not far from the properties of the ordinary hadrons and atoms, except that such a cluster has the mass dictated by the heavy family members.

3.2. *Dynamics of a heavy family baryons in our galaxy*

The density[8] of the dark matter ρ_{dm} in the Milky way can be evaluated from the measured rotation velocity (approximately independent of the distance r from the center of our galaxy) of stars and gas in our galaxy. Our Sun's velocity is $v_S \approx (170 - 270)$ km/s. ρ_{dm} is approximately spherically symmetric and proportional to $\frac{1}{r^2}$. At the position of our Sun ρ_{dm} is known within a factor of 10 ($\rho_0 \approx 0.3\,\mathrm{GeV}/(c^2\,\mathrm{cm}^3)$). We put $\rho_{dm} = \rho_0\,\varepsilon_\rho$, with $\frac{1}{3} < \varepsilon_\rho < 3$. The local velocity distribution of the dark matter cluster $\vec{v}_{dm\,i}$, in the velocity class i of clusters, can only be estimated, results depend strongly on the model.[8]

The velocity of the Earth around the center of the galaxy is equal to: $\vec{v}_E = \vec{v}_S + \vec{v}_{ES}$, with $v_{ES} = 30$ km/s and $\frac{\vec{v}_S \cdot \vec{v}_{ES}}{v_S v_{ES}} \approx \cos\theta, \theta = 60^0$. The dark matter cluster of the i- th velocity class hits the Earth with the velocity: $\vec{v}_{dmE\,i} = \vec{v}_{dm\,i} - \vec{v}_E$. Then the flux of our dark matter clusters hitting the Earth is: $\Phi_{dm} = \sum_i \frac{\rho_{dmi}}{m_{c_5}} |\vec{v}_{dm\,i} - \vec{v}_E|$, which (for $\frac{v_{ES}}{|\vec{v}_{dmi} - \vec{v}_S|}$ small) equals to $\Phi_{dm} \approx \sum_i \frac{\rho_{dmi}}{m_{c_5}} \{|\vec{v}_{dm\,i} - \vec{v}_S| - \vec{v}_{ES} \cdot \frac{\vec{v}_{dmi} - \vec{v}_S}{|\vec{v}_{dmi} - \vec{v}_S|}\}$. One can take approximately that $\sum_i |\vec{v_{dmi}} - \vec{v_S}|\,\rho_{dmi} = \varepsilon_{v_{dmS}}\,\varepsilon_\rho\, v_S\, \rho_0$, and further $\sum_i \vec{v}_{ES} \cdot \frac{\vec{v}_{dmi} - \vec{v}_S}{|\vec{v}_{dmi} - \vec{v}_S|} = v_{ES} \varepsilon_{v_{dmES}} \cos\theta\, \sin\omega t$. We estimate (due to experimental data and our theoretical evaluations) that $\frac{1}{3} < \varepsilon_{v_{dmS}} < 3$ and $\frac{1}{3} < \frac{\varepsilon_{v_{dmES}}}{\varepsilon_{v_{dmS}}} < 3$. This last term determines the annual modulations observed by DAMA.[9]

The cross section for our heavy dark matter baryon to elastically (the excited states of nuclei, which we shall treat, I and Ge, are at ≈ 50 keV or higher and are very narrow, while the average recoil energy of Iodine is expected to be 30 keV) scatter on an ordinary nucleus with A nucleons is $\sigma_A = \frac{1}{\pi\hbar^2} < |M_{c_5 A}| >^2\, m_A^2$. For our heavy dark matter cluster is m_A approximately the mass of the ordinary nucleus. In the case of a coherent scattering (if recognizing that $\lambda = \frac{h}{p_A}$ is for a nucleus large enough to make scattering coherent, when the mass of the cluster is 1 TeV or more and its velocity $\approx v_S$), the cross section is almost independent of the recoil velocity of the nucleus. For the case that the "nuclear force" as manifesting in the cross section $\pi\,(r_{c_5})^2$ brings the main contribution, the cross section is proportional to $(3A)^2$ (due to the square of the matrix element) times $(A)^2$ (due to the mass of the nuclei $m_A \approx 3A\, m_{q_1}$, with $m_{q_1}\, c^2 \approx \frac{1\text{GeV}}{3}$). When m_{q_5} is heavier than $10^4\,\text{TeV}/c^2$ (Table 1), the weak interaction dominates and σ_A is proportional to $(A-Z)^2\, A^2$, since to Z^0 boson exchange only neutron gives an appreciable contribution. Accordingly we have that $\sigma(A) \approx \sigma_0\, A^4\, \varepsilon_\sigma$, with $\sigma_0\, \varepsilon_\sigma$, which is $9\,\pi r_{c_5}^2\, \varepsilon_{\sigma_{nucl}}$, with $\frac{1}{30} < \varepsilon_{\sigma_{nucl}} < 30$ (taking into account the roughness with which we treat our heavy baryon's properties and the scattering procedure) when the "nuclear force" dominates, while $\sigma_0\, \varepsilon_\sigma$ is $(\frac{m_{n_1} G_F}{\sqrt{2}\pi}\, \frac{A-Z}{A})^2\, \varepsilon_{\sigma_{weak}}$ $(= (10^{-6}\, \frac{A-Z}{A}\,\text{fm})^2\, \varepsilon_{\sigma_{weak}})$, $\frac{1}{10} < \varepsilon_{\sigma_{weak}} < 1$ (the weak force is pretty accurately evaluated, but the way of averaging is not), when the weak interaction dominates.

3.3. *Direct measurements of the fifth family baryons as dark matter constituents*

Let us make a rough estimations of what the DAMA[9] and CDMS[10] experiments are measuring, provided that the dark matter clusters are made out of our heavy family quarks to see limitations these two experiments might put on properties of our heavy family members. We are completely aware of how rough our estimation is, yet we conclude that, since the number of measured events is proportional to $(m_{c_5})^{-3}$ for masses $\approx 10^4$ TeV or smaller (while for higher masses, when the weak interaction dominates, it is proportional to $(m_{c_5})^{-1}$) that even such rough estimations may in the case of our heavy baryons say, whether both experiments do at all measure our (any) heavy family clusters, if one experiment clearly sees the dark matter signals and the other does not (yet?).

Let N_A be the number of nuclei of a type A in the apparatus (of either DAMA,[9] which has $4 \cdot 10^{24}$ nuclei per kg of I, with $A_I = 127$, and Na, with $A_{Na} = 23$ (we shall neglect Na), or of CDMS,[10] which has $8.3 \cdot 10^{24}$ of Ge nuclei per kg, with $A_{Ge} \approx 73$). At velocities of a dark matter cluster $v_{dmE} \approx 200$ km/s are the $3A$ scatterers strongly bound in the nucleus, so that the whole nucleus with A nucleons elastically scatters on a heavy dark matter cluster. Then the number of events per second (R_A) taking place in N_A nuclei is due to the flux Φ_{dm} and the recognition that the cross section is at these energies almost independent of the velocity equal to

$$R_A = N_A\, \frac{\rho_0}{m_{c_5}}\, \sigma(A)\, v_S\, \varepsilon_{v_{dmS}}\, \varepsilon_\rho\, (1 + \frac{\varepsilon_{v_{dmES}}}{\varepsilon_{v_{dmS}}}\, \frac{v_{ES}}{v_S}\, \cos\theta\, \sin\omega t). \tag{12}$$

Let ΔR_A mean the amplitude of the annual modulation of R_A, $R_A(\omega t = \frac{\pi}{2}) - R_A(\omega t = 0) = N_A\, R_0\, A^4\, \frac{\varepsilon_{v_{dmES}}}{\varepsilon_{v_{dmS}}}\, \frac{v_{ES}}{v_S}\, \cos\theta$, where $R_0 = \sigma_0\, \frac{\rho_0}{m_{q_{c5}}}\, v_S\, \varepsilon$, and $\varepsilon = \varepsilon_\rho\, \varepsilon_{v_{dmES}} \varepsilon_\sigma$. Let $\frac{1}{300} < \varepsilon < 300$ demonstrates the uncertainties in the knowledge about the dark matter dynamics in our galaxy and our approximate treating of the dark matter properties. An experiment with N_A scatterers should measure $R_A \varepsilon_{cut\,A}$, with $\varepsilon_{cut\,A}$ determining the efficiency of a particular experiment to detect a dark matter cluster collision. For small enough $\frac{\varepsilon_{v_{dmES}}}{\varepsilon_{v_{dmS}}}\, \frac{v_{ES}}{v_S}\, \cos\theta$ it follows:

$$R_A\, \varepsilon_{cut\,A} \approx N_A\, R_0\, A^4\, \varepsilon_{cut\,A} = \Delta R_A \varepsilon_{cut\,A}\, \frac{\varepsilon_{v_{dmS}}}{\varepsilon_{v_{dmES}}}\, \frac{v_S}{v_{ES}\, \cos\theta}. \tag{13}$$

If DAMA[9] is measuring our heavy family baryons then $R_I \varepsilon_{cut\,dama} \approx \Delta R_{dama} \frac{\varepsilon_{v_{dmS}}}{\varepsilon_{v_{dmES}}}\, \frac{v_S}{v_{SE}\, \cos 60^0}$, with $\Delta R_{dama} \approx \Delta R_I\, \varepsilon_{cut\,dama}$. Most of unknowns about the dark matter properties, except the local velocity of our Sun, the cut off procedure ($\varepsilon_{cut\,dama}$) and $\frac{\varepsilon_{v_{dmS}}}{\varepsilon_{v_{dmES}}}$, are hidden in ΔR_{dama}. Taking for the Sun's velocity $v_S = 100, 170, 220, 270$ km/s, we find $\frac{v_S}{v_{SE}\, \cos\theta} = 7, 10, 14, 18$, respectively. DAMA/NaI,DAMA/LIBRA[9] publishes $\Delta R_{dama} = 0,052$ counts per day and per kg of NaI. Correspondingly is $R_I\, \varepsilon_{cut\,dama} = 0,052\, \frac{\varepsilon_{v_{dmS}}}{\varepsilon_{v_{dmES}}}\, \frac{v_S}{v_{SE}\, \cos\theta}$ counts per day and per kg. CDMS should then in 121 days with 1 kg of Ge ($A = 73$) detect $R_{Ge}\, \varepsilon_{cut\,cdms} \approx \frac{8.3}{4.0}\, (\frac{73}{127})^4\, \frac{\varepsilon_{cut\,cdms}}{\varepsilon_{cut\,dama}}\, \frac{\varepsilon_{v_{dmS}}}{\varepsilon_{v_{dmES}}}\, \frac{v_S}{v_{SE}\, \cos\theta}\; 0.052 \cdot 121$ events, which is for the above measured velocities equal to $(10, 16, 21, 25)\, \frac{\varepsilon_{cut\,cdms}}{\varepsilon_{cut\,dama}}\, \frac{\varepsilon_{v_{dmS}}}{\varepsilon_{v_{dmES}}}$. CDMS[10] has found no event.

The approximations we made might cause that the expected numbers $(10, 16, 21, 25)$ multiplied by $\frac{\varepsilon_{cut\,Ge}}{\varepsilon_{cut\,I}}\, \frac{\varepsilon_{v_{dmS}}}{\varepsilon_{v_{dmES}}}$ are too high (or too low!!) for a factor let us say 4 or 10. If in the near future CDMS (or some other experiment) will measure the above predicted events, then there might be heavy family clusters which form the dark matter. In this case the DAMA experiment puts the limit on our heavy family masses (Eq.(13)): We evaluate the lower limit for the mass $m_{q_5}\, c^2 > 200$ TeV. Observing that for $m_{q_5}\, c^2 > 10^4$ TeV the weak force starts to dominate, we estimate the upper limit $m_{q_5}\, c^2 < 10^5$ TeV. In the case that the weak interaction determines the n_5 cross section we find for the mass range 10 TeV $< m_{q_5}\, c^2 < 10^5$ TeV.

3.4. *Evolution of the abundance of the fifth family members in the universe*

To estimate the behaviour of our heavy family members in the expanding universe we need to know the particle—anti-particle asymmetry for the fifth family members, as well as their masses, and the way how do they form baryons. Let us simply assume that the fifth family members, q_5's and $\bar{q}_5$'s, e_5's, ν_5's (not necessarily all at the same time if they have different masses) start to decouple from the plasma, quarks at the temperature $T_1 \approx m_{q_5} c^2/k_b$, with k_b the Boltzmann constant, so that the quarks and anti-quarks form (recombine into) the baryons n_5 and $\bar{n}_5$. (The gluon interaction causes the annihilation of quarks and anti-quarks or forces quarks and anti-quarks to form baryons and anti-baryons, which are colour neutral possibly far before the colour phase transition at ≈ 0.3 GeV. We are studying these possibilities in more details in.[13])

To estimate the number of the fifth family quarks and anti-quarks clustered into n_5 and $\bar{n}_5$ we follow the ref.,[12] chapter 3. Let $\Omega_5 = \frac{\rho_{c_5}}{\rho_{cr}}$ ($\rho_{cr} = \frac{3H_0^2}{8\pi G}$, $H_0 \approx 10^{-10}$ year^{-1} and $G = \hbar c/(m_{Pl}^2)$, $m_{Pl}c^2 \approx 1.2 \cdot 10^{19}$ GeV) denote the today ratio of the density of the fifth family clusters and the critical density, estimated to be $\approx \frac{1}{3}$. It follows then for Ω_5

$$\Omega_5 = \frac{1}{\beta} \frac{T_1 k_B}{m_{c_5} c^2} \sqrt{g^*} \, (\frac{a(T^1)T^1}{a(T^0)T^0})^3 \sqrt{\frac{4\pi^3 G}{45(\hbar \, c)^3} \frac{(T_0 \, k_b)^3}{\rho_{cr} c^4}} \frac{1}{< \sigma_5 \, v/c >} \tag{14}$$

where we evaluated $\frac{T_1 k_B}{m_{c_5} c^2} \, (\frac{a(T^1)T^1}{a(T^0)T^0})^3 \approx 10^{-3}$, T_0 is the today's black body radiation temperature, $a(T^0) = 1$ and $a(T)$ is the metric tensor component in the expanding flat universe diag $g_{\mu\nu} = (1, -a(t)^2, -a(t)^2, -a(t)^2)$, $T = T(t)$, $\sqrt{g^*} \approx \sqrt{200}$ (g^* measures the number of degrees of freedom of four family fermions and all gauge fields at T_1), $0.1 < \beta < 10$ stays for uncertainty in the evaluation of $\frac{T_1 k_B}{m_{c_5} c^2}$ and $\frac{a(T^1)T^1}{a(T^0)T^0}$. Evaluating $\sqrt{\frac{4\pi^3 \, G}{45(\hbar c)^3} \frac{(T_0 \, k_b)^3}{\rho_{cr} c^4}} \approx 200 \, (10^{-7}\text{fm})^2$ we estimate (for $\frac{v}{c} \approx 1$) that the scattering cross section at the relativistic energies (when $k_b T^1 \approx m_{c_5} c^2$) is $(10^{-7}\text{fm})^2 < \sigma_5 < (10^{-6}\text{fm})^2$. Taking into account the relation for the relativistic scattering of quarks, with the one gluon exchange contribution dominating $\sigma = 8\,\pi (\frac{3\alpha_c(E) \, \hbar \, c}{E})^2$, we obtain the mass limit 10^2 TeV $< m_{q_5} \, c^2 < 10^4$ TeV.

4. Concluding Remarks

I presented in my talk very briefly the approach unifying spins and charges,[1–3] which is offering the way to explain the assumptions of the standard model of the electroweak and colour interactions, with the appearance of family included by proposing the mechanism for generating families. It is a simple starting Lagrange density with spinors which carry only two kinds of spins, no charges, and interact with vielbeins and the two kinds of spin connection fields, which manifests at observed energies all the observed properties of fermions and bosons.

Rough estimations made up to now predict the fourth family to be possibly seen at LHC and the stable fifth family, which is the candidate to form the Dark matter clusters. Predictions depend on the way of breaking the starting symmetries, and on the perturbative and nonperturbative effects, which follow the breaking. Accordingly future more sophisticated calculations will be very demanding. Our rough estimations predict the masses and the mixing matrices of the fourth family in agreement with the electroweak precision data analyses.[15]

Using the simple Bohr-like model to find the properties of baryons of the fifth family quarks, we evaluated the "nuclear" interaction of heavy baryons among themselves and with the ordinary nuclei, recognizing that the weak interaction dominates over the "nuclear interaction" for massive enough clusters ($m_{q_5} > 10^4$ TeV), while nonrelativistic clusters interact among themselves with the weak force only. Assuming that the DAMA and CDMS experiments[9,10] measure our fifth family baryons, we find the limit on the fifth family quark mass: $200\,\mathrm{TeV} < \mathrm{m_{q_5} c^2} < 10^5\,\mathrm{TeV}$. If the weak interaction determines the n_5 cross section we find: $10\mathrm{TeV} < m_{q_5}\, c^2 < 10^5$ TeV, which is as well the limit for m_{ν_5}. In this case is the estimated cross section for the dark matter cluster to (elastically, coherently and nonrelativisically) scatter on the nucleus determined on the lower mass limit by the "nuclear force" and on the higher mass limit by the weak force. The cosmological evolution suggests the mass limit 100 TeV $< m_{q_5}\, c^2 < \cdot 10^4$ TeV.

Our rough estimations predict that, if the DAMA experiments[9] observes the events due to our (any) heavy family members, the CDMS experiments[10] will observe a few events as well in the near future.

If a possible answer is a complicated mechanism (like in[16]), then since there are might be many possible complicated scenarios for the dark matter origin, it will be difficult to make the right choice among them, and we shall not find out very soon what is the dark matter constituted out of.

The fact that the fifth family baryons might form the dark matter does not contradict the measured (first family) baryon number and its ratio to the photon energy density as long as the fifth family members are heavy enough (> few TeV). Then they form neutral clusters far before the colour phase transition at around 1 GeV. Also the stable fifth family neutrino does not in this case contradict observations, either electroweak or cosmological.

Let me conclude this talk saying: If the approach unifying spins and charges is the right way beyond the standard model of the electroweak and colour interactions, then more than three families of quarks and leptons do exist, and the stable (with respect to the age of the universe) fifth family of quarks and leptons is the candidate to form the dark matter.

Acknowledgments

The author acknowledge very useful collaboration of the participants of the annual workshops entitled "What comes beyond the Standard models", starting in 1998, and in particular of H. B. Nielsen.

References

1. N.S. Mankoč Borštnik, Phys. Lett. **B 292** (1992) 25-29, J. Math. Phys. **34** (1993) 3731-3745, Modern Phys. Lett. **A 10** (1995) 587-595, Int. J. Theor. Phys. **40** (2001) 315-337.
2. A. Borštnik Bračič, N.S. Mankoč Borštnik, Phys Rev. **D 74** (2006) 073013-16.
3. G. Bregar, M. Breskvar, D. Lukman, N.S. Mankoč Borštnik, New J. of Phys. **10** (2008) 093002.
4. N. S. Mankoč Borštnik, hep-ph/0711.4681, 94-113.
5. N.S. Mankoč Borštnik, H.B. Nielsen, J. of Math. Phys. **43** (2002), (5782-5803), J. of Math. Phys. **44** (2003) 4817-4827.
6. N.S. Mankoč Borštnik, H.B. Nielsen, Phys. Lett. B **633** 771 (2006), hep-th/0311037, hep-th/0509101; Phys. Lett. **B** 10.1016 (2008).
7. M. Breskvar, D. Lukman, N.S. Mankoč Borštnik, hep-ph/0612250, p.25-50.
8. G. Bregar, N.S. Mankoč Borštnik, (sent to Phys. Rev. Lett. 14. Nov 2008), astro-ph¿arXiv: 0811.4106, arXiv:0812.0510 [hep-ph], 2-12.
9. R. Bernabei at al, Int.J.Mod.Phys. D13 (2004) 2127-2160, astro-ph/0501412, astro-ph/0804.2738v1.
10. Z. Ahmed et al., pre-print, astro-ph/0802.3530.
11. C. Amsler et al., Physics Letters B667, 1 (2008).
12. S. Dodelson, Modern Cosmology, Academic Press Elsevier 2003.
13. G. Bregar, M. Khlopov, N.S. Mankoč Borštnik, work in progress.
14. We thank cordially R. Bernabei (in particular) and J. Filippini for very informative discussions by emails and in private communication.
15. V.A. Novikov, L.B. Okun,A.N. Royanov, M.I. Vysotsky, *Phys. Lett.* **B 529**(2002) 111, hep-ph/0111028.
16. Maxim Yu. Khlopov, pre-print, astro-ph/0806.3581.

DARK ENERGY PERTURBATIONS AND A POSSIBLE SOLUTION TO THE COINCIDENCE PROBLEM

JAVIER GRANDE*, ANA PELINSON† and JOAN SOLÀ

High Energy Physics Group, Dept. ECM, and
Institut de Ciències del Cosmos, Univ. de Barcelona,
Av. Diagonal 647, E-08028 Barcelona, Catalonia, Spain
E-mails: jgrande@ecm.ub.es, apelinson@ecm.ub.es, sola@ifae.es

We analyze some generic properties of the dark energy (DE) perturbations, in the case of a self-conserved DE fluid. We also apply a simple test (the "F-test") to compare a model to the data on large scale structure (LSS) under the assumption of negligible DE perturbations. We exemplify our discussions by means of the ΛXCDM model, showing that it provides a viable solution to the cosmological coincidence problem.

Keywords: Dark energy; Cosmological perturbations; Renormalization group.

1. Introduction

In recent times Cosmology has become an accurately testable branch of Physics. Theoretical models can now be confronted with a large quantity of high-precision data coming from different sources, including studies of distant supernovae,[1] the anisotropies of the CMB[2] or the LSS of the Universe.[3] All these observations give strong support to the existence of DE, although the ultimate nature of this component remains a complete mystery.

Remarkably enough, the simplest DE candidate, namely a cosmological constant (CC) Λ, gives rise to a model (ΛCDM) in accurate agreement with all the currently available observational data. Moreover, a general prediction of quantum field theory (QFT) is the existence of a vacuum energy which would precisely take the form of a CC in the Einstein equations. However, the value predicted by the theory happens to be many orders of magnitude larger than the observed DE density. This fact, known as the "cosmological constant problem",[4] makes very unlikely the identification

*Speaker
†Present address: IAG, Univ. de São Paulo, Rua do Matão, CEP 05508-900, S.P., Brazil.

of DE with a strictly constant vacuum energy density, since that would require an extremely fine-tuned cancellation of the different contributions.

The CC problem could be alleviated if we allow the DE to be dynamical. The most popular models exploiting this idea are undoubtedly the scalar field models (XCDM).[5] These models, although well-motivated from the Particle Physics point of view, present two major drawbacks. First of all, the field should have an extremely tiny mass, $m_X \sim H_0 \sim 10^{-33}$ eV, which is even much smaller than the observed value of the mass scale associated to the DE ($\sqrt[4]{\rho_D} \sim 10^{-3}$ eV). And second, in this kind of models one implicitly assumes that the vacuum energy predicted by QFT cancels out for some reason, so the fine-tuning problems associated to the vacuum energy are not solved but simply obviated and traded for those of the scalar field itself. In short, the situation is as follows: on the one hand, from QFT we expect a vacuum energy contribution to the DE in the form of a CC; on the other, we have the popular and well-motivated scalar field models, which may serve to alleviate the CC problem due to their dynamical nature.

Therefore, it seems quite natural to study a more complete model in which the DE combines both ingredients, which we call the ΛXCDM model.[6] The new model presents additional advantages: e.g. Λ need not be constant, but may evolve with a renormalization group (RG) equation, as any other parameter in QFT. The other DE component, the "cosmon" X, need not be a fundamental field either; it could be e.g. an effective representation of dynamical fields of various sorts or even of higher order curvature terms in the action. In fact we do not have to assume anything about the nature of X: its dynamics is completely determined from that of Λ once we have a good ansatz for the latter, due to the fact that both components may exchange energy. In the original ΛXCDM model,[6] the result of these assumptions is a 3-parameter cosmological framework which incorporates the ΛCDM and XCDM models as special cases.

One of the most appealing features of the ΛXCDM model is that it provides a solution to the "cosmological coincidence problem",[4] i.e. the problem of explaining why the energy densities of matter and DE are currently of the same order. In the standard ΛCDM model, this fact is indeed a coincidence since the evolution of the two components is very different; namely, while the DE density remains constant, the matter density decays fast with the scale factor as a^{-3}. In contrast, in the ΛXCDM model, the ratio $r = \rho_D/\rho_M$ between the DE and matter densities may be bounded and not vary too far away from 1 for a significant fraction of the history of the Universe. It means that, for a very long time, ρ_D and ρ_M stay naturally of the same order as they are nowadays. Recently, a generalized version of the ΛXCDM model has been suggested in Ref. 7 with even more far reaching consequences, namely it is able to relax the value of the DE in the present Universe starting from an arbitrary value in the early epochs, i.e. it constitutes an interesting attempt to solve the old CC problem.

Whatever its nature, if the DE is not a strict CC, then, according to cosmological perturbation theory, it should fluctuate. In order to find out its impact on the LSS formation, we will discuss some generic properties of the DE perturbations, exemplifying them by means of the ΛXCDM model. We will also address the question of how the LSS data can be used to constrain a model. The information about LSS is encoded in the galaxy fluctuation power spectrum, $P_{GG}(k)$, which is determined observationally and must be reproduced by the predicted matter power spectrum $P(k) \equiv |\delta_M(k)|^2$ of the theoretical model. A first, and economical, approach to the problem is to simply neglect DE perturbations. Using the fact that the ΛCDM model provides a good fit to the data, we may take it as a reference and impose that the power spectrum of our model does not deviate by more than 10% from the ΛCDM value ("F-test"[8,9]). As we will see, this simple analysis may serve to strongly restrict the parameter space of a model. Nevertheless, it does not reflect some important features that only come to light when making a full study of the combined system of matter and DE perturbations. We will show that such a study[10] is useful not only to check the validity of the previous approach, in the allowed region of parameter space, but it can also help us to further constrain the physical region of that space.

The net result of our analysis of the DE perturbations and its implication on LSS formation is quite rewarding, as we are able to find a sizable region of the ΛXCDM parameter space where the model is in full agreement with LSS data, and other cosmological observations, while providing at the same time a plausible dynamical solution to the cosmological coincidence problem.

2. Dark Energy Perturbations

In this section, we discuss some general properties of the DE perturbations for models in which both matter and DE are self-conserved:

$$\rho_n' + \frac{3}{a}(1+\omega_n)\rho_n = 0 \tag{1}$$

Here a prime denotes differentiation with respect to the scale factor ($f' \equiv df/da$) and $n = M, D$ stands for each of the energy components, matter/radiation and DE. We take $\omega_M = 0$, since we are interested in studying the perturbations in the matter-dominated (MD) era, and we denote the equation of state (EOS) of the DE component as ω_e, where the subindex "e" serves us to remember that the EOS may be an effective one. Let us first introduce the basic equations for the fluctuations, which we derive following the standard approach.[11] For the background space-time we adopt the spatially flat FLRW metric, $ds^2 = dt^2 - a^2\delta_{ij}dx^i dx^j$. We perturb it

$$g_{\mu\nu} \rightarrow g_{\mu\nu} + h_{\mu\nu}\,, \tag{2}$$

keeping only the scalar part of the perturbation. In order to have uniquely defined fluctuations $h_{\mu\nu}$, a gauge choice is mandatory, i.e. we have to choose a specific coordinate system. Here we adopt the synchronous gauge,[11] for which $h_{00} = h_{0i} = 0$.

We should also perturb the energy-momentum tensor, considering thus perturbations on the density, pressure and 4-velocity of each fluid:

$$\rho_n \rightarrow \rho_n + \delta\rho_n\,, \qquad p_n \rightarrow p_n + \delta p_n\,, \qquad U_n^\mu \rightarrow U_n^\mu + \delta U_n^\mu\,. \tag{3}$$

The equations for the fluctuations are then obtained by perturbing the 00-component of the Einstein equations, $R_{\mu\nu} - g_{\mu\nu}R/2 = 8\pi G T_{\mu\nu}$, and the conservation law for the energy momentum-tensor, $\nabla_\mu T^\mu_\nu = 0$. At the end we obtain 5 equations depending on the following set of 7 variables:

$$\hat{h} \equiv \frac{\partial}{\partial t}\left(\frac{h_{ii}}{a^2}\right), \qquad \theta_n \equiv \nabla_\mu(\delta U_n^\mu) = \nabla_j(\delta U_n^j), \qquad \delta_n \equiv \frac{\delta\rho_n}{\rho_n}, \qquad \delta p_n \tag{4}$$

Therefore, in order to solve our system we need to give an expression for δp_D (since for the matter component, indeed $\delta p_M = 0$). In the case of adiabatic perturbations, we simply have $\delta p_D = c_a^2 \delta\rho_D$, where

$$c_a^2 = \frac{p_D'}{\rho_D'} = \omega_e - \frac{a}{3}\frac{\omega_e'}{(1+\omega_e)} \tag{5}$$

is the adiabatic speed of sound of the DE fluid. In general, however, there could be an entropy contribution to the pressure perturbation. In this case, the relation between δp_D and $\delta\rho_D$ in an arbitrary system of reference is given as follows[11]

$$\delta p_D = c_s^2 \delta\rho_D - a^3\,\rho_D'\,H(c_s^2 - c_a^2)\,\frac{\theta_D}{k^2}\,, \tag{6}$$

where c_s^2 is the rest-frame (or *effective*) speed of sound and k is the wave number, as we have moved to Fourier space. This expression is gauge-invariant, and thus it can be computed in any desired gauge, in particular in the synchronous one. When ρ_D is self-conserved, equation (1) holds for ρ_D and, in such case, (6) takes on the form

$$\delta p_D = c_s^2 \delta\rho_D + 3Ha^2(1+\omega_e)\rho_D(c_s^2 - c_a^2)\frac{\theta_D}{k^2}\,. \tag{7}$$

Finally, the equations for the perturbations read:

$$\hat{h}' + \frac{2}{a}\hat{h} - \frac{3H}{a}\tilde{\Omega}_M\delta_M = \frac{3H}{a}\tilde{\Omega}_D\left[(1+3c_s^2)\delta_D + 9a^2H(c_s^2 - c_a^2)\frac{\theta_D}{k^2}\right] \tag{8}$$

$$\delta_M' = -\frac{1}{aH}\left(\theta_M - \frac{\hat{h}}{2}\right) \tag{9}$$

$$\theta_M' = -\frac{2}{a}\theta_M \tag{10}$$

$$\delta_D' = -\frac{(1+\omega_e)}{aH}\left\{\left[1 + \frac{9a^2H^2(c_s^2 - c_a^2)}{k^2}\right]\theta_D - \frac{\hat{h}}{2}\right\} - \frac{3}{a}(c_s^2 - \omega_e)\delta_D \tag{11}$$

$$\theta_D' = -\frac{1}{a}\left(2 - 3c_s^2\right)\theta_D + \frac{k^2}{a^3H}\frac{c_s^2\delta_D}{(1+\omega_e)}, \tag{12}$$

where $\tilde{\Omega}_n(a) \equiv \Omega_n(a)H_0^2/H^2$ and $\Omega_n(a) \equiv \rho_n(a)/\rho_c^0 = 8\pi G\rho_n(a)/(3H_0^2)$. From these equations we get a second-order differential equation for δ_M:

$$\delta_M''(a) + \frac{3}{2}\left[1 - \omega_e(a)\tilde{\Omega}_D(a)\right]\frac{\delta_M'(a)}{a} - \frac{3}{2}\tilde{\Omega}_M(a)\frac{\delta_M(a)}{a^2} =$$
$$= \frac{3}{2}\tilde{\Omega}_D(a)\left[(1+3c_s^2)\frac{\delta_D(a)}{a^2} + 9H(a)\left(c_s^2 - c_a^2\right)\frac{\theta_D(a)}{k^2}\right]. \tag{13}$$

In order to study the properties of the DE perturbations, it is useful to write also a second-order differential equation for δ_D. This equation is much simpler if we use differentiation with respect to the conformal time η ($dt = ad\eta$) and work in the comoving gauge[a]. Defining $\mathcal{H} = (da/d\eta)/a$, it reads:[11]

$$\ddot{\Delta} - \left[3\left(2\omega_e - c_a^2\right) - 1\right]\mathcal{H}\dot{\Delta} + 3\left[\left(\frac{3}{2}\omega_e^2 - 4\omega_e - \frac{1}{2} + 3c_a^2\right)\mathcal{H}^2 + \frac{k^2}{3}c_s^2\right]\Delta = 0\,. \tag{14}$$

2.1. *Generic properties of the DE perturbations*

Once we have shown the basic equations, let us discuss some general properties of the DE perturbations. Looking at Eq. (14), we see that the coefficient

[a]Notice that gauge issues are unimportant for sub-Hubble perturbations, as the ones we study here, so the behavior of the perturbations will not depend on the chosen gauge.

of Δ presents two terms. If it is the second of these terms (the one proportional to k^2) that dominates, and forgetting for a moment about the term proportional to $\dot{\Delta}$, we are left with the equation of a harmonic oscillator. This defines the *sound horizon*, a "Jeans scale" for the DE,

$$\lambda_s = \left| \int_0^\eta c_s \mathrm{d}\eta \right| , \tag{15}$$

such that for scales well inside it, i.e. $l \sim k^{-1} \ll \lambda_s$:

$$\delta_D = C_1 e^{ic_s k\eta} + C_2 e^{-ic_s k\eta} , \tag{16}$$

where C_1 and C_2 are constants, and we have assumed constant c_s^2 for simplicity. Therefore, we see that:

- If $c_s^2 < 0$, the perturbations grow exponentially, situation which is unacceptable for structure formation. As long as ω_e is not varying too fast, $c_a^2 < 0$ [cf. Eq. (5)], so in general the perturbations cannot be adiabatic.
- If $c_s^2 > 0$, the perturbations oscillate. When we take into account the $\dot{\Delta}$ term, what we have is a damped harmonic oscillator, and thus the oscillations have decaying amplitude. As the matter perturbations grow typically as $\delta_M \sim a$, this ensures that $\delta_D/\delta_M \to 0$, i.e. that DE will be a smooth component, as usually assumed. Nevertheless, the larger the scale l or the smaller the speed of sound c_s^2, the more important DE perturbations are, because then $k^{-1} \ll \lambda_s$ is not such a good approximation.

Now the question is whether the scales relevant for the matter power spectrum are really inside the sound horizon or not. The linear regime of the power spectrum lies in the range $0.01h\mathrm{Mpc}^{-1} < k < 0.2h\mathrm{Mpc}^{-1}$ or, equivalently $(600H_0)^{-1} \lesssim \ell \lesssim (30H_0)^{-1}$. On the other hand, we expect[10] that at present $\lambda_s \sim c_s^2(H_0)^{-1}$. Thus we conclude that (at least for c_s^2 not too close to 0), the scales relevant for the observations of LSS are well below the sound horizon, and so the features previously described apply to them.

Inspection of Eqs. (11) and (12) reveals another important property of the DE perturbations: they diverge if the EOS of the DE acquires the value $\omega_e = -1$ (known as the "CC boundary"), i.e. if the DE changes from quintessence-like (QE) behavior ($-1/3 > \omega_e > -1$) to phantom ($\omega_e < -1$) or vice versa. Note that, even though c_a^2 diverges at the crossing [cf. (5)], $(1+\omega_e)c_a^2$ remains finite and, therefore, Eq. (11) is well-behaved. Thus, the problem lies exclusively in the $(1+\omega_e)$ factor in the denominator of (12) and only disappears for vanishing sound of speed c_s^2. One can argue that the physical source of momentum transfer is not θ_D but $\rho_D\mathcal{V}_D$, with $\mathcal{V}_D = \theta_D(1+\omega_e)$,[12] and hope to get rid of the divergence through such a redefinition of variables, but unfortunately this is not the case. Getting around this difficulty is not always possible, and in fact there is no way for a single scalar field (or single fluid) model to cross the CC boundary,[12] and even with two fields some very special conditions should be arranged. In the absence of a mechanism to avoid this singularity, we are forced to restrict our parameter space by removing the points that present such a crossing in the past.

3. The ΛXCDM Model

The properties discussed in the previous section apply in principle to any model in which the DE is self-conserved. The ΛXCDM model, introduced in Ref. 6 as a possible explanation to the cosmological coincidence problem, constitutes a non-trivial example of these kind of models. In it, the DE is a composite fluid, constituted by a variable CC and another generic component X, which can exchange energy with Λ:

$$\rho_D = \rho_\Lambda + \rho_X \,. \tag{17}$$

The evolution of Λ can be (as any parameter in QFT) tied to the RG in curved space-time:[13]

$$\frac{\mathrm{d}\rho_\Lambda}{\mathrm{d}\ln\mu} = \frac{3\nu}{4\pi} M_P^2 \mu^2 \longrightarrow \rho_\Lambda = \rho_\Lambda^0 + \frac{3\nu}{8\pi} M_P^2 (H^2 - H_0^2) \,, \tag{18}$$

where we identified μ (the energy scale associated to the RG in Cosmology) with the Hubble function H at any epoch and ν is a free, dimensionless, parameter related to the mass ratio (squared) of the heavy particles contributing to the running versus the Planck mass.[13] While the ultimate justification for this ansatz is the application of the RG method and the general considerations of covariance of the effective action in QFT in curved space-time, a more profound study is needed, see Refs. 14 and 10 (section VI) for a more detailed discussion. Interestingly enough, the evolution law (18) can be tested from different points of view,[15,16] including cosmological perturbations[17](see also Ref. 18 for related phenomenological studies).

It should be clear that, in spite of the dynamical nature of Λ, $\omega_\Lambda = -1$, and it is in this sense that we may call it a "cosmological constant". As for the X component, we assume that it has a constant EOS lying in the range $-1 - \delta < \omega_X < -1/3$ (where $\delta > 0$ is small). We need not make any assumption about the nature of the cosmon, since its evolution becomes determined by that of the CC through the energy conservation equation:

$$\rho_X' + \rho_\Lambda' + \frac{3}{a}(1 + \omega_X)\rho_X = 0 \,. \tag{19}$$

The solution of the model in the MD era can be found from Eqs. (18), (19) and the Friedmann equation:

$$H^2 = H_0^2[\Omega_M(a) + \Omega_D(a)]\,, \tag{20}$$

with $\Omega_M(a) = \Omega_M^0 a^{-3}$. For the normalized DE density, we find:

$$\Omega_D(a) = \frac{\Omega_\Lambda^0 - \nu}{1-\nu} + \frac{\epsilon\,\Omega_M^0\,a^{-3}}{w_X - \epsilon} + \left[\frac{1-\Omega_\Lambda^0}{1-\nu} - \frac{w_X\Omega_M^0}{w_X - \epsilon}\right] a^{-3(1+w_X-\epsilon)}\,, \tag{21}$$

where we have defined $\epsilon \equiv \nu(1+w_X)$. Assuming as a prior that $\Omega_D^0 = \Omega_X^0 + \Omega_\Lambda^0 \simeq 0.7$, we are left with 3 free parameters: ν, the parameter that controls the running of Λ; ω_X, the barotropic index of the X component; and Ω_Λ^0, the current energy density of the CC. Let us note that the model includes as special cases both the ΛCDM ($\nu = 0$, $\Omega_X^0 = 0$) and XCDM ($\nu = 0$, $\Omega_\Lambda^0 = 0$) models. The effective EOS parameter of the model,

$$\omega_e(a) = -1 - \frac{a}{3}\frac{1}{\Omega_D(a)}\frac{d\Omega_D(a)}{da}\,, \tag{22}$$

can present a variety of behaviors[6] compatible with $\omega_e(a_0) \simeq -1$ (the subindex 0 standing for the present value), as suggested by observations.[2]

3.1. *The coincidence problem*

In order to understand why the ΛXCDM model can provide an explanation for the coincidence problem, it is convenient to consider the ratio r between the DE and matter energy densities, which in the standard ΛCDM model reads:

$$r \equiv \frac{\Omega_D}{\Omega_M} = \frac{\Omega_\Lambda^0}{\Omega_M^0}a^3\,. \tag{23}$$

We see that r tends to zero in the past and grows unboundedly in the future. Only at the present time we have $r \simeq 1$. The unavoidable conclusion seems to be that we live in a very special moment, namely one very close to the time when the expansion of the Universe started to be accelerated.

In contrast, in the ΛXCDM, r reads as follows:

$$r = \frac{(\Omega_\Lambda^0 - \nu)a^3}{(1-\nu)\,\Omega_M^0} + \frac{\epsilon}{\omega_X - \epsilon} + \left[\frac{1-\Omega_\Lambda^0}{\Omega_M^0(1-\nu)} - \frac{\omega_X}{\omega_X - \epsilon}\right] a^{-3\,(\omega_X - \epsilon)}\,. \tag{24}$$

Such, more complex, structure allows for the existence of a maximum of this ratio in the future, which implies that r may be bounded and relatively small (not very different from 1), say $r \leq 10\,r_0$, for a very prolonged stretch of the history of the Universe. In this case, the value $r \sim 1$ would no longer be seen as special.

It is important to note that the ability to solve the coincidence problem is a very general feature of the model. In order to show that, let us recall that the solution to the coincidence problem is linked to the existence of a future stopping of the Universe expansion.[6] Now, from the Friedmann equation (20), it is clear that it is necessary that the DE density becomes negative for the expansion to stop. But this condition can be realized even in the simplest setups of the ΛXCDM model. Let us assume e.g. that $\nu = 0$, so there is no exchange of energy between the CC and the X component:

$$\Omega_D = \Omega_\Lambda^0 + \Omega_X^0 a^{-3(1+\omega_X)} . \tag{25}$$

In this case we have a truly constant Λ and the cosmon behaves effectively as a QE/phantom scalar field; Ω_D will eventually become negative if any of the following conditions is fulfilled:

$$\Omega_\Lambda^0 < 0 \text{ and } -1 < \omega_X < -1/3 \qquad \text{or} \qquad \Omega_X^0 < 0 \text{ and } \omega_X < -1 . \tag{26}$$

Let us also stress that in the ΛXCDM the individual components are not observable, the only thing we can measure is the total Ω_D. Therefore, there is no problem in having a negative value for Ω_Λ^0 or Ω_X^0, as long as $\Omega_D^0 = 0.7$. Remember also that X need not be a real fluid, its nature could be effective.

In Fig. 1a we show that there is a large 3D-volume of the parameter space for which this solution to the coincidence problem is possible (the projections of that volume onto three orthogonal planes are shown as the shaded regions in Fig. 1b, c, d.) All the points in it present a relatively low maximum of the ratio r ($r_{\rm max} \leq 10\, r_0$), ratio that, in addition, is small enough at the nucleosynthesis epoch ($r_N \lesssim 0.1$, where in this case Ω_M is the density of radiation and Ω_D is to be computed in the radiation-dominated era), so as to make sure that the predictions of the Big Bang model are not spoiled.

4. Perturbations in the ΛXCDM Model

In this section we address the problem of how the parameter space of a model can be constrained by means of LSS data, using as an example the ΛXCDM model. As we saw in Sec. 2.1, at the scales relevant to the linear part of the matter power spectrum the DE perturbations are expected to be negligible as compared to the matter ones. Thus, a reasonable approach is to simply neglect the DE perturbations from Eq. (13):

$$\delta_M''(a) + \frac{3}{2}\left[1 - \omega_e(a)\tilde{\Omega}_D(a)\right]\frac{\delta_M'(a)}{a} - \frac{3}{2}\,\tilde{\Omega}_M(a)\,\frac{\delta_M(a)}{a^2} = 0 , \tag{27}$$

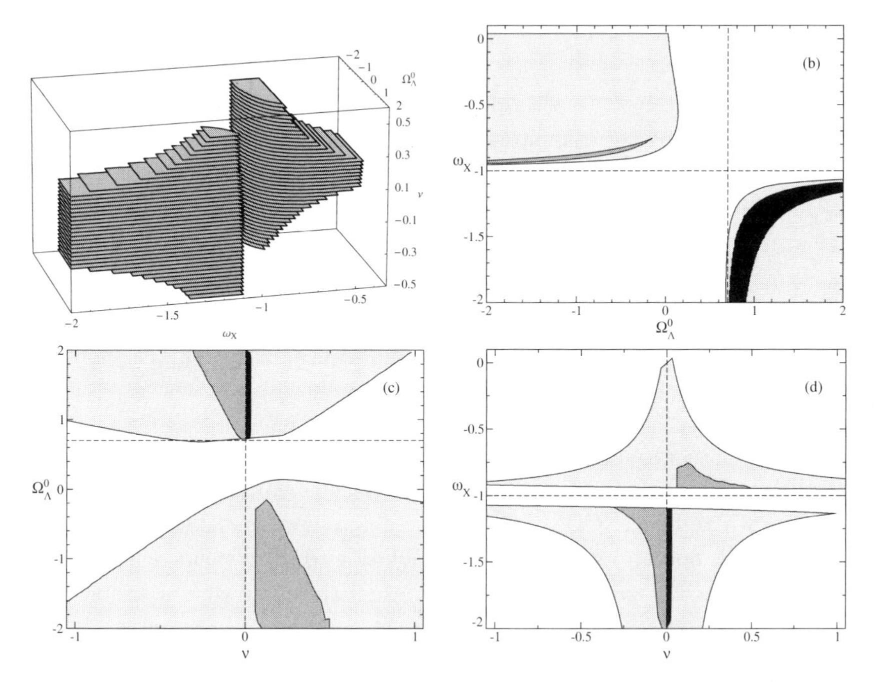

Fig. 1. (a) 3D volume constituted by the points of the ΛXCDM parameter space which provide a solution to the coincidence problem, presenting a low maximum of the ratio r (24), $r_{\max} \leq 10\, r_0$, and satisfy the nucleosynthesis bound $r_N \lesssim 0.1$ (see the text); (b), (c) and (d) Projections of the 3D volume in (a) onto the perpendicular planes $\nu = 0$, $\omega_e = 0$ and $\Omega^0_\Lambda = 0$ (all the shaded area). When we ask for the F-test (29) to be fulfilled and the current value of the EOS to be close to -1 (30), we are left with the medium and dark-shaded regions. Finally, by considering DE perturbations, we are forced to exclude the points for which the equations are ill-defined, i.e. those for which the EOS of the DE acquires the value -1 at some point in the past. By doing so we get the final physical parameter space of the ΛXCDM model (dark-shaded region).

and study the evolution of the perturbations from some initial scale factor $a = a_i$ in the MD era (where $\delta_M \sim a$) until the present time, $a_0 = 1$. As Eq. (27) does not depend on k, all the scales grow in the same fashion and we can characterize models by means of the "growth factor":

$$D(a) = \frac{\delta_M(a)}{\delta_M^{\rm CDM}(a_0)}\,, \tag{28}$$

whose present value, $D(a_0)$, compares the growth of the perturbations in the model considered to the growth in a pure cold dark matter (CDM) model. The parameter that measures the agreement between the observed galaxy distribution power spectrum, $P_{GG}(k)$, and the matter power spectrum of a model, $P(k) \equiv |\delta_M(k)|^2$, is the linear bias, which at the present time is defined as $b^2(a_0) = P_{GG}/P$. Most remarkably, the LSS data point to the value $b^2_\Lambda(a_0) = 1$, to within a 10% accuracy, for the ΛCDM model.[3] This suggests that the comparison to the ΛCDM can be a valid and more economical criterion for studying the viability of a model. In particular, we may require that any DE model should pass the following "F-test":[8]

$$|F| \equiv \left|1 - \frac{b^2(a_0)}{b^2_\Lambda(a_0)}\right| = \left|1 - \frac{P_\Lambda(a_0)}{P(a_0)}\right| = \left|1 - \frac{D^2_\Lambda(a_0)}{D^2(a_0)}\right| \leq 0.1 \,. \tag{29}$$

This was done for the ΛXCDM in Ref. 9, where, in addition, we imposed that the current value of the EOS parameter of the DE is close to -1:

$$|\omega_e(a_0) + 1| \leq 0.3 \,, \tag{30}$$

as suggested by recent observational limits[2,b]. As seen in Fig. 1b, c, d, there is still a sizable region of the parameter space (medium and dark-shaded regions) where the ΛXCDM satisfies these two new conditions and the nucleosynthesis bound, while providing a solution to the coincidence problem.

Neglecting DE perturbations provides us therefore with a simple and effective method to constrain the parameter space of a model. Although we expect it to be a reasonable approximation, we cannot be completely sure unless we perform a full analysis in which the DE fluctuations are also included. Such an analysis[10] implies an immediate and very important consequence. As discussed in Sec. 2.1, if the effective EOS of the model crosses the CC boundary ($\omega_e = -1$) at some point in the past, the perturbation equations will diverge. In the absence of a mechanism to get around this singularity (and indeed we cannot have it without a microscopic definition of the X component), we are forced to restrict our parameter space by removing the points that present such a crossing. This new constraint knocks out many of the points allowed by the previous simple analysis (we are left with the dark-shaded region in Fig. 1b, c, d) and so we end up with a rather definite prediction for the values of the parameters of the ΛXCDM model. It is worth noticing that only small (and positive) values of ν are allowed, $\nu \sim 10^{-2}$ at most, which is in very good agreement with theoretical expectations.[10] Another interesting consequence of the new constraint is that the effective EOS of the DE can be QE-like only,[10] i.e. $-1 < \omega_e < -1/3$.

We want to compare the matter power spectrum predicted by the ΛXCDM model, $P_{\Lambda X}(k)$, with the $P_{GG}(k)$[c] measured by the 2dFGRS col-

---

[b]Let us remark, though, that such limits on the EOS parameter are usually derived under the assumption of a constant ω_e and, therefore, do not strictly apply to our model.
[c]And also with the ΛCDM spectrum, $P_\Lambda(k)$, which provides a good fit to $P_{GG}(k)$

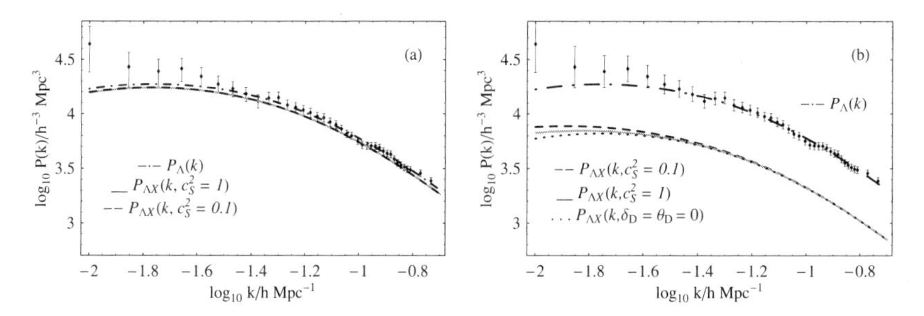

Fig. 2. The 2dFGRS observed galaxy power spectrum,[3] $P_{GG}(k)$ (points), and the ΛCDM power spectrum, $P_\Lambda(k)$ (dot-dashed line) versus the spectrum predicted by the ΛXCDM, $P_{\Lambda X}(k)$, for DE sound speeds $c_s^2 = 0.1$ (dashed line) and $c_s^2 = 1$ (solid/gray line): (a) for a set of parameters allowed by the analysis of Ref. 9 (in the dark-shaded region of Fig. 1b, c, d), $\Omega_\Lambda^0 = 0.8$, $\nu = \nu_0 \equiv 2.6 \times 10^{-2}$ and $w_X = -1.6$; (b) for a set of parameters satisfying all the conditions in that analysis but the F-test, $\Omega_\Lambda^0 = +0.35$, $\nu = -0.2$ and $w_X = -0.6$. In this case it is also shown the power spectrum obtained by neglecting DE perturbations (dotted line), which presents the same shape as $P_\Lambda(k)$.

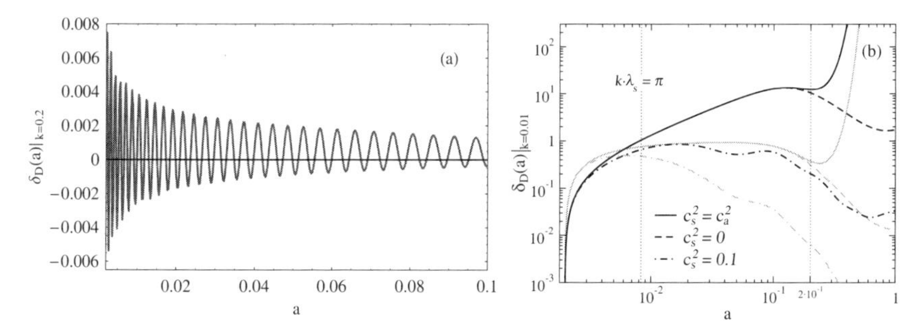

Fig. 3. (a) The ΛXCDM growth of DE perturbations for a small scale $k = 0.2$ (in units of $h\,\mathrm{Mpc}^{-1}$) and the same set of parameters assumed in Fig. 2a, and for DE sound speed $c_s^2 = 0.1$; (b) Evolution of the DE perturbations δ_D (black lines) for the same set of parameters as in (a), at the large scale $k = 0.01$ and for three different speeds of sound: $c_s^2 = c_a^2 < 0$ (solid line), $c_s^2 = 0$ (dashed line) and $c_s^2 = 0.1$ (dot-dashed line). The evolution of the ratio δ_D/δ_M is also shown (gray lines).

laboration.[3] The former can be found by evolving the perturbation equations (8)-(12) from $a = a_i$ to $a_0 = 1$, where now $a_i \ll 1$ is the scale factor at some time well after recombination. In order to set the initial conditions, we took into account that the DE does not begin to play an important role until very recently, so that the values of the metric and matter perturbations at $a = a_i$ should be the same for our model and for the ΛCDM model - the power spectrum $P_\Lambda(k)$ of the latter being available from standard analytical fits in the literature, see Ref. 10 and references therein. As for the

DE perturbations, given that we expect them to be negligible at any time, we assumed that they vanish at $a = a_i$.

The ΛXCDM power spectrum was calculated for two different fiducial sound speeds, $c_s^2 = 1$ and $c_s^2 = 0.1$ and several combinations of the parameters ν, ω_X and Ω_Λ^0. For values of the parameters *not* fulfilling the F-test (even though satisfying all the other conditions stated in Fig. 1) we obtain huge discrepancies (a large suppression) in the amount of growth with respect to the ΛCDM model (cf. Fig. 2b), as expected. This suppression is typical of the QE-like behavior [10] and occurs even if the DE perturbations are neglected (dotted line), in which case $P_{\Lambda X}(k)$ presents exactly the same shape as $P_\Lambda(k)$ (because then the k-dependence disappears from the equations, which reduce just to Eq. (27)). The effect of considering DE perturbations is only visible at large scales (small k), where they tend to compensate the aforementioned suppression. The smaller the speed of sound or the larger the scale, the more important the effect of DE perturbations, as expected from the general considerations of Sec. 2.1.

In contrast, in Fig. 2a we see that for values allowed by the F-test (and satisfying all the other constraints as well, i.e. lying in the dark-shaded region in Fig. 1b, c, d), $P_{\Lambda X}(k)$ is very similar to $P_\Lambda(k)$, with numerical results in very good agreement with those obtained through the F-method.[9,10] In particular, their shape is identical, indicating that DE perturbations do not play a role here. Indeed, in Fig. 3a we see that δ_D oscillates with decreasing amplitude, as predicted in Sec. 2.1. For positive sound speed, the perturbations get stabilized (and therefore the ratio δ_D/δ_M becomes negligible) once the sound horizon (15) is crossed, i.e. when $k\lambda_s = \pi$, as seen in Fig. 3b. Similarly, the adiabatic ($c_a^2 < 0$) perturbations begin their exponential growth once c_a^2 (which is negligible in the far past in the ΛXCDM model[6]), and thus the term proportional to k in (14), becomes important.

5. Conclusions

We have analyzed the behavior of the DE perturbations in models with self-conserved DE. We have exemplified them by means of the ΛXCDM model,[6] which is a non-trivial model of the cosmic evolution with a number of appealing properties. Unlike other proposed solutions to the coincidence problem (an incomplete list includes tracking scalar fields, interactive QE models, K-essence, Chaplygin gas, etc - see e.g. Ref. 10 and references therein), the ΛXCDM model accounts for the energy of vacuum through a (possibly running) Λ, giving allowance for other dynamical contributions, X, of general nature. The comparison of the ΛXCDM power spectrum to the LSS data, first by means of the F-test and then through a full analysis of the DE perturbations, resulted in a strong additional constraint on the parameter space of the model, hence increasing its predictive power and pinpointing a region where the ΛXCDM model provides a realistic solution to the coincidence problem, i.e. fully compatible with present observations.

Acknowledgments

Authors have been supported in part by MEC and FEDER under project FPA2007-66665 and also by DURSI Generalitat de Catalunya under project 2005SGR00564. We acknowledge the support from the Spanish Consolider-Ingenio 2010 program CPAN CSD2007-00042.

References

1. R. Knop *et al.*, *Astrophys. J.* **598** (2003) 102; A. Riess *et al. Astrophys. J.* **607** (2004) 665.
2. D.N. Spergel *et al.* [WMAP Collaboration], *Astrophys. J. Suppl.* **170** (2007) 377.
3. S. Cole et al., *Mon. Not. Roy. Astron. Soc.* **362** (2005) 505.
4. See e.g. S. Weinberg, *Rev. Mod. Phys.* **61** (1989) 1; T. Padmanabhan, *Phys. Rep.* **380** (2003) 235.
5. P.J.E. Peebles and B. Ratra, *Rev. Mod. Phys.* **75** (2003) 559.
6. J. Grande, J. Solà and H. Štefančić, JCAP **08** (2006) 011; *Phys. Lett.* **B645** (2007) 236; J. Phys. A 40 (2007) 6787.
7. F. Bauer, J. Solà, H. Štefančić, *Relaxing a large cosmological constant*, arXiv:0902.2215 [hep-th].
8. R. Opher and A. Pelinson, [arXiv:astro-ph/0703779].
9. J. Grande, R. Opher, A. Pelinson and J. Solà, JCAP **0712** (2007) 007.
10. J. Grande, A. Pelinson and J. Sola, Phys. Rev. D **79** (2009) 043006.
11. H. Kodama and M. Sasaki, *Prog. Theor. Phys. Supp.* **78** (1984) 1.
12. R.R. Caldwell and M. Doran, *Phys. Rev.* **D72** (2005) 043527.
13. I.L. Shapiro and J. Solà, JHEP 0202 (2002) 006; *Phys. Lett.* **B475B** (2000) 236.
14. I.L. Shapiro, J. Solà, *Can the cosmological 'constant' run? - It may run.*, arXiv:0808.0315 [hep-th].
15. I.L. Shapiro, J. Solà, C. España-Bonet and P. Ruiz-Lapuente, *Phys. Lett.* **B574** (2003) 149; JCAP **0402** (2004) 006; I.L. Shapiro and J. Solà, *Nucl. Phys. Proc. Supp.* **127** (2004) 71 [arXiv:hep-ph/0305279]; JHEP proc. AHEP2003/013, 2004, [arXiv:astro-ph/0401015].
16. J. Solà, H. Štefančić, *Mod. Phys. Lett.* **A21** (2006) 479; *Phys. Lett.* **B624** (2005) 147.
17. J.C. Fabris, I.L. Shapiro and J. Solà, JCAP **0702** (2007) 016.
18. S. Basilakos, *Cosmological implications and structure formation from a time varying vacuum*, arXiv:0903.0452 [astro-ph] (to appear in MNRAS).

PART X

Gravitation, Dark Energy, Dark Matter

DARK ENERGY WITHOUT DARK ENERGY: AVERAGE OBSERVATIONAL QUANTITIES

DAVID L. WILTSHIRE†

Department of Physics and Astronomy, University of Canterbury, Private Bag 4800, Christchurch 8140, New Zealand
†E-mail: David.Wiltshire@canterbury.ac.nz
http://www2.phys.canterbury.ac.nz/~dlw24/

I discuss potential observational tests of a "radically conservative" solution to the problem of dark energy in cosmology, in which the apparent acceleration of the universe is understood as a consequence of gravitational energy gradients that grow when spatial curvature gradients become significant with the nonlinear growth of cosmic structure. In particular, I discuss measures equivalent to the dark energy equation of state, baryon acoustic oscillation statistic D_V, $H(z)$, the $Om(z)$ diagnostic, an average inhomogeneity diagnostic, and the time–drift of cosmological redshifts.

Keywords: Dark energy; theoretical cosmology; observational cosmology.

1. Introduction

I will discuss some recent work[1] that further develops the cosmological model I presented at the previous conference in the Heidelberg series, Dark2007 in Sydney, which was discussed at length in the proceedings.[2] My theme is a radically conservative alternative to the standard homogeneous isotropic cosmology with vacuum energy or fluid–like dark energy.

Although the standard Lambda Cold Dark Matter (ΛCDM) model, provides a good fit to many tests, there are tensions between some tests, and also a number of puzzles and anomalies. Furthermore, at the present epoch the observed universe is only statistically homogeneous once one samples on scales of 150–300 Mpc. Below such scales it displays a web–like structure, dominated in volume by voids. Some 40%–50% of the volume of the present epoch universe is in voids with $\delta\rho/\rho \sim -1$ on scales of $30h^{-1}$ Mpc,[3] where h is the dimensionless parameter related to the Hubble constant by $H_0 = 100h\,\text{km sec}^{-1}\,\text{Mpc}^{-1}$. Once one also accounts for numerous minivoids, and perhaps also a few larger voids, then it appears that the present epoch universe is void-dominated. Clusters of galaxies are spread in sheets that surround these voids, and thin filaments that thread them.

One particular consequence of a matter distribution that is only statistically homogeneous, rather than exactly homogeneous, is that when the Einstein equations are averaged they do not evolve as a smooth Friedmann–Lemaître–Robertson–Walker (FLRW) geometry. Instead the Friedmann equations are supplemented by additional backreaction terms.[4] Whether or not one can fully explain the expansion history of the universe as a consequence of the growth of inhomogeneities and backreaction, without a fluid–like dark energy, is the subject of ongoing debate.[5]

Over the past two years I have developed a new physical interpretation of cosmological solutions within the Buchert averaging scheme.[6–8] I start by noting that in the presence of strong spatial curvature gradients, not only should the average evolution equations be replaced by equations with terms involving backreaction, but the physical interpretation of average quantities must also account for the differences between the local geometry and the average geometry. In other words, geometric variance can be just as important as geometric averaging when it comes to the physical interpretation of the expansion history of the universe.

I proceed from the fact that structure formation provides a natural division of scales in the observed universe. As observers in galaxies, we and the objects we observe in other galaxies are necessarily in bound structures, which formed from density perturbations that were greater than critical density. If we consider the evidence of the large scale structure surveys on the other hand, then the average location by volume in the present epoch universe is in a void, which is negatively curved. We can expect systematic differences in spatial curvature between the average mass environment, in bound structures, and the volume-average environment, in voids.

Spatial curvature gradients will in general give rise to gravitational energy gradients. Physically this can be understood in terms of a relative deceleration of expanding regions of different densities. Those in the denser region decelerate more and age less. Since we are dealing with weak fields the relative deceleration of the background is small. Nonetheless even if the relative deceleration is typically of order $10^{-10}\mathrm{ms}^{-2}$, cumulatively over the age of the universe it leads to significant clock rate variances.[8] I proceed from an ansatz that the variance in gravitational energy is correlated with the average spatial curvature in such a way as to implicitly solve the Sandage–de Vaucouleurs paradox that a statistically quiet, broadly isotropic, Hubble flow is observed deep below the scale of statistical homogeneity. Expanding regions of different densities are patched together so that the regionally measured expansion, in terms of the variation of the regional proper length, $\ell_r = \mathcal{V}^{1/3}$, with respect to proper time of isotropic observers (those who see an isotropic mean CMB), remains uniform. Although voids open up faster, so that their proper volume increases more quickly, on account of gravitational energy gradients the local clocks will also tick faster in a compensating manner.

Details of the fitting of local observables to average quantities for solutions to the Buchert formalism are described in detail in refs.[6,7] Negatively curved voids, and spatially flat expanding wall regions within which galaxy clusters are located, are combined in a Buchert average

$$f_{\rm v}(t) + f_{\rm w}(t) = 1, \tag{1}$$

where $f_{\rm w}(t) = f_{\rm wi} a_{\rm w}{}^3/\bar{a}^3$ is the *wall volume fraction* and $f_{\rm v}(t) = f_{\rm vi} a_{\rm v}{}^3/\bar{a}^3$ is the *void volume fraction*, $\mathcal{V} = \mathcal{V}_{\rm i}\bar{a}^3$ being the present horizon volume, and $f_{\rm wi}$, $f_{\rm vi}$ and $\mathcal{V}_{\rm i}$ initial values at last scattering. The time parameter, t, is the volume–average time parameter of the Buchert formalism, but does not coincide with that of local measurements in galaxies. In trying to fit a FLRW solution to the universe we attempt to match our local spatially flat wall geometry

$$ds^2_{fi} = -d\tau^2 + a_{\rm w}{}^2(\tau)\left[d\eta_{\rm w}^2 + \eta_{\rm w}^2 d\Omega^2\right] . \tag{2}$$

to the whole universe, when in reality the rods and clocks of ideal isotropic observers vary with gradients in spatial curvature and gravitational energy. By conformally matching radial null geodesics with those of the Buchert average solutions, the geometry (2) may be extended to cosmological scales as the dressed geometry

$$ds^2 = -d\tau^2 + a^2(\tau)\left[d\bar{\eta}^2 + r_{\rm w}^2(\bar{\eta},\tau)\, d\Omega^2\right] \tag{3}$$

where $a = \bar{\gamma}^{-1}\bar{a}$, $\bar{\gamma} = \frac{dt}{d\tau}$ is the relative lapse function between wall clocks and volume–average ones, $d\bar{\eta} = dt/\bar{a} = d\tau/a$, and $r_{\rm w} = \bar{\gamma}\,(1-f_{\rm v})^{1/3} f_{\rm wi}{}^{-1/3}\eta_{\rm w}(\bar{\eta},\tau)$, where $\eta_{\rm w}$ is given by integrating $d\eta_{\rm w} = f_{\rm wi}{}^{1/3} d\bar{\eta}/[\bar{\gamma}\,(1-f_{\rm v})^{1/3}]$ along null geodesics.

In addition to the bare cosmological parameters which describe the Buchert equations, one obtains dressed parameters relative to the geometry (3). For example, the dressed matter density parameter is $\Omega_M = \bar{\gamma}^3\bar{\Omega}_M$, where $\bar{\Omega}_M = 8\pi G\bar{\rho}_{M0}\bar{a}_0^3/(3\bar{H}^2\bar{a}^3)$ is the bare matter density parameter. The dressed parameters take numerical values close to the ones inferred in standard FLRW models.

2. Apparent Acceleration and Hubble Flow Variance

The gradient in gravitational energy and cumulative differences of clock rates between wall observers and volume average ones has important physical consequences. Using the exact solution obtained in ref.,[7] one finds that a volume average observer would infer an effective deceleration parameter $\bar{q} = -\ddot{\bar{a}}/(\bar{H}^2\bar{a}) = 2\,(1-f_v)^2/(2+f_v)^2$, which is always positive since there is no global acceleration. However, a wall observer infers a dressed deceleration parameter

$$q = -\frac{1}{H^2 a}\frac{\mathrm{d}^2 a}{\mathrm{d}\tau^2} = \frac{-\,(1-f_v)\,(8f_v{}^3+39f_v{}^2-12f_v-8)}{\left(4+f_v+4f_v{}^2\right)^2}\,, \tag{4}$$

where the dressed Hubble parameter is given by

$$H = a^{-1}\,\tfrac{\mathrm{d}}{\mathrm{d}\tau}\,a = \bar{\gamma}\bar{H} - \dot{\bar{\gamma}} = \bar{\gamma}\bar{H} - \bar{\gamma}^{-1}\,\tfrac{\mathrm{d}}{\mathrm{d}\tau}\,\bar{\gamma}\,. \tag{5}$$

At early times when $f_v \to 0$ the dressed and bare deceleration parameter both take the Einstein–de Sitter value $q \simeq \bar{q} \simeq \frac{1}{2}$. However, unlike the bare parameter which monotonically decreases to zero, the dressed parameter becomes negative when $f_v \simeq 0.59$ and $\bar{q} \to 0^-$ at late times. For the best-fit parameters[9] the apparent acceleration begins at a redshift $z \simeq 0.9$.

Cosmic acceleration is thus revealed as an apparent effect which arises due to the cumulative clock rate variance of wall observers relative to volume–average observers. It becomes significant only when the voids begin to dominate the universe by volume. Since the epoch of onset of apparent acceleration is directly related to the void fraction, f_v, this solves one cosmic coincidence problem.

In addition to apparent cosmic acceleration, a second important apparent effect will arise if one considers scales below that of statistical homogeneity. By any one set of clocks it will appear that voids expand faster than wall regions. Thus a wall observer will see galaxies on the far side of a dominant void of diameter $30h^{-1}$ Mpc to recede at a value greater than the dressed global average H_0, while galaxies within an ideal wall will recede at a rate less than H_0. Since the uniform bare rate $\bar{H}$ would also be the local value within an ideal wall, eq. (5) gives a measure of the variance in the apparent Hubble flow. The best fit parameters[9] give a dressed Hubble constant $H_0 = 61.7^{+1.2}_{-1.1}\,\mathrm{km\ sec^{-1}\,Mpc^{-1}}$, and a bare Hubble constant $\bar{H}_0 = 48.2^{+2.0}_{-2.4}\,\mathrm{km\ sec^{-1}\,Mpc^{-1}}$. The present epoch variance is 22%, and we can expect the Hubble constant to attain local maximum values of order $72.3\,\mathrm{km\ sec^{-1}\,Mpc^{-1}}$ when measured over local voids.

Since voids dominate the universe by volume at the present epoch, any observer in a galaxy in a typical wall region will measure locally higher values of the Hubble constant, with peak values of order 72 km $\text{sec}^{-1}\,\text{Mpc}^{-1}$ at the $30h^{-1}$ Mpc scale of the dominant voids. Over larger distances, as the line of sight intersects more walls as well as voids, a radially spherically symmetric average will give an average Hubble constant whose value decreases from the maximum at the $30h^{-1}$ Mpc scale to the dressed global average value, as the scale of homogeneity is approached at roughly the baryon acoustic oscillation (BAO) scale of $110h^{-1}$Mpc. This predicted effect could account for the Hubble bubble[10] and more detailed studies of the scale dependence of the local Hubble flow.[11]

In fact, the variance of the local Hubble flow below the scale of homogeneity should correlate strongly to observed structures in a manner which has no equivalent prediction in FLRW models.

3. Future Observational Tests

There are two types of potential cosmological tests that can be developed; those relating to scales below that of statistical homogeneity as discussed above, and those that relate to averages on our past light cone on scales much greater than the scale of statistical homogeneity. The second class of tests includes equivalents to all the standard cosmological tests of the standard model of a Newtonianly perturbed FLRW model. This second class of tests can be further divided into tests which just deal with the bulk cosmological averages (luminosity and angular diameter distances etc), and those that deal with the variance from the growth of structures (late epoch integrated Sachs–Wolfe effect, cosmic shear, redshift space distortions etc). Here I will concentrate solely on the simplest tests which are directly related to luminosity and angular diameter distance measures.

In the timescape cosmology we have an effective dressed luminosity distance

$$d_L = a_0(1+z)r_{\rm w}, \tag{6}$$

where $a_0 = \bar{\gamma}_0^{-1}\bar{a}_0$, and

$$r_{\rm w} = \bar{\gamma}\,(1-f_{\rm v})^{1/3} \int_t^{t_0} \frac{\mathrm{d}t'}{\bar{\gamma}(t')(1-f_{\rm v}(t'))^{1/3}\bar{a}(t')}\,. \tag{7}$$

We can also define an *effective angular diameter distance*, d_A, and an *effective comoving distance*, D, to a redshift z in the standard fashion

$$d_A = \frac{D}{1+z} = \frac{d_L}{(1+z)^2}\,. \tag{8}$$

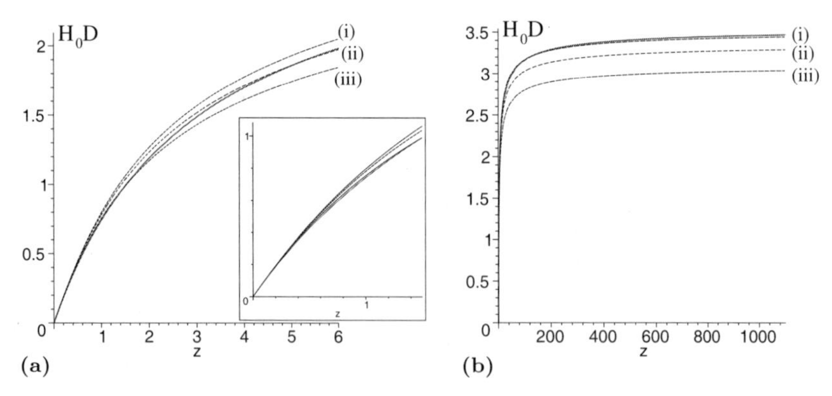

Fig. 1. The effective comoving distance $H_0D(z)$ is plotted for the best–fit timescape (TS) model, with $f_{v0} = 0.762$, (solid line); and for various spatially flat ΛCDM models (dashed lines). The parameters for the dashed lines are (i) $\Omega_{M0} = 0.249$ (best–fit to WMAP5 only[12]); (ii) $\Omega_{M0} = 0.279$ (joint best–fit to SneIa, BAO and WMAP5); (iii) $\Omega_{M0} = 0.34$ (best–fit to Riess07 SneIa only[13]). Panel (a) shows the redshift range $z < 6$, with an inset for $z < 1.5$, which is the range tested by current SneIa data. Panel (b) shows the range $z < 1100$ up to the surface of last scattering, tested by WMAP5.

A direct method of comparing the distance measures with those of homogeneous models with dark energy, is to observe that for a standard spatially flat cosmology with dark energy obeying an equation of state $P_D = w(z)\rho_D$, the quantity

$$H_0 D = \int_0^z \frac{\mathrm{d}z'}{\sqrt{\Omega_{M0}(1+z')^3 + \Omega_{D0} \exp\left[3\int_0^{z'} \frac{(1+w(z''))\mathrm{d}z''}{1+z''}\right]}}, \tag{9}$$

does not depend on the value of the Hubble constant, H_0, but only directly on $\Omega_{M0} = 1 - \Omega_{D0}$. Since the best-fit values of H_0 are potentially different for the different scenarios, a comparison of H_0D curves as a function of redshift for the timescape model versus the ΛCDM model gives a good indication of where the largest differences can be expected, independently of the value of H_0. Such a comparison is made in Fig. 1.

We see that as redshift increases the timescape model interpolates between ΛCDM models with different values of Ω_{M0}. For redshifts $z \lesssim 1.5$ $D_{\rm TS}$ is very close to $D_{\Lambda\rm CDM}$ for the parameter values $(\Omega_{M0}, \Omega_{\Lambda 0}) = (0.34, 0.66)$ (model (ii)) which best–fit the Riess07 supernovae (SneIa) data[13] only, by our own analysis. For very large redshifts that approach the surface of last scattering, $z \lesssim 1100$, on the other hand, $D_{\rm TS}$ very closely matches $D_{\Lambda\rm CDM}$ for the parameter values $(\Omega_{M0}, \Omega_{\Lambda 0}) = (0.249, 0.751)$ (model (i)) which best–fit WMAP5 only.[12] Over redshifts $2 \lesssim z \lesssim 10$, at which scales independent tests are conceivable, $D_{\rm TS}$ makes a transition over corresponding curves of $D_{\Lambda\rm CDM}$ with intermediate values of $(\Omega_{M0}, \Omega_{\Lambda 0})$. The $D_{\Lambda\rm CDM}$ curve for joint best fit parameters to SneIa, BAO measurements and WMAP5,[12] $(\Omega_{M0}, \Omega_{\Lambda 0}) = (0.279, 0.721)$ is best–matched over the range $5 \lesssim z \lesssim 6$, for example.

The difference of $D_{\rm TS}$ from any single $D_{\Lambda\rm CDM}$ curve are perhaps most pronounced only in the range $2 \lesssim z \lesssim 6$, which may be an optimal regime to probe in future experiments. Gamma–ray bursters (GRBs) now probe distances to redshifts $z \lesssim 8.3$, and could be very useful. There has already been much work deriving Hubble diagrams using GRBs. (See, e.g.,[14]) It would appear that more work needs to be done to nail down systematic uncertainties, but GRBs may provide a definitive test in future. An analysis of the timescape model Hubble diagram using 69 GRBs has just been performed by Schaefer,[15] who finds that it fits the data better than the concordance ΛCDM model, but not yet by a huge margin. As more data is accumulated, it should become possible to distinguish the models if the issues with the standardization of GRBs can be ironed out.

3.1. *The Effective "Equation of State"*

It should be noted that the shape of the $H_0 D$ curves depicted in Fig. 1 represent the observable quantity one is actually measuring when some researchers loosely talk about "measuring the equation of state". For spatially flat dark energy models, with $H_0 D$ given by (9), one finds that the function $w(z)$ appearing in the fluid equation of state $P_D = w(z)\rho_D$ is related to the first and second derivatives of (9) by

$$w(z) = \frac{\frac{2}{3}(1+z)D'^{-1}D'' + 1}{\Omega_{M0}(1+z)^3 H_0^2 D'^2 - 1} \tag{10}$$

where prime denotes a derivative with respect to z. Such a relation can be applied to observed distance measurements, regardless of whether the underlying cosmology has dark energy or not. Since it involves first and second derivatives of the observed quantities, it is actually much more difficult to determine observationally than directly fitting $H_0 D(z)$.

The equivalent of the "equation of state", $w(z)$, for the timescape model is plotted in Fig. 2. The fact that $w(z)$ is undefined at a particular redshift and changes sign through $\pm\infty$ simply reflects the fact that in (10) we are dividing by a quantity which goes to zero for the timescape model, even though the underlying curve of Fig. 1 is smooth. Since one is not dealing with a dark energy fluid in the present case, $w(z)$ simply has no physical

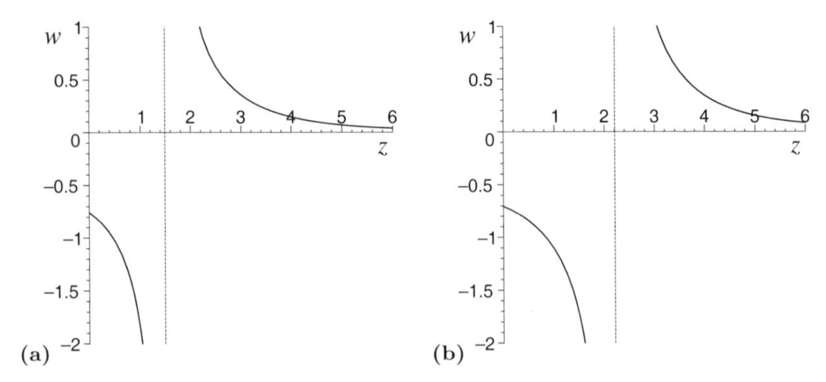

Fig. 2. The artificial equivalent of an equation of state constructed using the effective comoving distance (10), plotted for the timescape tracker solution with best–fit value $f_{v0} = 0.762$, and two different values of Ω_{M0}: **(a)** the canonical dressed value $\Omega_{M0} = \frac{1}{2}(1 - f_{v0})(2 + f_{v0}) = 0.33$; **(b)** $\Omega_{M0} = 0.279$.

meaning. Nonetheless, phenomenologically the results do agree with the usual inferences about w for fits of standard dark energy cosmologies to SneIa data. For the canonical model of Fig. 2(a) one finds that the average value of $w(z) \simeq -1$ on the range $z \lesssim 0.7$, while the average value of $w(z) < -1$ if the range of redshifts is extended to higher values. The fact that $w(z)$ is a different sign to the dark energy case for $z > 2$ is another way of viewing our statement above that the redshift range $2 \lesssim z \lesssim 6$ is optimal for discriminating model differences.

3.2. *The Alcock–Paczyński Test and Baryon Acoustic Oscillations*

Alcock and Paczyński devised a test[16] which relies on comparing the radial and transverse proper length scales of spherical standard volumes comoving with the Hubble flow. This test, which determines the function

$$f_{\mathrm{AP}} = \frac{1}{z}\left|\frac{\delta\theta}{\delta z}\right| = \frac{HD}{z}, \tag{11}$$

was originally conceived to distinguish FLRW models with a cosmological constant from those without a Λ term. The test is free from many evolutionary effects, but relies on one being able to remove systematic distortions due to peculiar velocities.

Current detections of the BAO scale in galaxy clustering statistics[17,18] can in fact be viewed as a variant of the Alcock–Paczyński test, as they make use of both the transverse and radial dilations of the fiducial comoving BAO

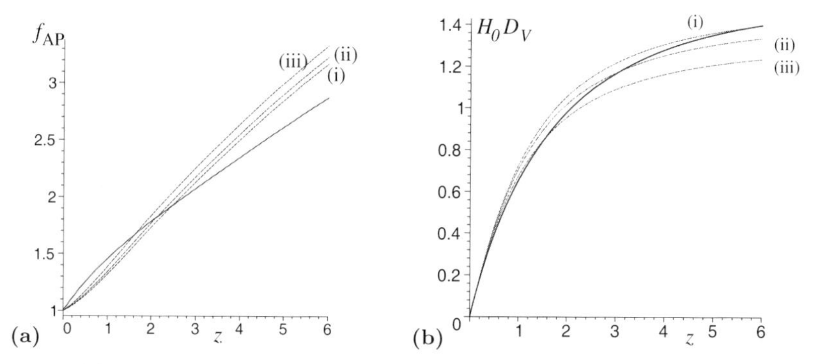

Fig. 3. **(a)** The Alcock–Paczyński test function $f_{\rm AP} = HD/z$; and **(b)** the BAO radial test function $H_0 D_V = H_0 D f_{\rm AP}^{-1/3}$. In each case the timescape model with $f_{v0} = 0.762$ (solid line) is compared to three spatially flat ΛCDM models with the same values of $(\Omega_{M0}, \Omega_{\Lambda 0})$ as in Fig. 1 (dashed lines).

scale to present a measure

$$D_V = \left[\frac{zD^2}{H(z)}\right]^{1/3} = D f_{\rm AP}^{-1/3}. \tag{12}$$

In Fig. 3 the Alcock–Paczyński test function (11) and BAO scale measure (12) of the timescape model are compared to that of spatially flat ΛCDM model with different values of $(\Omega_{\Lambda 0}, \Omega_{\Lambda 0})$. The curve for the timescape model has a distinctly different shape to those of the LCDM models, being convex. However, the extent to which the curves can be reliably distinguished would require detailed analysis based on the precision attainable with any particular experiment. Ideally, a model–independent determination of $H(z)$ would be required since $z f_{\rm AP} = H(z)D(z)$ for all the models under consideration.

3.3. *The* $H(z)$ *Measure*

Further observational diagnostics can be devised if the expansion rate $H(z)$ can be observationally determined as a function of redshift. Recently such a determination of $H(z)$ at $z = 0.24$ and $z = 0.43$ has been made using redshift space distortions of the BAO scale in the ΛCDM model.[19] This technique is of course model dependent, and the Kaiser effect would have to be re-examined in the timescape model before a direct comparison of observational results could be made. A model–independent measure of $H(z)$ is discussed in sec. 3.6.

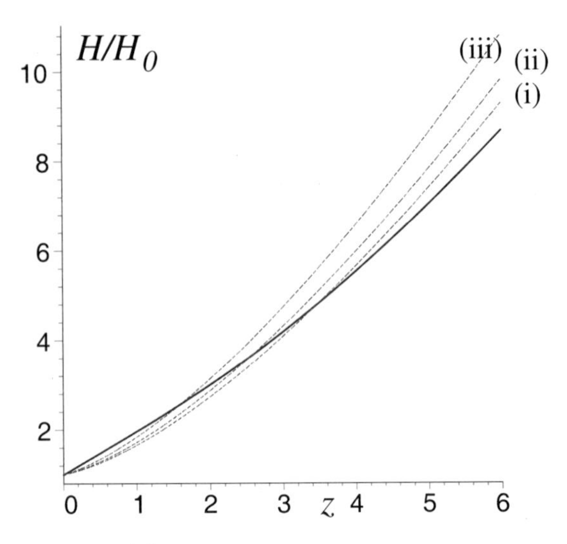

Fig. 4. The function $H_0^{-1}\frac{dz}{d\tau}$ for the timescape model with $f_{v0} = 0.762$ (solid line) is compared to $H_0^{-1}\frac{dz}{dt}$ for three spatially flat ΛCDM models with the same values of $(\Omega_{M0}, \Omega_{\Lambda 0})$ as in Fig. 1 (dashed lines).

In Fig. 4 we compare $H(z)/H_0$ for the timescape model to spatially flat ΛCDM models with the same parameters chosen in Fig. 1. The most notable feature is that the slope of $H(z)/H_0$ is less than in the ΛCDM case, as is to be expected for a model whose (dressed) deceleration parameter varies more slowly than for ΛCDM.

3.4. *The* $Om(z)$ *Measure*

Recently Sahni, Shafieloo and Starobinsky[20] have proposed a new diagnostic of dark energy, the function

$$Om(z) = \left[\frac{H^2(z)}{H_0^2} - 1\right]\left[(1+z)^3 - 1\right]^{-1}, \tag{13}$$

on account of the fact that it is equal to the constant present epoch matter density parameter, Ω_{M0}, for all redshifts for a spatially flat FLRW model with pressureless dust and a cosmological constant, but is not constant if the cosmological constant is replaced by other forms of dark energy. For general FLRW models, $H(z) = [D'(z)]^{-1}\sqrt{1 + \Omega_{k0} H_0^2 D^2(z)}$, which only involves a single derivatives of $D(z)$. Thus the diagnostic (13) is an easier to reconstruct observationally than the equation of state parameter, $w(z)$.

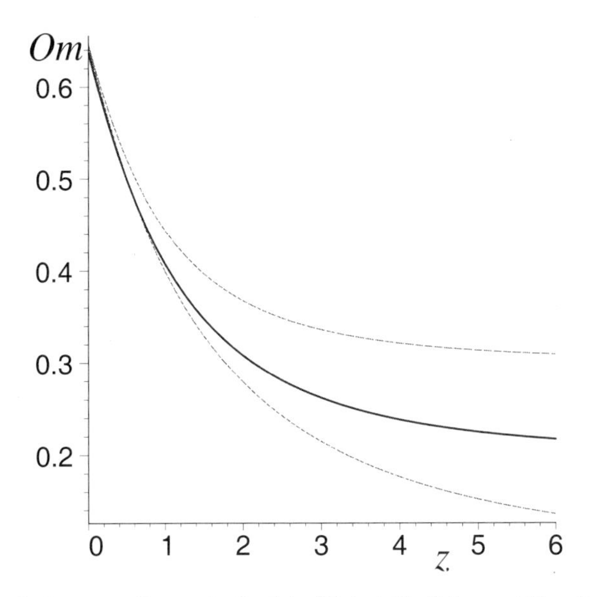

Fig. 5. The dark energy diagnostic $Om(z)$ of Sahni, Shafieloo and Starobinsky plotted for the timescape tracker solution with best–fit value $f_{v0} = 0.762$ (solid line), and 1σ limits (dashed lines) from ref.[9]

The quantity $Om(z)$ is readily calculated for the timescape model, and the result is displayed in Fig. 5. What is striking about Fig. 5, as compared to the curves for quintessence and phantom dark energy models as plotted in ref.,[20] is that the initial value

$$Om(0) = \tfrac{2}{3}\, H'|_0 = \frac{2(8f_{v0}^3 - 3f_{v0}^2 + 4)(2 + f_{v0})}{(4f_{v0}^2 + f_{v0} + 4)^2} \tag{14}$$

is substantially larger than in the spatially flat dark energy models. Furthermore, for the timescape model $Om(z)$ does not asymptote to the dressed density parameter Ω_{M0} in any redshift range. For quintessence models $Om(z) > \Omega_{M0}$, while for phantom models $Om(z) < \Omega_{M0}$, and in both cases $Om(z) \to \Omega_{M0}$ as $z \to \infty$. In the timescape model, $Om(z) > \Omega_{M0} \simeq 0.33$ for $z \lesssim 1.7$, while $Om(z) < \Omega_{M0}$ for $z \gtrsim 1.7$. It thus behaves more like a quintessence model for low z, in accordance with Fig. 2. However, the steeper slope and the completely different behaviour at large z mean the diagnostic is generally very different to that of typical dark energy models. For large z, $\bar{\Omega}_{M0} < Om(\infty) < \Omega_{M0}$, if $f_{v0} > 0.25$.

Interestingly enough, a recent analysis of SneIa, BAO and CMB data[21] for dark energy models with two different empirical fitting functions for

$w(z)$ gives an intercept $Om(0)$ which is larger than expected for typical quintessence or phantom energy models, and in the better fit of the two models the intercept (see Fig. 3 of ref.[21]) is close to the value $Om(0) = 0.638$ for the $f_{v0} = 0.762$ timescape model.

3.5. *Test of (in)Homogeneity*

Recently Clarkson, Bassett and Lu[22] have constructed what they call a "test of the Copernican principle" based on the observation that for homogeneous, isotropic models which obey the Friedmann equation, the present epoch curvature parameter, a constant, may be written as

$$\Omega_{k0} = \frac{[H(z)D'(z)]^2 - 1}{[H_0 D(z)]^2} \tag{15}$$

for all z, irrespective of the dark energy model or any other model parameters. Consequently, taking a further derivative, the quantity

$$\mathcal{C}(z) \equiv 1 + H^2(DD'' - D'^2) + HH'DD' \tag{16}$$

must be zero for all redshifts for any FLRW geometry.

A deviation of $\mathcal{C}(z)$ from zero, or of (15) from a constant value, would therefore mean that the assumption of homogeneity is violated. Although this only constitutes a test of the assumption of the Friedmann equation, i.e., of the Cosmological Principle rather than the broader Copernican Principle adopted in ref.,[6] the average inhomogeneity will give a clear and distinct prediction of a non-zero $\mathcal{C}(z)$ for the timescape model.

The functions (15) and (16) are computed in ref.[1] Observationally it is more feasible to fit (15) which involves one derivative less of redshift. In Fig. 6 we exhibit the function $\mathcal{B}(z) = [HD']^2 - 1$ from the numerator of (15) for the timescape model, as compared to two ΛCDM models with a small amount of spatial curvature. In the FLRW case $\mathcal{B}(z)$ is always a monotonic function whose sign is determined by that of Ω_{k0}. An open $\Lambda = 0$ universe with the same Ω_{M0} would have a monotonic function $\mathcal{B}(z)$ very much greater than that of the timescape model.

3.6. *Time Drift of Cosmological Redshifts*

For the purpose of the $Om(z)$ and (in)homogeneity tests considered in the last section, $H(z)$ must be observationally determined, and this is difficult to achieve in a model independent way. There is one way of achieving this, however, namely by measuring the time variation of the redshifts of different sources over a sufficiently long time interval,[23] as has been discussed

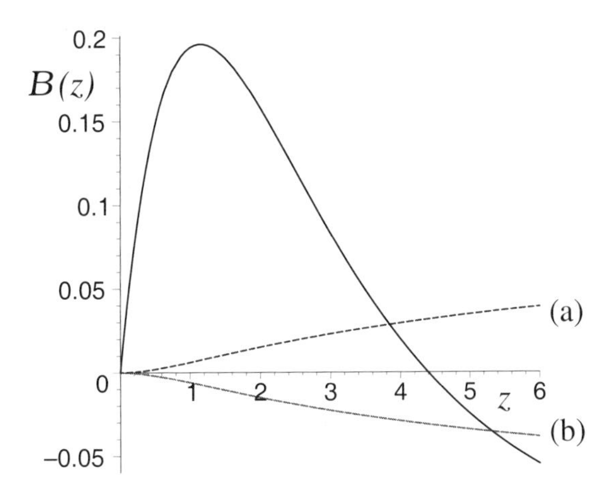

Fig. 6. The (in)homogeneity test function $\mathcal{B}(z) = [HD']^2 - 1$ is plotted for the timescape tracker solution with best–fit value $f_{v0} = 0.762$ (solid line), and compared to the equivalent curves $\mathcal{B} = \Omega_{k0}(H_0 D)^2$ for two different ΛCDM models with small curvature: **(a)** $\Omega_{M0} = 0.28$, $\Omega_{\Lambda 0} = 0.71$, $\Omega_{k0} = 0.01$; **(b)** $\Omega_{M0} = 0.28$, $\Omega_{\Lambda 0} = 0.73$, $\Omega_{k0} = -0.01$. A spatially flat FLRW model would have $\mathcal{B}(z) \equiv 0$.

recently by Uzan, Clarkson and Ellis.[24] Although the measurement is extremely challenging, it may be feasible over a 20 year period by precision measurements of the Lyman-α forest in the redshift range $2 < z < 5$ with the next generation of Extremely Large Telescopes.[25]

In ref.[1] an analytic expression for $H_0^{-1}\frac{dz}{d\tau}$ is determined, the derivative being with respect to wall time for observers in galaxies. The resulting function is displayed in Fig. 4 for the best-fit timescape model with $f_{v0} = 0.762$, where it is compared to the equivalent function for three different spatially flat ΛCDM models. What is notable is that the curve for the timescape model is considerably flatter than those of the ΛCDM models. This may be understood to arise from the fact that the magnitude of the apparent acceleration is considerably smaller in the timescape model, as compared to the magnitude of the acceleration in ΛCDM models. For models in which there is no apparent acceleration whatsoever, one finds that $H_0^{-1}\frac{dz}{d\tau}$ is always negative. If there is cosmic acceleration, real or apparent, at late epochs then $H_0^{-1}\frac{dz}{d\tau}$ will become positive at low redshifts, though at a somewhat larger redshift than that at which acceleration is deemed to have begun.

Fig. 4 demonstrates that a very clear signal of differences in the redshift time drift between the timescape model and ΛCDM models might be determined at low redshifts when $H_0^{-1}\frac{dz}{d\tau}$ should be positive. In particular, the magnitude of $H_0^{-1}\frac{dz}{d\tau}$ is considerably smaller for the timescape model as compared to ΛCDM models. Observationally, however, it is expected that measurements will be best determined for sources in the Lyman α forest in the range, $2 < z < 5$. At such redshifts the magnitude of the drift is somewhat more pronounced in the case of the ΛCDM models. For a source at $z = 4$, over a period of $\delta\tau = 10$ years we would have $\delta z = -3.3 \times 10^{-10}$ for the timescape model with $f_{v0} = 0.762$ and $H_0 = 61.7\,\text{km sec}^{-1}\,\text{Mpc}^{-1}$. By comparison, for a spatially flat ΛCDM model with $H_0 = 70.5\,\text{km sec}^{-1}\,\text{Mpc}^{-1}$ a source at $z = 4$ would over ten years give $\delta z = -4.7 \times 10^{-10}$ for $(\Omega_{M0}, \Omega_{\Lambda 0}) = (0.249, 0.751)$, and $\delta z = -7.0 \times 10^{-10}$ for $(\Omega_{M0}, \Omega_{\Lambda 0}) = (0.279, 0.721)$.

4. Discussion

The tests outlined here demonstrate several lines of investigation to distinguish the timescape model from models of homogeneous dark energy. The (in)homogeneity test of Bassett, Clarkson and Lu is definitive, since it tests the validity of the Friedmann equation directly.

In performing these tests, however, one must be very careful to ensure that data has not been reduced with built–in assumptions that use the Friedmann equation. For example, current estimates of the BAO scale such as that of Percival *et al.*[18] do not determine D_V directly from redshift and angular diameter measures, but first perform a Fourier space transformation to a power spectrum, assuming an FLRW cosmology.

Furthermore, in the case of SneIa, it has been recently claimed[26] that the timescape model does not compare favourably with the ΛCDM model when SneIa from the recent Union[27] and Constitution[28] compilations are used. However, the Union dataset and its extension use the SALT method to calibrate light curves. In this approach empirical light curve parameters and cosmological parameters – *assuming the Friedmann equation* – are simultaneously fit by analytic marginalisation before the raw apparent magnitudes are recalibrated. As Hicken *et al.* discuss,[28] a number of possible systematic discrepancies exist between data reduced by the SALT, SALT2, MLCS31 and MLCS17 techniques. In the case of the timescape model, for which the Friedmann equation does not apply, it turns out that these systematic differences lead to larger discrepancies in the determination of cosmological parameters from one method to another, as will be discussed in future work.[29]

The value of the dressed Hubble constant is also an observable quantity of considerable interest. A recent determination by Riess *et al.*[30] poses a challenge for the timescape model. However, in the presence of spatial curvature gradients a great deal of care must be taken, since in view of the expected Hubble bubble feature discussed in Sec. 2, estimates of H_0 made below the scale of statistical homogeneity will generally give higher values. The method that is used to anchor the Cepheid calibration in the SH_0ES survey[30] – namely a geometric maser distance to the relatively close galaxy NGC4258 – may in fact be the best way of determining whether a Hubble bubble feature exists, if sufficiently large numbers of maser distances can be determined within the scale of statistical homogeneity. The Megamaser project[31] may soon begin to provide the sort of data that is required.

Acknowledgements

I thank Prof. Remo Ruffini and ICRANet for support and hospitality while the work of ref.[1] was undertaken. This work was also partly supported by the Marsden fund of the Royal Society of New Zealand.

References

1. D.L. Wiltshire, in preparation.
2. D.L. Wiltshire, in *Dark Matter in Astroparticle and Particle Physics: Proceedings of the 6th International Heidelberg Conference*, eds H.V. Klapdor–Kleingrothaus and G.F. Lewis, (World Scientific, Singapore, 2008) pp. 565-596 [arXiv:0712.3984].
3. F. Hoyle and M.S. Vogeley, ApJ **566**, 641 (2002); ApJ **607**, 751 (2004).
4. T. Buchert, Gen. Relativ. Grav. **32**, 105 (2000); Gen. Relativ. Grav. **33**, 1381 (2001).
5. T. Buchert, Gen. Relativ. Grav. **40**, 467 (2008).
6. D.L. Wiltshire, New J. Phys. **9**, 377 (2007).
7. D.L. Wiltshire, Phys. Rev. Lett. **99**, 251101 (2007).
8. D.L. Wiltshire, Phys. Rev. **D 78**, 084032 (2008).
9. B.M. Leith, S.C.C. Ng and D.L. Wiltshire, ApJ **672**, L91 (2008).
10. S. Jha, A.G. Riess and R.P. Kirshner, ApJ **659**, 122 (2007).
11. N. Li and D.J. Schwarz, Phys. Rev. **D78**, 083531 (2008).
12. E. Komatsu *et al.*, Astrophys. J. Suppl. **180**, 330 (2009).
13. A.G. Riess *et al.*, ApJ **659**, 98 (2007).
14. B.E. Schaefer, ApJ **660**, 16 (2007); N. Liang, W. K. Xiao, Y. Liu and S.N. Zhang, ApJ **685**, 354 (2008). L. Amati, C. Guidorzi, F. Frontera, M. Della Valle, F. Finelli, R. Landi and E. Montanari, MNRAS **391**, 577 (2008); R. Tsutsui, T. Nakamura, D. Yonetoku, T. Murakami, Y. Kodama and K. Takahashi, arXiv:0810.1870.
15. B.E. Schaefer, in preparation.
16. C. Alcock and B. Paczyński, Nature **281**, 358 (1979).

17. D.J. Eisenstein *et al.*, ApJ **633** (2005) 560; S. Cole *et al.*, MNRAS **362** (2005) 505.
18. W.J. Percival *et al.*, MNRAS **381**, 1053 (2007).
19. E. Gaztañaga, A. Cabre and L. Hui, arXiv:0807.3551.
20. V. Sahni, A. Shafieloo and A. A. Starobinsky, Phys. Rev. **D 78**, 103502 (2008).
21. A. Shafieloo, V. Sahni and A.A. Starobinsky, arXiv:0903.5141.
22. C. Clarkson, B. Bassett and T.C. Lu, Phys. Rev. Lett. **101**, 011301 (2008).
23. A. Sandage, ApJ **136**, 319 (1962); G.C. McVittie, ApJ **136**, 334 (1962); A. Loeb, ApJ **499**, L111 (1998).
24. J.P. Uzan, C. Clarkson and G.F.R. Ellis, Phys. Rev. Lett. **100**, 191303 (2008).
25. P.S. Corasaniti, D. Huterer and A. Melchiorri, Phys. Rev. **D75**, 062001 (2007); J. Liske *et al.*, MNRAS **386**, 1192 (2008).
26. J. Kwan, M.J. Francis and G.F. Lewis, arXiv:0902.4249.
27. M. Kowalski *et al.*, ApJ **686**, 749 (2008).
28. M. Hicken *et al.*, arXiv:0901.4804.
29. P.R. Smale and D.L. Wiltshire, in preparation.
30. A.G. Riess *et al.*, ApJ **699**, 539 (2009).
31. M.J. Reid, J.A. Braatz, J.J. Condon, L. . Greenhill, C. Henkel and K.Y. Lo, ApJ **695**, 287 (2009).

SEARCH FOR QUANTUM GRAVITY SIGNATURE WITH PHOTONS FROM ASTROPHYSICAL SOURCES

A. JACHOLKOWSKA* and J. BOLMONT

LPNHE/CNRS/IN2P3, Universite Paris 6 and Paris 7,
Paris, 75252 cedex 05, France
**E-mail: Agnieszka.Jacholkowska@cern.ch*

The study of time-lags in light curves of cosmological sources such as Gamma-Ray Bursts and Active Galaxies, as a function of energy and redshift of the source, may lead to a detection of Lorentz Symmetry breaking or effects due to Quantum Gravity in extra-dimensions or Loop Quantum Gravity models. In this paper, the recent time-of-flight studies with photons published by the space and ground based experiments are reviewed. Various methods used in the time delay searches are described, and their performance discusseed. Since no significant time-lag value was found within experimental precision of the maesurements, the 95% Confidence Level limit on the Quantum Gravity scale of the order of $10^{16} GeV$ and $10^{18} GeV$ for the linear term in the standard photon dispersion relations, were established from the Gamma-Ray Burst and Active Galaxy data respectively.

Keywords: Lorentz Symmetry breaking; Quantum Gravity; Gamma Ray Bursts; Active Galaxies.

1. Introduction

A Quantum Gravity theory gives a unified picture based on the Quantum Mechanics and the General Relativity, thus leading to a common description of the four fundamental forces. The Quantum Gravity effects in the framework of the String Theory,[1] where the gravitation is considered as a gauge interaction, are resulting from a graviton-like exchange in a background classical space-time. In most of the String Theory models implying large extra-dimensions these effects would take place at the Planck scale, thus leading to no 'spontaneous' Lorentz Symmetry breaking, as it may happen in models with 'foamy' structure[1,2] of the quantum space-time. In this second class of models developed by,[3] photons propagate in vacuum which may exhibit a non-trivial refractive index due to its foamy structure on a characteristic scale approaching the Planck length or equivalently

Planck energy ($E_P = 1.22 \times 10^{19}$ GeV). This implies a light group velocity increasing as a function of energy of the subluminal photon, in contrast to the dispersion effects in any field theoretical vacuum or plasma. On the other hand, in models based on the General Relativity with Loop Quantum Gravity[4,5] which postulates discrete (cellular) space-time in the Planckian regime, the fluctuations would introduce perturbations to the propagation of photons. The luminous wave going through the discrete space matrix would feel an induced perturbation increasing with its decreasing wave-length or equivalently with increasing energy of photons. In consequence, the photons with different wave-length propagate with different velocity. As a result, one may expect a spontaneous violation of the Lorentz Symmetry at high energies to be the generic signature of the Quantum Gravity.

In four dimensions, the Quantum Gravity scale is supposed to be close to Planck mass and the standard photon dispersion relations up-to second order corrections in energy can be written as:

$$c^2p^2 = E^2(1 + \xi\frac{E}{M} + O(\frac{E^2}{M^2})) \text{ and } v(E) \approx c\left(1 - \xi\frac{E}{M}\right), \tag{1}$$

where ξ is a model parameter whose value is set to 1 in the following.[6]

From calculations of,[5] the modifications to the Maxwell equations in vacuum induced by Quantum Gravity effects provide helicity dependent dispersion relations, thus yielding parity violation and birefringent effects. In case of String Theory approach,[1] the dispersion relations come from the parity-conserving corrections to the Maxwell equations leading to a first order in $1/E_{QG}$ helicity independent effect, that is linear in the photon energy.

As suggested by[2] the tiny effects due to Quantum Gravity can add up to measurable time delays for photons from cosmological sources. The energy dispersion is best observed in sources that show fast flux variability, are at cosmological distances and are observed over a wide energy range. This is the case of Gamma Ray Bursts (GRBs) and Very High Energy (VHE) flares of Active Galactic Nuclei (AGNs), both types of sources being the preferred targets of these "time-of-flight" studies, which provide the least model dependent tests of the Lorentz Symmetry.

It should be noticed that the time-of-flight measurements are subject to a bias related to a potential dispersion introduced from the intrinsic source effects, which could cancel out or enhance the dispersion due to modifications to the speed of light. Till now, the most solid results on the Quantum Gravity scale of the order of $10^{16} GeV$ are provided by GRB observations with different redshifts as they take into account the possible time lags originating from source effects, resulting in limits of $\xi < 1300$.[7–9] The recent studies with flares of Mrk501[10] and PKS2155-304[11] provide results setting the Quantum Gravity scale between 0.2 $10^{18} GeV$ and 0.7 $10^{18} GeV$ for each source. It should be noticed here, that the obtained results concern only the linear term in Eq. (1).

2. Measurement of the Propagation Effects with Photons

When considering sources at the cosmological distances, the analyses of time lags as a function of redshift requires a correction due to the expansion of the universe, which depends on the cosmological model. Following the analysis of the BATSE data and recently of the HETE-2 and SWIFT GRB data, and within a framework of the Standard Cosmological Model[12] with flat expanding universe and a cosmological constant, the difference in the arrival time of two photons with energy difference ΔE is given by the formula:

$$\Delta t = \mathrm{H}_0^{-1} \frac{\Delta E}{\mathrm{E}_{\mathrm{QG}}} \int_0^z \frac{(1+z)\, dz}{h(z)}, \tag{2}$$

where

$$h(z) = \sqrt{\Omega_\Lambda + \Omega_\mathrm{M}(1+z)^3}. \tag{3}$$

and where Ω_m, Ω_Λ and H_0 are parameters of the Cosmological Standard Model ($\Omega_m = 0.3$, $\Omega_\Lambda = 0.7$ and $H_0 = 70$ km s^{-1} Mpc^{-1} from actually accepted values), and the ΔE is being the energy difference of the two photons. Then, the time-lag may decrease or increases with ΔE depending on the model in consideration. The term with the integral increases with the redshift and takes into account the fact that two photons travelling with two different speeds do not take the same time to cross the Universe and then do not see the same expansion. One also notes that for a given redshift, and a given value of ΔE, smaller Δt will lead to a more constraining limit on the Quantum Gravity scale M.

In order to probe the energy dependence of the velocity of light induced by Quantum Gravity, in most of the analyses, the time lags are studied as a function of the variable:

$$K_l(z) = \int_0^z \frac{(1+z)\, dz}{h(z)}, \tag{4}$$

except for nearby sources for which the Newtonian approximation applies.

Possible effects intrinsic to the astrophysical sources could also produce energy dependent time lags. The analysis as a function of K_l ensures, in principle, that the results are independent of such effects. In the first order of the dispersion relations, the evolution of the time-lags as a function of K_l can be written as:

$$< \Delta t > = \ \mathrm{a}\, K_l(z) + \mathrm{b}\,(1 + z), \tag{5}$$

where a and b parameters stand for extrinsic (Quantum Gravity) and intrinsic effects, respectively.

The goal of the 'time of flight' studies is to measure the energy dependence of time-lags. As proposed, two kind of variable extragalactic sources may be considered: the GRBs AGNs. However, both these sources could introduce intrinsic time lags in the measurements. Therefore, it is necessary to study the possible effects as a function of the redshift of the source. When this is not possible, the intrinsic (or source) effects are often assumed to be negligible. In the presented studies, this hypothesis was maintained.

There are various methods for the time-lag determination and subsequent derivation of the Quantum Gravity scale In order to measure a tiny deviation from its standard value of the light velocity and reach 10^{19} GeV domain, the following experimental conditions should be fulfilled: in case of the GRBs where keV/MeV photons are detected, the precision on Δ(t) of the order of 10^{-4} to 10^{-3}s is needed. For AGN flare detections, the energies of photons in GeV/TeV range imply Δ(t) of the order of a second.

The procedures in use to determine time-lags rely on different statistical treatements of the data:

- Cross Correlation Function (MCCF) in use for GRBs (BATSE[15,16]) and AGNs (H.E.S.S.[11])
- Energy Cost Function (ECF) for AGNs (MAGIC[10])
- Wavelet Transforms (CWT) for GRBs (BATSE, HETE2, SWIFT[7–9]) and AGNs (H.E.S.S.[11])
- Likelihood fits for GRBs (INTEGRAL[17] and AGN (MAGIC, H.E.S.S.)

For robust results, the use of at least two methods which probe different aspects of the light curves is recommended. As till now no significant time-lag was detected in any of the discussed below analyses, careful error calibration studies by the Monte Carlo simulations are mandatory for the extraction of the limits.

3. Data and Results

In this section the studies performed in view of search for Lorentz Symmetry breaking with GRBs and AGNs respectively are presented, combining best methods used in this field. The analyses performed by the High Energy Transient Explorer (HETE-2) and by the H.E.S.S. experiment are good examples of current studies in presently running experiments. The quality and significance of the results will be related to the acquired statistics and methods in use.

3.1. *Gamma Ray Bursts*

The GRBs are the most violent phenomena observed in the universe, detected as sudden and unpredictable bursts of hard X-rays and γ-rays, lasting tenths of seconds and presenting variable and unclassified light curves. The GRBs are of cosmological origin resulting from a death of massive stars or a collapse of compact binary objects. Their very high variability of the order of mili-seconds in the large energy range and distances going up to redshift values of 6, place them as excellent candidates for searches of the non-standard effects in the photon propagation.

A typical analysis in the discussed field has been performed by HETE-2 experiment. This mission was devoted to the study of GRBs using soft X-ray, medium X-ray, and γ-ray instruments mounted on a compact spacecraft. The analysis of the 15 GRBs with measured redshifts collected by HETE-2 mission in years 2001–2006 followed the procedure described in details in.[7,9] After the determination of the GRB interval time describing the start and the end of the burst, a cut above the background delimited the signal region to be studied in further analysis. The originality of the performed analysis was the choice of various energy ranges where the time-lags were computed. The study of various scenarios in energy was allowed by use of tagged photon data provided by the FREGATE sub-detector of the HETE-2 for each GRB. The time-lag calculation relied on 2 step procedure: de-noising of the light curves by a Discrete Wavelet Transform (DWT)[18] with a pre-selection of data in the studied time interval for each GRB and for each energy band, and search for the rapid variations (spikes) in the light curves for all energy bands using a Continuous Wavelet Transform (CWT).[19] As a result of these steps was a list of minima and maxima candidates, along with a coefficient characterizing their regularity (the Lipschitz coefficient), followed by the association in pairs of the found minima and maxima.

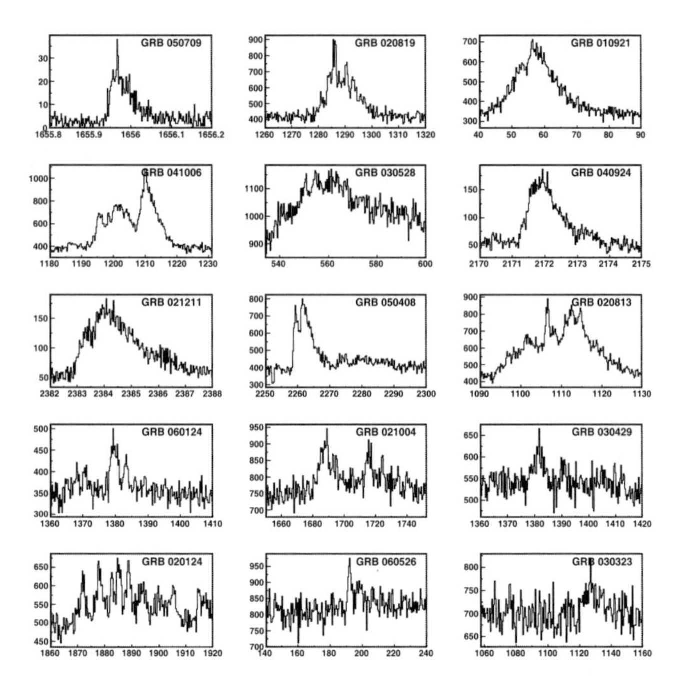

Fig. 1. Light curves of the 15 GRBs detected by HETE-2 in the energy range 6–400 keV. The bursts are sorted by increasing z from top left to bottom right. X-axes are graduated in seconds.

The average time lag of each GRB, the $< \Delta t >$, was then calculated and used later in the study of the Quantum Gravity model related to the foamy structure of the vacuum, as described in previous section. Finally, the evolution of the time lags as a function of the variable K_l leads to constraints on the minimal value of the Quantum Gravity scale M.

The light curves of the 15 GRBs are shown in figure 1. It appears clearly that the signal-to-noise ratio decreases for large redshifts. The large variety of light curve shapes is a characteristic feature of GRBs.

Figure 2 presents the evolution of $\chi^2(M)/\text{ndf}$ around its minimum $\chi^2_{min}(M)/\text{ndf}$ for maxima and minima together. All scenarios fulfill the condition $\chi^2_{min}/\text{ndf} \leq 2$. The two-parameter linear fits, as discussed in,[9] show a somewhat different behaviour in case of the maxima and the minima. However, no significant preference of any value of M is observed for most of the scenarios when considering both extrema. The 95% CL contours for a and b from 2-dimensional fit are presented in figure 3, showing that both parameters representing the time-lags expected from the propagation

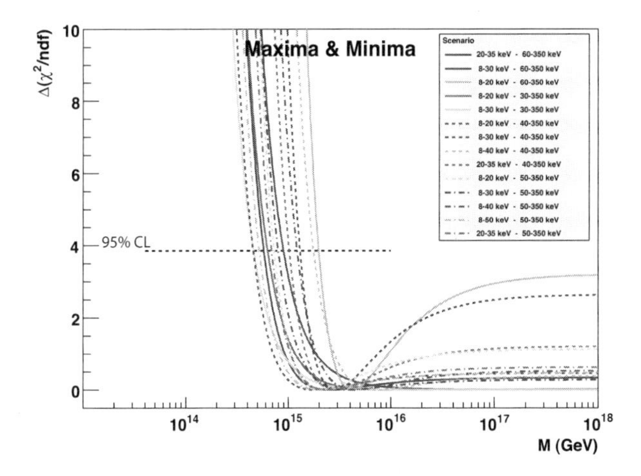

Fig. 2. Evolution of χ^2 function of M for maxima and for minima found by CCF procedure.

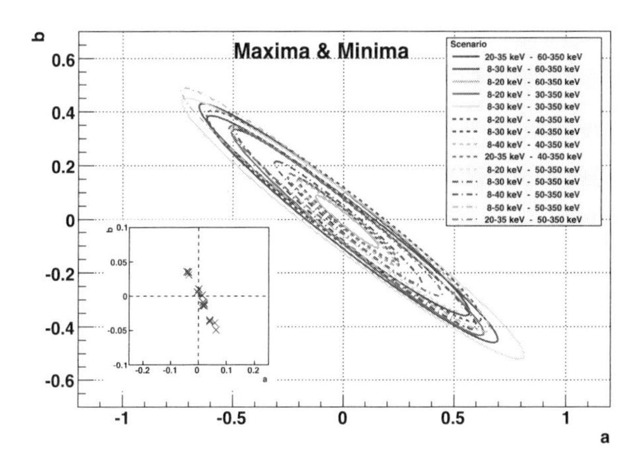

Fig. 3. 95% CL contours for a and b from the 2-parameter fits for the fourteen scenarios considered. The boxes at the bottom left of the plots show the position of contour centers.

and the source effects are strongly correlated. In this figure, the values of the offset parameter b are compatible with zero for all energy scenarios. Both fit results suggest no variation above $\pm 3\,\sigma$, so that in the following, the 95% CL lower limits on the Quantum Gravity scale were derived.

Concerning possible source effects, it has been known for a long time, that the peaks of the emission in GRB light curves are shorter and arrive earlier at higher energies.[20,21] These intrinsic lags, which have a sign opposite to the sign expected from Lorentz violation, have a broad dispersion of durations, which complicates the detection of the Lorentz Symmetry violation effects. Therefore the effect of the Lorentz Symmetry violation must be searched with a statistical study analyzing the average dependence of the lags with K_l, (redshift) and not in samples limited to one GRB. In this study, a universality of the intrinsic source time-lags has been assumed

In conlusion of the presented study of time lags with GRBs detected by HETE-2 and other satellite missions in the energy range of keV-MeV, a lower limit on the Quantum Gravity scale is placed at few $\times 10^{15}$ GeV. More recently, the Fermi collaboration[13] has published a value on Quantum Gravity scale just an order of magnitude below the Planck scale, from data taken with GBM and LAT detectors, when detecting a powerful GRB 080916C with redshift value about 4. This result relying on 1 photon at 13 GeV which determines the energy range, was derived by simple Δ(t) divided by distance formula. To progress further in this domain of energy, future detection of several GRBs by Fermi mission with known redshift and measurable time-lags are needed.

3.2. *Active Galaxies*

Blazars are variable AGNs, extragalactic sources producing γ-rays via the gravitational potential energy release of matter from an accretion disk surrounding a Super Massive Black Hole (SMBH). Beamed emission,large inferred luminosties, relativistic plasma jets pointing to the observer and the flux variations by large factors (flares) on hour scale in time, make them excellent objects for variability studies from radio to VHE frequencies. In addition, the Blazars are valuable transient candidates for searches of effects due to Lorentz Symmetry violation at Quantum Gravity scale. So far, 18 AGNs have been detected in VHE range with redshifts varying between 0.002 and 0.4 and here, the case of the observation of an exceptional flare of the blazar PKS 2155-304 (z = 0.116) by H.E.S.S. is described (more details are available elsewhere[11]).

In 2006, the H.E.S.S. experiment detected[14] an exceptional flare of this source, with a high flux (10 000 photons recorded in 1.5 hours) and a high variability (rise and fall times of $\sim$200 s) during the night of the 28th of July. The over-sampled light curve of the flare is shown by figure 4 in two different energy bands.

To measure the time lags between photons in two different light curves, two independant analyses were carried out, using two different methods:

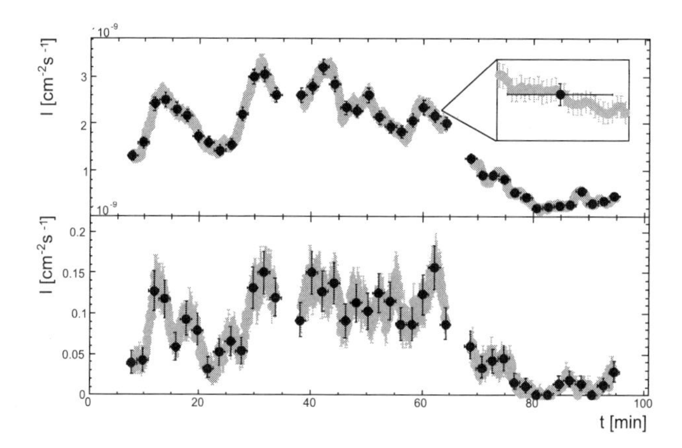

Fig. 4. Light curve of the PKS2155-304 flare during the night of July 28 in 2006. Top: 200-800 GeV. Bottom: >800 GeV. The original data points (in black) are binned in two-minute time intervals. The zero time point is set to MJD 53944.02. Gray points show the oversampled light curve, for which the two-minute bins are shifted in units of five seconds.

- The position of the maximum of the Modified Cross Correlation Function[15] (MCCF) which gives directly the value of the time lag. This method was applied to the oversampled light curves of figure 4 in the energy bands 200-800 GeV and >800 GeV (figure 5, left) which corresponds to a ΔE of ~1 TeV;
- The Continuous Wavelet Transform[19] (CWT) was used to locate the extrema of the light curves with great precision. An extremum of the low energy band is associated with an extremum in the high energy band to form a pair. A selection criteria were applied to reject fake extrema. The energy bands 210-250 GeV and >600 GeV were used, with a bin width of one minute and no oversampling of the light curves. These energy bands give a mean ΔE of ~0.92 TeV.

As shown on figure 5 (left), the MCCF is fitted with a gaussian plus a first degree polynomial to determine the position of the maximum. It gives $\tau_{\text{peak}} = 20$ s. In order to evaluate the uncertainties on this result, 10000 light curves were simulated in each energy band varying the flux within the error bars. For each pair of light curves, the MCCF was computed and its maximum was filled into the distribution shown on figure 5 (right). This distribution is slightly asymmetric, with a mean of 25 s and an RMS of 28 s. Another test was performed injecting a dispersion in the data showing

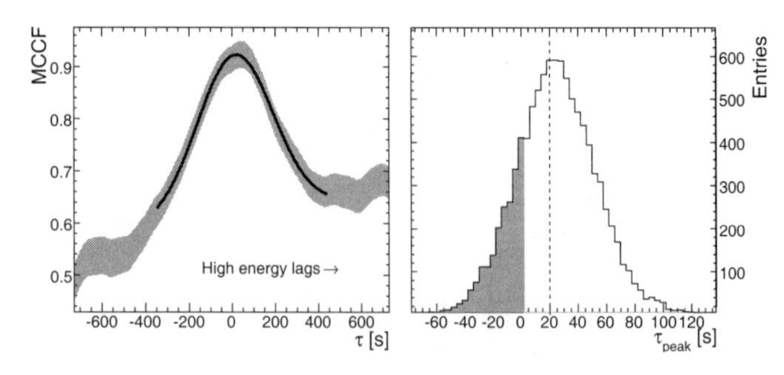

Fig. 5. Left: the MCCF obtained with the two light curves of Fig. 4. Right: the MCCF peak distribution (CCPD) obtained with 10000 realizations of the light curves.

Table 1. The results obtained with the two methods in the case of a linear dispersion in energy.

Method	Energy bands GeV	$< \Delta E >$ TeV	$(\Delta t/\Delta E)_{95\%\ \mathrm{CL}}$ s/TeV	$\mathrm{E}_{\mathrm{QG}\ 95\%\ \mathrm{CL}}$ GeV
MCCF	200 < E < 800 and E > 800	1.02	73	7.2×10^{17}
CWT	210 < E < 250 and E > 600	0.92	100	5.2×10^{17}

no significant deviation of the measured lag. Then, the obtained time-lag compatible with zero lead to a 95% CL upper limit on the linear dispersion of 73 s/TeV.

With the CWT method, two pairs of extrema were obtained giving a mean time delay of 27 seconds. A similar method as the one used for the CCF was used to determine the errors and they were found to be in a range between 30 and 36 seconds. A 95% confidence limit of 100 s/TeV was obtained for the linear correction to the dispersion relations.

The results obtained with the two methods are summarized in table 1. The MCCF leads to a limit of $\mathrm{E}_{\mathrm{QG}} > 7.2 \times 10^{17}$ GeV. The CWT gives a lower limit of $\mathrm{E}_{\mathrm{QG}} > 5.2 \times 10^{17}$ GeV, mainly due to a larger measured time-lag and a lower value of the $< \Delta E >$.

The presented analysis can be compared to the one performed by Whipple and MAGIC experiments with other AGN flares: Mkn 421 and Mkn 501. Figure 6 shows all the results obtained so far with AGNs. The Whipple collaboration has set a limit of 4×10^{16} GeV using a flare of Mkn 421 (z = 0.031) in 1996.[23] More recently, the MAGIC collaboration obtained a limit of 3.2×10^{17} GeV with a flare of Mkn 501 (z = 0.034).[10] The result obtained with H.E.S.S. are more constraining due to the fact that PKS 2155 is al-

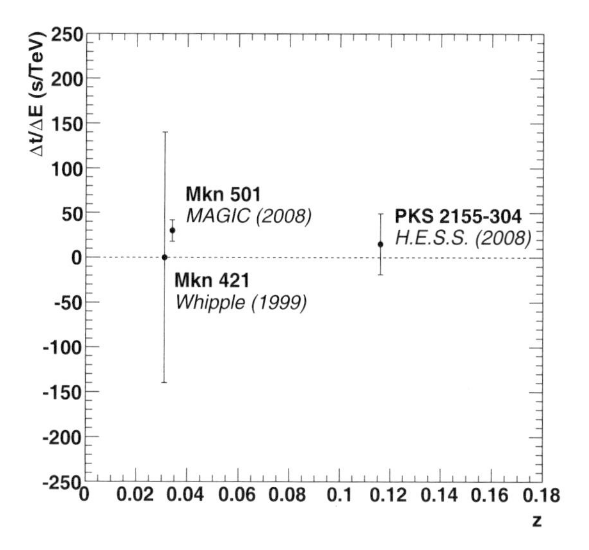

Fig. 6. The dispersion measured as a function of redshift. The results of the Whipple and MAGIC experiments are compared with the present H.E.S.S. result.

most four times more distant than Mkn 421 and Mkn 501 and the acquired statistics were higher by a factor of 10.

4. Summary and Discussion

Quantum Gravity phenomenology has known a growing interest in the past decade, especially since the time it was argued that Lorentz symmetry could be violated or at least deformed by the quantum nature of the space-time with measurable effects on photon propagation over large distances: the light group velocity could be modified.

The γ-ray dispersion studies in vacuum with GRBs and AGNs provide the cleanest way to look for Planck-scale modifications of the dispersion relations at the first order level. The results obtained with these sources are summarized in table 2.

As shown in the first part of the paper, the GRBs are good candidates for time of flight studies. The GRBs are easilly detected by satellite experiments at very high redshifts ($z < 6$) and up to a few hundreds of GeV ($\Delta E \sim 10$ GeV). Population studies have already been carried out[24] and lead to limits of the order of 10^{16} GeV. The best limit so far is $E_{QG} > 2 \times 10^{18}$ GeV. It has been obtained by Fermi experiment[13] with GRB 080916C ($z = 4.35$) using both GBM and LAT data ($\Delta E \sim 10$ GeV). However, this result

Table 2. A selection of limits obtained with various instruments.

Source(s)	Experiment(s)	Method	$E_{QG}(GeV)$
GRB 021206[22]	RHESSI	Fit + $\Delta(t)$	1.8×10^{17}
GRB 080916C[13]	Fermi (LAT + GBM)	$\Delta(t)$	1.5×10^{18}
9 GRBs[7]	BATSE + OSSE	Wavelets	6.0×10^{15}
15 GRBs[9]	HETE-2	Wavelets	4.0×10^{15}
17 GRBs[17]	INTEGRAL	Likelihood	1.5×10^{14}
35 GRBs[24]	BATSE/HETE-2 SWIFT	Wavelets	1.4×10^{16}
Mkn 421[23]	Whipple	Likelihood	0.6×10^{17}
Mkn 501[10]	MAGIC	ECF, Likelihood	2.6×10^{17}
PKS 2155-304[11]	H.E.S.S.	CCF, Wavelets	7.2×10^{17}

was obtained with only one GRB and a very limited photon statistics, therefore other observations and analyses are needed. On the other hand, the GRBs and the AGNs provide complementary analyses for the time-of-flight studies. AGN flares can be detected with high statistics with ground-based γ-ray telescopes, which give large values of $\Delta E \sim 1$ TeV. However, the absorption of the high energy photons by the Extra-galactic Background Light (EBL) limits the distance of the observed objects.

Till now no significant result on Lorentz Symmetry breaking has been obtained so far with neither AGNs nor GRBs for the linear and quadratic terms in the photon dispersion relations. The redshift dependences will be explored to distinguish between intrinsic effects to the source or induced by Quantum Gravity. Further observations of both a high number of GRBs and of AGN flares will be necessary to give robust conclusions on possible propagation effects. Present and future experiments such as Fermi, CTA[25] or AGIS[26] will greatly improve our capabilities in this area.

References

1. J.Ellis, N. E.Mavromatos and D. V.Nanopoulos, Microscopic recoil model for light-cone fluctuations in quantum gravity, Phys. Rev. D, **61**, 027503, (2000).
2. G. Amelino-Camelia *et al.*, Nature **395**, 525 (1998).
3. J. Ellis, N. E. Mavromatos and D. V. Nanopoulos, *Phys. Rev.* D **61**, 027503 (2002).
4. J. Alfaro *et al*, Phys. Rev. D **65**, 103509 (2002).
5. R. Gambini and J. Pullin *Phys. Rev.* D **59**, 124021 (1999).
6. G. Amelino-Camelia, arXiv:0806.0339.
7. J. Ellis *et al.*, A&A **402**, 409 (2003).
8. J. Ellis, N. E. Mavromatos, D. V. Nanopoulos, A. S. Sakharov and E. K. G Sarkisyan, *Astroparticle Physi cs* **25** 402 (2006).
9. J. Bolmont, A. Jacholkowska *et al.*, ApJ **676**, 532 (2008).

10. J. Albert *et al.* (MAGIC Collaboration) and J. Ellis *et al.*, Phys. Lett. B **668**, 253 (2008).
11. F. Aharonian *et al.* (HESS Collaboration), PRL **101**, 170402 (2008)
12. N. A. Bahcall, J. P. Ostriker, S. Perlmutter and P. J. Steinhardt, Science, **284**, 1481 (1999)..
13. A. A. Abdo *et al.*, Science Express, 02/19/2009.
14. F. Aharonian *et al.* (HESS Collaboration), ApJL **664**, L71 (2007).
15. T.-P. Li *et al.*, Chinese J. of Astronomy and Astrophys. **4**, 583 (2004).
16. J. P. Band, ApJ **486**, 928 (1997).
17. Lamon *et al.*., Gen. Rel. Grav. **40**, 1731 (2008).
18. D. Donoho and I. Johnstone, Biometrika, **81**, 425 (1994).
19. S. Mallat, *A Wavelet Tour of Signal Processing*, ed. Academic Press (1999); H. Bacry, *LastWave version 2.0.3* (2004).
20. E. E. Fenimore *et al.*, ApJ **448**, 1011 (1995).
21. J. P. Norris,ApJ **579**, 340 (2002).
22. S. Boggs *et al.*, ApJ **611**, L77 (2004).
23. S. D. Biller *et al.* (Whipple Collaboration), Phys. Rev. Lett. **83**, 2108 (1999).
24. See e.g. J. Ellis *et al.*, Astropart. Phys. **25**, 402 (2006), Astropart. Phys. **29**, 158 (2008),
25. `http://www.cta-observatory.org/`
26. `http://www.agis-observatory.org/`

CONSTRAINING DARK ENERGY AND GRAVITY WITH X-RAY GALAXY CLUSTERS

D. RAPETTI*, S. W. ALLEN, A. MANTZ, R. G. MORRIS

Kavli Institute for Particle Astrophysics and Cosmology, Stanford University, Stanford, California, 94305, USA
** E-mail: drapetti@slac.stanford.edu*

H. EBELING

Institute for Astronomy, 2680 Woodlawn Drive, Honolulu, HI 96822, USA

R. SCHMIDT

Astronomisches Rechen-Institut, Zentrum für Astronomie der Universität Heidelberg, Mönchhofstrasse 12-14, 69120 Heidelberg, Germany

A. C. FABIAN

Institute of Astronomy, Madingley Road, Cambridge CB3 0HA

Using two complementary X-ray galaxy cluster studies we present new constraints on dark energy and modified gravity. Using Chandra measurements of the X-ray gas mass fraction, $f_{\rm gas}$, in 42 hot, X-ray luminous, dynamically relaxed clusters spanning the redshift range $0.05 < z < 1.1$, we obtain a tight constraint on the mean matter density and a detection of the effects of dark energy on the distances to the clusters comparable in significance to recent type Ia supernovae (SNIa) studies. Using measurements of the growth of cosmic structure, as inferred from the observed evolution of the X-ray luminosity function (XLF) of clusters, we obtain the first precise determination of the dark energy equation of state and constrain departures from General Relativity (GR) on cosmological scales. For the latter, we employ the growth rate parameterization, $\Omega_{\rm m}(z)^{\gamma}$, for which GR predicts a growth index $\gamma \sim 0.55$. Combining the XLF with observations of the cosmic microwave background, SNIa, and $f_{\rm gas}$, to simultaneously constrain a flat cosmological constant (ΛCDM) background model, we obtain $\gamma = 0.51^{+0.16}_{-0.15}$, which is consistent with GR. Both the $f_{\rm gas}$ and XLF analyses provide strong, independent support to the GR+ΛCDM paradigm.

Keywords: Cosmology: theory; Cosmology: observations; dark energy; modified gravity; X-ray galaxy clusters.

1. Introduction

Precise cosmological measurements[4,6,7,10] indicate that the expansion history of the Universe has been accelerating at late times. Cosmic acceleration can be explained by either a new, exotic component, dark energy, or modifications of gravity at large scales. To distinguish between dark energy and modified gravity models, it is crucial to have measurements of both the expansion history and growth of cosmic structure. To determine the evolution of the background energy density, various techniques have been developed, such as distance measurements to X-ray luminous galaxy clusters[1–4] and type Ia supernovae (SNIa),[5,6] measurements of baryon acoustic oscillations in galaxy surveys,[7] and temperature and polarization anisotropy measurements in the cosmic microwave background (CMB).[8–10] Recently, independent constraints on dark energy[11,12] and modified gravity[14] models have been obtained from measurements of the growth of strutcure, as inferred from the observed X-ray luminosity function (XLF) of galaxy clusters.

Within the Markov Chain Monte Carlo (MCMC) code CosmoMC[13a], we have developed modules to analyse the $f_{\rm gas}$[4b] and XLF[11] data; and to investigate evolving dark energy[2,4] and modified gravity[14] models. In the following, we describe these analyses and models, and present constraints on these models from the $f_{\rm gas}$ and XLF data, and from combining $f_{\rm gas}$+SNIa+CMB+XLF data.

2. X-Ray Gas Mass Fraction Analysis

In the largest, hottest, most luminous and dynamically relaxed galaxy clusters in the Universe, the fraction of baryonic to total mass is expected to be an almost fair sample of the cosmic baryonic fraction $\Omega_{\rm b}/\Omega_{\rm m}$, where $\Omega_{\rm b}$ is the mean baryonic matter density and $\Omega_{\rm m}$ is the mean total matter density in units of the critical density.

In galaxy clusters, most of the baryons are in the form of X-ray emitting gas, and only a small fraction ($\sim 1/6$) are in the form of optically luminous material such as galaxies and intracluster light. Assuming hydrostatic equilibrium and a total mass profile such as that given by the NFW model,[15] the gas to total mass fraction, $f_{\rm gas}$, of a cluster can be determined from measurements of the X-ray gas density and temperature profiles.

[a]http://cosmologist.info/cosmomc/
[b]http://www.stanford.edu/~drapetti/fgas_module/

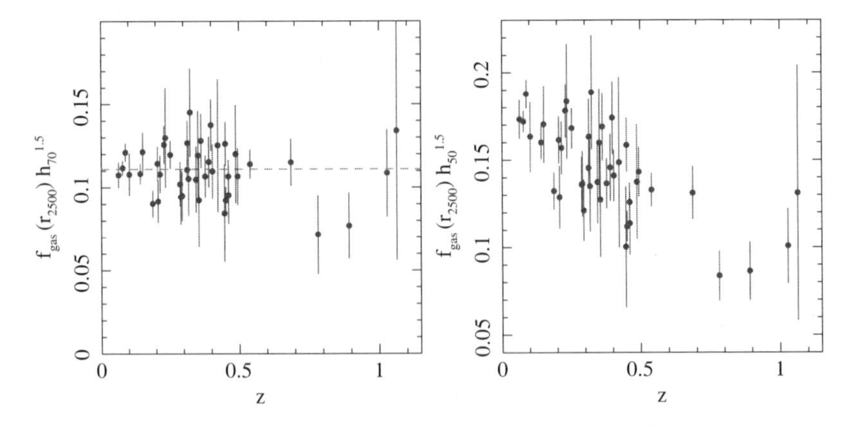

Fig. 1. The apparent variation of the $f_{\rm gas}$ measured within r_{2500} as a function of redshift,[4] using ΛCDM ($\Omega_{\rm m} = 0.3$, $\Omega_\Lambda = 0.7$, $h = 0.7$) (left panel) or SCDM ($\Omega_{\rm m} = 1.0$, $\Omega_\Lambda = 0.0$, $h = 0.5$) as the reference cosmology. The plotted error bars are root-mean-square 1σ uncertainties.

2.1. $f_{\rm Gas}$ *Measurements*

In Allen et al 2008[4] (hereafter A08), the $f_{\rm gas}$ measurements were made at the radius r_{2500}, corresponding to an angle θ_{2500}, for which the mean enclosed mass density is 2500 times the critical density for a given cosmology.

Figure 1 shows the $f_{\rm gas}$ measurements for the entire sample of 42 clusters spanning a redshift range $0.05 < z < 1.1$. The differences between the shapes of the $f_{\rm gas}(z)$ curves in the left and right panels of figure 1 reflect the dependence of the measured $f_{\rm gas}$ values on the assumed cosmology, i.e., on the angular diameter distances to the clusters. Non-radiative simulations of large clusters[16,17] suggest that $f_{\rm gas}$ should, in reality, be approximately constant with redshift. By simple inspection of figure 1 the ΛCDM is clearly favoured over the SCDM cosmology.

To determine constraints on cosmological parameters we fit a single, reference $f_{\rm gas}$ data set with a model that accounts for the expected apparent variation in $f_{\rm gas}(z)$ as the underlying cosmology is varied. In A08, a flat ΛCDM (cosmological constant plus cold dark matter) with $\Omega_{\rm m} = 0.3$ and $H_0 = 70$ km s^{-1}Mpc^{-1} was chosen as the reference cosmology.

2.2. $f_{\rm Gas}$ *Model*

In A08 the model fitted to the reference ΛCDM data is

$$f_{\rm gas}^{\Lambda{\rm CDM}}(z) = \frac{KA\gamma b(z)}{1+s(z)}\left(\frac{\Omega_{\rm b}}{\Omega_{\rm m}}\right)\left[\frac{d_{\rm A}^{\Lambda{\rm CDM}}(z)}{d_{\rm A}(z)}\right]^{1.5} \tag{1}$$

where $d_{\rm A}(z)$ and $d_{\rm A}^{\Lambda\rm CDM}(z)$ are the angular diameter distances to the clusters in the current test model and reference cosmologies. Here

$$d_A = \frac{c}{H_0(1+z)\sqrt{\Omega_{\rm k}}} \sinh\left(\sqrt{\Omega_{\rm k}} \int_0^z \frac{dz}{E(z)}\right), \tag{2}$$

where $\Omega_{\rm k}$ is the curvature density, $E(a)$ is defined as $H(a)^2 = H_0^2 E(a)^2$, $H(a)$ is the Hubble parameter and H_0 is its value today. See sections 3 and 7 for specific forms of $E(a)$ in dark energy and modified gravity models, respectively.

2.3. *Allowances for Systematic Uncertainties*

The factor A in equation 1 accounts for the change in angle subtended by r_{2500} as the underlying cosmology is varied:[4,18]

$$A = \left(\frac{\theta_{2500}^{\Lambda\rm CDM}}{\theta_{2500}}\right)^{\eta} \approx \left(\frac{H(z)d_{\rm A}(z)}{[H(z)d_{\rm A}(z)]^{\Lambda\rm CDM}}\right)^{\eta}, \tag{3}$$

where η is the logarithmic slope of the $f_{\rm gas}(r/r_{2500})$ data in the region of r_{2500}, as measured for the reference ΛCDM cosmology. For simplicity, in A08 we used the best-fit average slope of $\eta = 0.214 \pm 0.022$ determined from a fit to the whole sample and marginalize over it.

In equation 1 we also have the following parameters to account for systematic uncertainties: γ, s_0, $\alpha_{\rm s}$, b_0, $\alpha_{\rm b}$, K. γ models non-thermal pressure support in the clusters. Using hydrodynamical simulations,[19] a bias of ~ 9 per cent in $f_{\rm gas}$ at r_{2500} for relaxed clusters has been estimated. This bias originates primarily from subsonic motions in the intracluster gas. Non-thermal pressure support due to cosmic rays[20] and magnetic fields[21] is expected to be small. Based on these considerations, we adopted a uniform prior of $1.0 < \gamma < 1.1$. $s(z) = s_0(1+\alpha_{\rm s}z)$ models the baryonic mass fraction in stars. We included a 30 per cent Gaussian uncertainty on s_0 (derived from optical measurements), and a 20 per cent uniform prior on $\alpha_{\rm s}$.

The factor $b(z) = b_0(1+\alpha_{\rm b}z)$ is the 'depletion' or 'bias' factor by which the baryon fraction measured at r_{2500} is depleted with respect to the universal mean; such depletion is a natural consequence of the thermodynamic history of the gas. We included a conservative 20 per cent uniform prior on b_0, and allowed for a moderate evolution in $b(z)$ by using a uniform prior on $\alpha_{\rm b}$, $-0.1 < \alpha_{\rm b} < 0.1$. K is a 'calibration' constant that parameterizes residual uncertainty in the accuracy of the instrument calibration and X-ray modelling. A conservative 10 per cent Gaussian uncertainty was included.

3. Dark Energy Models

From the Friedmann equation, which relates the first time derivative of the scale factor $a = 1/(1+z)$, $(\dot{a}/a)^2 = H(a)^2$, to the total density, we have $E(a)^2 = \Omega_{\rm m} a^{-3} + \Omega_{\rm de} F(a) + \Omega_{\rm k} a^{-2}$; where $\Omega_{\rm de}$ is the dark energy density and $F(a)$ is its redshift dependence. (Note that this expression ignores the density contributions from radiation and relativistic matter, although they are included in the analysis.)

Our most general dark energy model has an evolving dark energy equation of state parameter as follows

$$w = \frac{w_{\rm et} z + w_0 z_{\rm t}}{z + z_{\rm t}} = \frac{w_{\rm et}(1-a)a_{\rm t} + w_0(1-a_{\rm t})a}{a(1-2a_{\rm t}) + a_{\rm t}}, \tag{4}$$

where w_0 and $w_{\rm et}$ are the equation of state at late (today) and early times, and $z_{\rm t}$ and $a_{\rm t}$ are the redshift and scale factor at the transition between the two.[2] We employ a uniform prior on the transition scale factor such that $0.5 < a_{\rm t} < 0.95$. As discussed by Rapetti et al (2005),[2] this model is both more general and more applicable to current data, which primarily constrain the properties of dark energy at redshifts $z < 1$, than models which impose a transition redshift $z = 1$, such as $w(a) = w_0 + w_a(1-a)$.[22,23] For the w parameterization of equation 4, $F(a) = a^{-3(1+w_{\rm et})} e^{-3(w_{\rm et}-w_0) g(a;a_{\rm t})}$ with

$$g(a;a_{\rm t}) = \left(\frac{1-a_{\rm t}}{1-2a_{\rm t}}\right) \ln\left(\frac{1-a_{\rm t}}{a(1-2a_{\rm t}) + a_{\rm t}}\right). \tag{5}$$

(Note that for ΛCDM the dark energy density is constant and $F = 1$.) In Rapetti et al (2007),[3] instead of relying on the Friedmann equation, we presented results from a purely kinematic modelling of the data, which is independent of the assumption of General Relativity.

4. Constraints on Dark Energy from $f_{\rm gas}$ Data

Using the $f_{\rm gas}$ method and the conservative allowances for systematic uncertainties discussed above, figures 2 and 3 show the constraints on various dark energy models that we obtained in A08. The left panel of figure 2 shows the constraints in the $\Omega_{\rm m}$, Ω_Λ plane, and the right panel those in the $\Omega_{\rm m}, w$ plane. In the left panel, a ΛCDM model is assumed, with the curvature included as a free parameter. In the right panel, a flat cosmology with a constant dark energy equation of state parameter, w, is assumed.

In figure 1, the $f_{\rm gas}$ constraints are shown as red contours, and independent constraints from SNIa and CMB data are shown as green and blue contours, respectively. The orange contours show the constraints obtained

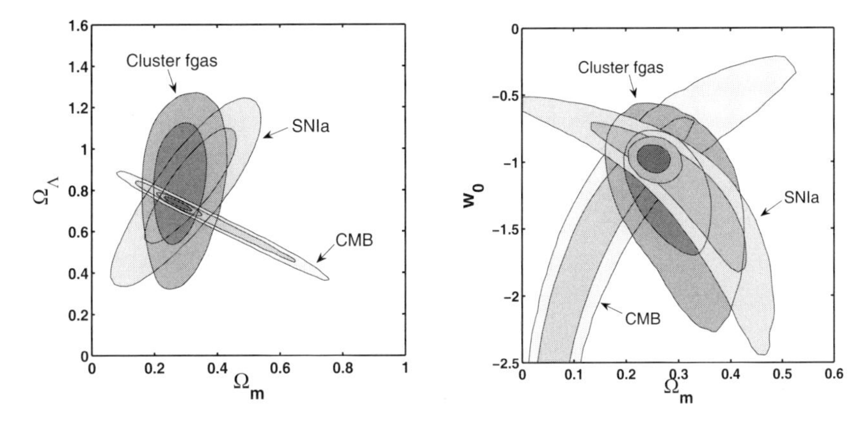

Fig. 2. The 68.3 and 95.4 per cent (1 and 2 σ) confidence constraints in the $\Omega_{\rm m}$, Ω_Λ plane (left panel) and $\Omega_{\rm m}, w$ plane (right panel) for the Chandra $f_{\rm gas}$ data[4] (red contours; standard priors on $\Omega_{\rm b}h^2$ and h are used). Also shown are the independent results obtained from CMB data[8] (blue contours) using a weak, uniform prior on h ($0.2 < h < 2$), and SNIa data (green contours; the results for the Davis et al 2007[5] compilation are shown). The inner, orange contours show the constraints obtained from all three data sets combined (no external priors on $\Omega_{\rm b}h^2$ and h are used).

from all three data sets combined. When the $f_{\rm gas}$ data is used alone, standard Gaussian priors on $\Omega_{\rm b}h^2$ and h are included: $\Omega_{\rm b}h^2 = 0.0214 \pm 0.0020$[24] and $h = 0.72 \pm 0.08$.[25] Using these priors, we measured $\Omega_{\rm m} = 0.27 \pm 0.06$ and $\Omega_\Lambda = 0.86 \pm 0.19$ (68 per cent confidence level), for the ΛCDM model, and $\Omega_{\rm m} = 0.28 \pm 0.06$ and $w = -1.14^{+0.27}_{-0.35}$, for the constant w model (wCDM). Combining all the three data sets, we obtained $\Omega_{\rm m} = 0.275 \pm 0.033$ and $\Omega_\Lambda = 0.735 \pm 0.023$, and $\Omega_{\rm m} = 0.253 \pm 0.021$ and $w = -0.98 \pm 0.07$.

Figure 3 shows the constraints on w_0 (solid, purple contours) and $w_{\rm et}$ (dashed, turquoise contours) obtained from the combined analysis of $f_{\rm gas}$+CMB+SNIa data, using equation 4 and assuming geometric flatness. This figure shows no evidence for evolution in the dark energy equation of state over the redshift range spanned by the data, obtaining $w_0 = -1.05^{+0.31}_{-0.26}$ and $w_{\rm et} = -0.83^{+0.48}_{-0.43}$ (right panel), which are both consistent with the cosmological constant model ($w = -1$).

Note, however, that when the SNIa data set of Riess et al (2007)[26] is used (right panel) there is a hint of evolution in $w(z)$. However, when the Davis et al (2007)[5] data set is used (left panel), which includes the high-quality, high-redshift HST supernovae from Riess et al (2007),[26] there is no hint of evolution, suggesting the possibility that the SNIa data set used in the right panel is affected by systematic uncertainties.

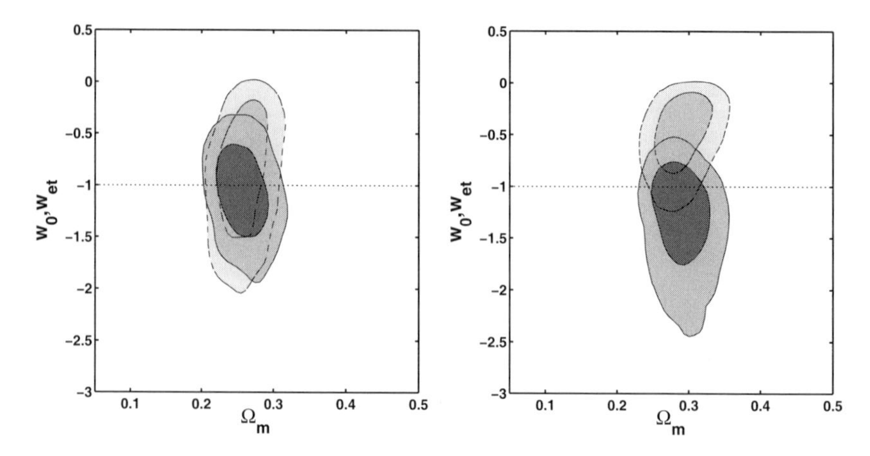

Fig. 3. The 68.3 and 95.4 per cent confidence limits in the ($\Omega_{\rm m}$;w_0,$w_{\rm et}$) plane (see equation 4) determined from the $f_{\rm gas}$+CMB+SNIa data,[4] with the transition scale factor marginalized over the range $0.5 < a_{\rm t} < 0.95$. The solid, purple contours show the results on ($\Omega_{\rm m}$,w_0). The dashed, turquoise contours show the results on ($\Omega_{\rm m}$,$w_{\rm et}$). The horizontal dotted line denotes the cosmological constant model ($w_0 = w_{\rm et} = -1$). The left and right panels show the results obtained for the two SNIa samples: (left panel) Davis et al (2007)[5] and (right panel) Riess et al (2007).[26] A flat geometry is assumed. The data provide no significant evidence for evolution in w and are consistent with the cosmological constant model.

5. X-Ray Luminosity Function Analysis

In Mantz et al (2008) (hereafter M08), we developed a full MCMC analysis that uses the observed X-ray luminosity function (XLF) of galaxy clusters to constrain cosmological parameters, simultaneously accounting for systematic uncertainties and X-ray survey biases. Using this analysis, in M08 we presented constraints on dark energy (see section 6), and in Rapetti et al (2009) (hereafter R09) we presented constraints on modified gravity (see section 9). For both analysis we employed data from the *ROSAT* Brightest Cluster Sample (BCS),[27] the *ROSAT*-ESO Flux Limited X-ray cluster sample (REFLEX),[28] and the MAssive Cluster Survey (MACS);[29] and in R09 we also employed the 400 square degrees *ROSAT* PSPC cluster survey (400sd)[30] (see figure 4).

In the XLF analysis, we compare X-ray flux-redshift data from the cluster samples to theoretical predictions. The relation between cluster mass and observed X-ray luminosity is calibrated using deeper pointed X-ray observations.[31]

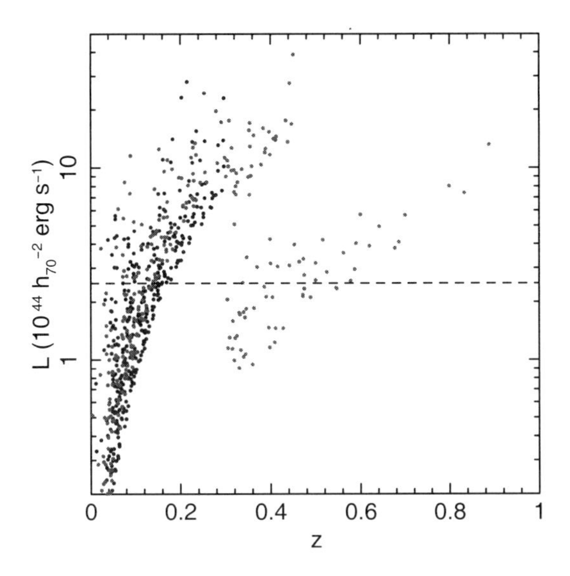

Fig. 4. Luminosity–redshift distribution of clusters in the BCS (blue), REFLEX (black), MACS (red) and 400sd (purple) surveys. Adopting a luminosity limit of $2.55 \times 10^{44} h_{70}^{-2}$ergs^{-1} (0.1 − 2.4keV) (dashed line), there are 78 clusters above a flux limit of 4.4×10^{-12}ergs^{-1}cm^{-2} from BCS, 130 above 3.0×10^{-12}ergs^{-1}cm^{-2} from REFLEX, 34 above 2×10^{-12}ergs^{-1}cm^{-2} from MACS, and 30 above 0.14×10^{-12}ergs^{-1}cm^{-2} from the 400sd. Error bars are not shown.

5.1. *Linear Theory*

The variance of the linearly evolved density field, smoothed by a spherical top-hat window of comoving radius R, enclosing a mass $M = 4\pi\rho_{\rm m}R^3/3$, is

$$\sigma^2(M,z) = \frac{1}{2\pi^2}\int_0^\infty k^2 P(k,z)|W_{\rm M}(k)|^2 dk\,, \tag{6}$$

where $W_{\rm M}(k)$ is the Fourier transform of the window function, and $P(k,z)$ is the linear matter power spectrum as a function of the wavenumber, k, and redshift, z: $P(k,z) \propto k^{n_{\rm s}} T^2(k,z_{\rm t})\,D(z)^2$. Here, $n_{\rm s}$ is the scalar spectral index of the primordial fluctuations, $T(k,z_{\rm t})$ is the matter transfer function at redshift $z_{\rm t}$, and $D(z) \equiv \delta(z)/\delta(z_{\rm t})$ is the growth factor of linear perturbations, normalized to unity at redshift $z_{\rm t}$. In R09 we chose $z_{\rm t} = 30$, well within the matter dominated era, and in M08 $z_{\rm t} = 0$ (today). The linear matter density contrast is defined as $\delta \equiv \delta\rho_{\rm m}/\rho_{\rm m}$, where $\rho_{\rm m}$ is the mean comoving matter density and $\delta\rho_{\rm m}$ a matter density fluctuation.

5.2. *Non-Linear Simulations*

Using large N-body simulations, it has been showed that the mass function of dark matter halos can be conveniently fitted by the expression[32]

$$f(\sigma^{-1}) \equiv \frac{M}{\rho_{\rm m}} \frac{dn(M,z)}{d\ln\sigma^{-1}} = A \exp\left(-|\ln\sigma^{-1} + B|^{\epsilon}\right), \qquad (7)$$

where $A = 0.316$, $B = 0.67$, and $\epsilon = 3.82$.[32] These fit values were determined using a spherical overdensity group finder, at 324 times the mean matter density. This formula is approximately 'universal' with respect to the cosmology assumed.[32–36] In M08 and R09 we used this formula to predict the number density of galaxy clusters, n, at a given M and z.

Despite the fact that equation 7 provides a useful approximation for a wide range of models, further simulations will be essential to determine its limitations for more complex tests of modified gravity and clustering dark energy models.

5.3. *Scaling Relation*

In both M08 and R09, we employed a power-law mass–luminosity relation, based on self–similar evolution[37] between the mass and X-ray luminosity, L, of the clusters, in which we included a conservative allowance for departures from self–similarity, encoded in the parameter ζ, $E(z)M_{\Delta} = M_0[L/E(z)]^{\beta}(1+z)^{\zeta}$; where M_{Δ} is the cluster mass defined at an overdensity of Δ with respect to the critical density, and $\log M_0$ and β are model parameters fitted in the MCMC analysis. We also assumed a log-normal intrinsic scatter in luminosity for a given mass, $\eta = \eta_0(1+\eta_{\rm z}z)$, which allows for possible evolution in the scatter.[38,39]

5.4. *Priors and Systematic Allowances*

The XLF analysis includes five nuisance parameters: A, ζ, $\eta_{\rm z}$, B, and $s_{\rm b}$. B and $s_{\rm b}$ (see details in M08) parameterize the bias, and the scatter in the bias, expected for hydrostatic mass measurements from X-ray data. We applied a Gaussian prior on A of 20 per cent, and for ζ and $\eta_{\rm z}$ we allowed up to $\sim$ 25 per cent change with respect to self–similar evolution out to $z = 1$, and $\sim$ 30 per cent evolution in the scatter to $z = 1$, respectively.

6. Constraints on Dark Energy from XLF Data

Figure 5 shows the constraints from the XLF analysis, as obtained in M08, in the $\Omega_{\rm m}, w$ and σ_8, w planes using standard priors on $\Omega_{\rm b}h^2$ and h. The

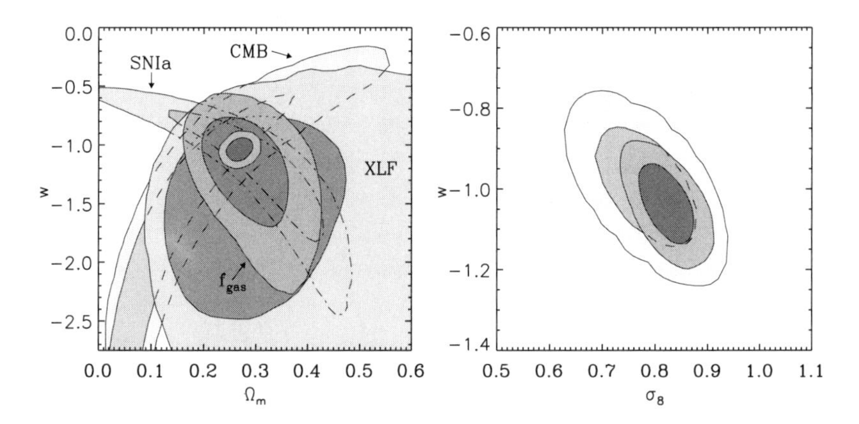

Fig. 5. The 68.3 and 95.4 per cent confidence constraints in the Ω_m, w plane (left panel) and σ_8, w (right panel) for a flat, constant w model using the XLF (purple contours).[11] Also shown are the independent constraints from figure 1, and the combination of all four (gold contours). In the right panel, the turquoise contours show the constraints obtained from a combined f_{gas}+CMB+SNIa analysis and the improved constraints obtained by combining these data with the XLF (gold contours).[11]

marginalized results from the XLF data are $\Omega_m = 0.24^{+0.15}_{-0.07}$, $\sigma_8 = 0.85^{+0.13}_{-0.20}$ and $w = -1.4^{+0.4}_{-0.7}$, and, from combining the XLF with f_{gas}+SNIa+CMB data, $\Omega_m = 0.269 \pm 0.016$, $\sigma_8 = 0.82 \pm 0.03$ and $w = -1.02 \pm 0.06$. The right panel of figure 5 shows that adding the XLF data to the combination of f_{gas}+SNIa+CMB data breaks the degeneracy between σ_8 and w, providing tighter constraints on dark energy.

7. Modified Gravity Model

Several authors[40–44] have parameterized the evolution of the growth rate as $f(a) \equiv d\ln\delta/d\ln a = \Omega_m(a)^\gamma$. Recasting this expression we have the differential equation

$$\frac{d\delta}{da} = \frac{\Omega_m(a)^\gamma}{a}\,\delta\,, \tag{8}$$

where γ is the growth index and $\Omega_m(a) = \Omega_m a^{-3}/E(a)^2$. For $\gamma \sim 0.55$, equation 8 accurately reproduces the evolution of δ obtained from the GR equation $\ddot{\delta} + 2H\dot{\delta} = 4G\pi\rho_m\delta$. Note that the growth rate $\Omega_m(a)^\gamma$ conveniently matches GR ($f = 1$) in the matter dominated era, and does not depend on k. Using $\delta(a)$ obtained from equation 8 we calculate $D(z)$, and thus $\sigma(M, z)$ from equation 6. This allows us to constrain γ using the XLF data.

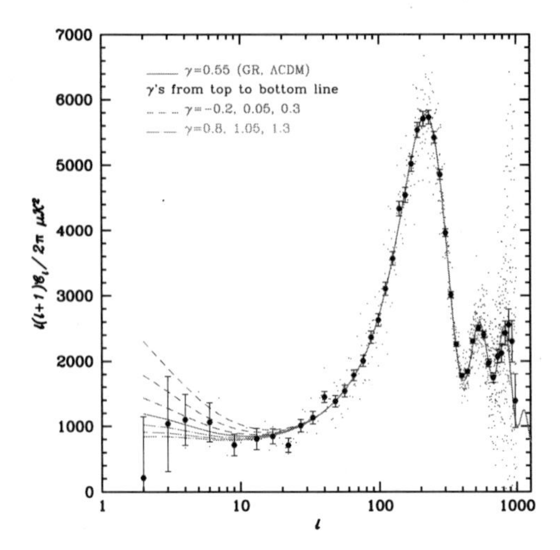

Fig. 6. CMB temperature anisotropy power spectra, C_l, for $\gamma = 0.55$ (GR; best-fit ΛCDM model from the five-year WMAP data[10]) (red, solid line), and for values of γ lower than GR (blue, dashed lines), and higher (magenta, dot-dashed lines).[14] The lines are equally spaced in γ. The data are more constraining for the lower values of γ, which produce larger changes in the ISW effect and significantly worse fits to the data. The dots and circles are the unbinned and binned WMAP5[10] data, respectively.

Using the phenomenological model of equation 8, in R09 we tested for departures from GR without adopting a particular, fully covariant modified gravity theory. In the absence of such an alternative gravity theory, we performed consistency tests using convenient parameterizations of the background expansion, within GR: flat ΛCDM, flat wCDM, and non-flat ΛCDM.

8. Integrated Sachs-Wolfe Effect

Through the integrated Sachs-Wolfe (ISW) effect, the low multipoles of the temperature anisotropy power spectrum of the CMB are sensitive to the growth of cosmic structure, and therefore to dark energy and modified gravity models.[45] The ISW effect arises when the gravitational potentials of large scale structures vary with time, yielding a net energetic contribution to the CMB photons crossing them.

Using the Poisson equation, we calculate the variation in time of the gravitational potential, ϕ, from the evolution of the matter density perturbations, $\delta\rho_{\rm m}$. Differentiating the Poisson equation with respect to conformal time t ('prime'), we have $\phi' = -(4\pi G/k^2)\partial(a^2\,\delta\rho_{\rm m})/\partial t$, which can be used to calculate the CMB temperature transfer function on large scales due to the ISW effect as[46]

$$\Delta_l^{\rm ISW}(k) = 2\int dt\, {\rm e}^{-\tau(t)}\phi' j_l\left[k(t-t_0)\right] , \tag{9}$$

where t_0 is the conformal time today, τ is the optical depth to the last scattering, and $j_l(x)$ is the spherical Bessel function for the multipole ℓ. Figure 6 shows the CMB temperature power spectrum we obtained using different values of γ.

9. Constraints on Modified Gravity

Using the XLF data to constrain the evolution of the density perturbations and the $f_{\rm gas}$+SNIa+CMB data to constrain the background evolution, we obtained (R09) the constraints in the $\Omega_{\rm m}, \gamma$ and w, γ planes shown in figure 7, and the marginalized 68 per cent confidence level constraints $\gamma = 0.51^{+0.16}_{-0.15}$ for flat ΛCDM, $\gamma = 0.44^{+0.17}_{-0.15}$ for flat wCDM, and $\gamma = 0.51^{+0.19}_{-0.16}$ for non-flat ΛCDM. We find no significant deviation from GR in any of these cases.

Note that the constraining power of the current ISW data is not competitive to that of the XLF data, which dominates our constraint on γ. However, the correlation of the ISW effect with galaxy surveys[47] may provide competitive, additional constraint on γ.

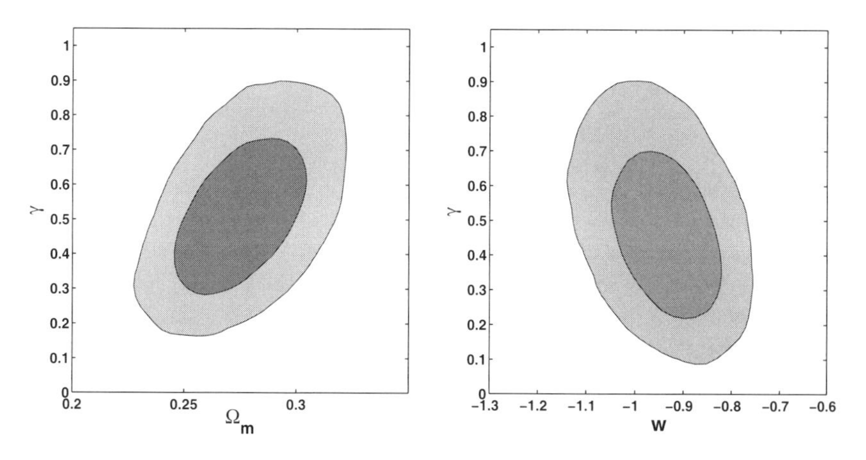

Fig. 7. 68 and 95 per cent confidence contours in the $\Omega_{\rm m}, \gamma$ plane (left panel) for the flat ΛCDM background model, and w, γ plane (right panel) for the flat wCDM model, from the combination of XLF, CMB, SNIa, and $f_{\rm gas}$ data.[14]

10. Conclusions

Using X-ray observations of galaxy clusters we have developed two independent and complementary analyses to measure the effects of cosmic acceleration in both the expansion history of the Universe, using $f_{\rm gas}$ data (A08), and the growth of cosmic structure, using XLF data. We have obtained the first precise constraints on dark energy (M08) and modified gravity on cosmological scales (R09) using the XLF analysis. Our work highlights the importance of cluster cosmology to investigate the nature of cosmic acceleration and to distinguish between dark energy and modified gravity models as its origin.

Acknowledgments

The computational analysis was carried out using the KIPAC XOC and Orange computer clusters at SLAC. We acknowledge support from the National Aeronautics and Space Administration through Chandra Award Numbers DD5-6031X and G08-9118X issued by the Chandra X-ray Observatory Center. This work was supported in part by the U.S. Department of Energy under contract number DE-AC02-76SF00515.

References

1. S. W. Allen, R. W. Schmidt, H. Ebeling, A. C. Fabian and L. van Speybroeck, *MNRAS* **353**, 457 (2004).
2. D. Rapetti, S. W. Allen and J. Weller, *MNRAS* **360**, 555 (2005).
3. D. Rapetti, S. W. Allen, M. A. Amin and R. D. Blandford, *MNRAS* **375**, 1510 (2007).
4. S. W. Allen, D. A. Rapetti, R. W. Schmidt, H. Ebeling, R. G. Morris and A. C. Fabian, *MNRAS* **383**, 879 (2008).
5. T. M. Davis *et al.*, *Astrophys. J.* **666**, 716 (2007).
6. M. Kowalski *et al.*, *Astrophys. J.* **686**, 749 (2008).
7. W. J. Percival *et al.*, *Mon. Not. Roy. Astron. Soc.* **381**, 1053 (2007).
8. D. N. Spergel *et al.*, *Astrophys. J. Suppl.* **170**, 377 (2007).
9. J. Dunkley *et al.*, *Astrophys. J. Suppl.* **180**, 306 (2009).
10. E. Komatsu *et al.*, *Astrophys. J. Suppl.* **180**, 330 (2009).
11. A. Mantz, S. W. Allen, H. Ebeling and D. Rapetti, *MNRAS* **387**, 1179 (2008).
12. A. Vikhlinin *et al.*, *Astrophys. J.* **692**, 1060 (2009).
13. A. Lewis and S. Bridle, *Phys. Rev.D* **D66**, 103511 (2002).
14. D. Rapetti, S. W. Allen, A. Mantz and H. Ebeling, *arXiv:0812.2259* (2008).
15. J. F. Navarro, C. S. Frenk and S. D. M. White, *ApJ* **490**, 493 (1997).
16. V. R. Eke, J. F. Navarro and C. S. Frenk, *Astrophys. J.* **503**, 569 (1998).
17. R. A. Crain, V. R. Eke, C. S. Frenk, A. J. Jenkins, M. I. G., N. J. F. and P. F. R., *MNRAS* **377**, 41 (2007).

18. D. Rapetti, S. W. Allen and A. Mantz, *MNRAS* **388**, 1265 (2008).
19. D. Nagai, A. Vikhlinin and A. V. Kravtsov, *Astrophys. J.* **655**, 98 (2007).
20. C. Pfrommer, T. A. Ensslin, V. Springel, M. Jubelgas and K. Dolag, *Mon. Not. Roy. Astron. Soc.* **378**, 385 (2007).
21. K. Dolag and S. Schindler, *Astr.Astrophys.* **364**, 491(December 2000).
22. M. Chevallier and D. Polarski, *Int. J. Mod. Phys.* **D10**, 213 (2001).
23. E. V. Linder, *Physical Review Letters* **90**, 091301 (2003).
24. D. Kirkman, D. Tytler, N. Suzuki, J. M. O'Meara and D. Lubin, *ApJ S.* **149**, 1 (2003).
25. W. Freedman *et al.*, *ApJ* **553**, 47 (2001).
26. A. G. Riess *et al.*, *ApJ* **659**, 98 (2007).
27. H. Ebeling *et al.*, *MNRAS* **301**, 881 (1998).
28. H. Böhringer *et al.*, *Astr.Astrophys.* **425**, 367 (2004).
29. H. Ebeling, A. C. Edge and J. P. Henry, *ApJ* **553**, 668 (2001).
30. R. A. Burenin, A. Vikhlinin, A. Hornstrup, H. Ebeling, H. Quintana and A. Mescheryakov, *ApJ S.* **172**, 561 (2007).
31. T. H. Reiprich and H. Böhringer, *ApJ* **567**, 716 (2002).
32. A. Jenkins, C. S. Frenk, S. D. M. White, J. M. Colberg, S. Cole, A. E. Evrard, H. M. P. Couchman and N. Yoshida, *MNRAS* **321**, 372 (2001).
33. A. E. Evrard *et al.*, *ApJ* **573**, 7 (2002).
34. M. Kuhlen, L. E. Strigari, A. R. Zentner, J. S. Bullock and J. R. Primack, *MNRAS* **357**, 387 (2005).
35. M. S. Warren, K. Abazajian, D. E. Holz and L. Teodoro, *Astrophys. J.* **646**, 881 (2006).
36. J. Tinker, A. V. Kravtsov, A. Klypin, K. Abazajian, M. Warren, G. Yepes, S. Gottlöber and D. E. Holz, *ApJ* **688**, 709 (2008).
37. G. L. Bryan and M. L. Norman, *Astrophys. J.* **495**, 80 (1998).
38. T. B. O'Hara, J. J. Mohr, J. J. Bialek and A. E. Evrard, *ApJ* **639**, 64 (2006).
39. Y. Chen, T. H. Reiprich, H. Böhringer, Y. Ikebe and Y.-Y. Zhang, *Astr.Astrophys.* **466**, 805 (2007).
40. P. J. E. Peebles, *The large-scale structure of the universe* (Princeton University Press, 1980).
41. O. Lahav, P. B. Lilje, J. R. Primack and M. J. Rees, *MNRAS* **251**, 128 (1991).
42. L.-M. Wang and P. J. Steinhardt, *Astrophys. J.* **508**, 483 (1998).
43. D. Huterer and E. V. Linder, *Phys. Rev.* **D75**, 023519 (2007).
44. E. V. Linder and R. N. Cahn, *Astropart. Phys.* **28**, 481 (2007).
45. W. Fang *et al.*, *Phys. Rev.* **D78**, 103509 (2008).
46. J. Weller and A. M. Lewis, *MNRAS* **346**, 987 (2003).
47. S. Ho, C. Hirata, N. Padmanabhan, U. Seljak and N. Bahcall, *Phys. Rev.D* **78**, 043519 (2008).

DARK MATTER OR MODIFIED DYNAMICS? HINTS FROM GALAXY KINEMATICS

G. GENTILE*

Institut d'Astronomie et d'Astrophysique, Université Libre de Bruxelles
1050 Brussels, Belgium
**E-mail: gianfranco.gentile@ugent.be*

I show two observational projects I am involved in, which are aimed at understanding better the existence and nature of dark matter, and also aimed at testing alternatives to galactic dark matter such as MOND (Modified Newtonian Dynamics). I present new HI observations of the nearby dwarf galaxy NGC 3741. This galaxy has an extremely extended HI disc (42 B-band exponential scalelengths). The distribution and kinematics are accurately derived by building model data cubes, which closely reproduce the observations. Mass modelling of the rotation curve shows that a cored dark matter halo or MOND provide very good fits, whereas Cold Dark Matter density profiles fail to fit the data. I also show new results about tidal dwarf galaxies, which within the CDM framework are expected to be dark matter-free but whose kinematics instead show a mass discrepancy, exactly of the magnitude that is expected in MOND (Modified Newtonian Dynamics).

Keywords: Spiral galaxies; Galaxy kinematics; dark matter.

1. Introduction

Rotation curves of spiral galaxies (the rotation velocity of gas and stars as a function from the distance of the galaxy centre) have been studied for several decades, and they are a useful tool to investigate the mass discrepancy in galaxies: the observed kinematics are different from the one that is expected from the observed baryonic distribution. There are two options: either an additional mass component is needed or one has to modify the law of gravity. In the recent years, rotation curves have been exploited to test the predictions of cosmological models of galaxy formation in the Universe, such as the currently favoured Λ Cold Dark Matter (ΛCDM). Most observations show that the baryons kinematically dominate the inner parts of rotation curves and that the "cuspy" halos predicted by ΛCDM (Navarro, Frenk & White 1996, Moore et al. 1999, Navarro et al. 2004) fail to reproduce observed rotation curves (see e.g. de Blok et al. 2001; de Blok & Bosma 2002; Gentile et al. 2004, 2005, 2007a,b; McGaugh et al. 2007).

MOND, the Modified Newtonian Dynamics (Milgrom 1983) is a successful prescription to predict the rotation curve of a spiral galaxy. In MOND, the mass discrepancy is not due to the dark matter halo, but instead it is the signature of a modification of Newtonian gravity to describe the observed kinematics at the low gravitational accelerations found in the outer parts of galaxies. In MOND, below a certain acceleration $a_0 \sim 10^{-8}$ cm s^{-2}, Newtonian gravity is no longer valid. Early concerns about the inconsistency of MOND with General Relativity are now overcome with the TeVeS theory (Bekenstein 2004).

In MOND, the actual gravitational acceleration g and the Newtonian one $g_{\rm N}$ are linked through:

$$g = \frac{g_{\rm N}}{\mu(g/a_0)} \tag{1}$$

where $\mu(x)$ is an interpolation function that regulates the transition between Newtonian regime and deep MOND regime. It tends to 1 for $g \gg a_0$ and it tends to g/a_0 for $g \ll a_0$. The "standard" functional form for $\mu(x)$ is:

$$\mu(x) = \frac{x}{\sqrt{1+x^2}} \tag{2}$$

Recently, Famaey & Binney (2005) proposed the following form of $\mu(x)$, which, contrary to Eqn. 2, Zhao & Famaey (2006) have shown to be consistent with the relativistic MOND theory of Bekenstein (2004):

$$\mu(x) = \frac{x}{1+x}. \tag{3}$$

MOND has an astonishing predictive power for the kinematical properties of galaxies: it fits the kinematics of small dwarf irregular galaxies (Gentile et al. 2007a,c), of our own Milky Way (Famaey & Binney 2005), of early-type spiral galaxies (Sanders & Noordermeer 2007), and of massive ellipticals (Milgrom & Sanders 2003). Also, it naturally explains observed tight scaling relations in spiral galaxies, such as the the mass discrepancy-acceleration relation and the baryonic Tully-Fisher relation (McGaugh 2004, 2005).

2. NGC 3741: An Extremely Extended Rotation Curve

The full description of the results is reported in Gentile et al. (2007a). NGC 3741 is a nearby dwarf irregular galaxy with an absolute B-band magnitude of -13.13. Its distance (Karachentsev et al. 2004), estimated through the tip of the Red-Giant Branch, is 3.0±0.3 Mpc.

The HI (neutral hydrogen) observations were performed with the WSRT (Westerbork Synthesis Radio Telescope) for 12 hours with the 'maxi-short' configuration. The final data cubes have a velocity resolution of 4.1 km s^{-1} and a spatial resolution of 15.2 × 11.7 arcsecs. The integrated HI flux is 59.6 Jy km s^{-1}, which corresponds to an HI mass of 1.3×10^8 $M_\odot$, and a total gas mass (including primordial Helium) of $M_{\rm gas} = 1.7 \times 10^8$ $M_\odot$.

The rotation curve was derived by creating models of the whole data cube, which account much better for resolution and projection effects than standard methods (see Gentile et al. 2004). Thanks to the exceptional extension of the gas disk (compared to the stellar disk), the rotation curve extends out to 6.6 kpc, which correspond to 42 B-band exponential scale lengths (see Fig. 1).

The observed rotation curve $V_{\rm obs}$ was decomposed into the stellar disk, gaseous disk and dark halo contributions $V_{\rm disk}$, $V_{\rm gas}$ and $V_{\rm halo}$ via

$$V_{\rm obs}^2(r) = V_{\rm disk}^2(r) + V_{\rm gas}^2(r) + V_{\rm halo}^2(r). \tag{4}$$

The shape of the stellar disk contribution is derived from the NIR photometry shown in Vaduvescu et al. (2005), and its amplitude is scaled using a constant mass-to-light (M/L) ratio. The gaseous disk contribution was derived from the HI observations presented here and in Gentile et al. (2007a).

For the dark matter halo, several possibilities were investigated. In numerous previous studies dark matter halos with a central constant-density core provided the best fits to the rotation curve. An example of such a cored halo is the Burkert halo (Burkert 1995); its density distribution is given by

$$\rho_{\rm Bur}(r) = \frac{\rho_0 r_{\rm core}^3}{(r + r_{\rm core})(r^2 + r_{\rm core}^2)} \tag{5}$$

where ρ_0 (the central density) and $r_{\rm core}$ (the core radius) are the two free parameters.

Then, another fit to the observed rotation curve was performed using the Navarro, Frenk and White halo (NFW), the result of an analytical fit to the density distribution of dark matter halos in CDM cosmological simulations,

$$\rho_{\rm NFW}(r) = \frac{\rho_{\rm s}}{(r/r_{\rm s})(1 + r/r_{\rm s})^2}, \tag{6}$$

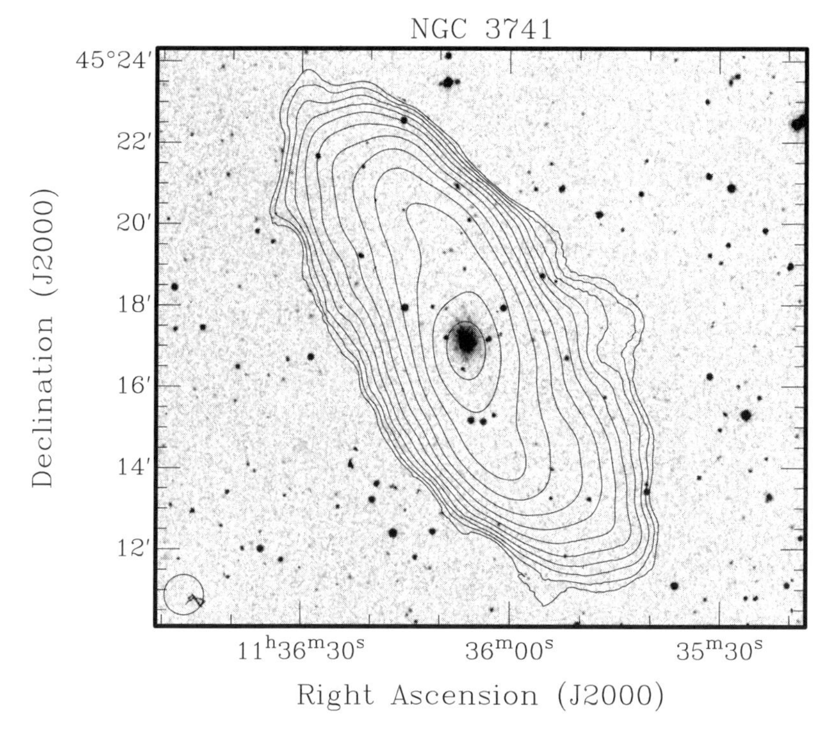

Fig. 1. HI column density map from the low-resolution data cube (contours) superimposed onto an optical image (DSS, greyscale). Contours are (1.5, 2.25, 3.375, ...) $\times$ 10^{19} atom cm^{-2}. The radio beam is shown in the bottom left corner.

where ρ_s and r_s are the characteristic density and scale of the halo. These two parameters are in principle independent, but they were shown to be correlated in simulations (e.g. Bullock et al. 2001, Wechsler et al. 2002), so it is actually a one-parameter family.

We also considered the so-called "gas scaling", where a fit to a rotation curve is made without a dark matter halo, by simply scaling up the contribution of the gaseous disk. This is linked to the hypothesis that large amounts of baryonic dark matter reside in the disk, in the form of e.g. cold clumps of molecular hydrogen (Pfenniger, Combes & Martinet 1994).

The results of the mass decompositions are shown in Fig. 2. The Burkert halo gives a very good-quality fit, with a core radius size of about five times the optical radius. The MOND fits, using any of the interpolation functions, are also extremely good, reproducing even the details in the rotation curve. The NFW fit is poor: the only way to try and reconcile this profile with the

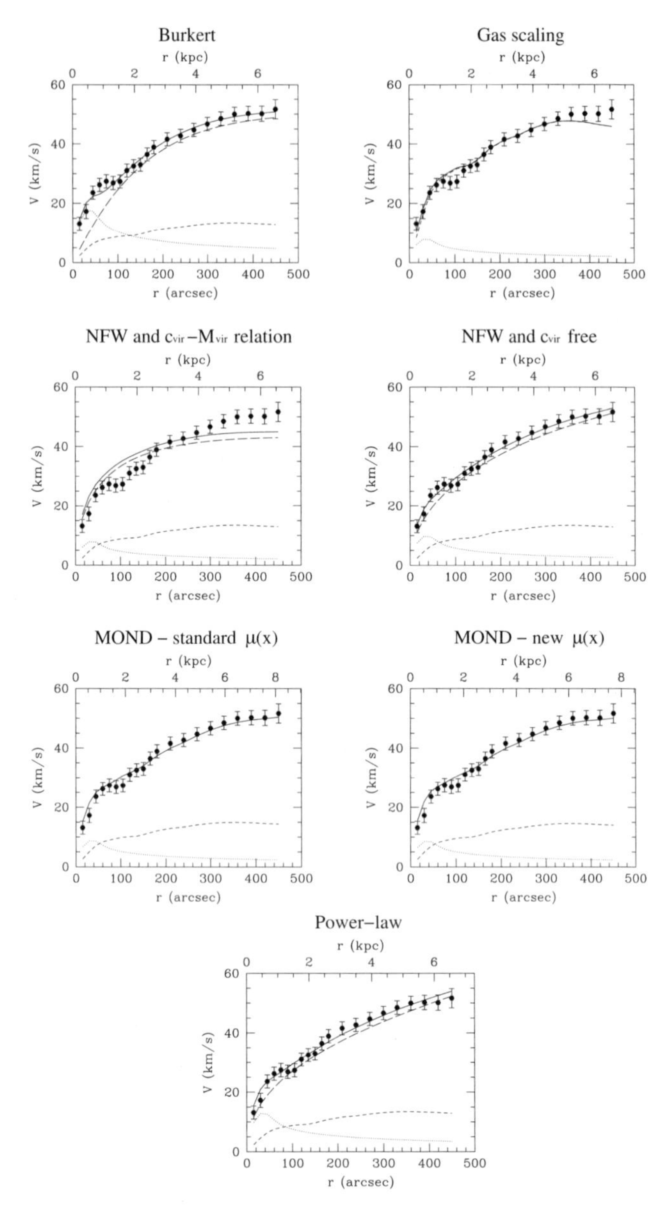

Fig. 2. Rotation curve fits. Dotted/short-dashed lines represent the stellar/gaseous disks, the long-dashed line depicts the dark matter halo, and the solid line is the best-fit to the rotation curve. The fit labelled "NFW and c_{vir}-M_{vir} relation" makes use of the relation between concentration and virial mass, and the fit called "MOND - new $\mu(x)$" uses the "simple" interpolation function for MOND (Famaey & Binney 2005).

observations is by relaxing the condition that the two structural parameters are correlated. By doing so, one finds a better fit, but at the expense of an unrealistically high virial mass and a concentration parameter which is at 2.5-σ from the predicted relation between the NFW structural parameters. The gas scaling fits also give a relatively good quality fit.

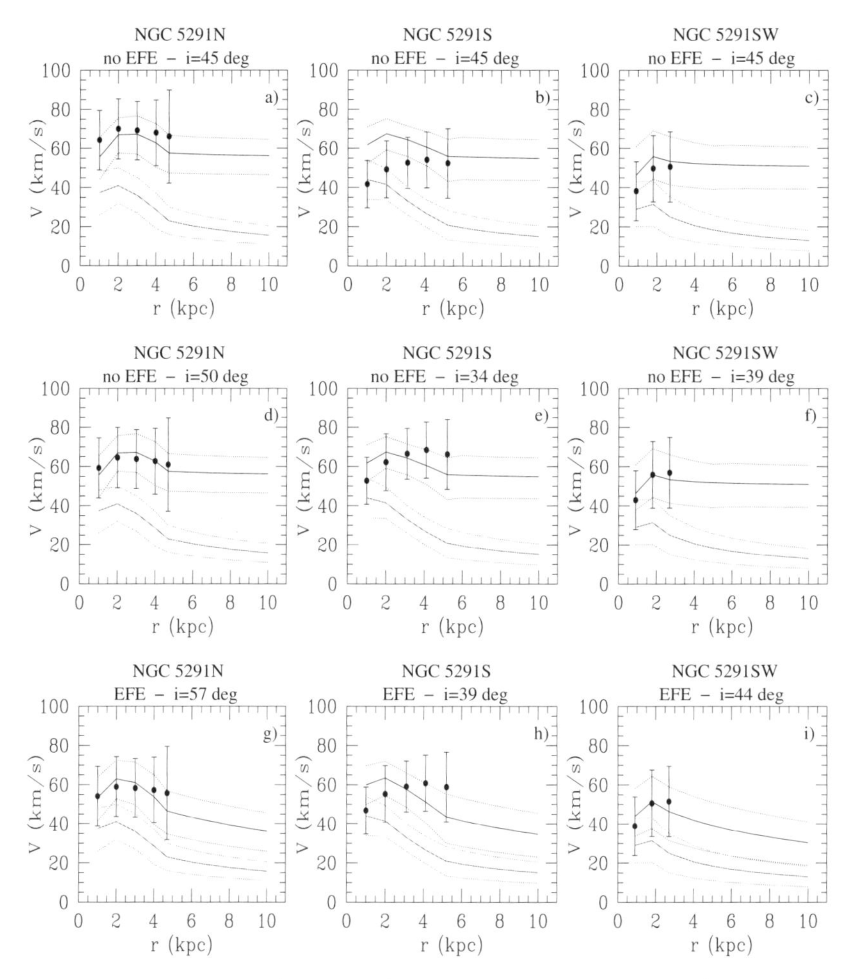

Fig. 3. Rotation curves (full circles) of the 3 tidal dwarf galaxies (Bournaud et al. 2007). The red curves are the Newtonian contribution of the baryons (and its uncertainty, indicated as dotted lines). The black curves are the MOND prediction and its uncertainty. The top panels assume (following Bournaud et al.) an inclination angle of 45 degrees. In the middle panels the inclination is a free parameter, and the bottom panels show the fits made with an estimate for the external field effect (EFE).

3. Tidal Dwarf Galaxies

The full description of the results is reported in Gentile et al. (2007c). Tidal dwarf galaxies (TDGs) form as a by-product of the collision of two disk galaxies. They form in the tidal tails, and they are made out of the material that was in the disks of the colliding galaxies, whereas dark matter is thought to reside in a halo. Hence, TDGs should be nearly devoid of collisionless dark matter (Barnes & Hernquist 1992), contrary to all other kinds of galaxies.

In a recent paper, Bournaud et al. (2007) analysed the rotation curves of 3 TDGs belonging to the NGC 5291 system of interacting galaxies. They find evidence for a mass discrepancy that is unexpected within the CDM framework, and they put forward the hypothesis of an (ad hoc) distribution of baryonic dark matter to explain the observations.

To make a rigorous fit of these galaxies in MOND, two important issues have to be taken into account: their inclination angle with respect to the line-of-sight and the EFE (external field effect) of MOND. Thus, we show here three fits for each one of the three TDGs: the first where the inclination is fixed at 45 degrees as in Bournaud et al. (2007); the second, where the inclination is left as a free parameter; the third, where we estimated the external field effect.

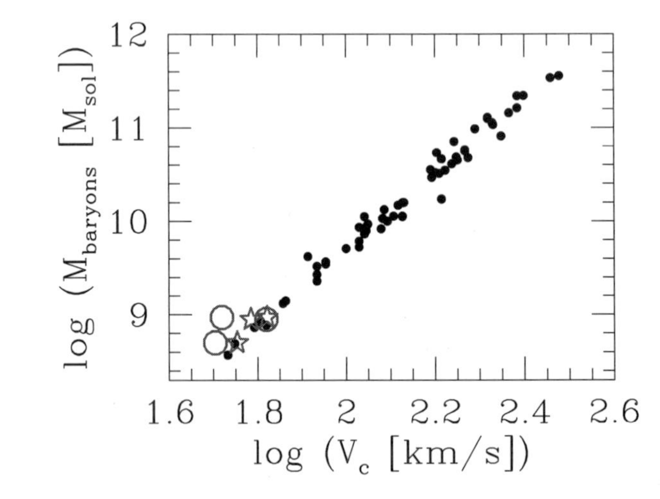

Fig. 4. Baryonic Tully-Fisher relation (baryonic mass versus a measure of the circular velocity). The small full circles are the disk galaxy data collected by McGaugh (2005). The 3 tidal dwarf galaxies studied here are shown as blue empty circles (when the inclination angle i is assumed to be 45°) and red stars (when i is a free parameter).

Fig. 3 shows the results of the fits. MOND gives very good fits to the rotation curves of these systems. In particular, the top panels show that even with *zero free parameters*, MOND can explain well the observed kinematics. Leaving the inclination as a free parameter obviously improves the quality of the fits, but the best-fit values of the inclination are not far from the initially assumed one. The external field effect does not affect our results significantly, but it might strongly affect the mass models if more extended data become available.

It is very important to notice that all three TDGs fall on the baryonic Tully-Fisher (BTF) relation (McGaugh 2005): the agreement is perfect for i as a free parameter and close to perfect for and inclination of $i = 45°$ (see Fig. 4). This shows that the kinematics of these three TDGs behave the same way as other galaxies, and is a very strong indication towards the fact that non-circular motions are negligible in these objects.

Acknowledgments

GG is a postdoctoral researcher of the FWO-Vlaanderen (Belgium)

References

1. Bekenstein, J. D. (2004) *Phys. Rev. D.* **70** 083509.
2. Bournaud, F., et al. (2007) *Science* **316** 1166.
3. Burkert, A. (1995) *ApJ* **447** L25.
4. Bullock, J.S., Kolatt, T.S., Rachel, Y.S., Somerville, S., Kravtsov, A.V., Klypin, A.A., Primack, J.R., Dekel, A. (2001) *MNRAS* **321** 559.
5. de Blok, W. J. G., McGaugh, S. S., Rubin, V. C. (2001) *AJ* **122** 2396.
6. de Blok, W. J. G., Bosma, A. (2002) *A&A* **385** 816.
7. Famaey, B., Binney, J. (2005) *MNRAS* **363** 603.
8. Gentile, G., Salucci, P., Klein, U., Vergani, D., Kalberla, P. (2004) *MNRAS* **351** 903.
9. Gentile, G., Burkert, A., Salucci, P., Klein, U., Walter, F. (2005) *ApJ* **634** L145.
10. Gentile, G., Salucci, P., Klein, U., Granato, G. L. (2007a) *MNRAS* **375** 199.
11. Gentile, G., Tonini, C., Salucci, P. (2007b) *A&A* **467** 925.
12. Gentile, G., Famaey, B., Combes, F., Kroupa, P., Zhao, H. S., Tiret, O. (2007c) *A&A* **472** L25.
13. Karachentsev, I. D., Karachentseva, V. E., Huchtmeier, W. K., Makarov, D. I. (2004) *AJ* **127** 2031.
14. McGaugh, S. S., de Blok, W. J. G., Schombert, J. M., Kuzio de Naray, R., Kim, J. H. (2007) *ApJ* **659** 149.
15. McGaugh, S. S. (2004) *ApJ* **609** 652.
16. McGaugh, S. S. (2005) *ApJ* **632** 859.
17. Milgrom, M. (1983) *ApJ* **270** 365.

18. Milgrom, M., Sanders, R. H. (2003) *ApJ* **599** L25.
19. Pfenniger, D., Combes, F., Martinet, L. (1994) *A&A* **285** 79.
20. Sanders, R. H., Noordermeer, E. (2007) *MNRAS* **379** 702.
21. Vaduvescu, O., McCall, M. L., Richer, M. G., Fingerhut, R. L. (2005) *AJ* **130** 1593.
22. Wechsler, R. H., Bullock, J. S., Primack, J. L., Kravtsov, A. V., Dekel, A. (2002) *ApJ* **568** 52.
23. Zhao, H. S., Famaey, B. (2006) *ApJ* **638** L9.

PART XI

Dark Matter and Non-standard Wigner Classes

QUANTUM FIELDS, DARK MATTER AND NON-STANDARD WIGNER CLASSES

A. B. GILLARD

Physics Department, University of Canterbury,
Private Bag 4800, Christchurch 8140, New Zealand
E-mail: abg22@student.canterbury.ac.nz

B. M. S. MARTIN

Mathematics & Statistics Department, University of Canterbury,
Private Bag 4800, Christchurch 8140, New Zealand
E-mail: B.Martin@math.canterbury.ac.nz

The Elko field of Ahluwalia and Grumiller is a quantum field for massive spin-1/2 particles. It has been suggested as a candidate for dark matter. We discuss our attempts to interpret the Elko field as a quantum field in the sense of Weinberg. Our work suggests that one should investigate quantum fields based on representations of the full Poincaré group which belong to one of the non-standard Wigner classes.

Keywords: Quantum field; Elko field; dark matter; non-standard Wigner class.

1. Introduction

In 2005 Ahluwalia and Grumiller introduced a new quantum field for massive spin-1/2 particles, which they called the *Elko* field.[1,2] They proposed the Elko field as a candidate for dark matter. It is natural to ask how the Elko field fits into Weinberg's formulation of quantum field theory (Ref. 3, Ch. 5). In this note we report on our recent investigations into this question. Forthcoming work[4] of Ahluwalia, Lee and Schritt also deals with aspects of this and related questions.

We begin by recalling the Elko field and its properties (Sec. 2). In Sec. 3 and Sec. 4 we briefly describe Weinberg's construction of quantum fields and compare the Elko field with the Dirac field. In the final section, Sec. 5, we discuss directions for future research.

2. Review of Elko Fields

We start with some notation. We denote the strict Lorentz and Poincaré groups by $\mathcal{L}_0$ and $\mathcal{P}_0$ respectively, and the full Lorentz and Poincaré groups — which include space inversion P and time reversal T — by $\mathcal{L}$ and $\mathcal{P}$ respectively. We represent elements of $\mathcal{L}_0$ by the symbol Λ, and elements of $\mathcal{P}_0$ by pairs (Λ, a), where a is a space-time translation.

The Elko field[a] is given by

$$\eta_i(x) = \int d^3\mathbf{p} \sum_\beta \left[e^{-ip_\mu x^\mu} \lambda_i^{\mathrm{S}}(\mathbf{p},\beta) c_\beta(\mathbf{p}) + e^{+ip_\mu x^\mu} \lambda_i^{\mathrm{A}}(\mathbf{p},\beta) d_\beta^\dagger(\mathbf{p}) \right], \quad (1)$$

where the index i ranges from 1 to 4, $c_\beta(\mathbf{p})$ is the annihilation operator for a certain species of particle and $d_\beta^\dagger(\mathbf{p})$ is the creation operator for the corresponding antiparticle. The index β takes two values $\pm$. The rest spinors are given up to proportionality by

$$\lambda^{\mathrm{S}}(\mathbf{0},+) = \begin{pmatrix} 0 \\ i \\ 1 \\ 0 \end{pmatrix}, \lambda^{\mathrm{S}}(\mathbf{0},-) = \begin{pmatrix} i \\ 0 \\ 0 \\ -1 \end{pmatrix}, \lambda^{\mathrm{A}}(\mathbf{0},+) = \begin{pmatrix} 0 \\ -i \\ 1 \\ 0 \end{pmatrix}, \lambda^{\mathrm{A}}(\mathbf{0},-) = \begin{pmatrix} -i \\ 0 \\ 0 \\ -1 \end{pmatrix} \quad (2)$$

and one obtains the spinors at nonzero momentum by multiplying the rest spinors by standard Lorentz boost matrices (Ref. 1, Sec. 3.3).

Recall the usual Dirac field [b]:

$$\psi_i(x) = \int d^3\mathbf{p} \sum_\sigma \left[e^{-ip_\mu x^\mu} u_i(\mathbf{p},\sigma) c_\sigma(\mathbf{p}) + e^{+ip_\mu x^\mu} v_i(\mathbf{p},\sigma) d_\sigma^\dagger(\mathbf{p}) \right], \quad (3)$$

where the rest spinors are given up to proportionality by

$$u(\mathbf{0},1/2) = \begin{pmatrix} 1 \\ 0 \\ 1 \\ 0 \end{pmatrix}, u(\mathbf{0},-1/2) = \begin{pmatrix} 0 \\ 1 \\ 0 \\ 1 \end{pmatrix}, v(\mathbf{0},1/2) = \begin{pmatrix} 0 \\ 1 \\ 0 \\ -1 \end{pmatrix}, v(\mathbf{0},-1/2) = \begin{pmatrix} -1 \\ 0 \\ 1 \\ 0 \end{pmatrix}. \quad (4)$$

The Dirac rest spinors are eigenspinors of the helicity operator. The starting point for the Elko construction is to choose rest spinors as in Eq. (2) which

[a] We have absorbed a p-dependent factor in the integrand into the definition of the spinors $\lambda(\mathbf{p},\beta)$, and we have done the same for the Dirac field below.

[b] See, e.g., Ref. 3, Eq. (5.5.34). We have changed the signs of the exponential factors to be consistent with Eq. (1): this amounts simply to adopting a different convention in the definition of how the translation operator acts on physical states.

are eigenspinors of the charge conjugation operator. These eigenspinors are not eigenspinors of the helicity operator: the top two and bottom two components have opposite helicities. For this reason Ahluwalia and Grumiller call β a dual helicity index.

Having defined the Elko field, Ahluwalia and Grumiller introduce a dual — which differs from the usual Dirac dual — on the space of spinors. They calculate the spin sums, the equation of motion and the propagator. The propagator turns out to be the Klein-Gordan propagator and the mass dimension of the Elko field is 1 (as opposed to the value 3/2 for the Dirac field). This implies that the Elko field cannot interact with the electromagnetic field and that interactions — ignoring gravity — with all standard model particles, except possibly the Higgs boson, are prohibited or suppressed. Hence the Elko field is a plausible candidate for a dark matter field.[c]

3. Weinberg's Definition of a Quantum Field

In his book Ref. 3, Weinberg provides a broad and coherent framework for introductory field theory based on a few basic symmetry principles. We briefly sketch an outline of his arguments and recall the relevant definitions. See Ref. 3, Ch. 5, for more details.

The ingredients we need are the following. We consider massive particles with positive energy, mass m and spin s. The little group SO(3) is a subgroup of $\mathcal{L}_0$. Let $R(\Lambda) = R_{\sigma\nu}(\Lambda)$ be the $2s+1$-dimensional irreducible representation of SO(3) corresponding to spin s. We construct a state space H as in Ref. 3, Ch. 2: the space of one-particle states is spanned by basis kets of the form $|p,\sigma\rangle$, having 4-momentum p and spin-z component σ in the rest frame. Using the matrices $R_{\sigma\nu}(\Lambda)$, we can construct an irreducible unitary representation $U(\Lambda, a)$ of $\mathcal{P}_0$ on H. Now let $D(\Lambda) = D_{ij}(\Lambda)$ be a t-dimensional representation of $\mathcal{L}_0$ for some $t \in \mathbb{N}$. Let $L(H)$ denote the space of linear operators from H to H. We define a *Weinberg quantum field based on the data* $(H, R(\Lambda), U(\Lambda, a), D(\Lambda))$ to be a collection of functions $\Psi(x) = (\Psi_i(x))_{1\leq i\leq t}$ from $\mathbb{R}^4$ to $L(H)$ such that for all $(\Lambda, a) \in \mathcal{P}_0$, we have

$$U(\Lambda,a)\Psi_i(x)U(\Lambda,a)^{-1} = \sum_j D_{ij}(\Lambda^{-1})\Psi_j(\Lambda x + a). \tag{5}$$

---

[c] The usual formalism of quantum field theory requires the fields to be local. The original version of the Elko field in Ref. 1 is not local. Recently Ahluwalia, Lee, Schritt and Watson discovered a slightly different field which does satisfy locality.[4] For simplicity we restrict ourselves to the original field in this note.

We say that a Weinberg quantum field — or more generally a collection of Weinberg quantum fields — is *local* if for any Ψ and Φ in the collection, for any indices i and j and for any $x, y \in \mathbb{R}^4$ such that $x - y$ is spacelike, the field components $\Psi_i(x)$ and $\Phi_j(y)$ commute (for bosons) or anti-commute (for fermions). Roughly speaking, Eq. (5) ensures that one can construct a Hamiltonian density $\mathcal{H}(x)$ from these fields which is a scalar under Poincaré transformations, and the extra requirement of locality ensures that the S-matrix obtained from $\mathcal{H}(x)$ transforms covariantly under Poincaré transformations.

It follows from quite general arguments that a Weinberg quantum field must be of the form[d]

$$\Psi_i(x) = \int \mathrm{d}^3\mathbf{p} \sum_{\sigma} \left[e^{-ip_\mu x^\mu} u_i(\mathbf{p}, \sigma) c_\sigma(\mathbf{p}) + e^{+ip_\mu x^\mu} v_i(\mathbf{p}, \sigma) d_\sigma^\dagger(\mathbf{p}) \right] . \quad (6)$$

Equation (5) implies that the spinors $u(\mathbf{p}, \sigma)$ and $v(\mathbf{p}, \sigma)$ for arbitrary $\mathbf{p}$ are completely determined by the rest spinors $u(\mathbf{0}, \sigma)$ and $v(\mathbf{0}, \sigma)$, and in many important cases, the values of $u(\mathbf{0}, \sigma)$ and $v(\mathbf{0}, \sigma)$ are also completely determined by Eq. (5) and the extra requirement of locality.

One advantage of Weinberg's approach is that physical insight falls out from the mathematical formalism: for instance, the Dirac equation — and the form of the Dirac field itself — can be derived from Eq. (5) and the requirement of locality (cf. Sec. 4), so we may view them simply as consequences of covariance under the Poincaré group. Similarly one can deduce the existence of anti-particles purely from the mathematical restrictions imposed by locality and Eq. (5) (see Ref. 3, Sec. 5.1).

4. The Dirac and Elko Fields in Weinberg's Formalism

Throughout this section, we fix $D(\Lambda) = D_{ij}(\Lambda)$ to be the usual $(1/2, 0) \oplus (0, 1/2)$ representation of $\mathcal{L}_0$. We wish to interpret the Elko field as a Weinberg quantum field $\Psi(x)$. To do this, we must identify β with the index σ in the construction of H above, for some suitable choice of H, $R(\Lambda)$ and $U(\Lambda, a)$. Consider the state space H and $U(\Lambda, a)$ constructed as above for $s = 1/2$, where $R_{\sigma\nu}(\Lambda)$ is chosen to be the usual spin-1/2 representation of SO(3); the index σ takes the values $\pm 1/2$. Using Eq. (5) and the assumption of locality, one obtains formulas for the rest spinors which involve constants

[d]Note that for each fixed x, i, $\mathbf{p}$ and σ, all of the quantities that appear inside the integral in Eq. (6) except for $c_\sigma(\mathbf{p})$ and $d_\sigma^\dagger(\mathbf{p})$ are c-numbers. Hence to evaluate the LHS of Eq. (5), one need only calculate $U(\Lambda, a) c_\sigma(\mathbf{p}) U(\Lambda, a)^{-1}$ and $U(\Lambda, a) d_\sigma^\dagger(\mathbf{p}) U(\Lambda, a)^{-1}$.

$c_\pm$, $d_\pm$. An argument involving parity conservation allows us to pin down the values of $c_\pm$ and $d_\pm$ up to an overall normalization (we return to this point in Sec. 5 below). We find that the rest spinors $u(\mathbf{0}, \sigma)$ and $v(\mathbf{0}, \sigma)$ are precisely the Dirac rest spinors given in Eq. (4) above, and $\Psi(x)$ is the Dirac field (Ref. 3, Sec. 5.5). Hence the Dirac field is the only Weinberg quantum field based on the data $(H, R(\Lambda), U(\Lambda, a), D(\Lambda))$. This shows that the Elko field cannot be a Weinberg quantum field based on this data. Indeed, Ahluwalia, Lee and Schritt have recently noted that the transformation properties of Elko spinors under rotations differ from the transformation rules that must be satisfied for a Weinberg spinor; the authors are grateful to them for communicating this observation to us.

We could instead have chosen the representation of SO(3) to be not the standard one $R(\Lambda)$ but another representation $R'(\Lambda)$ isomorphic to it. Then $R(\Lambda)$ and $R'(\Lambda)$ are related by a similarity transform: we have $R'(\Lambda) = SR(\Lambda)S^{-1}$ for some invertible linear transformation S. This change has no physical or mathematical significance, but it makes the resulting Weinberg quantum field look different. One can show by adapting the argument on pp220–1 of Ref. 3 that the rest spinors of a Weinberg quantum field based on $(H, R'(\Lambda), U(\Lambda, a), D(\Lambda))$ cannot take the form of the Elko rest spinors in Eq. (2), even if one does not assume parity conservation. Hence the Elko field cannot be a Weinberg quantum field based on $(H, R'(\Lambda), U(\Lambda, a), D(\Lambda))$. We will give full details of this calculation in forthcoming work.

5. Non-standard Wigner Classes

As we have seen, the most direct attempt to fit the Elko field into Weinberg's formalism fails. The next logical step, motivated by the discussion on p4, para. 1 of Ref. 1, is to consider Weinberg fields based on a state space H and an irreducible representation of the full Poincaré group $\mathcal{P}$ on H that belongs to one of the non-standard Wigner classes.[5] Fix $R(\Lambda)$ and let H be as in Sec. 3, with one-particle basis kets $|p, \sigma\rangle$. The unitary representation $U(\Lambda, a)$ of $\mathcal{P}_0$ extends to an (anti-)unitary representation $U(\Lambda, a)$ of the whole of $\mathcal{P}$; we abuse notation and denote this representation by $U(\Lambda, a)$ also. The operators $U(\mathsf{P})$ and $U(\mathsf{T})$ are unitary and anti-unitary respectively, and they act on the kets $|p, \sigma\rangle$ by $U(\mathsf{P})|p, \sigma\rangle = \eta_{\mathsf{P}}|\mathsf{P}p, \sigma\rangle$ and $U(\mathsf{T})|p, \sigma\rangle = \eta_{\mathsf{T}}(-1)^{s-\sigma}|\mathsf{P}p, -\sigma\rangle$ for some constants η_{P} and η_{T} (Ref. 3, Sec. 2.6). The representation $U(\Lambda, a)$ is said to belong to the standard Wigner class.

There are also three so-called non-standard Wigner classes of representations $U(\Lambda, a)$ of $\mathcal{P}$. Here the states of given 4-momentum p and spin-z

component σ become degenerate; the basis kets are labelled $|p, \sigma, \tau\rangle$, where τ is an extra index that breaks the degeneracy. Time reversal $U(\mathsf{T})$ couples states with different values of τ. The Weinberg quantum fields in these degenerate cases are not worked out in detail in Ref. 3 and we believe they are worth further study. Even if we cannot give an interpretation of the Elko field in this setting, perhaps there are other as yet unexplored Weinberg quantum fields that may be candidates for dark matter. Note that although the definition of a Weinberg field seems only to involve covariance under restricted Poincaré transformations, P and T play a crucial part (cf. the final step in the derivation of the Dirac field in Sec. 4).

We finish with some remarks on work of Lee and Wick which is relevant here. According to Ref. 6, if a field is local then the underlying representation of $\mathcal{P}$ must come from the standard Wigner class. There, however, they allow themselves the freedom to multiply the original $U(\mathsf{P})$ and $U(\mathsf{T})$ by symmetries of the internal state space. For the non-standard Wigner classes, where there are extra degrees of freedom coming from the index τ, one would expect there to be plenty of these internal symmetries above and beyond charge conjugation. A full study of the possible Weinberg fields would involve a systematic investigation of these internal symmetries.

Acknowledgments

Most of the ideas in this note had their roots in discussions of the authors with Dharamvir Ahluwalia, Cheng-Yang Lee, Dimitri Schritt and Thomas Watson, and we thank them for their contribution and for their comments on an earlier draft. The second author is grateful to Ahluwalia for introducing him to this area of research.

References

1. D. V. Ahluwalia and D. Grumiller, *Journal of Cosmology and Astroparticle Physics* **0507**, 1 (2005).
2. D. V. Ahluwalia and D. Grumiller, *Phys. Rev.* **D72**, 1 (2005).
3. S. Weinberg, *The Quantum Theory of Fields, Vol. I* (CUP, 1995).
4. D. V. Ahluwalia, C.-Y. Lee and D. Schritt, Local fermionic dark matter with mass dimension one, arXiv:0804.1854v4 [hep-th], (2008).
5. E. Wigner, Unitary representations of the inhomogeneous Lorentz group including reflections, in *Group Theoretical Concepts and Methods in Elementary Particle Physics*, (Gordon & Breach, New York, 1964).
6. T. D. Lee and G. C. Wick, *Phys. Rev.* **148**, 1385 (1966).

PART XII

Indirect Search for Dark Matter

DARK MATTER SEARCHES WITH IMAGING ATMOSPHERIC CHERENKOV TELESCOPES

E. MOULIN*
CEA -Saclay, IRFU,
Gif-sur-Yvette, 91191, France
**E-mail: emmanuel.moulin@cea.fr*

The annihilations of WIMPs produce high energy gamma-rays in the final state. These high energy gamma-rays may be detected by imaging atmospheric Cherenkov telescopes (IACTs). Amongst the plausible targets are the Galactic Center, the centre of galaxy clusters, dwarf Sphreroidal galaxies and substructures in Galactic haloes. I will review on the recent results from observations of ongoing IACTs.

Keywords: Dark matter; gamma-rays; Cherenkov telescopes.

1. Introduction

Cosmological and astrophysical probes suggest that ~23% of the Universe is composed of non-baryonic dark matter (DM), commonly assumed to be in the form of Weakly Interacting Massive Particles (WIMPs) arising in extensions of the Standard Model of Particle Physics (for reviews see, e.g.[1,2]). Amongst the most widely discussed DM candidates are the lightest neutralino in supersymmetric extensions of the Standard Model[3] and the first excitation of the Kaluza-Klein bosons (LKP) in universal extra dimension theories.[4–6]

The annihilation of WIMP pairs can produce in the final state a continuum of gamma-rays whose flux extends up to the DM particle mass, from the hadronization and decay of the cascading annihilation products. In supersymmetric models, the gamma-ray spectrum from neutralino annihilation is not uniquely determined and the branching ratios (BRs) of the open annihilation channels are not determined since the DM particle field content is not known *a priori*. In contrast, in Kaluza-Klein scenarios where the lightest Kaluza-Klein particle (LKP) is the first KK mode of the hypercharge gauge boson, the BRs of the annihilation channels can be computed given that the field content of the DM particle is known. The gamma-ray flux from annihilations of DM particles of mass m_{DM} accumulating in a spherical DM halo can be expressed in the form:

$$\frac{d\Phi(\Delta\Omega, E_\gamma)}{dE_\gamma} = \frac{1}{8\pi} \underbrace{\frac{\langle\sigma v\rangle}{m_{DM}^2} \frac{dN_\gamma}{dE_\gamma}}_{\text{Particle Physics}} \times \underbrace{\bar{J}(\Delta\Omega)\Delta\Omega}_{\text{Astrophysics}} \quad (1)$$

as a product of a particle physics component with an astrophysics component. The particle physics part contains $\langle\sigma v\rangle$, the velocity-weighted annihilation cross section, and dN_γ/dE_γ, the differential gamma-ray spectrum summed over the whole final states with their corresponding branching ratios. The astrophysical part corresponds to the line-of-sight-integrated squared density of the DM distribution J, averaged over the instrument solid angle $\Delta\Omega$ usually matching the angular resolution of the instrument:

$$J = \int_{l.o.s} \rho^2(r[s])ds \qquad \bar{J}(\Delta\Omega) = \frac{1}{\Delta\Omega}\int_{\Delta\Omega} PSF \times J\, d\Omega \quad (2)$$

where PSF stands for the point spread function of the instrument.

The annihilation rate being proportional to the square of the DM density integrated along the line of sight, regions with enhanced DM density are primary targets for indirect searches. Among them are the Galactic halo, external galaxies, galaxy clusters, substructures in galactic haloes, and the Galactic Center. We report here on recent results on dark matter searches with current IACTs such as H.E.S.S. and MAGIC, towards the Galactic Center, dwarf galaxies from the Local Group, Galactic globular clusters and DM mini-spikes around intermediate mass black holes (IMBHs).

2. The Galactic Center

H.E.S.S. observations towards the Galactic Center have revealed a bright pointlike gamma-ray source, HESS J1745-290,[7] coincident in position with the supermassive black hole Sgr A*, with a size lower than 15 pc. Diffuse emission along the Galactic plane has also been detected[8] and correlates well with the mass density of molecular clouds from the Central Molecular Zone, as traced by CS emission.[9] According to recent detailed studies,[10] the source position is located at an angular distance of $7.3'' \pm 8.7''_{\text{stat.}} \pm 8.5''_{\text{syst.}}$ from Sgr A*. The pointing accuracy allows to discard the association of the very high energy (VHE) emission with the center of the radio emission of the supernova remnant Sgr A East but the association with the pulsar wind nebula G359.95-0.04 can not be ruled out. From 2004 data set, the energy spectrum of the source is well fitted in the energy range 160 GeV - 30 TeV to a power-law spectrum $dN/dE \propto E^{-\Gamma}$ with a spectral index $\Gamma = 2.25 \pm 0.04_{stat.} \pm 0.1_{syst.}$. No deviation from a power-law is observed leading to an upper limit on the energy cut-off of 9 TeV (95% C.L.). The VHE emission from HESS J1745-290 does not show any significant periodicity or variability from 10 minutes to 1 year.[10] Besides plausible astrophysical origins (see e.g.[11] and references therein), an alternative explanation is the annihilation of DM in the central cusp of our Galaxy. The spectrum of HESS J1745-290 shows no indication for gamma-ray lines. The observed gamma-ray flux may also result from secondaries of DM annihilation. The left hand side of Fig. 2 shows the H.E.S.S. spectrum extending up to masses of about 10 TeV, which requires large neutralino masses (>10 TeV). They are unnatural in phenomenological MSSM scenarios. The Kaluza-Klein models provide harder spectra which still significantly deviate from the measured one. Non minimal version of the MSSM may yield flatter spectrum with mixed 70% $b\bar{b}$ and 30% $\tau^+\tau^-$ final states. Even this scenario does not fit to the measured spectrum. The hypothesis that the spectrum measured by H.E.S.S. originates only from DM particle annihilations is highly disfavored.

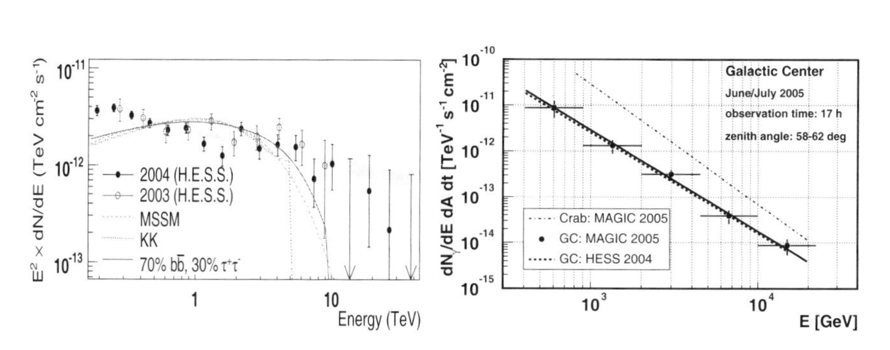

Fig. 1. Left : Spectral energy density E^2dN/dE of gamma-rays from HESS J1745-290 for 2003 (red empty circles) and 2004 (black filled circles) datasets of the H.E.S.S. observation of the Galactic Center. The shaded area shows the best power-law fit to the 2004 data points. The spectra expected from the annihilation of a MSSM-like 14 TeV neutralino (dashed green line), a 5 TeV KK DM particle (dotted pink line) and a 10 TeV DM particle annihilating into 70% $b\bar{b}$ and 30% $\tau^+\tau^-$ in final state (solid blue line) are presented. Right: Reconstructed VHE gamma-ray energy spectrum of the GC (statistical errors only) as measured by the MAGIC collaboration. The full line shows the result of a power-law fit to the data points. The dashed line shows the 2004 result of the HESS collaboration.[13] The dot-dashed line shows the energy spectrum of the Crab nebula as measured by MAGIC.

The MAGIC observations were carried out towards the Galactic Centre since 2004 and revealed a strong emission. The observed excess in the direction of the GC has a significance of 7.3 standard deviations[12] and is compatible with a pointlike source. Large zenith observation angle ($\geq 60°$) implies an energy threshold of $\sim$400 GeV. The source position and the flux level are consistent with the measurement of HESS[13] within errors. The right hand side of Fig. 1 shows the reconstructed VHE gamma-ray energy spectrum of the GC after the unfolding with the instrumental energy resolution. The differential flux can be well described by a power law of index $\Gamma = 2.2 \pm 0.2_{\text{stat.}} \pm 0.2_{\text{syst.}}$. The systematic error is estimated to be 35% in the flux level determination. The flux level is steady within errors in the time-scales explored within these observations, as well as in the two year time-span between the MAGIC and HESS observations. An interpretation in term of DM require a minimum value of mass higher than 10 TeV. Most probably, if DM signal exists is overcome by other astrophysical emitters.

3. Dwarf Galaxies from the Local Group

Dwarf spheroidal galaxies in the Local Group are considered as privileged targets for DM searches since they are among the most extreme DM-dominated environments. Measurements of roughly constant radial velocity dispersion of stars imply large mass-to-luminosity ratios. Nearby dwarfs are ideal astrophysical probes of the nature of DM as they usually consist of a stellar population with no hot or warm gas, no cosmic ray population and little dust. Indeed, these systems are expected to have a low intrinsic gamma-ray emission. This is in contrast with the Galactic Center where disentangling the dominant astrophysical signal from possible more exotic one is very challenging. Prior to the year 2000, the number of known satellites was eleven. With the Sloan Digital Sky Survey (SDSS), a population of ultra low-luminosity satellites has been unveiled, which roughly doubled the number of known satellites. Among them are Coma Berenices, Ursa Major II and Willman 1. IACTs have started observation campaigns on dwarf galaxies for a few years. Table 1 presents a possible list of the preferred targets for DM searches.

The star velocity dispersions in Draco reveal that this object is dominated by DM on all spatial scales and provide robust bounds on its DM profile, which thus decreases uncertainties on the astrophysical contribution to the gamma-ray flux. The MAGIC collaboration searched for a steady gamma-ray emission from the direction of Draco.[17] The analysis energy threshold after cuts is 140 GeV. No significant excess is found.

Table 1. A tentative list of preferred dwarf galaxies.

Name	Distance[a] (kpc)	Luminosity[a] (10^3 $L_\odot$)	M/L[b] ($M_\odot/L_\odot$)	Best positioned IACTs
Carina	101	430	40	HESS, CANGAROO
Coma Berenices	44	2.6	450	MAGIC, VERITAS
Draco	80	260	320	MAGIC, VERITAS
Fornax	138	15500	10	HESS, CANGAROO
Sculptor	79	2200	7	HESS, CANGAROO
Sagittarius	24	58000*	25	HESS, CANGAROO
Sextans	86	500	90	HESS, CANGAROO
Ursa Minor	66	290	580	MAGIC, VERITAS
Ursa Major II	32	2.8	1100	MAGIC, VERITAS
Willman 1	38	0.9	700	MAGIC, VERITAS

Note: [a]See Ref. 14. [b]See Ref. 15. *See Ref. 16.

For a power law with spectral index of 1.5, typical for a DM annihilation spectrum, and assuming a pointlike source, the 2σ upper limit is $\Phi(E > 140\text{GeV}) = 1.1 \times 10^{-11}\text{cm}^{-2}\text{s}^{-1}$. The measured flux upper limit is several orders of magnitude larger than predicted for the smooth DM distribution in mSUGRA models. The limit on the flux enhancement caused by high clumpy substructures or a black hole is around $O(10^3 - 10^9)$. The WHIPPLE collaboration has also observed Draco.[18] Fig. 3 shows the 95% C.L. exclusion curve on σv. The upper limit is at the level of 10^{-22} cm^3s^{-1} for 1 TeV neutralino annihilating with BRs of 90% in $b\bar{b}$ and 10% in $\tau^+\tau^-$.

The Sagittarius (Sgr) dwarf galaxy, one of the nearest Galaxy satellites of the Local Group, has been observed by H.E.S.S. since 2006. The annihilation signal from Sgr is expected to come from a region of 1.5 pc, which is much smaller than the H.E.S.S. point spread function. Thus, a pointlike signal has been searched for. No significant gamma-ray excess is detected at the nominal target position. A 95% C.L. upper limit on the gamma-ray flux is derived: $\Phi_\gamma^{95\%\text{C.L.}}(E_\gamma > 250\text{GeV}) = 3.6 \times 10^{-12}\text{cm}^{-2}\text{s}^{-1}$, assuming a power-law spectrum of spectral index 2.2.[19] A modelling of the Sgr DM halo has been carried out. Two models of the mass distribution for the DM halo have been studied: a cusped NFW profile and a cored isothermal profile, to emcompass a large class of profiles. The cored profile has a small core radius due to a cusp in the luminous profile. The value of the line-of-sight-integrated squared density is then found to be larger for the cored profile than for the NFW profile (see Ref. 19 for more details). The left hand side of Fig. 2 presents the constraints on the velocity-weighted annihilation cross section σv for a cusped NFW and cored profiles in the solid

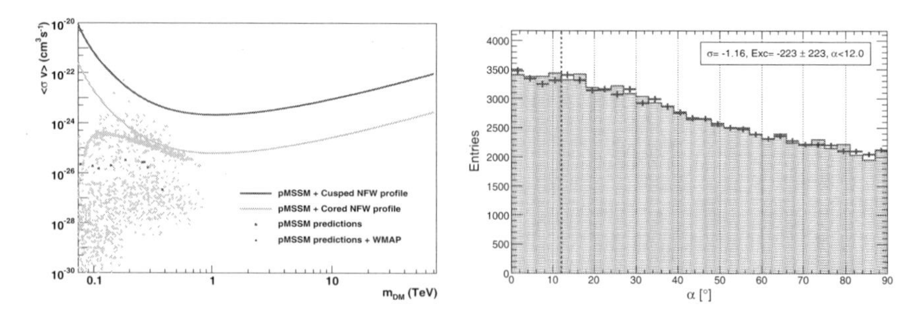

Fig. 2. Left: Upper limits at 95%C.L. on σv versus the neutralino mass for a cusped NFW and cored DM halo profile for Sgr. The predictions in pMSSM are also plotted with in addition those satisfying in addition the WMAP constraints on the cold DM density. Right: Willman 1 α-plot as seen by MAGIC in 15.5 hours above a fiducial energy threshold of 100 GeV. The red crosses represent the ON–data sample, the blue shaded region is the OFF–data sample normalized to the ON–data sample between $30° - 80°$. The vertical red dotted line represents the fiducial region $\alpha < 12°$ where the signal is expected.

angle integration region $\Delta\Omega = 2 \times 10^{-5}$ sr, for neutralino DM. Predictions for SUSY models are displayed. For a cusped NFW profile, H.E.S.S. does not set severe constraints on σv. For a cored profile, due to a higher central density, stronger constraints are derived and some pMSSM models can be excluded in the upper part of the scanned region. Although the nature of Canis Major is stilll debated, HESS observed this object and results on Canis Major[20] are presented in Ref. 21.

The recently discovered Willman 1 dwarf galaxy has been observed by MAGIC.[22] Willman 1 has a a total mass of $\sim 5\times 10^5$ $M_\odot$, which is about an order of magnitude smaller than those of the least massive satellite galaxies previously known. This object has one of the highest mass-to-luminosity ratio. No significant gamma-ray excess beyond 100 GeV above the background was observed in 15.5 hours of observation of the sky region around Willman 1. This is shown in the right hand side of Fig. 2, where the "α-plot" is reported [a]. The signal is searched with a cut slightly larger than for a pointlike source to take into account a possible source extension which makes a fiducial region to be $\alpha \leq 12°$. Flux upper limits of the order of 10^{-12} cm^{-2}s^{-1} for benchmark mSUGRA models are obtained.[22] Boost factors in flux of the order of 10^3 are required in the most optimistic scenarios, i.e. in

[a]The α-parameter is the angular distance between the shower image main axis and the line connecting the image barycenter and the camera center.

the funnel region or in region where the internal bremsstrahlung may play an important role. However, uncertainties on the DM distribution or the role of substructures may significantly reduce this boost.

4. Galactic Globular Clusters

Globular clusters are not believed to be DM-dominated objects. Their mass-to-luminosity ratios are well described by purely King profile which suggets no significant amount of DM. However, the formation of globular clusters fits in the hierarchical strucure formation scenario in which globular clusters may have formed in DM overdensities. During their evolution, they may have hold some DM in their central region. M15 is a nearby galactic globular cluster with a core radius of 0.2 pc and a central density of $\sim 10^7 M_\odot pc^{-3}$. Even if the mass-to-light ratio of M15 does not suggest a significant component of DM, the compact and dense core of M15 does not prevent from looking for a hypothetical DM signal. The presence of stars and DM in globular clusters may lead to a DM enhancement in their inner part. This is usually treated with the adiabatic contraction model.[23]

The Whipple collaboration obsered M15 in 2003. The modelling of DM profile assuming adiabatic contraction of DM from an initial NFW distribution leads to an enhancement in the astrophysical factor of $\sim 10^2$ to

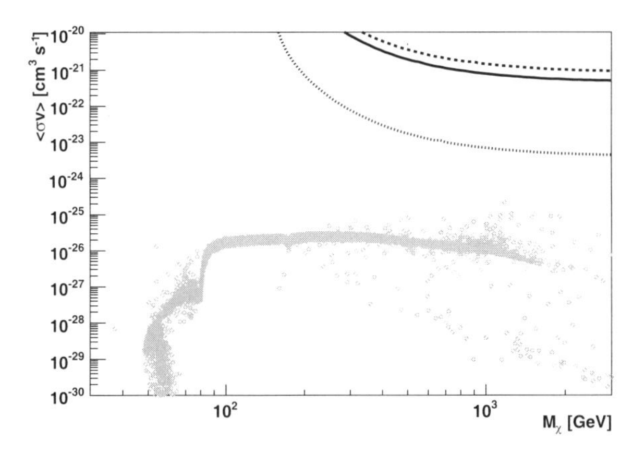

Fig. 3. Exclusion limit on σv as a function of the neutralino mass m_χ on M15 (dotted line) assuming an annihilation spectrum with 90% in $b\bar{b}$ and 10% in $\tau^+\tau^-$, and the NFW profile after adiabatic contraction. The solid line is the exclusion limit for Draco and the dashed line for Ursa Minor.

10^3. Fig. 3 shows the exclusion curve on σv (dotted line) versus the neutralino mass assuming a NFW profile after adiabatic contraction.[18] The limit reaches $\sim 10^{-24}$ cm^{3}s^{-1} for TeV neutralinos.

5. Galaxy Clusters

Astrophysical systems such as the VIRGO galaxy cluster[24] have been considered as target for DM annihilation searches. The elliptical galaxy M87 at the center of the VIRGO cluster has been observed by H.E.S.S. It is an active galaxy located at 16.3 Mpc which hosts a black hole of $3.2\times10^9 M_\odot$. From 2003 to 2006, H.E.S.S. detected a 13σ signal in 89 hour observation time at a location compatible with the nominal position of the nucleus of M87.[25] With the angular resolution of H.E.S.S., the VHE source is point-like, with an upper limit on the size of 3' at 99% confidence level. Given the distance of M87, this corresponds to a size of 13.7 kpc.

Fig. 4 shows the energy spectrum for 2004 and 2005 data. Both data sets are well fitted to a power-law spectrum with spectral indices $\Gamma = 2.62\pm0.35$ (2004) and $\Gamma = 2.22\pm0.25$ (2005). The 2005 average flux is found to be higher by a factor $\sim$5 to the average flux of 2004. The integrated flux above 730 GeV shows a yearly variability at the level of 3.2σ.[25] Variability at the day time scale has been observed in the 2005 data at the level of 4σ. This fast variability puts stringent constraints on the size of the VHE emission region. The position centered on M87 nucleus excludes the center of the

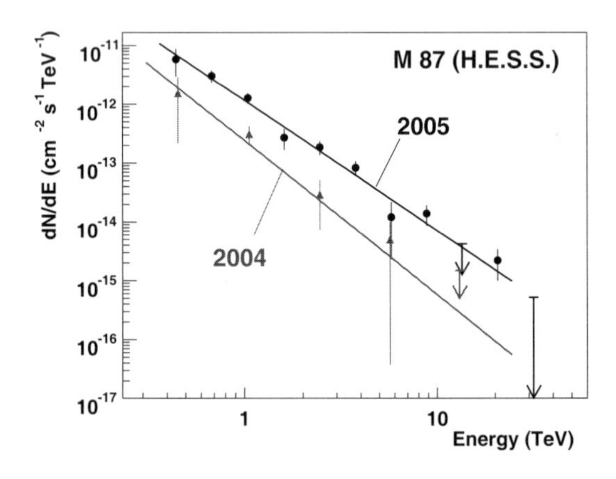

Fig. 4. Differential energy spectrum of M87 from $\sim$400 GeV up to 10 TeV from 2004 and 2005 data. Corresponding fits to a power-law function dN/dE E are plotted as solid lines. The data exhibit indices of $\Gamma = 2.62\pm0.35$ for 2004 and $\Gamma = 2.22\pm0.15$ for 2005.

VIRGO cluster and outer radio regions of M87 as the gamma-ray emission region. The observed variability at the scale of ~2 days requires a compact emission region below $50R_s$, where $R_s \sim 10^{-15}$ cm is the Schwarzschild radius of the M87 supermassive black hole.[25] The short and long term temporal variability observed with H.E.S.S. excludes the bulk of the TeV gamma-ray signal to be of DM origin.

6. Galactic Substructures: The Case for IMBHs

Mini-spikes around Intermediate Mass Black Holes have been recently proposed as promising targets for indirect dark matter detection.[26] The growth of massive black holes inevitably affects the surrounding DM distribution. The profile of the final DM overdensity, called mini-spike, depends on the initial distribution of DM, but also on astrophysical processes such as gravitational scattering of stars and mergers. Ignoring astrophysical effects, and assuming adiabatic growth of the black hole, if one starts from a NFW profile, a spike with a power-law index 7/3 is obtained, as relevant for the astrophysical formation scenario studied here characterized by black hole masses of $\sim 10^5$ $M_\odot$.[26] Mini-spikes might be detected as bright pointlike sources by current IACTs.

H.E.S.S. data collected between 2004 and 2007 during the Galactic plane survey have allowed to accurately map the Galactic plane between ±3° in galactic latitude and from -30° to 60° in galactic longitude with respect to the Galactic Center position. The study of the H.E.S.S. sensitivity in a large field of view to dark matter annihilations has been performed.[27] Fig. 5 shows the experimentally observed sensitivity map in the Galactic plane from Galactic longitudes l=-30° to l=+60° and Galactic latitudes b=-3° to b=+3°, for a DM particle of 500 GeV mass annihilating into the 100% BR $b\bar{b}$ channel. The H.E.S.S. sensitivity depends strongly on the exposure time and acceptance maps which are related to the choice of the pointing positions. The flux sensitivity varies along the latitude and longitude due to inhomogeneous coverage of the Galactic plane. In the band between -2° and 2° in Galactic latitude, a DM annihilation flux sensitivity at the level of 10^{-12} $cm^{-2}s^{-1}$ is achieved for a 500 GeV DM particle annihilating in the $b\bar{b}$ channel. Deeper observations of the Galactic Center and at Galactic longitude of ~20° allow the flux sensitivity to be of $\sim 5\times 10^{-13}$ $cm^{-2}s^{-1}$. For b≥2°, the sensitivity is deteriorated due to a weaker effective exposure. For b=0° and l=-0.5°, the flux sensitivity is $\sim 10^{-13}$ $cm^{-2}s^{-1}$ in $b\bar{b}$.

H.E.S.S. reached the required sensitivity to be able to test DM annihilations from mini-spikes in the context of one relatively favorable sce-

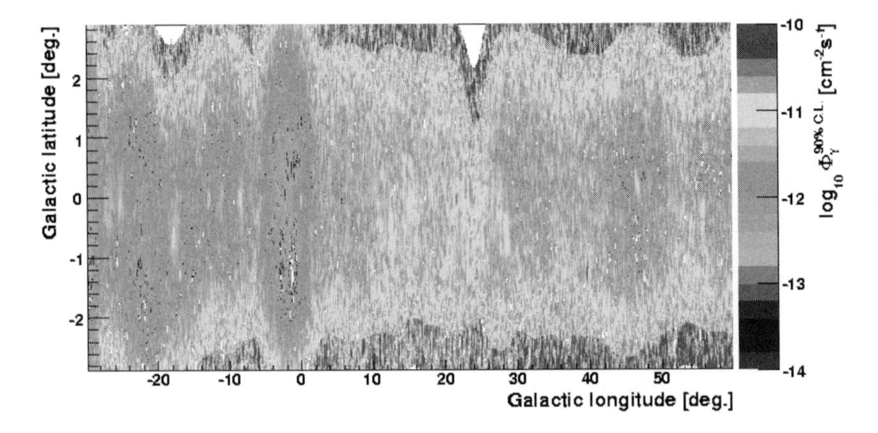

Fig. 5. H.E.S.S. sensitivity map in Galactic coordinates, i.e. 90% C.L. limit on the integrated gamma-ray flux above 100 GeV, for dark matter annihilation assuming a DM particle of mass $m_\chi = 500$ GeV and annihilation into the $b\bar{b}$ channel. The flux sensitivity is correlated to the exposure and acceptance maps. In the Galactic latitude band between -2° and 2°, the gamma-ray flux sensitivity reaches 10^{-12} cm^{-2}s^{-1}.

nario for IMBH formation and adiabatic growth of the DM halo around the black hole (e.g. scenario B of Ref. 26). H.E.S.S. observations (2004-2006) of the Galactic plane allowed to discover more than 20 very high energy gamma-ray sources .[28] Some of them have been identified owing to their counterparts at other wavelengths, but almost half of the sources have no obvious counterpart and are still unidentified.[29] An accurate reconstruction of their energy spectra shows that all the spectra are consistent with a pure power-law, spanning up to two orders of magnitude in energy above the energy threshold. None of them exhibits an energy cutoff, characteristic of DM annihilation spectra, in the energy range from ~100 GeV up to 10 TeV. Furthermore, the detailed study of their morphology[29] shows that all the sources have an intrinsic spatial extension greater than ~5 arcminutes, while mini-spikes are expected to be pointlike sources for H.E.S.S. No IMBH candidate has been detected so far by H.E.S.S. within the survey range. Based on the absence of plausible IMBH candidates in the H.E.S.S. data, constraints are derived on the scenario B of Ref. 26 for neutralino or LKP annihilations, shown as upper limits on σv.[27] Fig. 6 shows the exclusion limit at the 90% C.L. on σv as a function of the neutralino mass. The neutralino is assumed to annihilate into $b\bar{b}$ and $\tau^+\tau^-$ with 100% BR, respectively. Predictions for SUSY models are also displayed. The limits on σv are at the level of 10^{-28} cm^{3}s^{-1} for the $b\bar{b}$ channel for neutralino masses

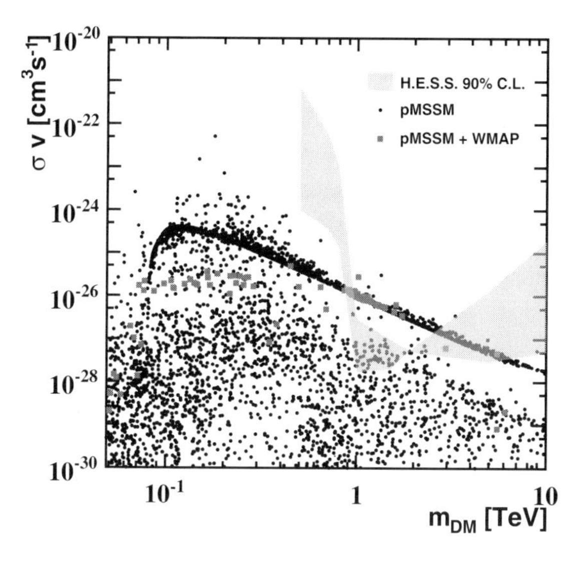

Fig. 6. Constraints on the IMBH gamma-ray production scenario for different neutralino parameters, shown as upper limits on σv as a function of the mass of the neutralino m_{DM}, but with a number of implicit assumptions about the IMBH initial mass function and halo profile (see Ref. 27 for details). For the scenario studied here, the probability of having no observable haloes in our Galaxy is 10% from Poisson statistics making these limits essentially 90% C. L. exclusion limits for this one particular (albeit optimistic) scenario (grey shaded area). The DM particle is assumed to be a neutralino annihilating into $b\bar{b}$ pairs or $\tau^+\tau^-$ pairs to encompass the softest and hardest annihilation spectra. The limit is derived from the H.E.S.S. flux sensitivity in the Galactic plane survey within the mini-spike scenario. SUSY models (black points) are plotted together with those satisfying the WMAP constraints on the DM particle relic density (magenta points).

in the TeV energy range. Limits are obtained one mini-spike scenario and constrain on the entire gamma-ray production scenario.

7. Conclusion

Dark matter searches will continue and searches with the phases 2 of H.E.S.S. and MAGIC will start soon. The phase 2 of H.E.S.S. will consist of a new large 28 m diameter telescope located at the center of the existing array, and will allow to lower the analysis energy threshold down to less than 50 GeV. The installation of the second telescope on the MAGIC site will allow to perform more sensitive searches by the use of the stereoscopic mode. The upcoming generation of IACTs, the Cherenkov Telescope Array, is in the design phase. This array composed of several tens of telescopes will permit to significantly improve the performances on both targeted DM searches and wide-field-of-view DM searches.

References

1. G. Bertone, D. Hooper and J. Silk, *Phys. Rept.* **405**, 279 (2005).
2. L. Bergstrom, *Rept. Prog. Phys.* **63**, 793 (2000).
3. G. Jungman, M. Kamionkowski and K. Griest, *Phys. Rept.* **267**, 195 (1996)
4. T. Appelquist, H.-C. Cheng and B. A. Dobrescu, Phys. Rev. D **64**, 035002 (2001).
5. H. Cheng, J. Feng and K. Matchev, *Phys. Rev. Lett.* **89**, 211301 (2002).
6. G. Servant and T. Tait, *Nucl. Phys.* **B 650**, 391 (2003).
7. F. Aharonian *et al.* [H.E.S.S. Collaboration], *Phys. Rev. Lett.* **97**, 221102 (2006) [Erratum-ibid. **97**, 249901 (2006)].
8. F. Aharonian *et al.* [H.E.S.S. Collaboration], *Nature* **439**, 695 (2006).
9. M. Tsuboi, H. Tsohihiro and N. Ukita, *Astrophys. J. Suppl.* 120 (2006) 675.
10. H.E.S.S. ICRC 2007 contributions, arXiv:0710.4057 [astro-ph].
11. F. Aharonian and A. Neronov, *Astrophys. J.* **619**, 306 (2005).
12. J. Albert *et al.* [MAGIC Collaboration], *Astrophys. J.* **638**, L101 (2006).
13. F. Aharonian *et al.* [HESS Collaboration], *Astron. Astrophys.* **425**, L13 (2004).
14. M. Mateo, *Ann. Rev. Astron. Astrophys.* **36**, 435 (1998).
15. J. D. Simon and M. Geha, *Astrophys. J.* **670**, 313 (2007).
16. A. Helmi and S. D. M. White, *Mon. Not. Roy. Astron. Soc.* **323**, 529 (2001).
17. J. Albert *et al.* [MAGIC Collaboration], *Astrophys. J.* **679**, 428 (2008).
18. M. Wood *et al.*, arXiv:0801.1708 [astro-ph].
19. F. Aharonian *et al.* [HESS Collaboration], *Astropart. Phys.* **29**, 55 (2008).
20. F. Aharonian *et al.* [HESS Collaboration], *Astrophys. J.* **691**, 175 (2009).
21. M. Vivier for the H.E.S.S. collaboration, *these proceedings.*
22. E. Aliu *et al.* [MAGIC Collaboration], arXiv:0810.3561 [astro-ph].
23. G. R. Blumenthal, S. M. Faber, R. Flores and J. R. Primack, *Astrophys. J.* **301**, 27 (1986).
24. E. A. Baltz, C. Briot, P. Salati, R. Taillet and J. Silk, *Phys. Rev.* **D 61**, 023514 (2000).
25. D. Berge *et al.*, *Science* **314**, 1424 (2006).
26. G. Bertone, A. R. Zentner and J. Silk, *Phys. Rev.* **D 72**, 103517 (2005).
27. F. Aharonian *et al.* [HESS Collaboration], *Phys. Rev.* **D 78**, 072008 (2008).
28. F. Aharonian *et al.* [HESS Collaboration], *Astrophys. J.* **636**, 777 (2006).
29. F. Aharonian *et al.* [HESS Collaboration], *Astron. Astrophys.* **477**, 353 (2008).

CONSTRAINTS ON DARK MATTER WIMPS MODELS WITH H.E.S.S. OBSERVATIONS OF THE CANIS MAJOR OVERDENSITY

M. VIVIER on behalf the H.E.S.S. collaboration

IRFU/DSM/CEA, CE Saclay, F-91191 Gif-sur-Yvette, Cedex, France
E-mail: matthieu.vivier@cea.fr

A search for a dark matter (DM) annihilation signal into γ-rays towards the direction of the Canis Major (CMa) overdensity has been performed with the HESS telescopes. The nature of CMa is still controversial and one scenario represents it as a dwarf galaxy, making it an interesting candidate for DM annihilation searches. A total of 9.6 hours of high quality data were collected with the H.E.S.S. array of Imaging Atmospheric Cherenkov Telescopes (IACTs) and no evidence for a very high energy γ-ray signal was found. Constraints on the velocity-weighted annihilation cross section $\langle \sigma v \rangle$ are calculated for specific WIMP scenarios, using a NFW model for the DM halo profile and taking advantage of numerical simulations of hierarchical structure formation. 95% C.L. exclusion limits of the order of 5×10^{-24} cm^3 s^{-1} are reached in the 500 GeV - 10 TeV WIMP mass range.

Keywords: γ-ray Astronomy; Dark Matter; Dwarf Galaxies.

1. Introduction

Many astrophysical objects, ranging from DM clumps to galaxy clusters are expected to lead to DM particle annihilation signals that are detectable with Imaging Atmospheric Cherenkov Telescopes (IACT). Regions of high concentration of DM are good candidates to search for such annihilations and the Galactic Centre (GC) was first considered. H.E.S.S. observations of the GC region[1,8] revealed a source of VHE γ-ray emission (HESS J1745-290) but ruled out the bulk of the signal as of DM origin.[5] There are also other candidates with high DM density in relative proximity that might lead to detectable DM annihilation signals. Satellite dwarf galaxies of the Milky Way (MW) such as Sagittarius, Draco or Canis Major are popular targets, owing to their relatively low astrophysical background.[20] A null result concerning the search for DM towards the Sagittarius dwarf spheroidal galaxy

(Sgr dSph) direction was published by the H.E.S.S. collaboration.[6] Constraints on the parameter spaces of two popular WIMPs models, namely the R-parity conserving Minimal Supersymmetric extension of the Standard Model (MSSM) and Kaluza-Klein (KK) scenarios with K-parity conservation, were derived. A null result was also established by the MAGIC collaboration[9] and the WHIPPLE collaboration,[38] when searching for a DM annihilation signal toward the Draco dwarf galaxy.

The present paper is based on a recently published analysis[7] and reports the search for a DM annihilation signal towards the direction of the CMa overdensity with the H.E.S.S. array of Cherenkov telescopes. The paper is organized as follows: in Sec 2 the controversial nature of the CMa overdensity is briefly discussed; in Sec 3 the analysis of the data is presented, while in Sec 4 the predictions for DM annihilation into γ-rays in the CMa overdensity are discussed. Constraints on the WIMP velocity-weighted annihilation rate are given.

2. A Galactic Warp or the Relic of a Dwarf Galaxy?

Since its discovery,[26] the nature of the Canis Major (CMa) overdensity is the subject of many discussions over whether it is a dwarf galaxy or simply a part of the warped Galactic disk. According to Ref. 28, the CMa overdensity simply reflects the warp and flare of the outer disk of the MW. A second scenario, which is of interest for the aim of this paper, considers this elliptical overdensity as the remnant of a disrupted dwarf galaxy that could have created the Monoceros "ring" structure.[26] Indeed, numerical simulations show that such a structure can be explained by an in-plane accretion event, in which the remnant of the dwarf galaxy would have an orbital plane close to the Galactic plane. The mass, luminosity and characteristic dimensions of CMa appear quite similar to those of the Sgr dwarf galaxy. As for many dSph, the CMa overdensity would thus be an interesting candidate for DM detection. In the remainder of the paper, the Canis Major object is assumed to be a dwarf galaxy.

The CMa overdensity is located towards the Galactic anti-centre direction at roughly 8 kpc from the sun[12] and is the closest observed dwarf galaxy. It is a very extended object (Δl = 12°, Δb = 10°) with a roundish core approximately centered at l = 240° and b = -8° according to various star surveys in this region.[25,26] In contrast to other dwarf galaxies, neither dispersion velocity measurements, nor luminosity profiles are available so that an accurate modelling of the CMa DM halo profile is not possible. However, there are enough constraints to estimate the expected γ-ray flux from DM particle annihilations in this object. The annihilation cross-section is given by the particle physics model (see section 4). As concerns the mass content of CMa, the narrow dispersion between the average mass values found for different dSph galaxies in the local group[27,37] is an indication that dSph's may possibly have a universal host halo mass.[17] The mass of the CMa dwarf galaxy can then be inferred to be in the same range as the Sgr dwarf galaxy and many other dSph's so that the CMa total mass would range between 10^8 and 10^9 $M_\odot$.[26] For instance, reference[20] gives a model where the CMa mass is taken as 3×10^8 $M_\odot$. The H.E.S.S. large FoV covers a large part of the CMa core, optimizing the chances to see a potential DM annihilation signal.

3. H.E.S.S. Observations and Analysis

3.1. *The H.E.S.S. Array of Imaging Atmospheric Cherenkov Telescopes*

H.E.S.S. is an array of four Imaging Atmospheric Cherenkov Telecopes (IACT's)[22] located in the Khomas Highland of Namibia at an altitude of 1800 m above sea level. The instrument uses the atmosphere as a calorimeter and images electromagnetic showers induced by TeV γ-rays. Each telescope collects the Cherenkov light radiated by particle cascades in the air showers using a large mirror area of 107 m^2 and a camera of 960 photomultiplier tubes (PMT's). The four telescopes are placed in a square formation with a side length of 120 m. This configuration allows for a accurate reconstruction of the direction and energy of the γ-rays using the stereoscopic technique. The cameras cover a total field of view of $5°$ in diameter. The energy threshold of the H.E.S.S. instrument is approximately 100 GeV at zenith and its sensitivity allows to detect fluxes larger than 2×10^{-13} cm^{-2} s^{-1} above 1 TeV in 25 hours. More details on the H.E.S.S. experiment can be found in Ref. 4.

3.2. *Data Processing*

Observations of the CMa dwarf galaxy with H.E.S.S. were carried out in November 2006 with pointing angles close to the zenith and extending up to $20°$. The nominal pointing direction was l = $240.15°$ and b = $-8.07°$ in Galactic coordinates. The data were taken in "wobble mode" with the telescope pointing typically shifted by $\pm 0.7°$ from the nominal target position. The dataset used for image analysis was selected using the standard quality criteria, excluding runs taken under bad or variable weather conditions.

Table 1. List of cuts used in the analysis.

Cut name	γ-event cut value
Combined cut	≤ 0.7
Image charge min.	≥ 60 photo-electrons
Reconstructed shower depth min.(rad. length)	-1
Reconstructed shower depth max. (rad. length)	4
Reconstructed nominal distance	$\leq 2.5°$
Reconstructed event telescope multiplicity	≥ 2

The CMa dataset amounts to 9.6 hours of live time after quality selection.

The data processing uses a combination of two techniques. The first technique computes the "Hillas geometrical moments" of the shower images to reconstruct shower geometry and energy, and to discriminate between γ-ray and hadronic events.[2] The second technique uses a semi-analytical model of air showers which predicts the expected intensity in each camera pixel.[30] The combination of these two techniques, referred hereafter as "Combined Hillas/Model analysis", uses a combined estimator (the so-called "Combined cut") and provides an improved background rejection. The background is estimated following the template background method.[32] Table 1 shows the different cut values used to select the γ-ray events. Events that pass the analysis cuts are labelled as "γ candidates" and are stored in the so-called γ candidate map $n_\gamma^{\text{candidate}}(l,b)$. Events that do not pass the analysis cuts are defined as "background events" and are stored in the so-called background map $n_{\text{bck}}(l,b)$.

The 2.5°×2.5° excess sky map is obtained by the following equation:

$$n_\gamma^{\text{excess}}(l,b) = n_\gamma^{\text{candidate}}(l,b) - \alpha(l,b) \times n_{\text{bck}}(l,b), \tag{1}$$

where $\alpha(l,b)$ refers to the template normalisation factor as described in Ref. 32. To search for a γ-ray signal, the raw fine-binned maps are integrated with a 0.1° radius around each point to match the H.E.S.S. angular resolution, resulting in new oversampled maps of γ-ray candidates and background events, and a corresponding γ-ray excess map. Using the prescription of Li and Ma[24] to derive the significance for each point of the oversampled map on the basis of the γ-ray candidate and background counts and the template normalization factor, no significant excess is found at the target position (Fig. 1, left panel). The distribution of significances for the entire map is shown in the right panel of Fig. 1 and is fully consistent of statistical fluctuations of the background signal. As the excess map does not show any signal, an upper limit on the number of γ-ray events for each point in the map can be derived using the method of Feldman

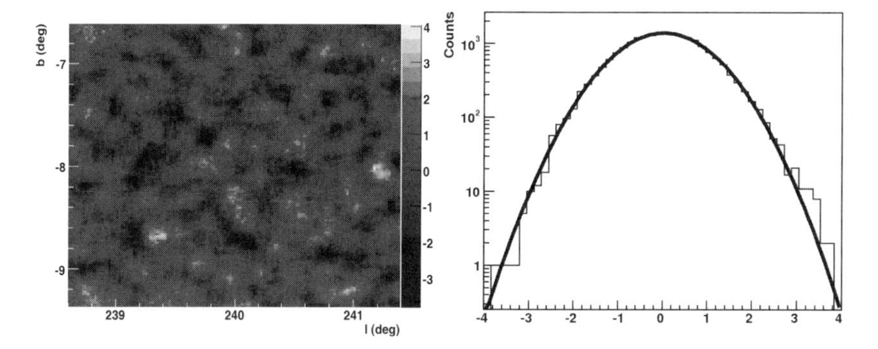

Fig. 1. Left panel: Significance map corresponding to the excess map computed in the analysis (see text), calculated according to the Li & Ma method.[24] Right panel: Significance distribution derived from the significance map. The solid line shows the Gaussian fit. The mean value is 0.01 ± 0.004 and the corresponding variance is 1.000 ± 0.005.

and Cousins.[21] The uncorrelated γ candidate and normalized background maps, plotted on a $0.2^\circ \times 0.2^\circ$ grid to have bins not smaller than the H.E.S.S. angular resolution, are used for the upper limits calculations.

4. Predictions for Dark Matter Annihilations in the Canis Major Overdensity

The DM particles are expected to annihilate into a continuum of γ-rays through various processes such as the hadronization of quark final states, hadronic decay of τ leptons and subsequent decay of mesons. Two DM candidates are commonly discussed in literature: the so-called neutralino arising in supersymmetric extensions of the standard model (SUSY),[23] and the first excitation of the hypercharge gauge boson in Universal Extra Dimension theories (UED) called the $B^{(1)}$ particle.[33] Typical masses for these DM candidates range from 50 GeV to several TeV. The value of the annihilation cross-section is constrained to give a thermal relic abundance of WIMPs that is in agreement with the WMAP+SDSS derived value.[36] The velocity-weighted annihilation cross-sections can be as low as 10^{-30} cm^3 s^{-1}, for scenarios involving co-annihilations processes, and be as high as 10^{-25} cm^3 s^{-1}.

The expected flux ϕ_γ of γ-rays from WIMP annihilations occuring in a spherical dark halo is commonly written as a product of a particle physics term ($\mathrm{d}\Phi^{\mathrm{PP}}/\mathrm{d}E_\gamma$) and an astrophysics term (f^{AP}):

$$\phi_\gamma = \frac{\mathrm{d}\Phi^{\mathrm{PP}}}{\mathrm{d}E_\gamma} \times f^{\mathrm{AP}} \tag{2}$$

The velocity-weighted cross-section for WIMP annihilation $\langle\sigma v\rangle$ and the WIMP mass are fixed to compute the particle physics term in Eq.2:

$$\frac{\mathrm{d}\Phi^{\mathrm{PP}}}{\mathrm{d}E_\gamma} = \frac{\langle\sigma v\rangle}{4\pi m_{\mathrm{DM}}^2}\left(\frac{\mathrm{d}N}{\mathrm{d}E_\gamma}\right)_{\mathrm{DM}}, \tag{3}$$

where $(\mathrm{d}N/\mathrm{d}E_\gamma)_{\mathrm{DM}}$ is the γ-ray spectrum originating for DM particle annihilation. The shape of the continuum γ-ray spectrum predicted in the framework of the phenomenological Minimal Supersymmetric extension of the Standard Model (pMSSM) depends on the model in a complicated way. A simplified parametrization of this shape, for higgsino-like neutralinos mainly annihilating via pairs of W and Z gauge bosons, was taken from Ref. 13. In the case of KK $\mathrm{B}^{(1)}$ particle annihilations, the branching ratios to final states are independent of the WIMP mass. The differential photon continuum has been simulated with the PYTHIA package[34] using branching ratios from Ref. 33.

The astrophysics term f^{AP} illustrates the DM concentration dependency of the expected γ-ray flux toward the pointed source and is given by:

$$f^{\mathrm{AP}} = \int_{\Delta\Omega}\int_{\mathrm{los}} \rho^2(l)\mathrm{d}l\mathrm{d}\Omega, \tag{4}$$

where $\rho(l)$ is the mass density profile of the CMa dwarf galaxy and $\Delta\Omega$ the detection solid angle ($\Delta\Omega = 10^{-5}$ sr, corresponding to the integration radius of $0.1°$).

4.1. *Model of the Canis Major Dark Matter Halo within the ΛCDM Cosmology*

The estimate of the astrophysical term f^{AP} relies on the modelling of the CMa DM mass distribution. Observationally, the DM mass content of dSph galaxies can be derived using velocity dispersion measurements of their stellar population as well as their luminosity profile. The comparison between models and observations can constrain the parameters of their assumed density profiles. In the case of the CMa dSph, the lack of available observational data prevents the modelling of its density profile in the same way as in the literature.[6,16,20]

In the absence of observational data, a standard cusped NFW halo[29] was assumed to model the CMa dwarf mass distribution:

$$\rho_{\text{cusped}}(r) = \frac{\rho_0}{\frac{r}{r_s}(1+\frac{r}{r_s})^2}, \tag{5}$$

where ρ_0 is the overall normalisation and $\mathrm{r_s}$ the scale radius. The parameters ρ_0 and $\mathrm{r_s}$ determining the shape of the profile as well as the halo virial mass M_{vir} are found by solving a system of three equations:

$$M_{\text{vir}} = \int_0^{R_{\text{vir}}} \rho_{\text{cusped}}(r) \mathrm{d}^3\vec{r} \tag{6}$$

$$M_{\text{vir}} = \frac{4\pi}{3}\rho_{200} \times R_{\text{vir}}^3 \tag{7}$$

$$C_{\text{vir}}(M_{\text{vir}}, z) = \frac{c_0}{1+z} \times \left(\frac{M_{\text{vir}}}{10^{14}h^{-1}M_\odot}\right)^\alpha, \tag{8}$$

where $\mathrm{R_{vir}}$ is the halo virial radius. The virial radius is computed given the virial mass $\mathrm{M_{vir}}$ and is defined as the radius within which the mean density equals ρ_{200} ($\rho_{200} = 200 \times \rho_c$, where ρ_c is the critical density of the universe).

Eq. 8 relates the concentration parameter $\mathrm{C_{vir}}$ to the virial mass. The concentration parameter is defined as the ratio between the virial radius and the scale radius $\mathrm{R_{vir}/r_s}$ in the case of a NFW profile. The relation between C_{vir} and M_{vir} is not well-known and was studied in various simulations of structure formation. Eq. 8 is used following the halo concentration fit of[18] which is in good agreement with most of N-body simulations proposed in the literature (see Ref. 15 and Ref. 19 as examples). In Eq. 8, z denotes the redshift and h the present day normalized Hubble expansion rate. c_0 and α are the parameters of the halo concentration fit and depend on the cosmological scenario ($c_0 = 9.6$ and $\alpha = -0.1$ in a ΛCDM cosmology).

The CMa dSph galaxy is located close to the Galactic disk and suffers from strong tidal disruptions. A reasonably good estimator of its total dark halo mass is then the mass enclosed inside its tidal radius rather than its virial mass. The tidal radius of the CMa dwarf galaxy is calculated via the Roche criterion:

$$\frac{M_{\text{dSph}}(r_t)}{r_t^3} = \frac{M_{\text{MW}}(d-r_t)}{(d-r_t)^3}, \tag{9}$$

where d is the distance of CMa to the center of the MW. $\mathrm{M_{MW}(r)}$ denotes the mass of the MW galaxy enclosed in a sphere of radius r. A NFW profile for the MW halo is considered with a concentration parameter equal to 10 and a virial mass of 10^{12} $\mathrm{M_\odot}$. The total mass of the dSph galaxy is computed by iterative tidal stripping, inserting first M_{vir} in Eq. 9 and computing successively the total halo mass (using Eq. 6) and the tidal radius (using Eq. 9), until the convergence of the procedure is reached. The question is now whether or not tidal forces significantly remodel the internal structure of tidally affected dSph. Discrepant results have been reported in the literature regarding this question.[31,35] Here, it is assumed that tidal forces do not affect the inner part of the density profile so that the initial halo structural parameters are kept constant during the stripping procedure. The remaining mass is typically found to be an order of magnitude lower than the virial mass.

According to our model, the values of f^{AP} range from f^{AP} ~2.3 10^{23} GeV^2 cm^{-5} for a halo mass of 10^6 $M_\odot$ to f^{AP} ~1.2 10^{25} GeV^2 cm^{-5} for a halo mass of 10^{10} $M_\odot$.

4.2. *Sensitivity to the Annihilation Cross-Section of WIMP Candidates*

In this part, the CMa total mass is fixed to be 3×10^8 $M_\odot$, which is the mass quoted by Ref. 20. In that case, the value of the astrophysical factor is f^{AP} =2.2 10^{24} GeV^2 cm^{-5}. Limits on the velocity-weighted annihilation cross-section $\langle\sigma v\rangle^{95\%C.L}$ can then be derived as a function of the DM particle mass in the framework of SUSY and KK models. The SUSY parameters were computed with the micrOMEGAs v1.37 software package.[11] The left panel of Fig. 2 shows the H.E.S.S. exclusion limits on the velocity weighted cross-section. The black points illustrate the computed pMSSM scenarios and the red points represent those satisfying the WMAP+SDSS constraints on the CDM relic density $\Omega_{CDM}h^2$.[36] $\Omega_{CDM}h^2$ is allowed to range between 0.09 and 0.11. The H.E.S.S. observations of the CMa dSph allows to exclude velocity weighted cross-sections of the order of 5×10^{-24} cm^3 s^{-1}, comparable with those derived for the Sgr dSph modelled with a cusped NFW profile. The limits obtained are an order of magnitude larger than the velocity-weighted annihilation cross sections of higgsino-like neutralinos.

In the case of KK scenarios, predictions for the velocity-weighted cross-section are computed with the formula given in.[10] The expression of $\langle\sigma v\rangle$ is inversely proportional to the squared mass of the lightest Kaluza-Klein (LKP) particle, namely the $B^{(1)}$ particle. Considered KK models that reproduce the CDM relic measured by WMAP and SDSS require a LKP mass ranging from 0.7 TeV to 1 TeV. The right panel of Fig. 2 shows the H.E.S.S. limits obtained within these models. The H.E.S.S. observations do not constrain the KK velocity weighted cross-section.

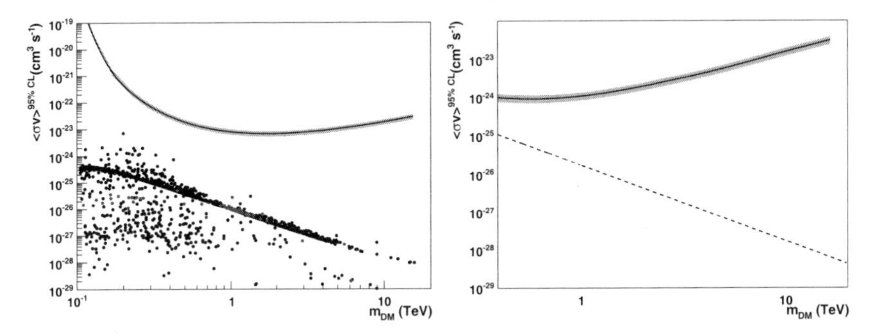

Fig. 2. Upper limits at 95% CL on the velocity weighted cross-section as a function the DM particle mass in the case of pMSSM (left panel) and KK (right panel) scenarios, for an assumed CMa total mass of 3 × 10^8 $M_\odot$. The shaded area represents the 1σ error bars on $\langle\sigma v\rangle^{95\%C.L}$ (see text for details). Left: The pMSSM models are represented by black points, and those giving a CDM relic density in agreement with the measured WMAP+SDSS value are illustrated by red points. Right: The KK models are represented by the black dashed line, and those verifying the WMAP+SDSS constraint on $\Omega_{CDM}h^2$ are labelled in red.

5. Conclusions

The CMa overdensity is the subject of many debates over whether it is a dwarf galaxy or the warp and flare of the Galactic outer disk. Considering the first scenario, its relative proximity makes it potentially the best region for searches of a DM annihilation signal. However, the lack of observational data prevents the precise modelling of its density profile. Assuming a NFW profile and a mass content of 3 × 10^8 $M_\odot$ within its tidal radius, typical of dwarf galaxies, H.E.S.S. is close to exclude a few pMMSM scenarios with higgsino-like neutralinos, but does not reach the necessary sensitivity to test models compatible with the WMPA+SDSS constraint on the CDM relic density. In the case of DM made of $B^{(1)}$ particle from KK models with extra dimensions, no constraints are obtained.

Acknowledgments

The support of the Namibian authorities and of the University of Namibia in facilitating the construction and operation of H.E.S.S. is gratefully acknowledged, as is the support by the German Ministry for Education and Research (BMBF), the Max Planck Society, the French Ministry for Research, the CNRS-IN2P3 and the Astroparticle Interdisciplinary Programme of the CNRS, the U.K. Particle Physics and Astronomy Research Council

(PPARC), the IPNP of the Charles University, the Polish Ministry of Science and Higher Education, the South African Department of Science and Technology and National Research Foundation, and by the University of Namibia. We appreciate the excellent work of the technical support staff in Berlin, Durham, Hamburg, Heidelberg, Palaiseau, Paris, Saclay, and in Namibia in the construction and operation of the equipment.

References

1. Aharonian F., et al. (2004) *A&A* **425** L13.
2. Aharonian F., et al. (2005) *A&A* **430** 865.
3. Aharonian F., et al. (2006a) *Science* **314** 1424.
4. Aharonian F., et al. (2006b) *A&A* **457** 899.
5. Aharonian F., et al. (2006c) *Phys. Rev. Lett.* **97** 221102.
6. Aharonian F., et al. (2008) *Astropart. Phys.* **29** 55.
7. Aharonian F., et al. (2009a) *ApJ* **691** 175.
8. Aharonian F., et al. (2009b) *accepted in A&A*.
9. Albert J., et al. (2007) *preprint astro-ph/*07112574.
10. Baltz E., Hooper D. (2005) *JCAP* **0507** 001.
11. Belanger G., Boudjema F., Pukhov A., Semenov A. (2004) *preprint hep-ph/* **0405253**.
12. Bellazinni M., et al. (2004) *MNRAS* **354** 1263.
13. Bergström L., Ullio P., Buckley J. (1998) *Astropart. Phys.* **9** 137.
14. Bertone G., Hooper D., Silk J. (2005) *Phys. Rept.* **405** 279.
15. Bullock J., et al. (2001) *MNRAS* **321** 559.
16. Colafrancesco S., Profumo S., Ullio P. (2007) *Phys. Rev.* **D75** 023513.
17. Dekel A., Silk J. (1986) *ApJ* **303** 39.
18. Dolag K., et al. (2004) *A&A* **416** 853.
19. Eke V.R., Navarro J.F., Steinmetz M. (2001) *ApJ* **554** 114.
20. Evans N.W., Ferrer F., Sarkar S. (2004) *Phys. Rev.* **D69** 123501.
21. Feldman G., Cousins R. (1998) *Phys. Rev.* **D57** 3873.
22. Hofmann W., et al. (2003) *in Proc. of the 28th ICRC (Tsubuka)* **Vol.1** p.2811.
23. Jungman G., Kamionkowski K., Griest K. (1996) *Phys. Rept.* **276** 195.
24. Li T., Ma Y. (1983) *ApJ* **272** 317.
25. Martinez-Delgado D., et al. (2004) *preprint (astro-ph/* **0410611**).
26. Martin N.F., et al. (2004) *MNRAS* **348** 12.
27. Mateo M. (1998) *Annu. Rev. Astron. Astrophys.* **36** 435.
28. Momany Y., et al. (2006) *A&A* **451** 515.
29. Navarro J., Frenk C., White S. (1997) *ApJ* **490** 493.
30. de Naurois M., et al. (2003) *in Proc. of the 28th ICRC (Tsubuka)* **Vol.5** p.2907.
31. Reed D., et al. (2005) *MNRAS* **357** 82.
32. Rowell G.P. (2003) *A&A* **410** 389.
33. Servant G., Tait T. (2003) *Nucl. Phys.* **B650** 391.

34. Sjöstrand T., Lönnblad L., Mrenna S., Skands P. (2003) *PYTHIA 6.3, preprint (hep-ph/* **0308153**.
35. Stoehr F., et al. (2002) *MNRAS* **335** L84.
36. Tegmark M., et al. (2006) *Phys. Rev.* **D74** 123507.
37. Walker M.G., et al. (2007) *ApJ* **667** L53.
38. Wood M., et al. (2007) *preprint (astro-ph/* **08011708**.

RECENT RESULTS AND STATUS OF ICECUBE

S. H. SEO

Department of Physics, Stockholm University,
Stockholm, 106 91, Sweden
E-mail: seo@physto.se
www.su.se
for the IceCube Collaboration *

IceCube is a neutrino telescope currently under construction at the geographical S. Pole. It will be a cubic kilometer size by 2011 when complete. So far IceCube has been successful in both deploying strings and taking data with its partial detector together with its predecessor, AMANDA. Its performance was well verified as it was originally designed. Here we present some interesting recent results from IceCube and AMANDA: point source search, GRB080319B, indirect dark matter search, magnetic monopole search and search for violation of Lorentz invariance. The IceCube deep core which will consist of 6 special strings is expected to improve low energy physics of IceCube such as indirect dark matter search.

Keywords: Neutrino telescope; IceCube; AMANDA; deep core; South Pole.

1. Introduction

Owing to the success of AMANDA (Antarctic Muon And Neutrino Detection Array), IceCube was conceived to extend AMANDA to a cubic kilometer scale (~80 times bigger than AMANDA in fiducial volume) to explore the high energy (1 TeV ~ 1 PeV) neutrino universe. With a better detector technology as well as the bigger size, IceCube is expected to improve time and directional resolutions as well as event rates compared to AMANDA.

Since 2004 IceCube has been successfully taking data at the S. Pole with its partial detector (~73% as of 2009). We have obtained some interesting preliminary physics results using the data taken with the partial IceCube detectors, and 7 (or less) years of AMANDA data.

Here we would like to describe our detector first, and then to present some recent physics results followed by our future plan.

*http://www.icecube.wisc.edu/

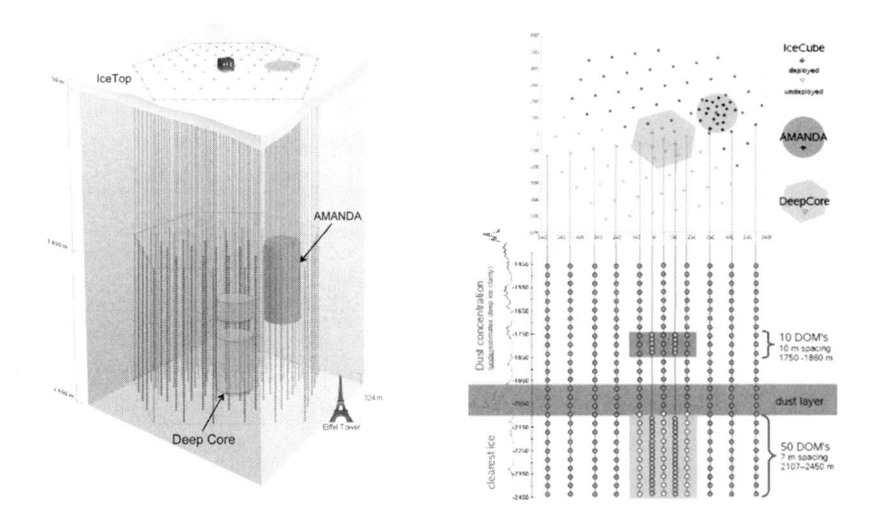

Fig. 1. The IceCube detector (left) and schematics of the deep core (right).

2. The IceCube Detector

As shown in Fig. 1 (left) IceCube consists of 3 sub-detector parts: ice-top surface array, in-ice detector, and the predecessor, AMANDA [a].

Ice-top is a cosmic ray air shower array of which main goals are to study the compositions and the energy spectrum of cosmic rays. The complete ice-top will consist of 320 Digital Optical Modules (DOM), distributed in 160 frozen water tanks (or in 80 stations with 2 side-by-side tanks each) with 1 high gain and 1 low gain DOMs per tank.

The complete in-ice detector will consist of 4800 DOMs equally distributed in 80 vertical strings (60 DOMs per string) alligned with ice-top stations. Each DOM in the same string is separated by 17 m and each string is separted by 125 m. A single DOM is a fundemental building block of the IceCube detector and operates independently. Currently, however, it is required to meet a local conincidence condition in any DOM pair to reduce noise hits. A DOM consists of a PMT (Hamamatsu: 25.4 cm, 10 stage), a main board, a flasher board and a PMT base, all of which are housed in a 33 cm diameter benthosphere. In the main board, there are two types of waveform digitizers: Analog Transient Waveform Digitizer (ATWD) and Fast Analog to Digital Converter (FADC). ATWD gives waveform information

[a]AMANDA is scheduled to stop taking data as of April, 2009. The deep core in Fig. 1 will be discussed in Section 4.

in detail for a short time period (~450 nsec). On the other hand, FADC gives less precise waveform information but for a longer time period (~6 μsec). These waveform digitizers convert analog signals to digital ones *in situ*, which makes IceCube more advanced technologically than AMANDA. More technical details on the IceCube DOM are described in Ref. 1.

2.1. *Detector Resolutions*

Owing to the *in situ* waveform digitizers as well as the bigger detection volume, the performance of IceCube is expected to be better than AMANDA in terms of time and directional resolutions which are very important for searching for high energy neutrino sources. For track-like events angular resolution improves from $2 \sim 3°$ to $1°$ at about 1 TeV and time resolution improves from $5 \sim 7$ nsec to 3 nsec. Energy resolution of the track-like events is about 0.4 in $\log_{10}$(E/Gev) in both AMANDA and IceCube. The same studies for cascade-like events which are much harder to detect than track-like events, are actively on-going.

2.1.1. *Shadow of the Moon*

The Moon can block cosmic rays from reaching the Earth. This known effect can be used for calibration, and was observed in the first 3 months of data taken with the IceCube 40 strings (IC-40) by observing 4.2σ deficit of the cosmic ray muons within ~0.7° from the direction of the moon, as shown in Fig. 2 (preliminary). The ultimate goal of this measurement is to calibrate our detector in pointing resolution as well as absolute pointing.

2.2. *Background Rates*

Main backgrounds for IceCube signal events are atmospheric neutrinos and muons from cosmic ray interactions in the atmosphere. At the depth of IceCube, downward-going muon flux is about 10^6 times more than atmospheric neutrino flux. The downward-going muons, however, can be avoided by looking only upward-going events, but the atmospheric neutrinos generate an irreducible background.

Table 1 [b] shows these atmospheric muon and neutrino rates in different IceCube configurations. With IC-22 (IC-40) strings we detected (expect) ~550 (~1000) muons per second and ~28 (~110) atmospheric neutrinos

[b]The IC-40 and IC-80 rates are expected ones.

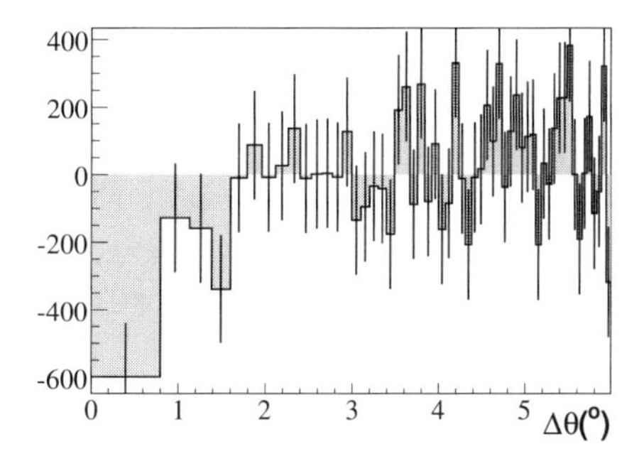

Fig. 2. The x-axis shows $\Delta\theta$, the space angle between the moon direction and reconstructed angle, and the y-axis is the difference between the expected and observed cosmic ray muons. The 4.2σ deficit (~600 events) of the cosmic ray muons in the direction of the moon with IceCube 40 strings was observed within ~7° in $\Delta\theta$ (preliminary).

Table 1. Event rates in different IceCube configurations.

Year	# Strings	Run Length	CR μ rate	atm. ν rate
2006	IC-9	137 days	~80 Hz	~1.7/day
2007	IC-22	319 days	~550 Hz	~28/day
2008	IC-40	1 year	~1000 Hz	~110/day
2011	IC-80	≥10 years	~1650 Hz	~220/day

per day. With IC-80 we expect to detect ~1650 muons per second and ~220 atmospheric neutrinos per day.

3. Recent Physics Results

IceCube physics reach is broad from astronomy to particle physics. In the following subsections, however, only some recent preliminary physics results from AMANDA and IceCube are discussed.

3.1. *AMANDA and IceCube-22 Sky Maps*

Fig. 3 shows the northern sky map using AMANDA data taken for 7 years (equivalent to 3.8 years livetime).[2] A maximum significance of 3.38σ was found at the position of 54° declination and 11.4° right ascension. However, 95% of randomized data sets also had a maximum significance of 3.38σ or greater, which implies the hottest spot is not statistically significant. The results of the point source search with the IceCube 22-string detector (275.7 days livetime) are shown in Fig. 4. The largest deviation from background

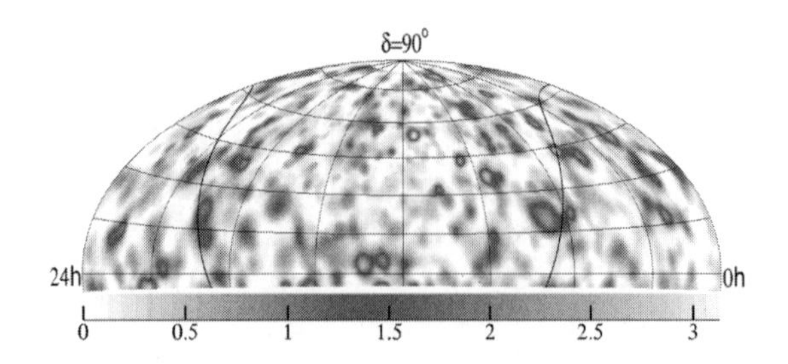

Fig. 3. The AMANDA northern sky map using data taken for 3.8 years livetime.

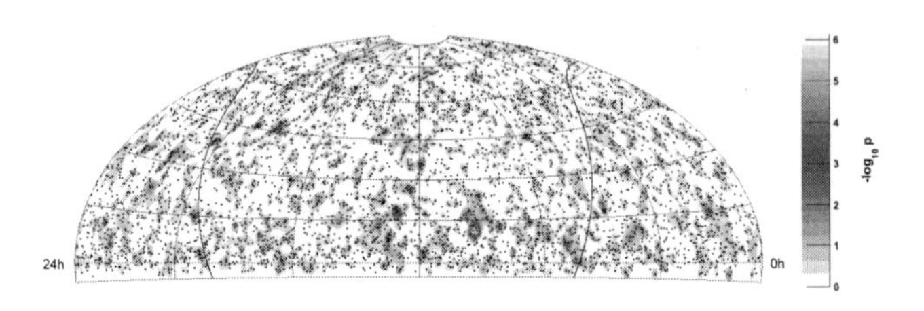

Fig. 4. The IC-22 northern sky map (5114 ν_μ candidates in 275.7 days livetime of data).

is located at 11.4° declination and 153.4° right ascension, with pre-trial significance of 4.8σ. Accounting for all trials in the analysis, the final p-value for this result is 1.34%, consistent with the null hypothesis of background-only events at the 2.2σ level (if the p-value is expressed as the one-sided tail of a Gaussian distribution).

3.2. *Gamma Ray Bursts*

Gamma Ray Bursts (GRBs) are one of the most energetic astronomical phenomena and may produce very high energy neutrinos as well as photons. The most brightest optical GRB ever observed in March 19th, 2008 (06:12:49 UT) for about 70 sec at 217.9° declination and 36.3° right ascention.[3,4] This GRB080319B was so bright that one was able to see it with ones bare eyes. IceCube 9 strings were taking data at that time[c] but our analysis[5] did not find any muon neutrinos from this source so that the 90% CL upper limit was set, as shown in Fig. 5. The expected neutrino fluence in Fig. 5 was obtained using measured burst parameters (gamma-ray spec-

[c]In March, 2008 IceCube took nominal data with 22 strings (with 40 strings from April) but unfortunately IceCube was on a special test mode taking data with only 9 strings at the time of GRB080319B occurrence.

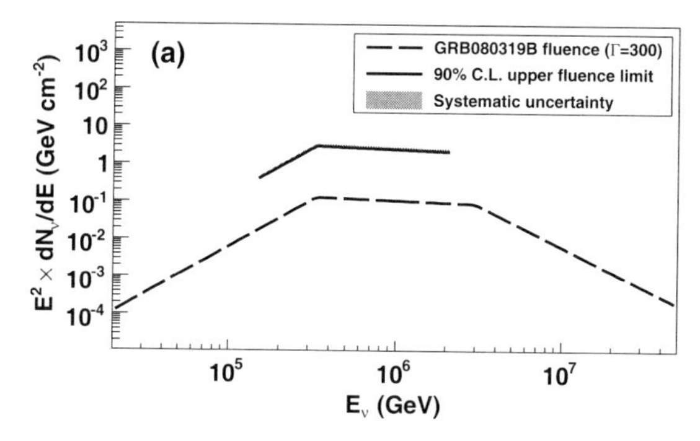

Fig. 5. The 90% CL upper limit (line) on the muon neutrino fluence from GRB080319B using the data taken with the IceCube 9 strings and the expected neutrino fluence (dashed line) from GRB080319B (see text).

trum and redshift) and assuming the Lorentz boost factor (Γ) of 300, which results in ~0.1 expected ν_μ events in IC-9 from the GRB080319B.[5] There are also on-going coincident analyses of GRBs using IC-22 and IC-40 with the GRBs reported by Swift, Fermi, etc.

3.3. *Astrophysical Neutrinos*

Astrophysical neutrinos are expected to provide us with some clues to understand our universe better and thus pursued by neutrino telescopes. So far no neutrino telescope has observed them and therefore set only upper limits. The Fig. 6 shows the upper limits (sensitivities) on astrophysical neutrino point source flux measured (studied) by several experiments. The current best limit was set by AMANDA with the 3.8 years livetime of data as $10^{-11} \sim 10^{-10}$ TeV cm^{-2} s^{-1} (black triangle in Fig. 6). The sensitivity of IC-80 with 1 year livetime of data is expected to be about an order of magnitude better than that of the AMANDA limit.

3.4. *Atmospheric Neutrinos*

In IceCube atmospheric neutrinos are main background in searching for astrophysical neutrinos but at the same time can play an important role in checking the detector performance and calibration. With the same 7 years of AMANDA data, atmospheric neutrino flux was measured,[22] as shown in Fig. 7 (red band) together with SuperK's measurement[6] (blue band). These two measurements are consistent with the two popular atmospheric neutrino models[7,8] in low (SuperK) and high (AMANDA) energy regions, but can not differenciate the two models.

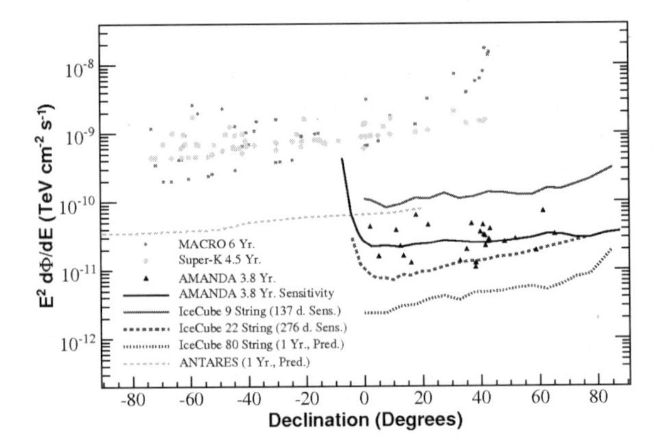

Fig. 6. Astrophysical neutrino point source flux limits and sensitivities by various experiments.

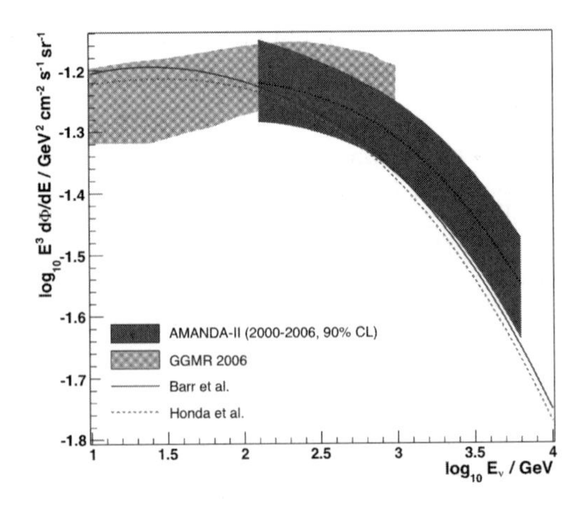

Fig. 7. The 90% confidence belt (red band) on atmospheric neutrino flux measured by using AMANDA 3.8 years livetime of data and SuperK's 90% confidence belt (blue band). These measurements are consistent with the two atmospheric neutrino models (blue line and red dotted line).

3.5. *Solar WIMP Search*

Indirect dark matter search is one of the main research topics in IceCube. There are still many viable candidates[9] of dark matter in the market currently and one of the most favored one is WIMP[10] (Weakly Interacting Massive Particle), especially SUSY neutralinos.[11] An analysis with IceCube 22 strings (104 days livetime of data) did not find any ν_μ from neutralino

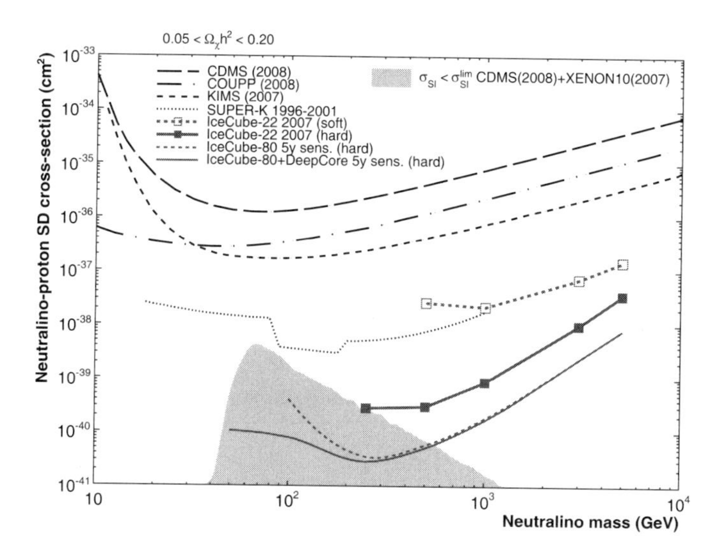

Fig. 8. The 90% CL upper limits on SD cross-section of neutralino and proton scattering for hard (W^+W^-) channel (red line with squares) and soft ($b\bar{b}$) channel (red dotted line with squares) using 104 days livetime of IC-22 data. SuperK 90% CL limit on soft channel (black dotted line) is also shown for comparison. Conservative sensitivities of IC-80 with (red line) and without (red dotted line) deep core (see later section) are also shown.

annihilations in the Sun (in the neutralino mass of 250 to 5000 GeV) but was able to set the current best upper limit (90% CL) on the muon flux from the Sun. The muon flux can be converted[12] to neutralino and proton scattering cross-section in the Sun and we obtained the best upper limit on the spin-dependent (SD) cross-section, $10^{-3} \sim 10^{-1}$ pb, in the hard channel (W^+W^-), as shown in Fig. 8. This is about 1 to 3 orders of magnitude better than the current best limit, ~1 pb, from the direct experiments like KIMS[14] and COUPP.[15] Using IC-80 with (red line) and without (red dotted line) deep core (5 years livetime of data), we expect to reach upto 10^{-4} pb in the SD cross-section. More details on IC-22 WIMP search can be found in Ref. 13, and WIMP searches with AMANDA/IceCube can be found in A. Rizzo's contribution in the same proceedings.

3.6. *Monopole Search*

According to Rubakov-Callan mechanism,[16,17] cross-section of magnetic monopoles from the Grand Unification Theory (GUT) on a proton target can be enhanced to at the level of strong interaction cross-section. These GUT monopoles which are very heavy ($10^{13} \sim 10^{17}$ GeV) and thus slowly

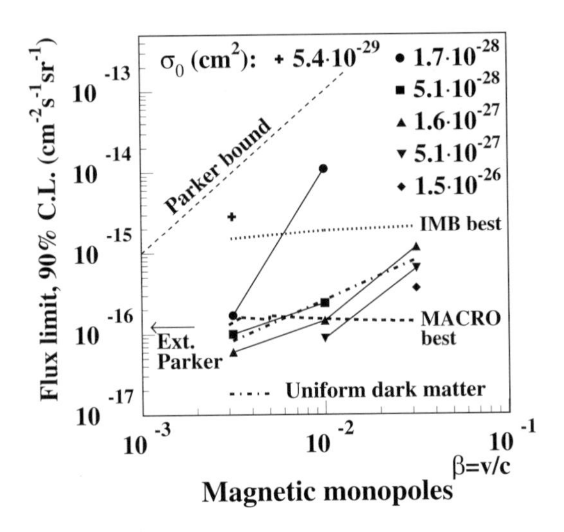

Fig. 9. Preliminary upper limits (90% CL) on sub-relativistic magnetic monopole flux as a function of β for 4 different strong interaction cross-sections.

moving, could be detected in AMANDA/IceCube if the Rubakov-Callan mechanism is true. Using the data taken with AMANDA (64 days livetime of data from 2001), 90% CL upper limits were set on the sub-relativistic magnetic monopole flux as a function of their velocity (β) for 4 different strong interaction cross-sections, as shown in Fig. 9. More details on this analysis is found in Ref. 18. A preliminary upper limit (90% CL) on relativistic monopole[19] flux with AMANDA (154 days livetime of data) was also obtained as $\sim 4 \cdot 10^{-16}$ cm^{-2}s^{-1}sr^{-1} at $\beta = 1$.[20]

3.7. *Search for Violation of Lorentz Invariance*

One of the many features in Violation of Lorentz Invariance (VLI) is neutrino oscillation which is different from the usual mass-induced oscillations, being generated instead from different neutrino maximum-velocity eigenstates.[21] The same 7 years of AMANDA data (3.8 years livetim) showed no evidence for ν_μ oscillation due to VLI.[22] A 90% CL upper limit was set on the velocity-splitting parameter $\delta c/c$ of 2.8 10^{-27} at the maximum mixing angle of $\sin^2 2\xi = 1$. Fig. 10 shows several limits (90%, 95% and 99% CLs from AMANDA and 90% CL from SuperK + K2K[23]) on $\delta c/c$ as a function of VLI mixing angle (ξ). In the same plot IceCube 10 year sensitivity (90% CL) with IC-80 is expected to improve about an order of magnitude at the maximum mixing angle.

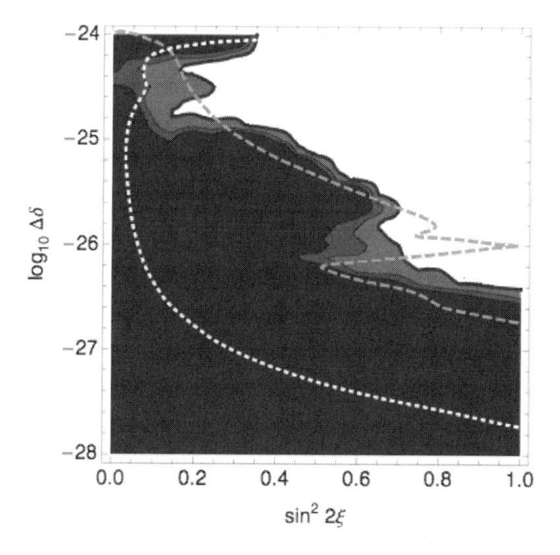

Fig. 10. Preliminary CL upper limits (90%, 95% and 99% from darker to lighter blue lines) on Violation of Lorentz Invariance using 3.8 years livetime of AMANDA data. Also shown are IC-80 90% sensitivity (yellow dotted line) for 10 years of data, and SuperK + K2K 90% contour (red dotted line).

4. Current Status of Construction and Future of IceCube

As of 2009 IceCube has deployed 59 strings (~73%) and aims to finish the construction by 2011 with the originally planned 80 strings plus 6 more special strings (60 DOMs/string) surrounding a central IceCube string, as shown in Fig.1. The purpose of the 6 special strings is to improve low energy event detection efficiency. Thus the DOM spacing is much denser (7 m intra-string and 72 m inter-string) for the bottom-positioned 50 DOMs in each string, which will be placed below a big dust layer where the ice is the clearest ($\lambda_{sca} \sim 40$ m and $\lambda_{abs} \sim 220$ m at 400 nm wavelength of light)[d] The top-positioned 10 DOMs will be placed above the big dust layer to be used as a veto. In addition to the denser spacing of the DOMs the PMTs used in the deep core strings have higher quantum efficiency (~35% more light collection). Several low energy physics topics in IceCube, e.g., WIMP search and atmospheric neutrino oscillation[24,25] etc..., will benefit from the deep core. Fig. 11 shows the effective areas for muon neutrinos with (blue) and without (pink) the deep core. The effective area in the low energy region

[d]In IceCube the average values of these parameters are $\lambda_{sca} \sim 20$ m and $\lambda_{abs} \sim 110$ m at 400 nm wavelength of light.

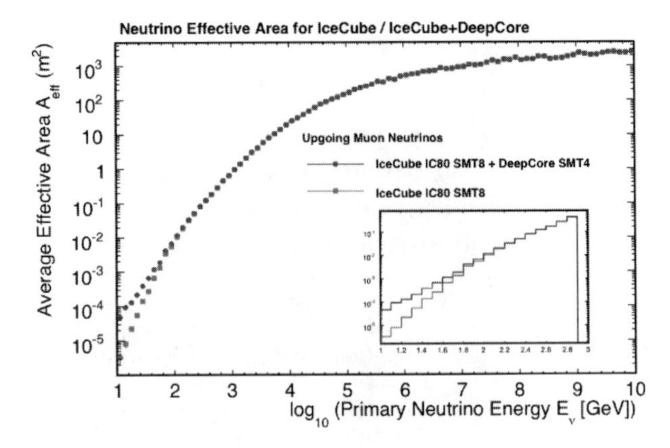

Fig. 11. Muon neutrino effective areas for IC-80 only (pink) and for IC-80 with the deep core (blue).

(below 100 GeV) has visibly improved. As discussed in the earlier section, the SD cross-section of neutralino-proton scattering is improved by adding the deep core for neutralino mass below 200 GeV (Fig. 8). The deep core together with the nearest 7 IceCube strings can also be utilized to look for the southern sky including the galactic center, by using the rest of IceCube DOMs as a downward-going muon veto.

5. Conclusion

IceCube has been successful in both deploying strings and taking data since 2004 at the S. Pole. IceCube construction will be finished by 2011 with the 80 strings plus the 6 special deep core strings. Very interesting physics results has already been obtained with data taken with IceCube partial detectors as well as AMANDA 7 years (or less) of data even though no evidence of astrophysical neutrino has been found yet. With the complete detector, however, our results are expected to be much more improved. In addition to that, the new deep core strings will play a critical role in low energy physics topics (WIMP search, atmospheric neutrino oscillation etc...) in IceCube. High energy extension of IceCube strings which is not presented here, is also being discussed within the IceCube collaboration.

References

1. R. Abbasi *et al.*, *NIM A* **601**, 294 (2009).
2. R. Abbasi *et al.*, *Phys. Rev. D* **79**, 062001 (2009).
3. J. S. Bloom *et al.*, *Astrophys. J.* **691**, 723, (2009).

4. S. Dado *et al.*, *arXiv:0804.0621* [astro-ph] (2008).
5. R. Abbasi *et al.*, *arXiv:0902.0131 [astro-ph]* (2009).
6. M. C. González-García *et al.*, *JHEP* **0610**, 075 (2006)
7. G. Barr *et al.*, *Phys. Rev. D*, **70**, 023006 (2004).
8. M. Honda *et al.*, *Phys. Rev. D*, **75**, 043006 (2007).
9. M. Taoso *et al.*, *JCAP* **0803**, 022 (2008).
10. V. Rubin and W. K. Ford, *Astrophys. J.* **159**, 379 (1970).
11. M. Drees and M. M. Nojiri, *Phys. Rev. D.* **47**, 499 (1993).
12. G. Wikström and J. Edsjö, *arXiv:0903.2986* [astro-ph] (2009).
13. R. Abbasi *et al.*, *arXiv:0902.2460 [astro-ph.CO]* (2009).
14. H. S. Lee *et al.*, *Phys. Rev. Lett.* **99**, 091301 (2007).
15. E. Benke *et al.*, *Science* **319**, 933 (2008).
16. V. A. Rubakov, *JETP Lett* **12**, 644 (1981)
17. C. G. Callan, *Phys. Rev. D.* **26**, 2058 (1982)
18. A. Pohl, Search for Subrelativistic Particles with the AMANDA Neutrino Telescope, PhD Thesis, Uppsala Universitet, (Uppsala, Sweden, 2009)
19. S. D. Wick *et al.*, *Astropart. Phys.* **9**, 663 (2003).
20. M. Ackermann *et al.*, *The 30th International Cosmic Ray Conference, arXiv:0711.0353 [astro-ph]*, 139 (2007).
21. D. Mattingly, *Living Rev. Rel.* **8**, 5 (2005).
22. R. Abbasi *et al.*, *arXiv:0902.0675 [astro-ph.HE]* (2009).
23. M. C. González-García and M. Maltoni, *Phys. Rev. D* **70**, 033010 (2004).
24. C. Rott, *arXiv:0810.3698 [astro-ph]*, (2008).
25. O. Mena *et al.*, *Phys. Rev. D* **78**, 093003 (2008).

SEARCH FOR DARK MATTER WITH AMANDA AND ICECUBE DETECTORS

A. RIZZO*
(for the IceCube Collaboration)
Interuniversity Institute for High Energies, Vrije Universiteit Brussel,
Brussels, B-1050, Belgium
**E-mail: alfio.rizzo@vub.ac.be*
w3.iihe.ac.be

If non-baryonic dark matter exists in the form of neutralinos, a neutrino flux is expected from the decay of neutralino pair annihilation products inside heavy celestial bodies. Data taken with the AMANDA (Antarctica Muon and Neutrino Detector Array) neutrino telescope located at the South Pole have been used in a search for this indirect dark matter signal. We present result from searches for neutralinos accumulated in the Sun and in the centre of the Earth, using the data taken up to 2003.

The IceCube neutrino detector is being deployed at the South Pole since 2006. This cubic kilometer observatory with 80 strings of 60 photomultipliers will be completed in 2011. The data taken in 2007 with 22 strings have been used in the search for neutrino signal from neutralinos in the Sun. Preliminary results of this analysis will be shown. The planned IceCube detector will be complemented with a dense inner core, Deep Core, to improve the sensitivity in the GeV-TeV energy domain. We will also discuss the expected performance of the combined IceCube - Deep Core detector in relation to dark matter searches.

Keywords: Dark Matter; Neutralino; Neutrino Telescopes; AMANDA/IceCube; South Pole.

1. Introduction

The existence of dark matter (DM) was first inferred from astrophysical observations that probe gravitational potentials.[1–3] One possible scenario to explain the mismatch between the required mass, needed to supply the derived gravitational potential, and the observed mass is represented by the Cold Dark Matter (CDM) model, i.e. non-relativistic massive particles (GeV or heavier) produced in the Big Bang.[4] This CDM makes up 23% of the energy density of the Universe, according to the WMAP measurements of the temperature anisotropies in the Cosmic Microwave Background, in combination with data on the Hubble expansion and the density fluctuations in the Universe.[5] The thermally averaged cross section of the DM particles at the freeze-out temperature explains why the DM can only have weak and gravitational interactions. Therefore, the DM particles are generically called WIMPs, Weakly Interacting Massive Particles.[6,7] One of the most popular and widely studied WIMP candidates is the lightest supersymmetric neutralino $\tilde{\chi}_1^0$ (or simply χ). In the Minimal Supersymmetric extension of the Standard Model (MSSM),[8] where the multiplicative quantum number R-parity is conserved, the neutralino is the mixture of the superpartners of the B and W_3 gauge bosons and the neutral Higgs bosons, H_1^0 and H_2^0. The attractiveness of this candidate stems from the fact that it is electrically neutral, and thus it does not interact electromagnetically, is stable, and can only disappear via pair annihilation (it is a Majorana particle) or coannihilation with the next-to-lightest supersymmetric particle; therefore the neutralinos may have survived since the Big Bang.[9] Consequently, relic neutralinos in the galactic halo will pass through massive bodies like the Sun or the Earth, where they can lose energy by scattering off nuclei. Over time, the neutralinos concentrate near the centres of these celestial bodies and annihilate producing Standard Model particles. The products of these annihilations will, in general, decay and produce neutrinos. The latter will be able to escape and would potentially be visible in a high energy neutrino telescope at the surface of the Earth as an excess over the atmospheric neutrino flux.[10]

2. The AMANDA and IceCube Detectors

AMANDA-II, the final stage of the AMANDA detector, has been operating since January 2000 with 677 optical modules (OMs) attached to 19 strings.[11] Most of the OMs are located between 1500 m and 2000 m below the South Pole ice surface; each OM consists of an eight inch photomultiplier tube (PMT) enclosed in a glass sphere. The previous configuration of AMANDA called AMANDA-B10, the inner core of 302 OMs on 10 strings, operated between 1997 and 1999. Muons from charged-current neutrino interactions near or inside the array can be detected by the Cherenkov light they produce when traversing the ice. The PMT signals are amplified and sent to a majority logic trigger. Since 2001, an additional lower multiplicity trigger (referred to as "string trigger") is operational, which exploits not only the temporal information but also the space topology of the hit pattern. A detailed description of the reconstruction techniques used in AMANDA is given in Ref. 12. IceCube builds on the successful deployment and operation of the AMANDA telescope. In addition to its larger volume (about a cubic kilometer), IceCube differs from AMANDA in a significant way since the signals are digitized inside each optical module, to minimize loss of information from degradation of analog signals sent over long distances.[13] The IceCube neutrino observatory, the completion of which is planned for 2011, will consist of 4800 Digital Optical Modules (DOMs) installed on 80 strings between 1450 m and 2450 m below the surface.[13,14] This In-Ice array is complemented by the IceTop surface shower array that can be used for calibration, for veto of certain backgrounds generated by air showers and for the study of cosmic rays. The current (2009) IceCube configuration consists of 59 strings (referred to as "IceCube-59"), one of which is part of the IceCube Deep Core (see Sec. 4.2). The 2008 configuration consisted of 40 strings (referred to as "IceCube-40"), while the 2007 configuration consisted of 22 strings (referred to as "IceCube-22").

3. Dark Matter Searches with Neutrino Detectors

Two annihilation channels were considered in our analyses for each neutralino mass tested (50, 100, 250, 500, 1000, 3000 and 5000 GeV/c^2), $\chi\chi \rightarrow b\bar{b}$ (referred to as the "soft" channel) and $\chi\chi \rightarrow W^+W^-$ (referred to as the "hard" channel, replaced with $\tau^+\tau^-$ in case of $m_\chi < m_W$). The simulated zenith angular range for the Solar WIMP analyses (see Sec. 3.1 and Sec. 3.2), was restricted to [90°,113°], since the muons induced by neutralinos can be distinguished from the dominant atmospheric muon background only when the Sun is seen below the horizon at the South Pole. The zenith angle for the Earth WIMP analysis (see Sec. 3.3) was simulated close to 180°, taking into account that light neutralinos annihilate less closely to the centre of the Earth than heavy neutralinos. The simulation of the solar neutralino-induced neutrino signal was performed using the programme called `WIMPSIM`.[15] For the Earth WIMP analysis an older version of the code, called `WIMPSIMP`,[16] was used. The background for this search is represented by upward-going atmospheric neutrinos and mis-reconstructed downward-going atmospheric muons. The atmospheric neutrino flux was simulated according to Ref. 17 with `ANIS`,[18] with the energy between 10 GeV and 10 PeV and zenith angles between 70° and 180°. The simulation of the atmospheric muons was made with `CORSIKA`[19] using the South Pole atmosphere parameters and protons as primaries. The sample was generated isotropically between zenith angles 0° and 85°, and with proton energies, E_p, between 600 GeV and 10^{11} GeV, with a differential energy distribution $\propto$ to $E_p^{-2.7}$. Several quality criteria on the reconstructed track were adopted in order to select the events, and the cuts on the variables used in these searches have been optimised independently for each mass and channel, in order to exploit the differences in the resulting muon energy spectra.

3.1. *To catch a neutralino from the Sun with AMANDA*

The data set used contained events collected in 150.4 days of effective live-time during 2003. The event selection was optimised in a "blind" way by explicitly excluding the time stamp of the event, and thus the Sun direction, in the development of the selection criteria.[20] The first filtering steps resulted in a rejection of 99% of the data (mostly downward-going muons) while the simulated signal was kept between 55% - 90%. The last step was a cut on a multivariate classifier (Boosted Decision Tree) output. This final event selection keeps 7% - 37% of the simulated signal, while the atmospheric muon background is reduced by a factor of $\sim 10^7$. The remaining data were used for the hypothesis test, where the hyphotesis that the data contain background-only is tested versus the hypothesis that the data contain signal and background. The pdf of the space angle distribution between the reconstructed track and the scrambled position of the Sun, i.e. the estimated background $f_{\rm B}(\psi)$, can be constructed from data; while the signal pdf, $f_{\rm S}(\psi)$, can be estimated from simulation. The likelihood of the presence of μ signal events in an experiment that observes $n_{\rm obs}$ events can be then calculated: $\mathcal{L}(\mu) = \prod_{i=1}^{n_{\rm obs}} f(\psi_i|\mu)$, where $f(\psi|\mu) = \frac{\mu}{n_{\rm obs}} f_{\rm S}(\psi) + \left(1 - \frac{\mu}{n_{\rm obs}}\right) f_{\rm B}(\psi)$ is the combined pdf. Defining $\xi = \frac{\mu}{n_{\rm obs}}$, we can construct the logarithm of the likelihood ratio $R(\xi) = \log\left(\prod_{i=1}^{n_{\rm obs}} \frac{f_\xi(\psi_i)}{f_{\xi_{\rm best}}(\psi_i)}\right)$, where $\xi_{\rm best}$ represents the physically allowed (non-negative) maximum likelihood signal content.[21] The confidence interval, at significance $\alpha = 0.9$, can be then derived for the presence of a signal content $\mu^{90\%}$. Based on this we can calculate the neutrino to muon conversion rate $\Gamma^{90\%}_{\nu\to\mu} = \frac{\mu^{90\%}}{V_{\rm eff}\cdot T_{\rm live}}$, where $V_{\rm eff}$ is the effective volume and $T_{\rm live}$ is the experimental live-time. Since no evidence of a statistically significant excess of events from the Sun direction was found after un-blinding, the upper limits on the muon flux at the Earth, for an energy threshold of 1 GeV, were calculated: $\phi_\mu^{90\%}(E_\mu \geq E_{\rm thr}) = \frac{\Gamma_A^{90\%}}{4\pi r_\odot^2}\int_{E_{\rm thr}}^{\infty} dE_\mu \frac{dN}{dE_\mu}$, where $r_\odot$ is the distance to the Sun, $\frac{dN}{dE_\mu}$ the muon energy spectrum at the detector and $\Gamma_A^{90\%}$ the annihilation rate in the Sun (the conversion from $\Gamma^{90\%}_{\nu\to\mu}$ to $\Gamma_A^{90\%}$ and then to $\phi_\mu^{90\%}$ is done using the code described in Ref. 22).

The AMANDA limits on the muon flux from neutralino annihilation in

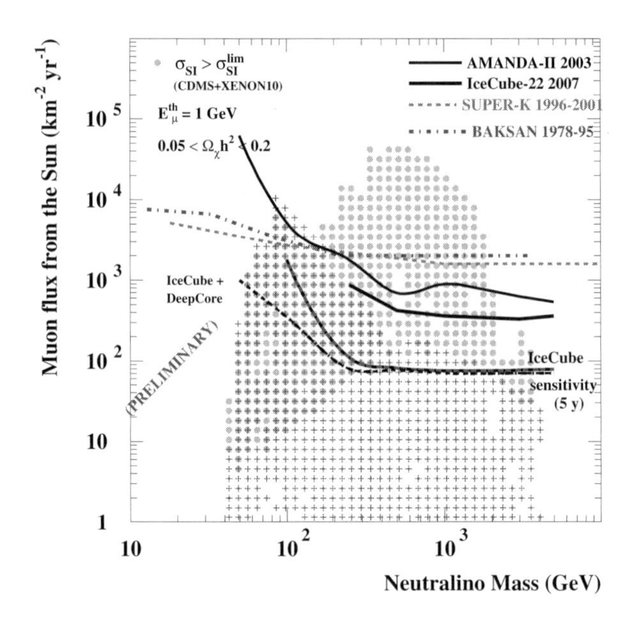

Fig. 1. AMANDA (systematics not included) and IceCube upper limits at 90% CL on the muon flux from neutralino annihilation ("hard" channel) in the centre of the Sun, as function of the neutralino mass, compared with other searches in the phase space of relevant neutralino models (+) and those disfavoured by CDMS[28] and XENON10[29] (•).

the "hard" channels in the Sun, as a function on the neutralino mass, compared to other indirect limits[23–25] and to theoretical predictions from MSSM models are shown in Fig. 1. Systematic errors are not yet included in this analysis. Ongoing analyses[26,27] with more statistics will include studies on systematic errors.

3.2. *To catch a neutralino from the Sun with IceCube-22*

The data set used in this analysis corresponds to 104.26 days of livetime during 2007 and is focused on five neutralino masses ($m_\chi \geq 250$ GeV/c^2). The analysis is performed bearing in mind the "blindness" criterion as described in Sec. 3.1. The data were processed through four filtering levels in order to reduce the downward-going muon background. These levels consist of a series of cuts on the zenith angle and on some variables that give a certain rank to the reconstructed track, in order to get a good rejection of the mis-reconstructed background. The final level was a cut on the product of two multivariate classifiers (Support Vector Machines) outputs,[33] trained

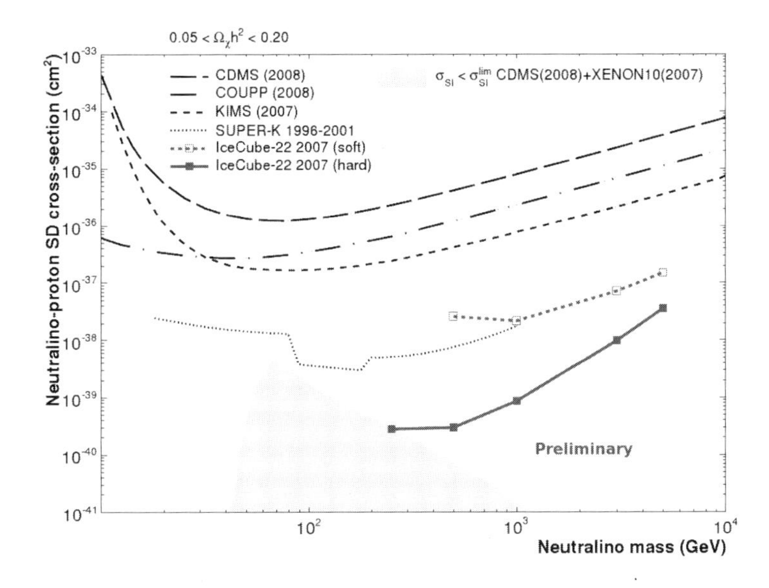

Fig. 2. IceCube-22 upper limits ("soft" and "hard" channels at 90% CL) on σ^{SD} adjusted for systematic effects, as a function of neutralino mass. The shaded area represents MSSM models not disfavoured by direct searches[29,30] based on σ^{SI} . The limits from other experiments[23,30–32] are shown for comparison.

with 6 parameters each. To evaluate the signal content in the final event sample, hypothesis testing was performed in a way similar to that described in Sec. 3.1. No excess of neutrino events from the Sun was observed, thus the 90% confidence upper limits on the muon flux at the Earth were calculated. If we assume equilibrium between the capture and the annihilation rate in the Sun, then the annihilation rate is directly proportional to the cross-section. Assuming also a local WIMP density of 0.3 GeV/cm^3 and a Maxwellian velocity distribution with a dispersion of 270 km/s, the limits on the spin dependent cross-section σ^{SD} were calculated[34] (by assuming $\sigma^{SI} = 0$ and neglecting planetary effects). To estimate the systematic errors on the calculation of the signal effective volume, some simulation studies were performed. They showed a dependence on the WIMP model, which ranged from $\pm 19\%$ to $\pm 24\%$. Figure 1 shows the limits on $\phi_\mu^{90\%}$ as function of m_χ, compared to the AMANDA results and to other indirect experiments,[23–25] and to theoretical predictions from MSSM. Figure 2 shows the IceCube-22 limits on σ^{SD} with the bound from other experiments[23,30–32] along with theoretical predictions from the MSSM.

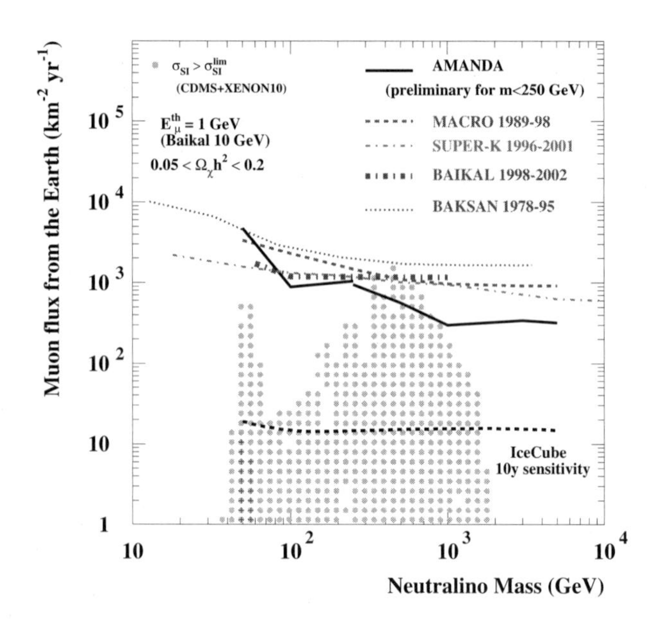

Fig. 3. AMANDA and IceCube upper limits (90% CL) on the muon flux from neutralino annihilation ("hard" channel) in the centre of the Earth, as function of the neutralino mass, compared with other searches in the phase space of relevant neutralino models (+) and those disfavoured by CDMS[28] and XENON10[29] (•).

3.3. *To catch a neutralino from the Earth with AMANDA*

The data collected during the 2001–2003 period correspond to 688.0 days of live-time of the detector. Besides the standard multiplicity trigger, this search[35] makes use of the low energy string trigger, and focuses on low mass WIMP models ($m_\chi \leq 250$ GeV$/c^2$) in order to achieve an improvement with respect to a previous analysis with AMANDA-B10.[36] The "blind" procedure for this search consists in developing the analysis on only 20% of the data, and then deriving the results using the remaining 80% data sample. Because of an upgrade of the data acquisition system (DAQ) and change in the string trigger settings between 2001 and 2002, the optimisation was performed separately for 2001 and for the two combined years 2002–2003. The sample, which is dominated by mis-reconstructed downward-going atmospheric muons, was reduced by a method which consisted of a series of sequential cuts on the variables that form the selection criteria of the analysis, thus maximising $\varepsilon_{\rm S} \times (1 - \varepsilon_{\rm B})$ (with ε being the signal and background efficiencies). At the final stage the entire data (dominated by the atmospheric neutrinos) of the three years were combined to apply the final cut selection on the zenith angle. Without a significant excess of vertical tracks being observed, the upper limits on the muon flux from the centre of the Earth were calculated for each of the neutralino masses and annihilation channels. The systemaic error studies on the calculation of the signal effective volume, through some Monte Carlo simulations, showed a WIMP model dependence which ranges from 20% to 34%. Figure 3 shows the AMANDA muon flux limits ("hard" channel) compared with other indirect searches[23–25] and with theoretical predictions from MSSM models.

4. IceCube Prospects

4.1. *Expected sensitivity for IceCube-80*

The sensitivity of the completed IceCube detector (IceCube-80) for neutrinos from WIMP annihilations in the Sun and in the Earth has also been studied by Monte Carlo. Cuts on the quality of the reconstruction have been applied. Using an effective volume of 1 km^3 and a live-time of 10 years for the Earth and of 5 years for the Sun, the derived sensitivity on the muon flux from the Sun and the Earth (Fig. 1 and Fig. 3) have been estimated. The role of IceCube in the searches for supersymmetric dark matter is expected to be complementary to that of the direct detection experiments. As a matter of fact, models with significant spin-dependent neutralino-proton couplings may be very important for neutralino capture in the Sun,[37] and IceCube should play a notable role putting some relevant constrains. Also, it should be stressed that direct and indirect searches probe different epochs of the history of the solar system and different parts of the neutralino velocity distribution.

4.2. *IceCube Deep Core*

To improve the sensitivity for low energy neutrinos, such as those originating from WIMPs, the so called IceCube Deep Core (IC-DC) project was approved and its first string was successfuly deployed during the season 2008-09. The design of the IC-DC consists of six dense strings centered around one of the central IceCube strings, with several advantages compared with AMANDA: 50 out of the 60 DOMs on a IC-DC string will be installed in the deep clear ice between 2107 – 2450 m, below the existing dust-layer, thus improving the reconstruction efficiency and angular resolution due to the longer scattering length of the Cherenkov light. The top 36 DOMs and three layers of the nearest IceCube strings will form the so called "Veto Volume", required to reject the bulk of the downward-going muon background. The IC-DC will give the possibility to observe neutrinos from above the horizon thus permitting the increase of the exposure time for neutrinos from dark matter annihilations up to the entire year.[38] The expected sensitivity on the muon flux, for 5 years of live-time, from WIMPs annihilating in the Sun is showed in Fig. 1.

5. Summary and Outlook

Figure 1 and Fig. 3 show the AMANDA and IceCube upper limits (rescaled to a common muon threshold of 1 GeV) on the muon flux from neutralino annihilation in the Sun and in the Earth respectively, together with the results from other indirect searches. The cosmologically relevant MSSM models are also shown, allowed (+) and disfavoured (•) by the direct search results from CDMS[28] and XENON10.[29] Compared to the previously published AMANDA results from searches for neutralinos in the Earth,[36] the analysis with 2001–2003 data benefits from the larger detector volume and the addition of the string trigger with its lower energy threshold. Good results were achieved also for the solar neutralino analysis of 2003 AMANDA data, thanks to the increased detector exposure, improved reconstruction techniques and the string trigger. The neutralino searches with the AMANDA detector will be continued with a larger set of data from 2001–2006. The solar neutralino analysis with IceCube-22 has put the most stringent limits so far on neutralino annihilation in the Sun and on the spin-dependent WIMP-proton cross section for neutralino masses above 250 GeV (see Fig. 2). IceCube will continue to take data during its construction and increase its sensitivity for solar WIMPs by several factors in the forthcoming years. Furthemore the IceCube Deep Core will improve the performance at low energy, and make it possible to explore the mass region down to 50 GeV/c^2.

References

1. F. Zwicky, *Helv. Phys. Acta.* **6**, p. 110 (1933).
2. W. J. G. de Blok *et al.*, *astro-ph/0103102* (2001).
3. K. G. Begeman, A. H. Broeils and R. H. Sander, *Monthly Notices of the Royal Astronomical Society* **249**, p. 523 (1991).
4. L. Wang *et al.*, *Astrophysical Journal* **530**, p. 17 (2000).
5. D. N. Spergel *et al.*, *Astrophys. J. Suppl. Ser.* **170**, p. 377 (2007).
6. G. Jungman, M. Kamionkowski and K. Griest, *Phys. Rept.* **267**, p. 195 (1996).
7. G. Bertone, D. Hooper and J. Silk, *hep-ph/0404175* (2004).
8. J. Ellis *et al.*, *Nucl. Phys. B* **238**, p. 453 (1984).

9. L. Bergström, *Rep. Prog. Phys.* **63**, p. 793 (2000).
10. J. Edsjö, Aspects of neutrino detection of neutralino dark matter, PhD thesis, Uppsala University1997.
11. E. Andres *et al.*, *Nature* **410**, p. 441 (2001).
12. J. Ahrens *et al.*, *Nucl. Instrum. Meth. A* **524**, p. 169 (2004).
13. A. Achterberg *et al.*, *Astropart. Phys.* **26**, p. 155 (2006).
14. J. Ahrens *et al.*, *IceCube Preliminary Design Document*, tech. rep., The University of Wisconsin, Madison, Wisconsin (2001), `http://www.icecube.wisc.edu/science/publications/pdd`.
15. M. Blennow, J. Edsjö and T. Ohlsson, *hep-ph/0709.3898* (2007).
16. L. Bergström, J. Edsjö and P. Gondolo, *Phys. Rev. D* **58**, p. 103519 (1998).
17. P. Lipari, *Astropart. Phys.* **1**, p. 195 (1993).
18. A. Gazizov and M. Kowalski, *astro-ph/0406439* (2004).
19. D. Heck *et al.*, *A Monte Carlo code to simulate extensive air showers*, tech. rep., Forschungszentrum Karlsruhe Report FZKA, 6019 (1998).
20. T. Burgess, A search for solar neutralino dark matter with the AMANDA-II neutrino telescope, PhD thesis, Stockholm University2008.
21. G. J. Feldman and R. D. Cousins, *Phys. Rev. D* **57**, p. 3873 (1998).
22. P. Gondolo *et al.*, *Journ. of Cosm. and Astropart. Phys.* **0407**, p. 008 (2004).
23. S. Desai *et al.*, *Phys. Rev. D* **70**, p. 083523 (2004).
24. M. Ambrosio *et al.*, *Phys. Rev. D* **60**, p. 082002 (1999).
25. M. M. Boliev *et al.*, Baksan neutralino search, in *Dark Matter in Astro- and Particle Physics, Dark '96*, eds. H. V. Klapdor-Kleingrothaus and Y. Ramachers (World Scientific, 1997).
26. D. Hubert, Search with the AMANDA detector for neutralino dark matter in the centre of the sun, PhD thesis, Vrije Universiteit BrusselMay 2009.
27. A. Rizzo, Results of the search of neutralino dark matter with 6 years of data taken with the AMANDA-II neutrino telescope, PhD thesis, Vrije Universiteit Brussel2009. in preparation.
28. D. S. Akerib *et al.*, *Phys. Rev. Lett.* **96**, p. 011302 (2006).
29. J. Angle *et al.*, *Phys. Rev. Lett.* **100**, p. 021303 (2008).
30. Z. Ahmed *et al.*, *Phys. Rev. Lett.* **102**, p. 011301 (2009).
31. E. Behnke *et al.*, *Science* **319**, p. 933 (2008).
32. H. S. Lee *et al.*, *Phys. Rev. Lett.* **99**, p. 091301 (2007).
33. R. Abbasi *et al.*, *astro-ph/0902.2460* (2009), submitted to PRL.
34. G. Wikström and J. Edsjö, *astro-ph/0903.2986v1* (2009).
35. A. Davour, Search for low mass WIMPs with the AMANDA neutrino telescope, PhD thesis, Uppsala University2007.
36. A. Achterberg *et al.*, *Astropart. Phys.* **26**, p. 129 (2006).
37. F. Halzen and D. Hooper, *Phys. Rev. D* **73**, p. 123507 (2006).
38. E. Resconi (for the IceCube Collaboration), *astro-ph/0807.3891v1* (2008).

INDIRECT SEARCH FOR DARK MATTER WITH THE ANTARES NEUTRINO TELESCOPE

H. MOTZ for the ANTARES Collaboration

Erlangen Centre for Astroparticle Physics, Erwin Rommel Strasse 1,
Erlangen, 91058, Germany
E-mail: Holger.Motz@physik.uni-erlangen.de

ANTARES (Astronomy with a Neutrino Telescope and Abyss environmental RESearch) is currently the largest neutrino detector on the Northern Hemisphere. The detector consists of twelve lines, carrying 885 ten-inch photomultipliers in total, placed at a depth of about 2480 meters in the Mediterranean Sea near Toulon, France. The PMTs detect Cherenkov light emitted by muons from neutrino charged current interactions in the surrounding seawater and the rock below. The neutrinos momentum is transferred to the muons allowing for reconstruction of the neutrinos direction. The goals of ANTARES are among others the search for astrophysical neutrino point sources and for neutrinos produced in self-annihilation of dark matter particles. A likely source of the latter type of neutrino emission would be the Sun, where dark matter particles from the galactic halo are expected to accumulate. ANTARES is taking data with its full twelve line configuration since May 2008, and has been before in a five and ten line setup for more than a year. First results on the search for dark matter annihilation in the Sun, and their interpretation in the framework of mSugra are presented, as well as sensitivity studies on Dark Matter search with the full ANTARES detector and the future large undersea KM3NeT neutrino telescope.

Keywords: Dark matter; neutrino astronomy; supersymmetry.

1. The ANTARES Detector

ANTARES is a deep-sea water Cherenkov detector located in the Mediterranean Sea near Toulon, Southern France at a depth of 2475 metres. It consists of twelve strings (also called 'lines') of 480 metres length fixed to the seabed by anchors and straightened by buoys. Each string holds 25 triplets ('storeys') of 10-inch photomultipliers (PMT) in pressure resistent glass spheres ('Optical Modules'), vertically spaced by 14.5 m. An additional Instumentation Line (IL) is equipped with environment monitoring devices. In addition, the IL as well as the topmost part of Line 12 contain the hydrophones of the AMADEUS[1] project instead of PMTs. The 885 PMTs are used to detect Cherenkov light from neutrino induced muons and cascades. The detection principle is to reconstruct the direction and position of the muon tracks and hadronic cascades from timing and position of their Cherenkov light emissions, which requires a precise knowledge of the position and orientation of each PMT. As flexible structures the strings move with the sea current and the attached storeys are shifted, tilted and rotated with respect to the straight line case without torque. Therefore ANTARES features a positioning system[2] consisting of a tiltmeter/compass board on each storey to measure its orientation and five hydrophones per line on selected storeys to triangulate their position from the propagation times of acoustic signals emitted by tranceivers at the string anchor stations. Light flashes from four LED-beacons installed on each line are used for timing calibration of the PMTs. Data transmission and power supply is provided to the lines via a 40 km long electro-optical cable linking the detector to the shore station located in La Seyne-sur-Mer (France). This main cable is terminated off-shore by a Junction Box on which each line is individually connected by an interlink cable with a submarine vehicle. The construction of ANTARES was performed in several steps. The first detector line (Line 1) was deployed on February 14th 2006, and connected to the junction box on March 2nd. The second line was connected on September 21st 2006. Due to ambiguities in the reconstructed geometry of an event measured with one and two lines respectively, direction reconstruction of muons and neutrinos was restricted to the Zenith angle with these setups.[3] From January 29th 2007 five detector lines were operational, providing full functionality. The data taken in this period has been analysed and results are presented in this paper. The connection of five more lines and the IL was performed on December 7th 2007 and the last two lines were added May 30th 2008, marking the completion of the ANTARES detector.

2. Performance of the ANTARES Detector

The detector energy threshold for reconstructed muons is approximately 10 GeV. The effective area increases sharply with energy. Neutrinos from annihilation of SUSY dark matter are generally of low energy, so the sensitivity of ANTARES is determined by its neutrino effective area in the regime of 10 GeV to 400 GeV, which is depicted in figure 1.

Calculation of the effective area takes into account a background rate of 60 kHz for each PMT. The light causing optical background rates is produced by bioluminescent organisms and Cherenkov photons due to ^{40}K

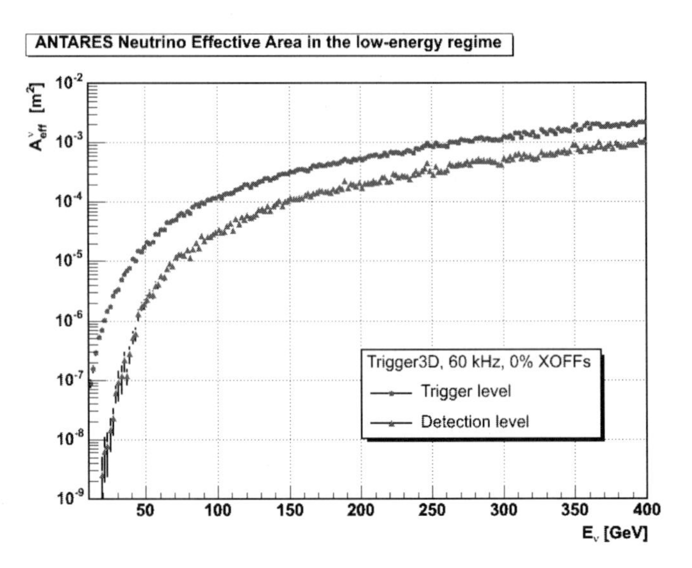

Fig. 1. ANTARES effective area up to 400 GeV neutrino energy, at trigger level (blue) and for fully reconstructed events with quality cuts applied (red).

beta decays. This background poses a challenge for reconstructing the muon tracks, but the coincident Cherenkov photons caused by β-decay of ^{40}K are used to check the efficiency and time calibration of the Optical Modules. The distribution of time differences of background hits on the Optical Modules of one storey displayed in figure 2 features a gaussian peak of 4 ns width of genuine coincident photons above a pedestal caused by random coincidences from ^{40}K and bioluminescence. The rate for genuine coincident photons is predicted by Monte-Carlo to be 18 Hz, with a 4 Hz systematic uncertainty, to which the average value of the measured rates of 14 Hz, shown in figure 3, is compared. With a perfect timing calibration, the peak from coincidences should be at zero, making the deviation from zero a tool to monitor the quality of the timing calibration. The RMS of the distribution of peak offsets in figure 3 is found to be 0.65 ns, indicating that the time calibration is within the required precision.

Unlike the background rate from ^{40}K-decay, which is constant, the background contribution from bioluminescence shows variation on short time scales, i.e. featuring bursts of a few seconds, as well as on long timescales as presented in figure 4.

According to simulation studies, the expected angular resolution of the ANTARES detector is better than 0.3 degree for neutrino events with an

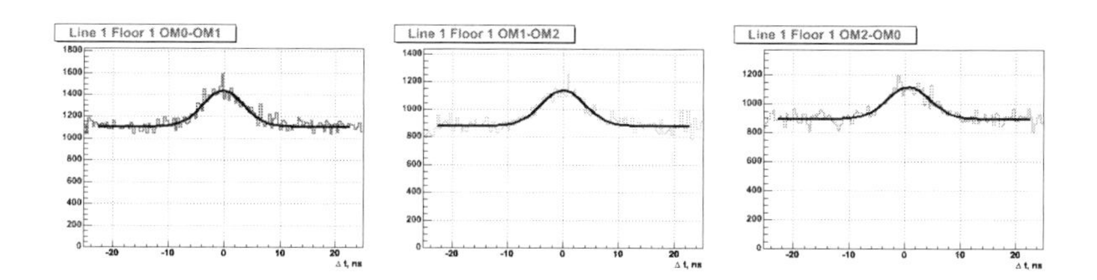

Fig. 2. Examples of time difference distributions of background hits on Optical Modules of Storey 1 of Line 1.

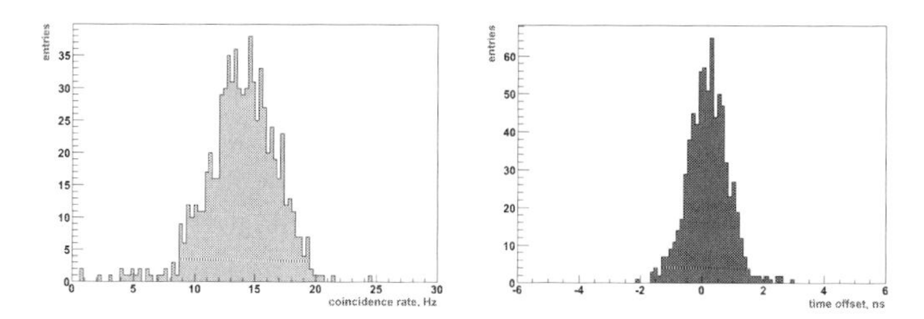

Fig. 3. Distribution of the coincidence rate measured with pairs of Optical Modules (left, mean value 14 Hz, RMS 2.9 Hz)) and distribution of peak offsets (right, RMS 0.65 ns).

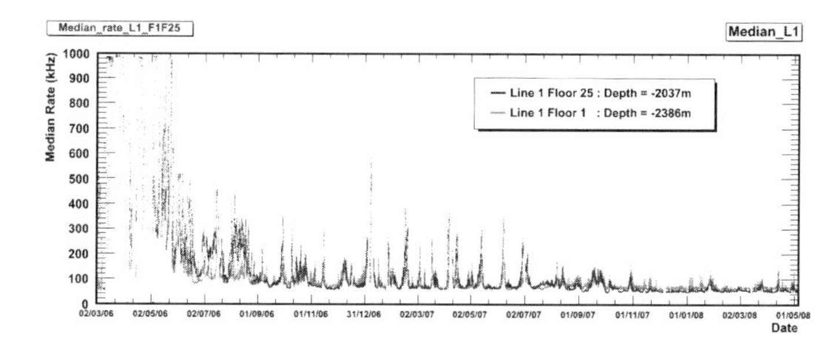

Fig. 4. Mean background rate measured on the lowest and highest storey of Line 1 from March 2006 till May 2008.

energy above a few TeV. At lower energy the kinematic angle between the neutrino and the induced muon dominates.

3. KM3NeT

The KM3NeT project[4] aims at building a neutrino telescope of cubic-kilometre scale in the Mediterranean Sea. During the Design Study project phase of the project, several detector designs are explored to an extent which allows to predict their capabilities concerning dark matter indirect detection. One possible detector configuration consists of 225 lines, arranged in a cubic layout, carrying 36 optical modules each. An optical module consists of 21 three-inch PMTs in a glass sphere which also houses read-out electronics and calibration devices. The sensitivity calculations presented here are based on this detector layout.

4. Indirect Dark Matter Search with Neutrino Telescopes

Indirect search for Dark Matter is one of the aims of ANTARES and KM3NeT. Supersymmetry features a Dark Matter candidate in the form of the neutralino χ which could be the Lightest Supersymmetric Particle (LSP) and thus stable due to conservation of R-Parity. As the lightest mass-Eigenstate of the superpartners of the neutral gauge- and Higgs-bosons, the neutralino would only interact through gravity and the weak force, and, with a mass in the range of GeV to TeV, fulfill the requirements to be the cold dark matter (CDM) predicted by the cosmological standard model. The primary method for indirect Dark Matter search with neutrino telescopes is the detection of neutrinos created in R-Parity conserving self annihilation of neutralinos into standard model particles. If CDM is composed of neutralinos, an excess in high energy neutrinos from the core of massive stellar objects, like the Sun, is expected.

As outlined in figure 5 neutralinos from the galactic dark matter halo accumulate in the Sun by elastic scattering. The resulting high neutralino den-

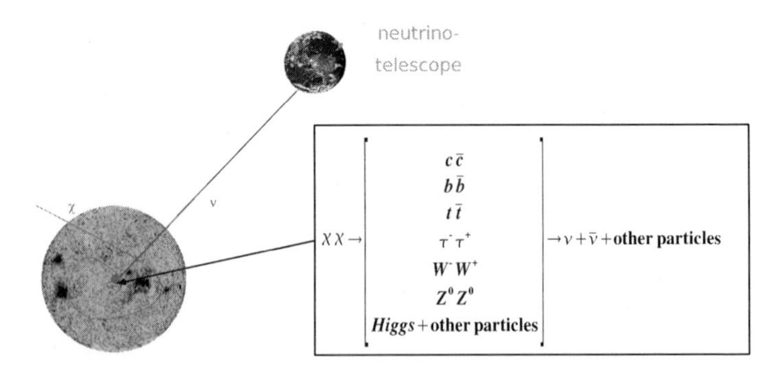

Fig. 5. Principle method of indirect dark matter search with neutrino telescopes.

sity leads to an increased self-annihilation rate in equilibrium with the capture rate.[5] From the decay of the primary annihilation products high energy neutrinos are produced which leave the Sun and propagate to Earth, where they can be detected. Another possible source is the galactic centre,[6] but the expected flux is highly dependent of the shape of the galactic dark matter halo, while for the position of the Solar System most proposed halo c onfigurations feature a local Dark Matter density of approximately 0.3 GeV/c^2 per cm^3.[7] While this leads to less astrophysics-dependent predictions for neutralino capture in the Sun, the capture rate inside the Earth is highly dependent on the velocity distribution of neutralinos already bound to the Solar System.[8]

5. Sensitivity to Neutrinos from Neutralino Annihilations

The expected neutrino flux from neutralino annihilation in the Sun depends in the mSugra scenario on the four parameters m_0, $m_{1/2}$, A_0, $\tan(\beta)$ and $\mathrm{sign}(\mu)$. From those parameters defined at the GUT energy-scale, the properties of the supersymmetric particle spectrum including the neutralino at the electroweak energy scale are calculated using renormalization group equations (RGE). The expected neutrino flux was calculated for approximately four million parameter sets with a modified version of DarkSUSY 4.1[9] using a random walk method to scan the regions of the parameter space[10] allowed by theoretical and experimental constraints and to highlight models predicting a neutralino relic density in close agreement with the one derived from the WMAP measurement[11] of the CMB. The RGE-code ISASUGRA[12] was used for the calculation of the supersymmetric particle spectrum and the halo model of Navarro, Frenk and White[13] was assumed. The flux calculation takes into account the effect of absorption and oscillations of neutrinos inside the Sun, numerically calculated by tracking the neutrinos using path-ordered propagators,[14] as well as during their propagation through vacuum from Sun to Earth. The mixing angles and mass-differences of the LMA oscillation scenario[15] were applied. Figure 6 shows the neutrino and antineutrino flux on the Earth integrated above a threshold energy of 10 GeV as a function of the neutralino mass for a wide class of mSugra models.

Based on the effective area and the neutrino flux an estimated detection rate for ANTARES, shown in figure 7, is calculated. C onsidering the irreducible background coming from atmospheric neutrinos and another 10% of that flux due to misreconstructed atmospheric muons, a sensitivity is derived according to the unified approach method of Feldman and Cousins.[16]

Thus, assuming that only the average background rate will be measured, an achievable upper limit for three years of data taking with the complete ANTARES detector can be derived and compared to the detection rate predicted for each individual mSugra model from the scan of the parameter space as shown in figure 7. Since the angular resolution of ANTARES at

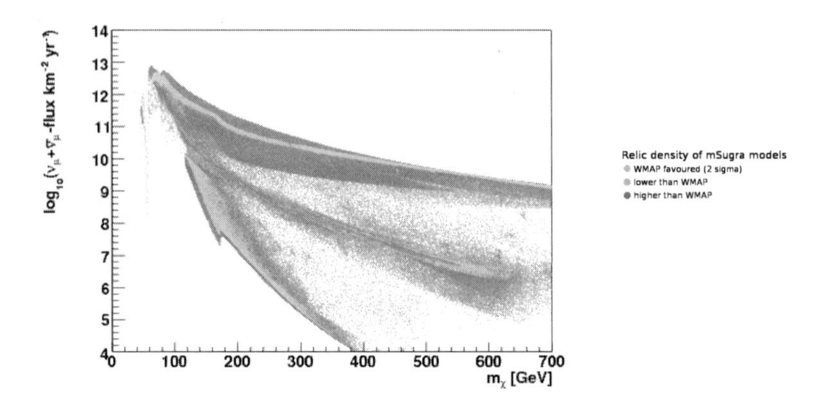

Fig. 6. Integrated neutrino and antineutrino flux above 10 GeV neutrino energy per km^2 and year versus neutralino mass m_χ. The relic density of green coloured models is within the 2 σ region of the WMAP results, blue represents lower relic density, red higher.

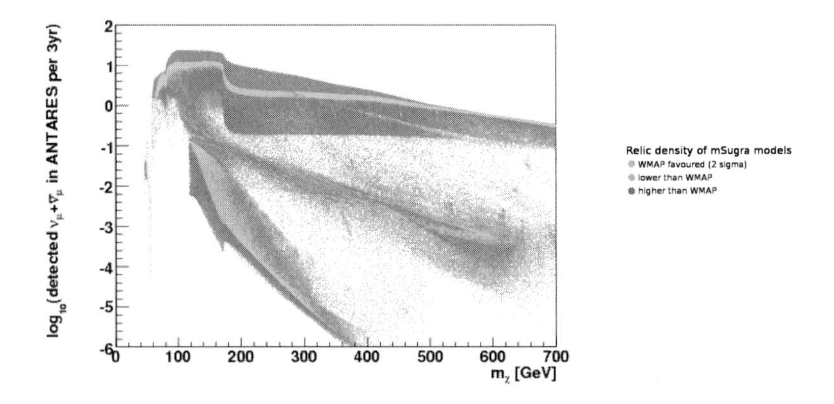

Fig. 7. Detection rate resulting from the flux shown in figure 6 in ANTARES for three years of data taking. The relic density of green coloured models is within the 2 σ region of the WMAP results, blue represents lower relic density, red higher.

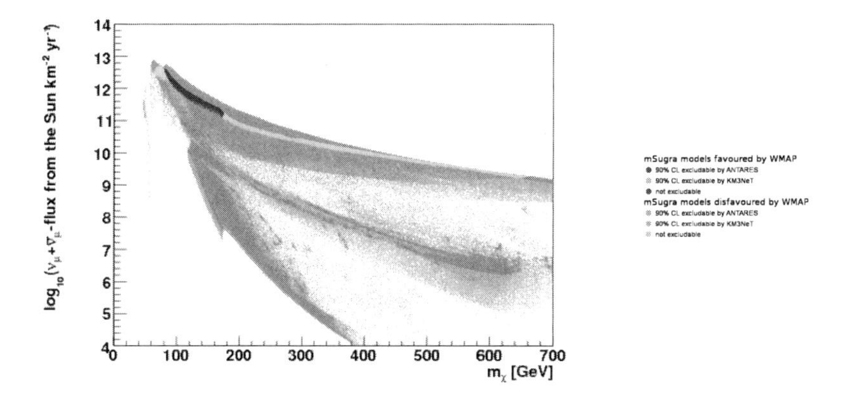

Fig. 8. Sensitivity of ANTARES and KM3NeT to neutrinos from annihilation of mSugra dark matter on a plot showing the $\nu_\mu + \bar{\nu_\mu}$-flux integrated above 10 GeV E_ν (top right) versus neutralino mass. Blue indicates models within the sensitivity of ANTARES, green models KM3NeT is sensitive to and red those outside the sensitivity of both experiments. Brightly coloured are models with a relic density predicted within the 2 σ WMAP-region, shaded ones those outside.

low energies is dominated by the kinematic angle between the neutrino and the muon generated from it, a search cone of three degree radius around the Sun was assumed, corresponding to an averaged expected background of 6.95 neutrinos in three years.

The sensitivity of ANTARES is sufficient to put constraints on parts of the so-called Focus-Point region[17] of the mSugra parameter space, where the neutralino is mainly Higgsino, for which a high neutrino flux and hard neutrino energy spectrum are expected. The analogous calculations for the proposed configuration of the KM3NeT detector indicate that such a detector can be sensitive to most of this region, which corresponds to the uppermost branch of the neutrino flux indicated in figure 8. The sensitivity of ANTARES and KM3NeT is also shown in figure 9 as a function of the mSugra parameters directly.

To compare this model-dependent sensitivity with limits from direct detection experiments, the WIMP-nucleon scattering cross-sections were calculated and compared to recent limits obtained by the CDMS[18] and XENON10[19] experiments on spin-independent scattering as well as the sensitivity prediction for EdelweissII, shown in figure 10. Spin-dependent scattering experiments do not yet reach the sensitivity required to probe the mSugra parameter space.

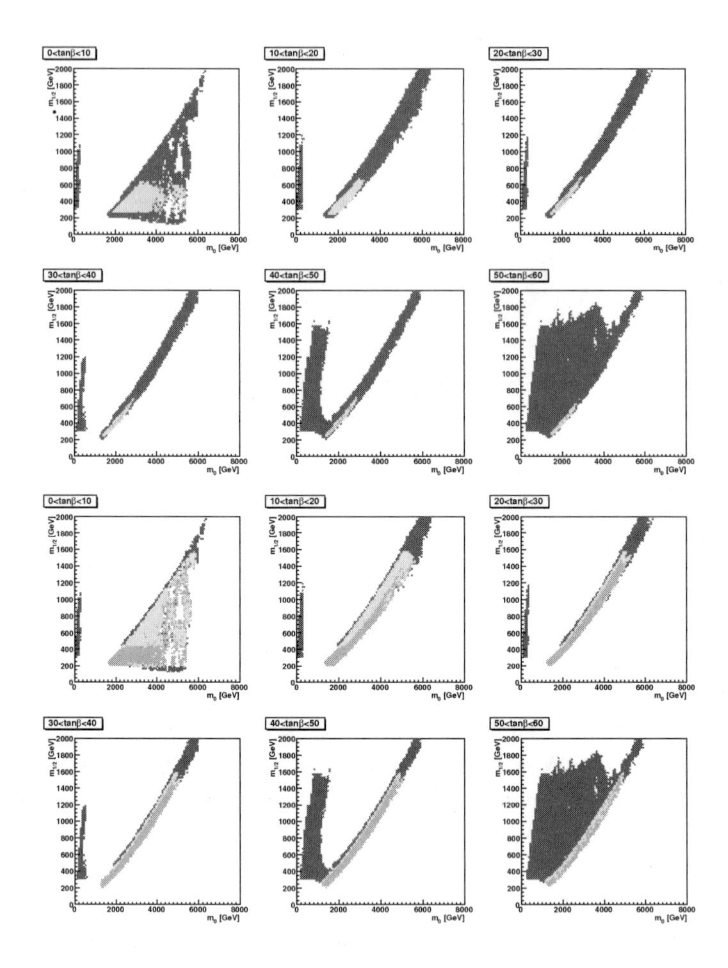

Fig. 9. Sensitivity of ANTARES (top) and KM3NeT (bottom) in the mSugra parameter space for different intervalls of $\tan(\beta)$. A_0 varies between -3 m_0 and 3 m_0. Green indicates that the considered experiment has a sensitivity allowing for exclusion of all models of the given $\tan(\beta)$ intervall; yellow regions: only some models can be excluded; red regions: none.

6. Muon and Neutrino Distributions from ANTARES 5-Line Data

Atmospheric muons, due to the approximately 10^5 times higher rate in ANTARES, compared to atmospheric neutrinos, represent an opportunity to check the systematics of the detector. From data taken in 2007 with five detector lines in operation, their measured distribution in Azimuth and Zenith angle was compared to the Monte Carlo predictions. As figure 11

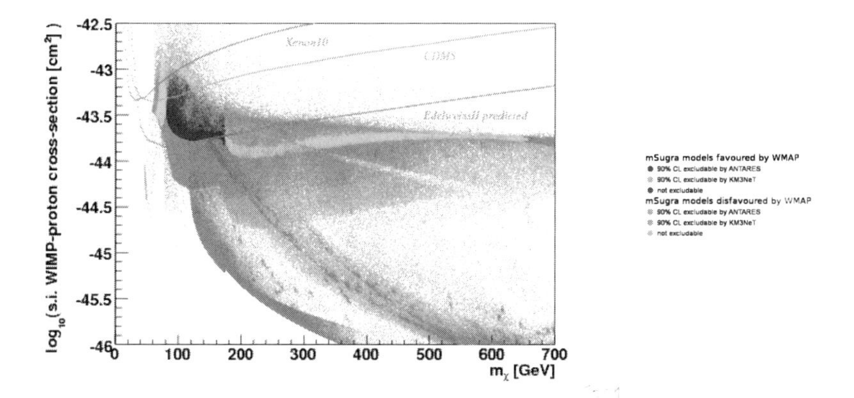

Fig. 10. Sensitivity of ANTARES and KM3NeT to neutrinos from annihilation of mSugra dark matter on a plot showing the spin independent WIMP-proton cross-section versus neutralino with limits set by the CDMS and XENON experiments and the sensitivity prediction for EDELWEISS II. Blue indicates models within the sensitivity of ANTARES, green models KM3NeT is senistive to and red those outside the sensitivity of both experiments. Brightly coloured are models within the 2 σ WMAP-region, shaded ones those outside.

shows, the data is within the uncertainties from modeling the cosmic ray flux and the hadronic showers from which the muons originate, as well as within those from detector properties, e.g. the exact angular acceptance and quantum efficiency of the PMTs. The events which are reconstructed as upward-going represent the neutrino candidates. From the 5-line period 185 such events were reconstructed in a data sample of 164 days effective live time and compared to the expected flux of cosmic ray induced atmospheric neutrinos. As shown in figure 12 the zenith and azimuth distributions are compatible with the prediction of the background simulation.

7. Limit on Muon and Neutrino Flux from Dark Matter Annihilation in the Sun

The neutrino events measured during the 5-line period with an effective lifetime of 68.4 days, reduced from the full 164 days due to trigger dead-time and the condition that the Sun has to be below the horizon, allow to look for a possible excess in the neutrino and muon flux in the direction of the Sun. Figure 13 shows the number of observed events inside a search cone centered towards the direction of the Sun, compared to the exected background as a function of the half cone opening angle. The background expected from Monte-Carlo simulation is in good agreement with the estimation obtained

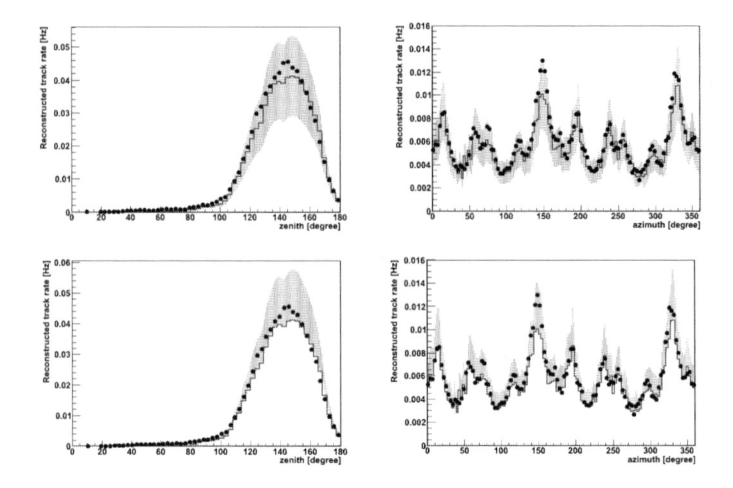

Fig. 11. Zenith (left) and Azimuth (right) distributions of atmospheric muons, measured (black dots) and predicted by Monte Carlo (lines) with uncertainty band due to detector properties (blue/top) and theoretical modeling (red/bottom).

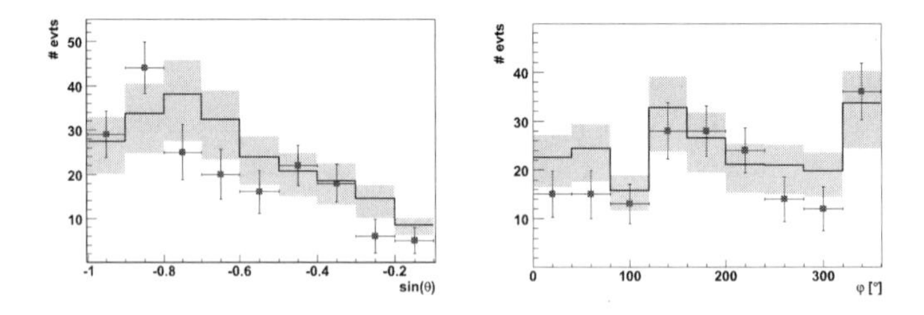

Fig. 12. Zenith (left) and Azimuth (right) distributions of neutrino candidates, measured (red dots) with statistical error and predicted by Monte Carlo (black line) and uncertainty band due to detector properties (grey).

on the data sample by randomizing the direction of the measured upward-going events. The limit on observed events given by the unified-approach-method of Feldman and Cousins depending on the search cone size is also shown in figure 13.

From Monte-Carlo studies the optimal search cone was estimated for a hard (W-boson) and soft (b-quark) annihilation spectrum and depending on neutralino mass, before analysing the data. With the optimal search cone size, limits on the neutrino flux above an energy threshold energy of 10 GeV for the hard and soft channel have been calculated, as displayed in figure 14. To compare with the limits published by other indirect detection

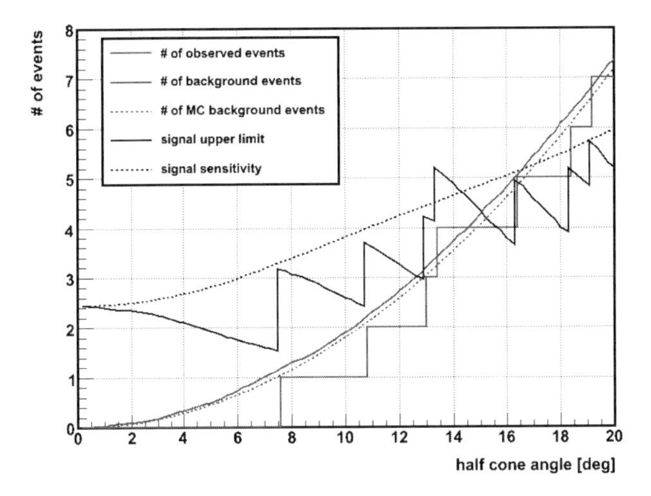

Fig. 13. Number of neutrino events reconstructed with the 5-line ANTARES detector during an effective lifetime of 68.4 days as a function of the angular distance to the direction of the Sun, and corresponding limit.

experiments,[20] the corresponding muon flux above a muon energy threshold of 1 GeV was calculated. The limits on this quantity are displayed in figure 15.

8. Conclusion

The ANTARES neutrino telescope, which was completed in May 2008, measured sucessfully distributions of atmospheric muons and neutrinos in accordance with Monte Carlo predictions, during the building stage in 2007 when five detector-lines were operational. From this data a limit on a possible neutrino flux excess from Dark Matter annihilations in the Sun was derived. Sensitivity calculations for the full twelve-line ANTARES detector show that within the mSugra theory some part of the Focus-Point region will be accessible in three years of taking data with ANTARES, while the future KM3NeT detector will be sensitive to most of the Focus-Point region.

Acknowledgments

We acknowledge support of this work by the BMBF (Project Number 05CN5WE1/7).

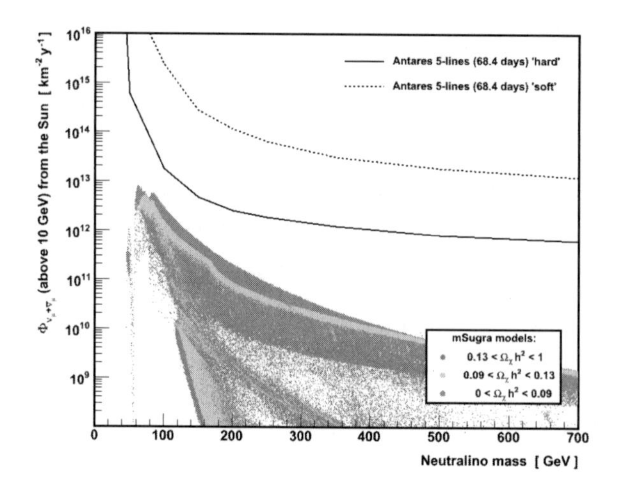

Fig. 14. Limit on neutrino flux from the Sun above a neutrino energy of 10 GeV from ANTARES 5-Line period (effective lifetime of 68.4 days) in comparison to the expected flux from mSugra Dark Matter annihilations. The relic density of green coloured models is within the 2 σ region of the WMAP results, blue represents lower relic density, red higher.

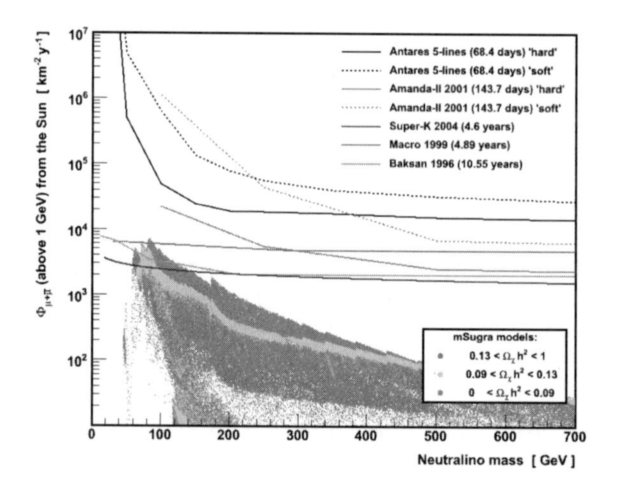

Fig. 15. Limit on muon flux from the Sun above a muon energy of 1 GeV from ANTARES 5-Line period (effective lifetime of 68.4 days) and limits set by other experiments in comparison to the expected flux from mSugra Dark Matter annihilations. The relic density of green coloured models is within the 2 σ region of the WMAP results, blue represents lower relic density, red higher.

References

1. K. Graf *et al.*, *J. Phys. Conf. Ser.* **60**, 296 (2007).
2. M. Ardid, Positioning system of the ANTARES neutrino telescope, Prepared for the VLVnT08 Workshop, April 21.-24. 2008 , Toulon, (2008).
3. M. Ageron *et al.*, Performance of the First ANTARES Detector Line, astro-ph/0812.2095, (2008).
4. P. Bagley *et al.*, KM3NeT - Conceptual Design for a Deep Sea Research Infrastructure Incorporating a Very Large Volume Neutrino Telescope in the Mediterranean Sea (2008), available at www.km3net.org.
5. A. Gould, *Astrophys. J.* **388**, 338(April 1992).
6. P. Gondolo and J. Silk, *Phys. Rev. Lett.* **83**, 1719(Aug 1999).
7. M. Kamionkowski and A. Kinkhabwala, *Phys. Rev.* **D57**, 3256 (1998).
8. L. Bergstrom, T. Damour, J. Edsjo, L. M. Krauss and P. Ullio, *JHEP* **08**, p. 010 (1999).
9. P. Gondolo *et al.*, *JCAP* **0407**, p. 008 (2004).
10. E. A. Baltz and P. Gondolo, *JHEP* **10**, p. 052 (2004).
11. D. N. Spergel *et al.*, *Astrophys. J. Suppl.* **148**, p. 175 (2003).
12. F. E. Paige, S. D. Protopescu, H. Baer and X. Tata, Isajet 7.69: A monte carlo event generator for p p, anti-p p, and e+ e- reactions, hep-ph/0312045, (2003).
13. J. F. Navarro, C. S. Frenk and S. D. M. White, *Astrophys. J.* **490**, 493 (1997).
14. T. Ohlsson and H. Snellman, *Eur. Phys. J.* **C20**, 507 (2001).
15. M. Maltoni, T. Schwetz, M. A. Tortola and J. W. F. Valle, *New J. Phys.* **6**, p. 122 (2004).
16. G. J. Feldman and R. D. Cousins, *Phys. Rev.* **D57**, 3873 (1998).
17. J. L. Feng, K. T. Matchev and T. Moroi, *Phys. Rev.* **D61**, p. 075005 (2000).
18. Z. Ahmed *et al.*, *Phys. Rev. Lett.* **102**, p. 011301 (2009).
19. J. Angle *et al.*, *Phys. Rev. Lett.* **100**, p. 021303 (2008).
20. S. Desai *et al.*, *Phys. Rev.* **D70**, p. 083523 (2004).

PART XIII

Direct Search for Cold Dark Matter II

HOW PRECISELY COULD WE IDENTIFY WIMPS MODEL–INDEPENDENTLY WITH DIRECT DARK MATTER DETECTION EXPERIMENTS

MANUEL DREES

Physikalisches Institut and Bethe Center of Theoretical Physics
Universität Bonn, D-53115 Bonn, Germany
E-mail: drees@th.physik.uni-bonn.de

CHUNG-LIN SHAN*

School of Physics and Astronomy, Seoul Nat'l Univ., Seoul 151-747, Republic of Korea
E-mail: cshan@hep1.snu.ac.kr

In this talk we present data analysis methods for reconstructing the mass and couplings of Weakly Interacting Massive Particles (WIMPs) by using directly future experimental data (i.e., measured recoil energies) from direct Dark Matter detection. These methods are independent of the model of Galactic halo as well as of WIMPs. The basic ideas of these methods and the feasibility and uncertainties of applying them to direct detection experiments with the next generation detectors will be discussed.

Keywords: Dark Matter; WIMP; direct detection; direct detection simulation.

1. Introduction

Weakly Interacting Massive Particles (WIMPs) χ arising in several extensions of the Standard Model of electroweak interactions with masses roughly between 10 GeV and a few TeV are one of the leading candidates for Dark Matter.[1,2] Currently, the most promising method to detect different WIMP candidates is the direct detection of the recoil energy deposited by elastic scattering of ambient WIMPs on the target nuclei.[3,4] The differential event rate for elastic WIMP–nucleus scattering is given by:[1]

$$\frac{dR}{dQ} = \left(\frac{\rho_0 \sigma_0}{2 m_\chi m_{\rm r,N}^2} \right) F^2(Q) \int_{v_{\rm min}}^{v_{\rm max}} \left[\frac{f_1(v)}{v} \right] dv \,. \tag{1}$$

*Speaker

Here R is the event rate, i.e., the number of events per unit time and unit mass of detector material, Q is the energy deposited in the detector, ρ_0 is the WIMP density near the Earth, σ_0 is the total cross section ignoring the form factor suppression, $F(Q)$ is the elastic nuclear form factor, $f_1(v)$ is the one–dimensional velocity distribution function of incident WIMPs, v is the absolute value of the WIMP velocity in the laboratory frame. The reduced mass $m_{\rm r,N}$ is defined by $m_{\rm r,N} \equiv m_\chi m_{\rm N}/(m_\chi + m_{\rm N})$, where m_χ is the WIMP mass and $m_{\rm N}$ that of the target nucleus. Finally, $v_{\rm min} = \alpha\sqrt{Q}$ with $\alpha \equiv \sqrt{m_{\rm N}/2m_{\rm r,N}^2}$ is the minimal incoming velocity of incident WIMPs that can deposit the energy Q in the detector, and $v_{\rm max}$ is related to the escape velocity from our Galaxy at the position of the Solar system.

The total WIMP–nucleus cross section σ_0 in Eq.(1) depends on the nature of the WIMP couplings on nucleons. Generally speaking, one has to distinguish spin–independent (SI) and spin–dependent (SD) couplings. Through e.g., squark and Higgs exchanges with quarks, Majorana WIMPs e.g., neutralinos in the supersymmetric models, can have a SI scalar interaction with nuclei:[1,2]

$$\sigma_0^{\rm SI} = A^2 \left(\frac{m_{\rm r,N}}{m_{\rm r,p}}\right)^2 \sigma_{\chi{\rm p}}^{\rm SI}\,, \qquad \sigma_{\chi{\rm p}}^{\rm SI} = \left(\frac{4}{\pi}\right) m_{\rm r,p}^2 |f_{\rm p}|^2\,, \tag{2}$$

where A is the atomic number of target nucleus, $m_{\rm r,p}$ is the reduced mass of WIMPs and protons, and $f_{\rm p}$ is the effective $\chi\chi{\rm pp}$ four–point coupling. Note here that the approximation $f_{\rm n} \simeq f_{\rm p}$ predicted in most theoretical models has been adopted and the tiny mass difference between a proton and a neutron has been neglected.

Meanwhile, through e.g., squark and Z boson exchanges with quarks, WIMPs can couple to the spin of the target nuclei. The total cross section for the spin coupling can be expressed as[1,2]

$$\sigma_0^{\rm SD} = \left(\frac{32}{\pi}\right) G_F^2\, m_{\rm r,N}^2 \left(\frac{J+1}{J}\right) \Big[\langle S_{\rm p}\rangle a_{\rm p} + \langle S_{\rm n}\rangle a_{\rm n}\Big]^2\,, \tag{3a}$$

and

$$\sigma_{\chi({\rm p,n})}^{\rm SD} = \left(\frac{24}{\pi}\right) G_F^2\, m_{\rm r,p}^2\, |a_{({\rm p,n})}|^2\,. \tag{3b}$$

Here G_F is the Fermi constant, J is the total spin of the target nucleus, $\langle S_{({\rm p,n})}\rangle$ are the expectation values of the proton and the neutron group spins, and $a_{({\rm p,n})}$ is the effective SD WIMP coupling to protons and neutrons.

2. Determining the WIMP Mass

It has been found that the one–dimensional velocity distribution function of incident WIMPs, $f_1(v)$, can be solved analytically from Eq.(1) directly[5] and, consequently, its generalized moments can be estimated by[6]

$$\begin{aligned}
&\langle v^n \rangle (v(Q_{\rm min}), v(Q_{\rm max})) \\
&= \int_{v(Q_{\rm min})}^{v(Q_{\rm max})} v^n f_1(v)\, dv \\
&= \alpha^n \left[\frac{2Q_{\rm min}^{(n+1)/2} r(Q_{\rm min})/F^2(Q_{\rm min}) + (n+1) I_n(Q_{\rm min}, Q_{\rm max})}{2Q_{\rm min}^{1/2} r(Q_{\rm min})/F^2(Q_{\rm min}) + I_0(Q_{\rm min}, Q_{\rm max})} \right] . \qquad (4)
\end{aligned}$$

Here $v(Q) = \alpha\sqrt{Q}$, $Q_{(\rm min,max)}$ are the minimal and maximal cut–off energies of the experimental data set, respectively, $r(Q_{\rm min}) \equiv (dR/dQ)_{Q=Q_{\rm min}}$ is an estimated value of the scattering spectrum at $Q = Q_{\rm min}$, and $I_n(Q_{\rm min}, Q_{\rm max})$ can be estimated through the sum:

$$I_n(Q_{\rm min}, Q_{\rm max}) = \sum_a \frac{Q_a^{(n-1)/2}}{F^2(Q_a)} , \qquad (5)$$

where the sum runs over all events in the data set between $Q_{\rm min}$ and $Q_{\rm max}$.

By requiring that the values of a given moment of $f_1(v)$ estimated by Eq.(4) from two detectors with different target nuclei, X and Y, agree, a general expression for determining m_χ appearing in the prefactor α^n on the right–hand side of Eq.(4) has been found as:[7]

$$m_\chi|_{\langle v^n \rangle} = \frac{\sqrt{m_X m_Y} - m_X(\mathcal{R}_{n,X}/\mathcal{R}_{n,Y})}{\mathcal{R}_{n,X}/\mathcal{R}_{n,Y} - \sqrt{m_X/m_Y}} , \qquad (6)$$

where

$$\mathcal{R}_{n,X} \equiv \left[\frac{2Q_{{\rm min},X}^{(n+1)/2} r_X(Q_{{\rm min},X})/F_X^2(Q_{{\rm min},X}) + (n+1) I_{n,X}}{2Q_{{\rm min},X}^{1/2} r_X(Q_{{\rm min},X})/F_X^2(Q_{{\rm min},X}) + I_{0,X}} \right]^{1/n} , \qquad (7)$$

and $\mathcal{R}_{n,Y}$ can be defined analogously. Here $n \neq 0$, $m_{(X,Y)}$ and $F_{(X,Y)}(Q)$ are the masses and the form factors of the nucleus X and Y, respectively. Note that, since the general moments of $f_1(v)$ estimated by Eq.(4) are independent of the WIMP–nucleus cross section σ_0, the estimator (6) of m_χ can be used either for SI or for SD scattering.

Additionally, since in most theoretical models the SI WIMP–nucleus cross section given in Eq.(2) dominates,[1,2] and on the right–hand side of Eq.(1) is in fact the minus–first moment of $f_1(v)$, which can be estimated by Eq.(4) with $n = -1$, one can find that[6]

$$\rho_0 |f_{\rm p}|^2 = \frac{\pi}{4\sqrt{2}} \left(\frac{m_\chi + m_{\rm N}}{\mathcal{E} A^2 \sqrt{m_{\rm N}}} \right) \left[\frac{2 Q_{\rm min}^{1/2} r(Q_{\rm min})}{F^2(Q_{\rm min})} + I_0 \right] . \tag{8}$$

Here $\mathcal{E}$ is the exposure of the experiment which relates the actual counting rate to the normalized rate in Eq.(1). Since the unknown factor $\rho_0 |f_{\rm p}|^2$ on the left–hand side above is identical for different targets, it leads to a second expression for determining m_χ:[6]

$$m_\chi|_\sigma = \frac{(m_X/m_Y)^{5/2} m_Y - m_X (\mathcal{R}_{\sigma,X}/\mathcal{R}_{\sigma,Y})}{\mathcal{R}_{\sigma,X}/\mathcal{R}_{\sigma,Y} - (m_X/m_Y)^{5/2}} . \tag{9}$$

Here $m_{(X,Y)} \propto A_{(X,Y)}$ has been assumed,

$$\mathcal{R}_{\sigma,X} \equiv \frac{1}{\mathcal{E}_X} \left[\frac{2 Q_{{\rm min},X}^{1/2} r_X(Q_{{\rm min},X})}{F_X^2(Q_{{\rm min},X})} + I_{0,X} \right] , \tag{10}$$

and similarly for $\mathcal{R}_{\sigma,Y}$.

In order to yield the best–fit WIMP mass as well as its statistical error by combining the estimators for different n in Eq.(6) with each other and with the estimator in Eq.(9), a χ^2 function has been introduced[6]

$$\chi^2(m_\chi) = \sum_{i,j} (f_{i,X} - f_{i,Y}) \mathcal{C}_{ij}^{-1} (f_{j,X} - f_{j,Y}) , \tag{11}$$

where

$$f_{i,X} \equiv \left(\frac{\alpha_X \mathcal{R}_{i,X}}{300 \ {\rm km/s}} \right)^i , \qquad \text{for } i = -1,\ 1,\ 2,\ \ldots,\ n_{\rm max}, \tag{12a}$$

and

$$f_{n_{\rm max}+1,X} \equiv \frac{A_X^2}{\mathcal{R}_{\sigma,X}} \left(\frac{\sqrt{m_X}}{m_\chi + m_X} \right) ; \tag{12b}$$

the other $n_{\rm max} + 2$ functions $f_{i,Y}$ can be defined analogously. Here $n_{\rm max}$ determines the highest moment of $f_1(v)$ that is included in the fit. The f_i are normalized such that they are dimensionless and very roughly of order unity. Note that the first $n_{\rm max} + 1$ fit functions depend on m_χ through the overall factor α and that m_χ in Eqs.(12a) and (12b) is now a fit parameter, which may differ from the true value of the WIMP mass. Moreover, $\mathcal{C}$ is the total covariance matrix. Since the X and Y quantities are statistically completely independent, $\mathcal{C}$ can be written as a sum of two terms:

$$\mathcal{C}_{ij} = {\rm cov}(f_{i,X}, f_{j,X}) + {\rm cov}(f_{i,Y}, f_{j,Y}) . \tag{13}$$

Finally, since we require that, from two experiments with different target nuclei, the values of a given moment of the WIMP velocity distribution estimated by Eq.(4) should agree, this means that the upper cuts on $f_1(v)$ in two data sets should be (approximately) equal†. This requires that[6]

$$Q_{\mathrm{max},Y} = \left(\frac{\alpha_X}{\alpha_Y}\right)^2 Q_{\mathrm{max},X} \,. \tag{14}$$

Note that α is a function of the true WIMP mass. Thus this relation for matching optimal cut–off energies can be used only if m_χ is already known. One possibility to overcome this problem is to fix the cut–off energy of the experiment with the heavier target, minimize the $\chi^2(m_\chi)$ function defined in Eq.(11), and estimate the cut–off energy for the lighter nucleus by Eq.(14) algorithmically.[6]

As demonstration we show some numerical results for the reconstructed WIMP mass based on Monte Carlo simulations. The upper and lower bounds on the reconstructed WIMP mass are estimated from the requirement that χ^2 exceeds its minimum by 1. ^{28}Si and ^{76}Ge have been chosen as two target nuclei. The scattering cross section has been assumed to be dominated by spin–independent interactions. The theoretically predicted recoil spectrum for the shifted Maxwellian velocity distribution ($v_0 = 220$ km/s, $v_e = 231$ km/s)[1,2,5] with the Woods-Saxon elastic form factor[1,2,8] have been used. The threshold energies of two experiments have been assumed to be negligible and the maximal cut–off energies are set as 100 keV. $2 \times 5{,}000$ experiments with 50 events on average before cuts from each experiment have been simulated. In order to avoid large contributions from very few events in the high energy range to the higher moments,[5] only the moments up to $n_{\mathrm{max}} = 2$ have been included in the χ^2 fit.

In Fig. 1 the dotted (green) curves show the median reconstructed WIMP mass and its 1σ upper and lower bounds for the case that both Q_{max} have been fixed to 100 keV. This causes a systematic *underestimate* of the reconstructed WIMP mass for input WIMP masses $\gtrsim 100$ GeV.[7] The solid (black) curves have been obtained by using Eq.(14) for matching the cut–off energy $Q_{\mathrm{max,Si}}$ perfectly with $Q_{\mathrm{max,Ge}} = 100$ keV and the true (input) WIMP mass, whereas the dashed (red) curves show the case that $Q_{\mathrm{max,Ge}} = 100$ keV, and $Q_{\mathrm{max,Si}}$ has been determined by minimizing $\chi^2(m_\chi)$. As shown here, with only 50 events from one experiment, the algorithmic process seems already to work pretty well for WIMP masses

†Here the threshold energies have been assumed to be negligibly small.

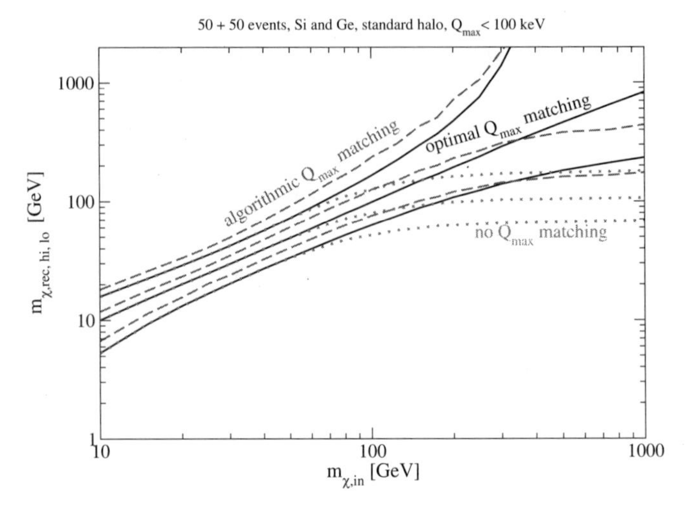

Fig. 1. Results for the median reconstructed WIMP mass as well as its 1σ statistical error interval based on the χ^2–fit in Eq.(11). See the text for further details.

up to $\sim$ 500 GeV. Though for $m_\chi \lesssim 100$ GeV m_χ determined in this way *overestimates* its true value by 15 to 20%, the true WIMP mass always lies within the median limits of the 1σ statistical error interval up to even $m_\chi = 1$ TeV.[6]

On the other hand, in order to study the statistical fluctuation of the reconstructed WIMP mass by algorithmic $Q_{\rm max}$ matching in the simulated experiments, an estimator δm has been introduced as[6]

$$\delta m = \begin{cases} 1 + \dfrac{m_{\chi,\rm lo1} - m_{\chi,\rm in}}{m_{\chi,\rm lo1} - m_{\chi,\rm lo2}}\,, & \text{if } m_{\chi,\rm in} \le m_{\chi,\rm lo1}\,; \\[2ex] \dfrac{m_{\chi,\rm rec} - m_{\chi,\rm in}}{m_{\chi,\rm rec} - m_{\chi,\rm lo1}}\,, & \text{if } m_{\chi,\rm lo1} < m_{\chi,\rm in} < m_{\chi,\rm rec}\,; \\[2ex] \dfrac{m_{\chi,\rm rec} - m_{\chi,\rm in}}{m_{\chi,\rm hi1} - m_{\chi,\rm rec}}\,, & \text{if } m_{\chi,\rm rec} < m_{\chi,\rm in} < m_{\chi,\rm hi1}\,; \\[2ex] \dfrac{m_{\chi,\rm hi1} - m_{\chi,\rm in}}{m_{\chi,\rm hi2} - m_{\chi,\rm hi1}} - 1\,, & \text{if } m_{\chi,\rm in} \ge m_{\chi,\rm hi1}\,. \end{cases} \tag{15}$$

Here $m_{\chi,\rm in}$ is the true (input) WIMP mass, $m_{\chi,\rm rec}$ its reconstructed value, $m_{\chi,\rm lo1(2)}$ are the 1 (2) σ lower bounds satisfying $\chi^2(m_{\chi,\rm lo(1,2)}) = \chi^2(m_{\chi,\rm rec}) + 1\,(4)$, and $m_{\chi,\rm hi1(2)}$ are the corresponding $1\,(2)\,\sigma$ upper bounds.

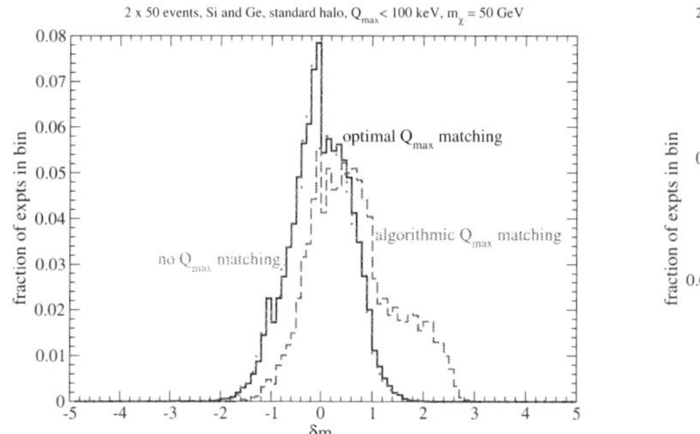

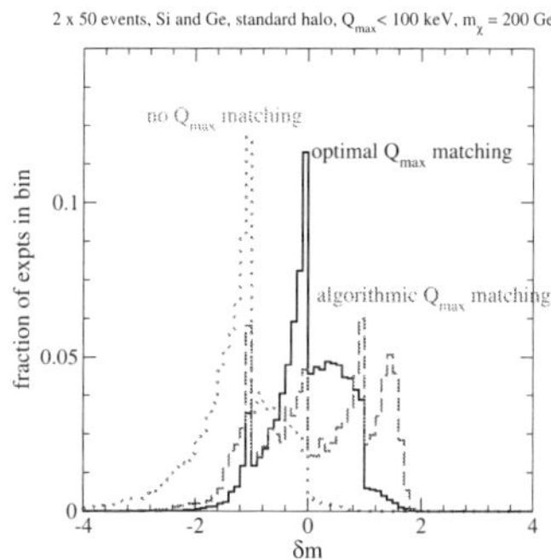

Fig. 2. Normalized distribution of the estimator δm defined in Eq.(15) for WIMP masses of 50 GeV (left) and 200 GeV (right). Parameters and notations are as in Fig. 1. Here the bins at $\delta m = \pm 5$ are overflow bins, i.e., they also contain all experiments with $|\delta m| \geq 5$.

Figures 2 show the distribution of the estimator δm calculated from 5,000 simulated experiments for WIMP masses of 50 GeV (left) and 200 GeV (right). For the lighter WIMP mass, simply fixing both $Q_{\rm max}$ values to 100 keV still works fine (the dotted (green) curves in Fig. 1). However, the distributions for both fixed $Q_{\rm max}$ and optimal $Q_{\rm max}$ matching show already an asymmetry of the statistical uncertainties with $m_{\chi,{\rm hi1}} - m_{\chi,{\rm rec}} > m_{\chi,{\rm rec}} - m_{\chi,{\rm lo1}}$. The overestimate of light WIMP masses reconstructed by algorithmic $Q_{\rm max}$ matching shown in Fig. 1 is also reflected by the dashed (red) histogram here, which has significantly more entries at positive values than at negative values. Moreover, these distributions also indicate that the statistical uncertainties estimated by minimizing $\chi^2(m_\chi)$ are in fact overestimated, since nearly 90% of the simulated experiments have $|\delta m| \leq 1$,[6] much more than $\sim$ 68% of the experiments, which a usual 1σ error interval should contain.

For the heavier WIMP mass of 200 GeV, as shown in the right frame of Figs. 2, the situation becomes less favorable. While the distributions for both fixed $Q_{\rm max}$ and optimal $Q_{\rm max}$ matching look more non–Gaussian but more concentrated on the median values, the distribution for algorithmic $Q_{\rm max}$ matching spreads out in the range $-1 < \delta m < 2$. It has even been observed that, for larger samples (e.g., with 500 events on average) the outspread distribution becomes broader.[6] Hence, the statistical fluctuation by the algorithmic procedure for determining $Q_{\rm max}$ of the experiment with the lighter target nucleus by minimizing χ^2 could be problematic for the determination of m_χ if WIMPs are heavy.

3. Estimating the SI WIMP–Proton Coupling

As shown in the previous section, by combining two experimental data sets, one can estimate the WIMP mass m_χ without knowing the WIMP–nucleus cross section σ_0. Conversely, by using Eq.(8), one can also estimate the SI WIMP–proton coupling, $|f_{\rm p}|^2$, from experimental data directly *without* knowing the WIMP mass.[9]

In Eq.(8) the WIMP mass m_χ on the right–hand side can be determined by the method described in Sec. 2, $r(Q_{\rm min})$ and I_0 can also be estimated from one of the two data sets used for determining m_χ or from a third experiment. However, due to the degeneracy between the local WIMP density ρ_0 and the coupling $|f_{\rm p}|^2$, one *cannot* estimate each one of them without making some assumptions. The simplest way is making an assumption for the local WIMP density ρ_0.

Figure 3 shows the reconstructed SI WIMP–proton coupling as a function of the input WIMP mass.. The WIMP mass has again been reconstructed with ^{28}Si and ^{76}Ge. In order to avoid complicated calculations of the correlation between the error on the reconstructed m_χ and that on the estimator of I_0, a second, independent data set with Ge has been chosen as the third target for estimating I_0. Parameters are as in Fig. 1, except that the SI WIMP–proton cross section has been set as 10^{-8} pb.

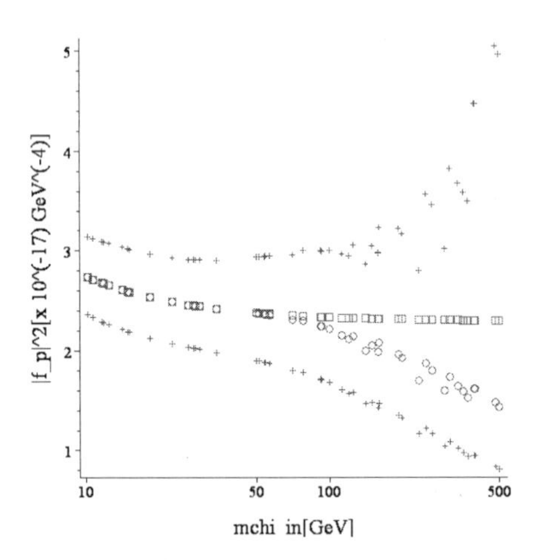

Fig. 3. The reconstructed SI WIMP–proton coupling as a function of the input WIMP mass. The (red) squares indicate the input WIMP masses and the true values of the coupling. The (blue) circles and the (blue) crosses indicate the reconstructed couplings and their 1σ statistical errors. Parameters are as in Fig. 1, in addition $\sigma_{\chi{\rm p}}^{\rm SI}$ has been set as 10^{-8} pb. See the text for further details.

It can be seen that the reconstructed $|f_{\rm p}|^2$ are *underestimated* for WIMP masses $\gtrsim$ 100 GeV. This systematic deviation is caused mainly by the underestimate of I_0. However, in spite of this systematic deviation the true value of $|f_{\rm p}|^2$ always lies within the 1σ statistical error interval. Moreover, for a WIMP mass of 100 GeV, one could in principle already estimate the SI WIMP–proton coupling with a statistical uncertainty of only $\sim 15\%$ with just 50 events from each experiment. Note that this is much smaller than the systematic uncertainty of the local Dark Matter density (of a factor of 2 or even larger).

4. Determining Ratios of WIMP–Nucleon Cross Sections

So far we have discussed only the case that the spin–independent WIMP–nucleus interaction dominates. In this section we turn to consider the case of the spin–dependent cross section as well as of a general combination of these two cross sections.

4.1. *Determining the $a_{\rm n}/a_{\rm p}$ ratio*

Consider at first the case that the SD WIMP–nucleus interaction dominates. By substituting $\sigma_0^{\rm SD}$ in Eq.(3a) and $\langle v^{-1}\rangle$ estimated by Eq.(4) into Eq.(1) and combining two data sets with different target nuclei, an expression for the ratio between two SD WIMP-nucleon couplings can be given as

$$\left(\frac{a_{\rm n}}{a_{\rm p}}\right)^{\rm SD}_{\pm,n} = -\frac{\langle S_{\rm p}\rangle_X \mathcal{R}_{J,n,Y} \pm \langle S_{\rm p}\rangle_Y \mathcal{R}_{J,n,X}}{\langle S_{\rm n}\rangle_X \mathcal{R}_{J,n,Y} \pm \langle S_{\rm n}\rangle_Y \mathcal{R}_{J,n,X}}, \tag{16}$$

with

$$\mathcal{R}_{J,n,X} \equiv \left[\left(\frac{J_X}{J_X+1}\right)\frac{\mathcal{R}_{\sigma,X}}{\mathcal{R}_{n,X}}\right]^{1/2}, \tag{17}$$

and similarly for $\mathcal{R}_{J,n,Y}$, where $n \neq 0$. Note that $a_{\rm n}/a_{\rm p}$ can be estimated from experimental data directly through estimating $\mathcal{R}_{n,X}$, $\mathcal{R}_{\sigma,X}$ and two Y terms by Eqs.(7) and (10)[‡] *without* knowing the WIMP mass.

Because the couplings in Eq.(3a) are squared, we have two solutions for $a_{\rm n}/a_{\rm p}$ here; if exact "theory" values for $\mathcal{R}_{J,n,(X,Y)}$ are taken, these solutions coincide for $a_{\rm n}/a_{\rm p} = -\langle S_{\rm p}\rangle_X/\langle S_{\rm n}\rangle_X$ and $-\langle S_{\rm p}\rangle_Y/\langle S_{\rm n}\rangle_Y$, which depends only on the properties of target nuclei[§]. Moreover, one of these

[‡]Note that the form factor $F^2(Q)$ here must be chosen for the SD cross section.

[§]Some relevant spin values of the nuclei used for our simulations shown in this paper are given in Table. 1.

Table 1. List of the relevant spin values of the nuclei used for simulations shown in this paper (Data from Ref. 10).

nucleus	Z	J	$\langle S_{\rm p}\rangle$	$\langle S_{\rm n}\rangle$	$-\langle S_{\rm p}\rangle/\langle S_{\rm n}\rangle$
^{17}O	8	5/2	0	0.495	0
^{23}Na	11	3/2	0.248	0.020	−12.40
^{37}Cl	17	3/2	−0.058	0.050	1.16
^{73}Ge	32	9/2	0.030	0.378	−0.079

two solutions has a pole at the middle of two intersections, which depends simply on the signs of $\langle S_{\rm n}\rangle_X$ and $\langle S_{\rm n}\rangle_Y$: since $\mathcal{R}_{J,n,X}$ and $\mathcal{R}_{J,n,Y}$ are always positive, if both of $\langle S_{\rm n}\rangle_X$ and $\langle S_{\rm n}\rangle_Y$ are positive or negative, the "minus" solution $(a_{\rm n}/a_{\rm p})^{\rm SD}_{-,n}$ will diverge and the "plus" solution $(a_{\rm n}/a_{\rm p})^{\rm SD}_{+,n}$ will be the "inside" solution, which has a smaller statistical uncertainty (see Figs. 4); in contrast, if the signs of $\langle S_{\rm n}\rangle_X$ and $\langle S_{\rm n}\rangle_Y$ are opposite, the "minus" solution $(a_{\rm n}/a_{\rm p})^{\rm SD}_{-,n}$ will be the "inside" solution.

Figures 4 show the reconstructed $(a_{\rm n}/a_{\rm p})^{\rm SD}$ estimated by Eq.(16) with $n = 1$ as functions of the true (input) $a_{\rm n}/a_{\rm p}$ for a WIMP mass of 100 GeV (left) and as functions of the input WIMP masses for $a_{\rm n}/a_{\rm p} = 0.7$ (right), respectively. The shifted Maxwellian velocity distribution with a form factor calculated in the thin-shell approximation for the SD cross section[4,11] has been used. Parameters are as earlier, except that the minimal cut-off energy has been increased to 5 keV for both experiments. Here we have chosen

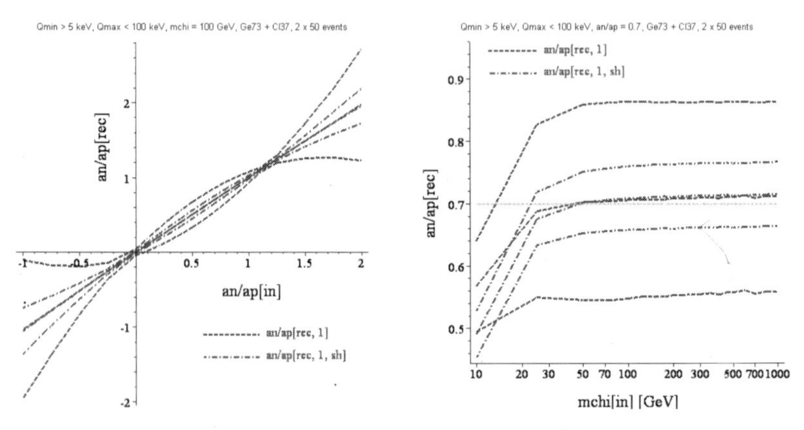

Fig. 4. Preliminary results for the reconstructed $(a_{\rm n}/a_{\rm p})^{\rm SD}$ estimated by Eq.(16) with $n = 1$ as functions of the true (input) $a_{\rm n}/a_{\rm p}$ (left frame, for a WIMP mass of 100 GeV) and as functions of the input WIMP mass m_χ (right frame, for $a_{\rm n}/a_{\rm p} = 0.7$), respectively. See the text for further details.

^{73}Ge and ^{37}Cl as two target nuclei in order to test the range of interest $0 \leq a_\text{n}/a_\text{p} \leq 1$;[12,13] and $(a_\text{n}/a_\text{p})^\text{SD}_{+/-,n}$ are thus the inside/outside solutions.

For estimating the statistical error on a_n/a_p, one needs to estimate the counting rate at the threshold energy, $r(Q_\text{min})$, and its statistical error, $\sigma(r(Q_\text{min}))$. It has been found that, instead of Q_min, one can estimate the counting rate and its statistical error at the shifted point $Q_{s,1}$ (from the central point of the first bin, Q_1):[5]

$$Q_{s,1} = Q_1 + \frac{1}{k_1} \ln \left[\frac{\sinh(k_1 b_1/2)}{k_1 b_1/2} \right], \tag{18}$$

where k_1 is the logarithmic slope of the reconstructed recoil spectrum in the first Q–bin and b_1 is the bin width. We see in the right frame of Figs. 4 very clearly that, for WIMP masses $\gtrsim 30$ GeV, the 1σ statistical error estimated with $Q_{s,1}$ (the dash–dotted (red) lines, labeled with "sh") is $\sim 7\%$, only 1/3 of the error estimated with Q_min (the dashed (blue) lines).

One more advantage with using $Q_{s,1}$ instead of Q_min is that the statistical error on a_n/a_p estimated with different n (namely with different moments of the WIMP velocity distribution) at $Q = Q_{s,1}$ are almost equal. Therefore, since

$$\mathcal{R}_{J,-1,X} = \left[\left(\frac{J_X}{J_X+1} \right) \frac{2\, r_X(Q_{X,s,1})}{\mathcal{E}_X F_X^2(Q_{X,s,1})} \right]^{1/2}, \tag{19}$$

one needs thus only events in the low energy range (~ 20 events between 5 and 15 keV in our simulations) for estimating a_n/a_p.

4.2. *Determining the $\sigma^\text{SD}_{\chi\text{p/n}}/\sigma^\text{SI}_{\chi\text{p}}$ ratios*

Now let us combine WIMP–nucleus scattering induced by both SI and SD interactions given in Eqs.(2) and (3a) (with the corresponding form factors). By modifying $F^2(Q)$ and I_n in the estimator (4) of the moments of the WIMP velocity distribution, the ratio of the SD WIMP-proton cross section to the SI one can be solved analytically as[¶, ‖]

$$\frac{\sigma^\text{SD}_{\chi\text{p}}}{\sigma^\text{SI}_{\chi\text{p}}} = -\frac{F^2_{\text{SI},X}(Q_{\text{min},X})\mathcal{R}_{m,Y} - F^2_{\text{SI},Y}(Q_{\text{min},Y})\mathcal{R}_{m,X}}{\mathcal{C}_{\text{p},X} F^2_{\text{SD},X}(Q_{\text{min},X})\mathcal{R}_{m,Y} - \mathcal{C}_{\text{p},Y} F^2_{\text{SD},Y}(Q_{\text{min},Y})\mathcal{R}_{m,X}}. \tag{20}$$

[¶] In this section we consider only the case of $\sigma^\text{SD}_{\chi\text{p}}$, but all formulae given here can be applied straightforwardly to the case of $\sigma^\text{SD}_{\chi\text{n}}$ by exchanging n $\leftrightarrow$ p.

[‖] Q_min appearing in this section can be replaced by $Q_{s,1}$ everywhere.

Here

$$\mathcal{R}_{m,X} \equiv \frac{r_X(Q_{\min,X})}{\mathcal{E}_X m_X^2}\,, \tag{21}$$

and

$$\mathcal{C}_{\mathrm{p},X} \equiv \frac{4}{3}\left(\frac{J_X+1}{J_X}\right)\left[\frac{\langle S_\mathrm{p}\rangle_X + \langle S_\mathrm{n}\rangle_X (a_\mathrm{n}/a_\mathrm{p})}{A_X}\right]^2; \tag{22}$$

$\mathcal{R}_{m,Y}$ and $\mathcal{C}_{\mathrm{p},Y}$ can be defined analogously. Note that a "minus $(-)$" sign appears in the expression (20).

By introducing a third target having *only* the SI interaction with WIMPs, $a_\mathrm{n}/a_\mathrm{p}$ appearing in $\mathcal{C}_{\mathrm{p},X}$ and $\mathcal{C}_{\mathrm{p},Y}$ can again be solved analytically as

$$\left(\frac{a_\mathrm{n}}{a_\mathrm{p}}\right)_{\pm}^{\mathrm{SI+SD}} = \frac{-\left(c_{\mathrm{p},X} s_{\mathrm{n/p},X} - c_{\mathrm{p},Y} s_{\mathrm{n/p},Y}\right) \pm \sqrt{c_{\mathrm{p},X} c_{\mathrm{p},Y}}\,\left|s_{\mathrm{n/p},X} - s_{\mathrm{n/p},Y}\right|}{c_{\mathrm{p},X} s_{\mathrm{n/p},X}^2 - c_{\mathrm{p},Y} s_{\mathrm{n/p},Y}^2}\,. \tag{23}$$

Here

$$c_{\mathrm{p},X} \equiv \frac{4}{3}\left(\frac{J_X+1}{J_X}\right)\left(\frac{\langle S_\mathrm{p}\rangle_X}{A_X}\right)^2 F_{\mathrm{SD},X}^2(Q_{\min,X}) \times \left[F_{\mathrm{SI},Z}^2(Q_{\min,Z})\left(\frac{\mathcal{R}_{m,Y}}{\mathcal{R}_{m,Z}}\right) - F_{\mathrm{SI},Y}^2(Q_{\min,Y})\right], \tag{24}$$

$c_{\mathrm{p},Y}$ can be obtained by simply exchanging $X \leftrightarrow Y$, and $s_{\mathrm{n/p}} \equiv \langle S_\mathrm{n}\rangle/\langle S_\mathrm{p}\rangle$. However, in order to reduce the statistical uncertainties contributed from estimate of $a_\mathrm{n}/a_\mathrm{p}$ involved in $\mathcal{C}_{\mathrm{p},X}$ and $\mathcal{C}_{\mathrm{p},Y}$, one can use one target with the SD sensitivity (almost) only to protons or to neutrons combined with another one with only the SI sensitivity. For this case $\mathcal{C}_{\mathrm{p},X}$ is independent of $a_\mathrm{n}/a_\mathrm{p}$ and the expression (20) for $\sigma_{\chi\mathrm{p}}^{\mathrm{SD}}/\sigma_{\chi\mathrm{p}}^{\mathrm{SI}}$ can be reduced to

$$\frac{\sigma_{\chi\mathrm{p}}^{\mathrm{SD}}}{\sigma_{\chi\mathrm{p}}^{\mathrm{SI}}} = -\frac{F_{\mathrm{SI},X}^2(Q_{\mathrm{thre},X})\mathcal{R}_{m,Y} - F_{\mathrm{SI},Y}^2(Q_{\mathrm{thre},Y})\mathcal{R}_{m,X}}{\mathcal{C}_{\mathrm{p},X} F_{\mathrm{SD},X}^2(Q_{\mathrm{thre},X})\mathcal{R}_{m,Y}}\,. \tag{25}$$

Figures 5 show the reconstructed $\sigma_{\chi\mathrm{p}}^{\mathrm{SD}}/\sigma_{\chi\mathrm{p}}^{\mathrm{SI}}$ (left) and $\sigma_{\chi\mathrm{n}}^{\mathrm{SD}}/\sigma_{\chi\mathrm{p}}^{\mathrm{SI}}$ (right) estimated by Eqs.(20) and (25) as functions of the true (input) $a_\mathrm{n}/a_\mathrm{p}$, respectively. Besides $^{73}\mathrm{Ge}$ and $^{37}\mathrm{Cl}$, $^{28}\mathrm{Si}$ has been chosen as the third target for estimating $a_\mathrm{n}/a_\mathrm{p}$ by Eq.(23); whereas $^{76}\mathrm{Ge}$ has been chosen as the second target having only the SI interaction with WIMPs and combined with $^{23}\mathrm{Na}$ (for $\sigma_{\chi\mathrm{p}}^{\mathrm{SD}}/\sigma_{\chi\mathrm{p}}^{\mathrm{SI}}$) and $^{17}\mathrm{O}$ (for $\sigma_{\chi\mathrm{n}}^{\mathrm{SD}}/\sigma_{\chi\mathrm{p}}^{\mathrm{SI}}$) for using Eq.(25). We see here that, since the SD WIMP–nucleus interaction doesn't dominate for our simulation setup, $\sigma_{\chi\mathrm{p}}^{\mathrm{SD}}/\sigma_{\chi\mathrm{p}}^{\mathrm{SI}}$ estimated by Eq.(20) has two discontinuities around the intersections at $a_\mathrm{n}/a_\mathrm{p} = -0.079$ and especially at $a_\mathrm{n}/a_\mathrm{p} = 1.16$,

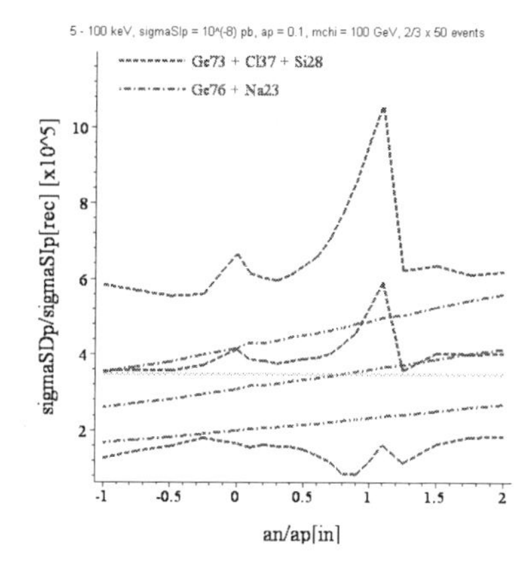

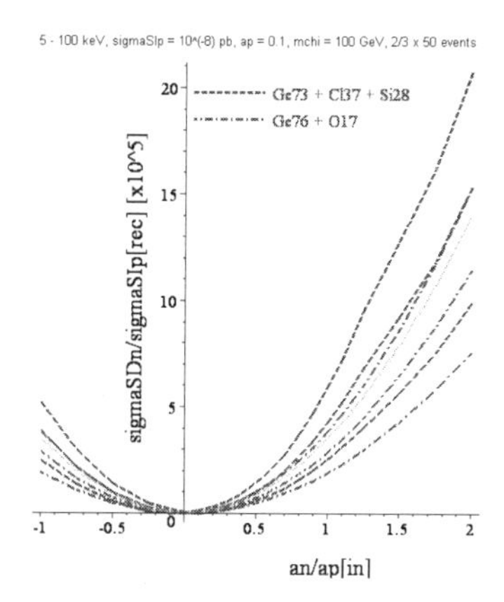

Fig. 5. Preliminary results for the reconstructed $\sigma_{\chi\mathrm{p}}^{\mathrm{SD}}/\sigma_{\chi\mathrm{p}}^{\mathrm{SI}}$ (left) and $\sigma_{\chi\mathrm{n}}^{\mathrm{SD}}/\sigma_{\chi\mathrm{p}}^{\mathrm{SI}}$ (right) as functions of the true (input) $a_{\mathrm{n}}/a_{\mathrm{p}}$, respectively. The dashed (blue) curves indicate the values estimated by Eq.(20) with $a_{\mathrm{n}}/a_{\mathrm{p}}$ estimated by Eq.(23); whereas the dash–dotted (red) curves indicate the values estimated by Eqs.(25). $\sigma_{\chi\mathrm{p}}^{\mathrm{SI}}$ and a_{p} have been set as 10^{-8} pb and 0.1, respectively. The other parameters are as in Figs. 4. Note that, since we fix $\sigma_{\chi\mathrm{p}}^{\mathrm{SI}}$ and a_{p}, $\sigma_{\chi\mathrm{p}}^{\mathrm{SD}}/\sigma_{\chi\mathrm{p}}^{\mathrm{SI}}$ shown here is a constant, whereas $\sigma_{\chi\mathrm{n}}^{\mathrm{SD}}/\sigma_{\chi\mathrm{p}}^{\mathrm{SI}} \propto a_{\mathrm{n}}^2$ a parabola.

the intersection determined by the $-\langle S_{\mathrm{p}}\rangle/\langle S_{\mathrm{n}}\rangle$ value of ^{37}Cl. However, from two experiments with only $\sim$ 20 events in the low energy range, one could in principle already estimate $\sigma_{\chi\mathrm{p}}^{\mathrm{SD}}/\sigma_{\chi\mathrm{p}}^{\mathrm{SI}}$ and $\sigma_{\chi\mathrm{n}}^{\mathrm{SD}}/\sigma_{\chi\mathrm{p}}^{\mathrm{SI}}$ by using Eq.(25) with a statistical uncertainties of $\sim$ 35%.

5. Summary and Conclusions

In this article we described model–independent methods for determining the WIMP mass and their couplings on nucleons by using future experimental data from direct Dark Matter detection. The main focus is *how well* we could extract the nature of WIMPs with *positive* signals and *which problems* we could meet by applying these methods to (real) data analysis.

In Secs. 2 and 3 we discussed the determinations of the WIMP mass and its spin–independent coupling on protons. If WIMPs are light ($m_\chi \simeq 50$ GeV), with $\mathcal{O}(50)$ events from one experiment, their mass and SI coupling could be estimated with errors of $\sim$ 35% and $\sim$ 15%, respectively. However, in case WIMPs are heavy ($m_\chi \gtrsim$ 200 GeV), the statistical fluctuation by the algorithmic procedure for matching the maximal cut–off energies of the experiments could be problematic for estimating their mass, and thereby their SI coupling.

In Sec. 4 we turned to consider the spin–dependent interaction. The simulations show pretty small statistical uncertainties. Moreover, differing from the traditional method for constraining the SD WIMP–nucleon couplings,[10,14–16] we do not make any assumptions on ρ_0, $f_1(v)$, and m_χ. The price one has to pay for this is that positive signals in at least two different data sets with different target nuclei are required. In addition, without independent knowledge of ρ_0, one can only determine ratios of cross sections.

In summary, once two (or more) experiments measure WIMP events, the methods presented here could in principle help us to extract the nature of halo WIMPs. This information will allow us not only to constrain the parameter space in different extensions of the Standard Model, but also to confirm or exclude some candidates for WIMP Dark Matter.[17,18]

Acknowledgments

This work was partially supported by the Marie Curie Training Research Network "UniverseNet" under contract no. MRTN-CT-2006-035863, by the European Network of Theoretical Astroparticle Physics ENTApP ILIAS/N6 under contract no. RII3-CT-2004-506222, as well as by the BK21 Frontier Physics Research Division under project no. BA06A1102 of Korea Research Foundation.

References

1. G. Jungman, M. Kamionkowski, and K. Griest, *Phys. Rep.* **267**, 195 (1996).
2. G. Bertone, D. Hooper, and J. Silk, *Phys. Rep.* **405**, 279 (2005).
3. P. F. Smith and J. D. Lewin, *Phys. Rep.* **187**, 203 (1990).
4. J. D. Lewin and P. F. Smith, *Astropart. Phys.* **6**, 87 (1996).
5. M. Drees and C. L. Shan, *J. Cosmol. Astropart. Phys.* **0706**, 011 (2007).
6. M. Drees and C. L. Shan, *J. Cosmol. Astropart. Phys.* **0806**, 012 (2008).
7. C. L. Shan and M. Drees, `arXiv:0710.4296 [hep-ph]` (2007).
8. J. Engel, *Phys. Lett.* **B 264**, 114 (1991).
9. M. Drees and C. L. Shan, `arXiv:0809.2441 [hep-ph]` (2008).
10. F. Giuliani and T. A. Girard, *Phys. Rev.* **D 71**, 123503 (2005).
11. H. V. Klapdor-Kleingrothaus, I. V. Krivosheina, and C. Tomei, *Phys. Lett.* **B 609**, 226 (2005).
12. V. A. Bednyakov, *Phys. Atom. Nucl.* **67**, 1931 (2004).
13. J. Ellis, K. A. Olive, and C. Savage, *Phys. Rev.* **D 77**, 065026 (2008).
14. D. R. Tovey *et al.*, *Phys. Lett.* **B 488**, 17 (2000).
15. F. Giuliani, *Phys. Rev. Lett.* **93**, 161301 (2004).
16. T. A. Girard and F. Giuliani, *Phys. Rev.* **D 75**, 043512 (2007).
17. G. Bertone *et al.*, *Phys. Rev. Lett.* **99**, 151301 (2007).
18. V. Barger, W. Y. Keung, and G. Shaughnessy, *Phys. Rev.* **D 78**, 056007 (2008).

HOT ON THE TAIL OF THE ELUSIVE WIMP: DIRECT DETECTION DARK MATTER SEARCHES ENTER THE 21st CENTURY

T. SAAB*

Department of Physics, University of Florida,
Gainesville, FL 32611, USA
**E-mail: tsaab@phys.ufl.edu*
http://www.phys.ufl.edu/

(On behalf of the CDMS Collaboration)
http://cdms.berkeley.edu/cdms_collab.html

It is now well established that roughly 25% of the total mass-energy density of the Universe is in the form of non-relativistic particles referred to as Dark Matter. This paper will review various ongoing Dark Matter searches and the variety of techniques and implementation used to both detect the rare Dark Matter interactions as well as reject the vast number of background events, with focus given to the recent results obtained in this past year. Eight years into the 21st century, the field Dark Matter searches is very active, with many new experimental results, and some intriguing hints.

Keywords: Dark Matter; WIMP; Direct Detection; Superconducting detectors; CDMS; Background.

1. Introduction

While hints for the existence of an unobserved form of matter on extra-galactic scales have been around for ~70 years, over the last ten years mounting evidence has made the case for non-baryonic particle dark matter on size scales, from galactic to cosmological, inexorable.[1,2] Weakly Interacting Massive Particles (WIMPs), whose origin lies in supersymmetric extensions to the Standard Model (SUSY), are a generic class of particles having a mass in range of few $GeV/c^2 - TeV/c^2$, a very low scattering cross-section with Standard Model particles, and which can satisfy the constraints on dark matter.[3]

2. Principles of Direct Detection

Dark matter direct detection searches are based on the elastic scattering interaction between WIMPs and atomic nuclei. The differential recoil energy spectrum of such an interaction between a WIMP of mass m_χ and a nucleus of mass m_N is described by equation 1:[3,4]

$$\frac{dR}{dQ} = \frac{\sigma_0 \rho_0}{2 m_\chi m_r^2} F^2(Q) T(Q) \tag{1}$$

Where ρ_0 is the WIMP density in the local neighborhood of the galactic halo, and σ_0 is the elastic scattering cross-section between the WIMP and nucleus, m_r is the WIMP-nucleus reduced mass $m_r = \left(\frac{m_\chi m_N}{m_\chi + m_N}\right)$, $F(Q)$ is the nuclear form factor, and $T(Q)$ is the WIMP velocity distribution. For spin-independent scattering, σ_0 is approximately proportional to the WIMP-proton scattering cross-section $\sigma_{\chi-p}$: $\sigma_0 \propto A^2 \sigma_{\chi-p}$. This relationship emphasizes the dependence of the interaction rate on the size of the nucleus A. The resulting spectra are roughly falling exponentials with an average energy of $\approx 10\,\text{keV}$. The advantage of the A^2 scaling of the scattering cross-section is somewhat mitigated by the form factor $F(Q)$ of the heavier nuclei, leading to a sharper falloff at high momentum transfers.

The interaction rates, ranging from of 10^{-5} to 10^{-4} events/kg/keV/day, are orders of magnitude lower than the background rates achieved in the cleanest detector materials, typically ~ 1 event/kg/kev/day. The ability to distinguish potential dark matter events from background interactions, therefore, becomes essential for a successful direct detection experiment.

2.1. *Discriminating Dark Matter Signals from Backgrounds*

Several physical phenomena can be exploited to discriminate between signal and background events. The discrimination is mostly based on the statistical properties of a large number of signal (and background) interactions, however, there also exist some techniques which permit the identification of the interaction on an event by event basis.

(1) Temporal variation in the signal and background event rates, generally referred to as the seasonal or annual modulation effect. This method of discrimination is based on the variation in the WIMP event rates and spectrum as the relative motion between the laboratory frame of reference and the WIMP rest frame varies along the earth's orbit of the sun,[5] whereas background sources are not expected to exhibit such a variation. Under the assumption of a non-rotating WIMP halo the

event rate is expected to exhibit maxima/minima in June/December, with an amplitude of few percent.

(2) Directional variation in the recoiling nucleus. The strong correlation between the direction of the incoming WIMP and the recoiling nucleus means that the majority of signal events, as seen from the laboratory frame, should point in the direction of the WIMP wind[5] (defined by the motion of the solar system through the galactic halo). Background events are not expected to exhibit a non-uniform directionality, or none that is correlated with the relative directions of the laboratory frame and the WIMP wind, which varies on a 24 hour time scale as the earth rotates on its axis. Employing such a discrimination technique requires detectors capable of reconstructing the tracks of individual nuclei.

(3) Variation in detector based response to signal and background events. In the keV momentum transfer range, signal events are due to WIMP interactions with atomic nuclei, whereas the majority of the backgrounds is due to electromagnetic interactions with the atomic electrons. For a given energy, a recoiling nucleus travels a much smaller distance in the detector than a recoiling electron. Therefore, an interacting WIMP results in a larger local deposited energy density than that of a background interaction, which can lead to a variation in the overall detector response. Detection techniques that are sensitive to this effect allow for the identification and rejection of electromagnetic backgrounds on an event to event basis. Background events resulting in nuclear-recoils, primarily due to neutron interactions within the detector, however cannot be rejected with this technique.

(4) Operation at a deep underground site to reduce the flux of energetic neutrons. Regardless of the experimental technique, reducing the rate of background event is essential for all dark matter experiments. Energetic neutrons, which are created by the interaction of cosmic-ray muons within the materials and structures surrounding the experiment, lead to interactions in a detector that are indistinguishable from a WIMP signal. A large overburden of rock results in the attenuation of the muon flux, and subsequently the neutron background rates. Since the nature of the overburden above any given laboratory varies, the depths are typically quoted in a normalized unit based on the equivalent depth of a water column. This is referred to as "meters water equivalent (m.w.e.)". Fig. 1 shows the relative muon and neutron fluxes at a selection of underground labs.

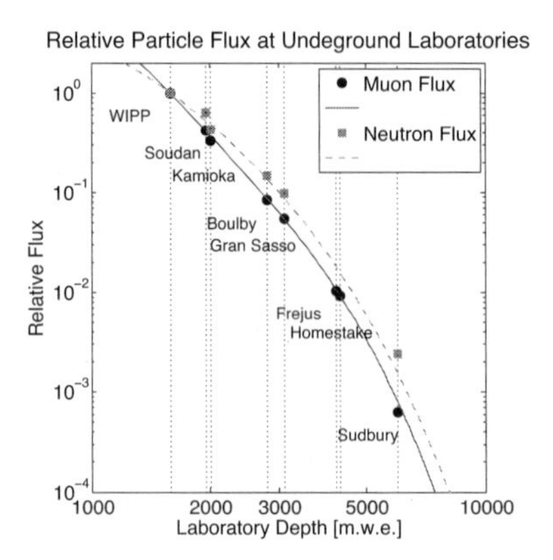

Fig. 1. Normalized flux of cosmic ray muons (squares) and the resulting energetic neutrons (circles) at various underground laboratories.[6] The squares are measured values, and lines is a functional fit. The neutron data are based on a Monte-Carlo simulation: the line is the total flux, whereas the circles are the flux above 10 MeV.

In the remainder of this article I will review a variety of experimental techniques and implementations which are currently active in WIMP searches. This list is a small sampling of the variety of experimental approaches with a focus on recent developments in terms of detector technology. A comprehensive survey of direct detection experiments can be found in references 7–11. The reader is also directed to Ref. 12 for a review of dark matter candidates.

3. Direct Detection Techniques and Experiments

I refer you articles in these proceedings by P. Belli, I. Krivosheina, and N. Smith which discuss the experimental techniques employed by the DAMA-LIBRA, GENIUS-TF, and ZEPLIN-III experiments respectively.

3.1. *Gaseous Detectors*

Gaseous dark matter detectors allow the possibility of reconstructing the track of a recoiling nucleus. As mentioned in section 2.1 this information can be very useful in distinguishing a dark matter signal from among the ever-present background. The disadvantage of this approach, however, is the relatively small target mass, and thus exposure, that can be readily achieved.

The DRIFT (Directional Recoil Identification From Tracks) experiment employs a negative on time projection chamber filled with CS_2 gas as a

Dark Matter detector. An interaction occurring within the detector results in a recoiling nucleus and the creation of electron-ion pairs along the track. DRIFT is able to do discriminate between different classes of events, such as: electron-recoils, nuclear-recoils, and α particles, based on the track morphology. Namely, nuclear-recoils (in the 10 keV range) have short tracks, within a few mm, and are expected to deposit a signal on one anode wires. α particles have a longer track length, more than tens of mm, and result in a signal seen in multiple anode wires. Electron-recoil events have a very long track length results in a small signal spread over multiple anode wires. The DRIFT-IIa detector, which has a $1\,\text{m}^3$ instrumented volume, containing 167 g of CS_2 gas at a pressure of 40 Torr, has a threshold of $\approx$ 1000 ion-pairs, corresponding to a $\approx 20\,\text{keV}$ S nuclear-recoil.

The experiment, located at the Boulby underground laboratory at a depth of 3300 m.w.e., has collected a data with exposures totaling 6 kg-days without a neutron moderator and 10.2 kg-days with the detector enclosed within a neutron moderator. The electron-recoil rejection was measured to be 8×10^6 while maintaining a 94% acceptance for nuclear-recoil in the range of 1000-6000 ion-pairs.[13]

3.2. *Superheated Liquid Detectors*

Superheated liquids as a dark matter detection target offer some very appealing features. A phase transition occurs in a superheated liquid occurs when a bubble in the liquid exceeds a critical radius r_c. Although such a process can occur spontaneously, nucleation of such bubbles can be induced by the deposition of energy in a small volume leading to local heating of the liquid. By tuning the thermodynamic parameters of the superheated liquid, such as temperature and pressure, the minimum energy threshold as well as the deposited energy density required to induce bubble formation can be varied. This mechanism leads to a mode of operation in which the detector is only sensitive to nuclear events (such as the potential dark matter signal), while being highly insensitive to electron-recoil backgrounds.

The COUPP (Chicagoland Observatory for Underground Particle Physics) experiment uses CF_3I as the target liquid. The experiment is sensitive to both spin-independent and spin-dependent WIMP interactions due to the presence of the F ($J = 1/2$) and I ($A = 126.9$) target nuclei. Operated as a bubble chamber, the COUPP detector uses two cameras (viewing orthogonal planes) to reconstruct the 3 dimensional position of all nuclear-recoil interactions above the threshold energy. An electron-background sensitivity of better than 1 in 10^9 was measured in a prototype detector.[14]

Subsequently, a 2 kg detector that has been operated at a 300 m.w.e. deep site on the Fermi National Accelerator Laboratory grounds. A total exposure of $\sim$ 250 kg-day was accumulated over a period spanning 2005-2006. COUPP is currently constructing a 20 and 60 kg version of the detectors.

3.3. *Cryogenic Detectors*

Cryogenic detection techniques offer new ways for determining the properties of an interaction in the detector, may not otherwise not be available at room temperature. At sub-K temperatures, the following physical channels contain information relevant for direct detection experiments:

- The entire recoil energy of an event (independent of the nature of the recoil) can be determined from the response of the thermal and athermal phonon populations.
- Charge carriers can be drifted and collected with electric fields of only a few V/cm. This makes it possible to perform an ionization measurement overwhelming the signal in the phonon channel.[15,16]
- Physical effects that only exist below a certain critical temperature lead to novel readout techniques. For example:
  - Semiconductor thermistors and superconducting transition-edge sensors allow the measurement of an event's recoil energy with excellent resolution.
  - Superconducting kinetic induction devices can be used to instrument and readout 100's to 1000's kg of detector mass.
  - Vibration wire resonators in a superfluid permit the measurement of recoil energy in a ^{3}He target.

Unlike the superheated liquids, and to some extend the gaseous time projection chambers, cryogenic detectors are sensitive to all particle interactions. This means that electron-recoil backgrounds constitute the main source of the interaction rate. The ability to distinguish electron-recoils from nuclear-recoils on an event by event basis becomes paramount and permits the rejection of the great majority of such backgrounds.

The CDMS II experiment uses semiconducting Ge as well as Si crystals as the dark matter detectors.[17] A fraction of the energy deposited in the detector by a particle interaction results in the creation of electron-hole pairs, with the remainder of the energy creating a population of high frequency athermal phonons. An electric field of 3 V/cm, across the crystal, causes the electrons and holes to drift to opposing electrodes, where they are subsequently measured with a charge amplifier. The energy in the phonons system is measured by four Transition-Edge Sensors that are photolithographically patterned on the surface into independent quadrants, leading to a signal that is proportional to the recoil energy.[18,19]

The ratio of the ionization to phonon signals (referred to as ionization yield) permits the rejection of electron recoils[20] with an efficiency of better than 1 in 10^4.[17] Events occurring within $\sim 10\,\mu$m of the detector surfaces, have a diminished ionization response resulting in a degradation of the electron-recoil identification to 99.79%. The information in the time structure of the phonon pulse, however, can be used to identify events occurring near the surface versus those occurring in the bulk, and thus improve the surface electron rejection, resulting in an overall electron recoil rejection of better than 1 in 10^6.

For the analysis described of this data, a timing parameter was developed to maximize the rejection of surface electron recoils while maintaining a large acceptance for nuclear recoils. The timing parameter is defined as the sum of the rise-time of the largest phonon pulse and the delay of the pulse relative to the prompt ionization pulse. Figure 2(a) shows the distribution of bulk electron recoils (γs from a ^{133}Ba calibration source), surface electron recoils (βs ejected from the surfaces closest to the detectors during a calibration with a ^{133}Ba source), and bulk nuclear recoils (neutrons from a ^{252}Cf calibration source) in the yield-timing parameter space illustrating the power of the combination of ionization yield and phonon timing parameters at rejecting electron recoil backgrounds.

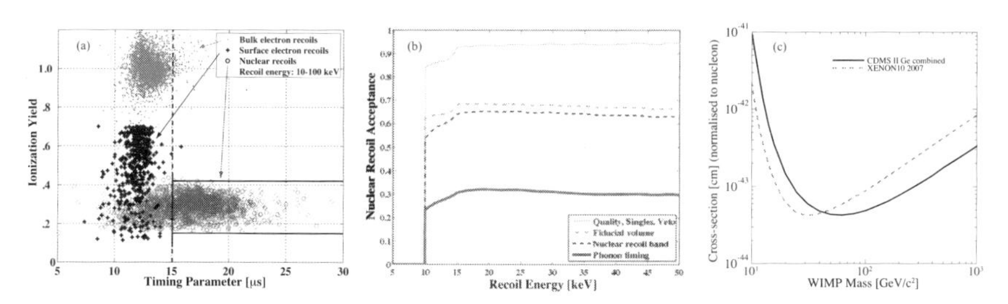

Fig. 2. (a) Ionization yield vs timing parameter for calibration data in a Ge detector. The vertical dashed line indicates the minimum timing parameter allowed for candidate dark matter events in this detector, and the box shows the approximate signal region. (b) Acceptance efficiency for the WIMP signal selection cuts. The efficiencies are averaged over all 15 detectors and weighted by their individual livetimes. The four curves represent cumulative efficiencies, beginning with events and culminating with the final efficiency used in the analysis. (c) Spin-dependent WIMP-neutron cross-section upper limits (90% C.L.) versus WIMP mass. The current CDMS limit (lower solid line) verifies much of the cross-section range covered by XENON10[21] (dashed line). Data for (c) courtesy of 22.

Despite their intrinsic rejection capability, it is necessary to reduce the overall background flux incident on the detectors. This is achieved with a of passive Pb shielding to reduce electromagnetic backgrounds (such as gamma rays) as well as a combination of passive polyethylene shielding, an active scintillator veto, and situating the experiment at an underground depth of 2090 meters-water-equivalent to reduce the flux of neutron backgrounds. It is important to note that neutron interactions, which result in a nuclear recoil, cannot be distinguished from the WIMP signal on an event by event basis.

4. CDMS Data Analysis

The data presented here is from a data set taken between October 2006 and March 2007, referred to as Run 123 and is reported in in more detail in 23,24. For the purposes of minimizing systematic uncertainties a total of 15 Ge detectors (3.75 kg) which were used for this analysis.

4.1. *Data Cuts*

The analysis cuts were developed on calibration data taken with (^{133}Ba) γ and (^{252}Cf) neutron sources. Alternating events from the ^{133}Ba data were separated into two statistically independent samples one of which was used to develop the analysis cuts and the other to determine their efficiencies in an unbiased manner. After selecting events taken during period with nominal detector performance (referred to as data quality cuts) the following cuts were used to define the WIMP signal region:

- Fiducial volume cut: Events in the outer part of each detector, based on the partitioning of energy between the two concentric charge electrodes, were rejected.
- Single scatter cut: Events were required to deposit energy $> 4\sigma$ above mean noise in only one detector (since the probability of a multiply interacting WIMPs is vanishingly small). All 30 detectors contributed to active vetoing of multiple scatter events at all times.
- Veto cut: Events were required to show no activity in the surrounding scintillator veto shield during a 200 μs window around the trigger in order to reject events associated with cosmic ray muons or radioactive sources external to the cryostat.
- Surface event cut: Events wer required to exceed a minimum value of the timing parameter, which was determined independently for each detector.

- Nuclear recoil yield cut: Events were required to lie within 2σ region of the nuclear recoil distribution in ionization yield, as determined by neutron calibration.
- Nuclear recoil timing cut: Events were required to lie within 4σ of the mean of the neutron recoil distribution in timing parameter, as determined by neutron calibration.

Figure 2(b) shows the measured cut acceptance efficiency as a function of energy. Prior to the cuts being applied the exposure of this data was 397.8 kg-days, which was reduced to 121.3 kg-days after thes cuts (averaged over recoil energies 10–100 keV, weighted for a WIMP mass of 60 GeV). The β and neutron background leakage rates were estimated to be $0.6^{+0.5}_{-0.3}(stat.)^{+0.3}_{-0.2}(syst.)$ and < 0.1 events respectively.

4.2. *Results*

Once all the analysis cuts were fixed and the background leakage were estimated the nuclear recoil region in the dark matter data was unmasked. There were zero events in the WIMP signal region. Given the total exposure and the observed number of events a Poisson 90%C.L. upper limit on the spin-independent WIMP-nucleon cross section is derived under the assumption of a standard non-rotating galactic halo.[4] The exclusion curve, derived from the Optimum Interval method,[25] for all the CDMS data taken so far is shown in figure 2(c) as the solid curve with a minimum at $4.6 \times 10^{-44}\,\text{cm}^2$ at for a WIMP mass of 60 GeV, a factor of ~3 stricter than our previously published limit. The major result from the run, however, is the zero observed events. While a potential WIMP signal would have been very exciting, this run demonstrated the ability of the CDMS detector discrimination to limit backgrounds to less than one event over the run. This is important for maintaining a high dark matter discovery potential.

5. Future SuperCDMS Detectors

The sights of the CDMS collaborations are set on achieving an experimental sensitivity beyond $1 \times 10^{-46}\,\text{cm}^2$ with a 100-1000 kg scale experiment. Simply increasing the number of detectors is not very feasible approach, since the thermal, and mechanical difficulties which arise with instrumenting and cooling down $\sim$ a thousand detectors are non-trivial. The path towards larger mass experiments, therefore, requires larger mass detectors and a larger fraction of *fiducial volume* per detector.

We are currently in the midst of the transition from the CDMS II experiment to the SuperCDMS experiment. The SuperCDMS detectors will consist of 3 inch diameter by 1 inch thick crystals (a factor of 2.5 increase in mass) for a total experimental payload of $\sim$ 15 kg. The volume increase by itself translates into a factor of 2.5 reduction in the the rate of surface backgrounds, such as the low yield electron recoil events. The detectors will use the same type of crystal, i.e. same dislocation densities of 100–1000 cm^{-3}, as the CDMS detectors. This value of dislocation density has been found to be optimal for the operation of ionization detectors.

Beyond the SuperCDMS design, however, we are investigating larger diameter ($\sim$ 6 inch) detectors. Crystals of such diameters cannot be grown with the optimal dislocation densities. In fact, the fabrication process produces dislocation free crystals which cannot be used as ionization detectors, at liquid nitrogen temperatures, due to the resulting large number of point defect trapping sites.[26,27] Since the CDMS detectors are operated at mK temperatures, we expected that the point defect trapping sites can be neutralized with the same techniques used with the CDMS II detectors.[28] Due to the low thermal energy available at mK temperatures, the traps would subsequently remain neutralized, allowing an ionization measurement to be performed. A small dislocation free crystal, operated as a test device, showed that full charge collection efficiency can be achieved for a field of a few V/cm, similar to that used in the current CDMS detectors. This demonstration has given us confidence that proceeding with 6 inch (or larger) diameter crystals should pose no intrinsic impediment to building a successful large mass dark matter detector.

Acknowledgments

The CDMS collaboration gratefully acknowledges Patrizia Meunier, Daniel Callahan, Pat Castle, Dave Hale, Susanne Kyre, Bruce Lambin and Wayne Johnson for their contributions. This work is supported in part by the National Science Foundation (Grant Nos. AST-9978911, PHY-0542066, PHY-0503729, PHY-0503629, PHY-0503641, PHY-0504224 and PHY-0705052), by the Department of Energy (Contracts DE-AC03-76SF00098, DE-FG02-91ER40688, DE-FG03-90ER40569, and DE-FG03-91ER40618), by the Swiss National Foundation (SNF Grant No. 20-118119), and by NSERC Canada (Grant SAPIN 341314-07).

References

1. D. Clowe, M. Bradač, A. H. Gonzalez, M. Markevitch, S. W. Randall, C. Jones and D. Zaritsky, *Astrophysical Journal* **648**, L109(September 2006).

2. D. N. Spergel *et al.*, *Astrophysical Journal Supplement Series* **170**, 377(June 2007).
3. G. Jungman *et al.*, *Physics Reports* **267**, p. 195(Mar 1996).
4. J. D. Lewin and P. F. Smith, *Astroparticle Physics* **6**, p. 87(Dec 1996).
5. D. N. Spergel, *Physical Review D* **37**, 1353(March 1988).
6. D. M. Mei and A. Hime, *Physical Review D* **73**, p. 53004(Mar 2006), (c) 2006: The American Physical Society.
7. R. J. Gaitskell, *Annual Review of Nuclear and Particle Science* **54**, 315(December 2004).
8. G. Chardin, *Dark matter direct detection*, Jan 2005), pp. 313–357.
9. A. Morales, *Nuclear Physics B Proceedings Supplements* **114**, p. 39(Jan 2003).
10. L. Mosca, *Nuclear Physics B Proceedings Supplements* **114**, p. 59(Jan 2003).
11. T. Saab, *Modern Physics Letters A* **23**, p. 457(Jan 2008), (c) 2008: World Scientific Publishing Company.
12. G. Bertone, D. Hooper and J. Silk, *Phys Rep* **405**, 279(Jan 2005).
13. S. Burgos *et al.*, *Astroparticle Physics* **28**, p. 409(Dec 2007).
14. W. J. Bolte *et al.*, *Nuclear Instruments and Methods in Physics Research Section A* **577**, p. 569(Jul 2007).
15. P. N. Luke, *Journal of Applied Physics* **64**, 6858(December 1988).
16. B. Neganov and V. Trofimov, *JTEP Letters* **28**, p. 328 (1978).
17. D. S. Akerib *et al.*, *Physical Review D* **72**, 052009(September 2005).
18. K. D. Irwin *et al.*, *Review of Scientific Instruments* **66**, p. 5322(Nov 1995).
19. K. D. Irwin, *Applied Physics Letters* **66**, p. 1998(Apr 1995).
20. T. Shutt *et al.*, *Physical Review Letters* **69**, p. 3425(Dec 1992).
21. J. Angle *et al.*, *Physical Review Letters* **100**, p. 21303(Jan 2008), (c) 2008: The American Physical Society.
22. R. Gaitskell, V. Mandic and J. Filippini, (2008) `http://dmtools.brown.edu`.
23. Z. Ahmed *et al.*, *Physical Review Letters* **102**, p. 11301(Jan 2009), (c) 2009: The American Physical Society.
24. T. Saab, *SUPERSYMMETRY AND THE UNIFICATION OF FUNDAMENTAL INTERACTIONS. AIP Conference Proceedings* **1078**, p. 545(Nov 2008), (c) 2008: American Institute of Physics.
25. S. Yellin, *Physical Review D* **66**, p. 32005(Aug 2002), (c) 2002: The American Physical Society.
26. W. C. Dash, *Journal of Applied Physics* **30**, 459(April 1959).
27. E. E. Haller *et al.*, *Advances in Physics* **30**, 93(October 1981).
28. S. T. E., *A Dark Matter Detector Based in Simultaneous Measurement of Phonons and Ionization at 20 mK* (Ph.D. Thesis, UC Berkeley, 1993).

THE ZEPLIN-III VETO DETECTOR

P. R. SCOVELL*
on behalf of the ZEPLIN-III Collaboration
School of Physics and Astronomy, University of Edinburgh, Edinburgh, EH9 3JZ, United Kingdom
**E-mail: p.scovell@ed.ac.uk*

An active veto detector to complement the ZEPLIN-III two phase Xenon, direct dark matter device is described. The design consists of 52 plastic scintillator segments, individually read out by high efficiency photomultipliers, coupled to a Gd loaded passive polypropylene shield. Experimental work was performed to determine the plastic scintillator characteristics which were used to inform a complete end-to-end Monte Carlo simulation of the expected performance of the new instrument, both operating alone and as an active veto detector for ZEPLIN-III. The veto device will be capable of tagging over 65% of expected coincident nuclear recoil events in the energy range of interest in ZEPLIN-III, and over 14% for gamma ray rejection (gamma and neutron rate is predicted by simulation), while contributing no significant additional background. In addition it will also provide valuable diagnostic capabilities. The inclusion of the veto to ZEPLIN-III will aid to significantly improve the sensitivity to spin independent WIMP-nucleon cross sections $\sim 10^{-9}$ pb.

Keywords: Dark matter; ZEPLIN-III; Liquid Xenon; Radiation detectors; WIMPs.

1. Introduction

The ZEPLIN-III detector[1] is a low background, high sensitivity 12 kg, two phase xenon time projection chamber developed to observe low energy nuclear elastic recoils resulting from the scattering of weakly interacting massive particles (WIMPs) from Xenon nuclei. The ZEPLIN-III detector aspires to achieve a WIMP-neutron capture cross section $\sim 10^{-9}$ pb. To reach the sensitivities required a number of extreme measures must be implemented to remove background sources of radiation. Such measures include working 1100m underground at Boulby mine to reduce the cosmic muon flux, using high purity components of low radionuclide content, and surrounding the target by a high efficiency veto to identify neutron events. ZEPLIN-III will be retrofitted with a veto system in 2009 which will serve both as a diagnostic device and will increase background rejection capability. The veto will be capable of identifying background events that cause energy depositions within the ZEPLIN-III active volume, in coincidence with the veto detector. Events that occur in coincidence will be indicative of non WIMP events as WIMPs are highly unlikely to interact within a detector system twice; any event that does so may be discarded. The active veto is part of a larger upgrade including the addition of new custom built ultra-low background photomultiplier tubes (PMTs). ZEPLIN-III has recently completed its first phase of the project achieving a WIMP-neutron capture cross section of 7.7×10^{-8} pb[2].

2. Veto Design

The veto was developed using a combination of Monte Carlo simulations and laboratory testing at ITEP (Moscow), the University of Edinburgh and Boulby mine. Materials for the veto were selected by testing each component's radiological impact using a dedicated low background hyperpure germanium (HPGe) detector located at Boulby mine[3]. The veto uses 1060 kg of plastic scintillator cut into 52 blocks, arranged in a barrel formation around the ZEPLIN-III detector to provide $> 3\pi$ sr coverage (figure 1). The scintillator material is polystyrene based UPS-923A (p-terphenyl 2%; POPOP 0.02%) produced by Amcrys-H, Kharkov, Ukraine. The walls of the veto are comprised of 32 standing parallelepiped segments (length 985mm, parallel sides of 128 mm and 156 mm and a thickness of 150 mm) in a barrel like formation with inner and outer diameters of 130 mm and 160 mm respectively. The remaining 20 sections extend over the full diameter of the barrel sections, above ZEPLIN-III. These sections include 4 different shapes all 150 mm thick ranging in length from 515 mm to 800 mm in length. Optically coupled to each plastic segment are 3″ ETL 9302KB PMTS for readout. The polystyrene material emits scintillation light at a maximum emission wavelength of 420 nm. The density of UPS-923A is $1.06 g/cm^3$ and has a refractive index of 1.52. The scintillator material operates as the active component of the veto system which works conjointly with a passive component of 15cm thick gadolinium loaded polypropylene shielding. The active plastic is positioned outside the inner Gd-loaded polypropylene, which is fashioned in a cylindrical geometry to maximise the efficiency.

Fig. 1. Veto design. The veto consists of 32 trapezoid scintillator blocks for the base section with the roof section comprising of 20 cuboid scintillator blocks. All PMTs are attached to the far end of each block to minimise radiological impact on the ZEPLIN-III data.

3. Experimental and Simulation Work for the Veto

3.1. *Experimental Work Characterising Scintillator Material*

Given below are details of experimental work performed in ITEP and the University of Edinburgh to characterise the attenuation length and light yield of the scintillator material.

The performance of each of the 52 pieces has been individually assessed as follows. For each fully wrapped plastic scintillator section, spectra of the PMT response to a ^{22}Na gamma ray calibration source were recorded for each of six positions along the length. For these measurements, to ensure consistency, the same PMT, voltage divider network, preamplifier and operating bias was used. Furthermore, taking advantage of the characteristic back-to-back 511 keV gamma rays produced in the annihilation of positrons from the ^{22}Na beta decay, effective collimation was achieved using a coincidence requirement with a 3" diameter 3" long NaI scintillation detector, in the geometry show in (figure 2). A typical spectrum obtained is shown in

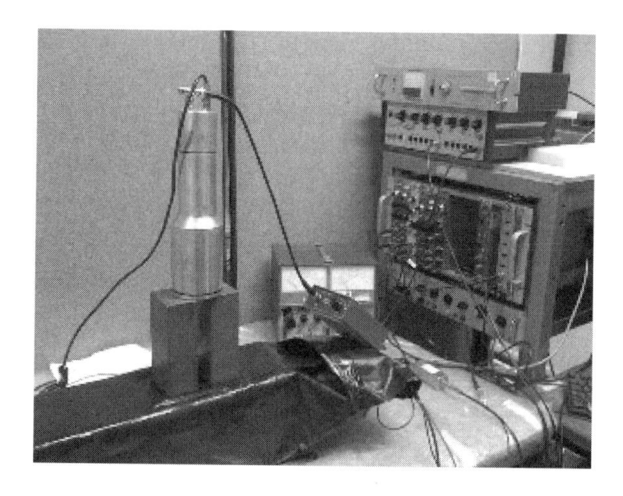

Fig. 2. The NaI detector set up in coincidence with the slab measurement.

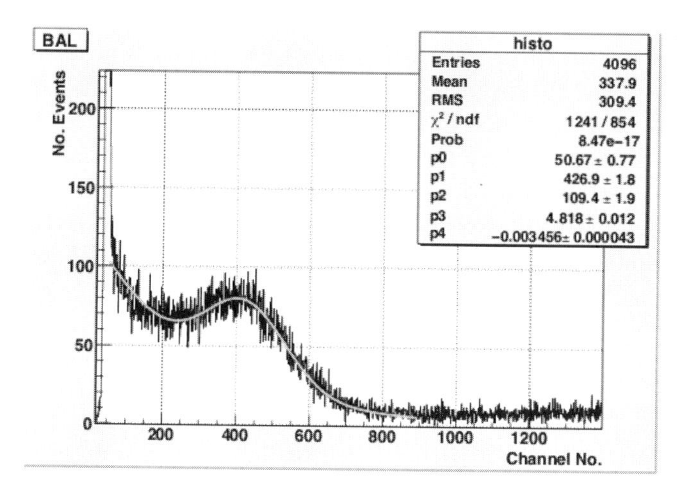

Fig. 3. A graph of the ^{22}Na peak for a fixed distance along a single slab, used to calculate its BAL.

(figure 3), revealing the characteristic Compton edge feature as expected. A Gaussian fitted to the Compton edge together with an exponential to describe the background, was used to characterise the 511 keV peak. The movement of the centroid of the peak as a function of the position of the source along the length was used as the indicator of the technical attenuation length (TAL).

Whereas the bulk attenuation length (BAL) describes the attenuation of photons due to the scintillator only, the TAL includes the losses due to imperfect reflections on surfaces and the effects of geometery. In the case of the units measured here, a highly specularly reflective mirror surface has also been placed at the end away from the PMT. Consequently, to first order, one expects a the response as a function of distance from the PMT to vary as the sum of two exponentials, one describing the reduction due to the separation of PMT and source position, and which describes the reduction due to the distance between the PMT and the image of the source in the mirror. The exact function used was

$$S(x) = Ae^{-x/T} + Ae^{-(l-x)/T}$$

where $S(x)$ is the centroid of the measured Compton edge, x is the distance from the PMT, l is the length of the scintillator section (the distance between the PMT and the mirror), T is the TAL and A is an arbitrary scaling constant. In principle, one could use different scaling for the term describing the response due to reflections from the mirror, effectively accounting for imperfect reflections, but the number of data points available did not justify an additional free parameter. The measurements of TAL have been complemeted with cosmic ray muon measurements in which the NaI coincidence detector was replaced by two small plastic scintillator detectors placed above and below the test piece. In this case the spectra obtained are well described and fitted by a Landau distribution, and again the response as a function of measurement position determined. Consistent results were found for the TAL measured either with the ^{22}Na source or with cosmic rays.

3.2. *Calculating Component Radio-Contamination*

The cosmic ray muon flux is reduced by a factor of $\sim 10^6$ to a level of $(4.09 \pm 0.15) \times 10^{-8}$ muons cm$^{-2}s^{-1}$ at a vertical depth of 1070m (2805m water-equivalent shielding)[4]. The resultant neutrons from cosmic ray muon spallation and secondary cascades lead to a single scattering neutron event rate in ZEPLIN-III of <1 event/year for a nuclear recoil energy above 10 keV within the central 8 kg of xenon[5]. Thus, the most important radiological background that must be considered comes from the local environment of the laboratory and the instruments themeselves. Consequently, all components proposed for use in the veto have been assessed for their radiological content. The main contributions come from U and Th, which produce background neutrons via (α,n) reactions on other materials and a small contribution from fission, and from K, which via beta-decay of ^{40}K populating the 1.461 keV excited state in ^{40}Ca leads to a gamma-ray background. To assess the overall neutron and gamma-ray environment that the instruments will be exposed to, additional measurements of other components have also been considered, for example the ZEPLIN-III instrument PMTs and the cavern rock. Two principal methods have been employed to determine the radioactivity content of components.

3.2.1. *Direct measurements of radioactivity of components*

Direct measurements of the gamma-ray emission from candidate components has been made using a HPGe detector located in a dedicated low background counting facility at the Boulby mine. Details of this work have been presented elsewhere [3] and only a summary is provided here. The HPGe detector head was encased in an inner copper and outer lead castle leaving a 30 cm^{-3} test cell. The ambient gamma-ray flux experienced by the HPGe detector in the abscence of a test sample was then assessed through an approximately week-long background run. A sample of material was placed close to the head and further data accrued over a period of several days. The background subtracted spectra were then analysed, typically showing a net excess in yield at energies characteristic of decays corresponding the the presence of certain isotopes. To assess the quantity of that isotope that must have been present in order to generate the measured excess in the background-subtracted spectrum, a GEANT4 Monte Carlo of the low background counting set up was performed. Inputs to the Monte Carlo included accurate geometry of the detector, the test chamber and the surrounding laboratory, and the efficiency of the HPGe detector. Comparison of Monte Carlo to data included experimental factors such as dead time and run duration.

Typically, prominent lines in the measured spectra were seen corresponding to the U, Th and K decays of interest. Sensitivities at the parts per billion level for U and Th, and parts per million level for K were achieved using this technique. Components tested included the scintillator plastic, the PMTs, the PMT bases, voltage divider networks and preamplifier electronics, cabling, connectors, black outer wrapping, PTFE sheet inner wrapping, adhesive PTFE tape, and the steel supports for the Pb shielding. The conversion between gamma-rays emitted from an isotope and the source contamination level is given using, for 1ppb U, 1ppb Th, and 1ppm K, 2310, 958, and 278 gamma-rays per kg per day, respectively.

3.2.2. *Mass spectrometry measurements of the radioactivity of components*

To supplement the gamma-ray measurements, and to provide a cross check of their accuracy, a subset of components have had their U, Th and K content measured by a commerial company employing mass spectroscopy technique. These results are in good agreement with results obtained using the HPGe method.

4. Simulating the Full Veto Array

GEANT4 is utilised to model the full veto array of 52 segments around the ZEPLIN-III detector in a complete end-to-end program culminating in simulated data sets from both detectors simultaneously within a given run. The ZEPLIN-III simulation has been fully described in [6]. The additions to the model include the individual veto crystals with complete wrapping, veto PMTs, PMT bases and cups, and the passive Gd loaded hydrocarbon shielding. The entire assembly is surrounded by Pb shielding and this castle is placed within a cavern cut from rock salt and with the dimensions of the Boulby underground laboratory. Three metres of rock surrounding the cavern is retained since neutrons generated from deeper than 3m do not contribute to the flux through the rock face, whereas for gamma rays a 25cm depth is found to be sufficient. A screen-shot of some of the major components of the ZEPLIN-III simulation including the full veto array is shown in (figure 4).

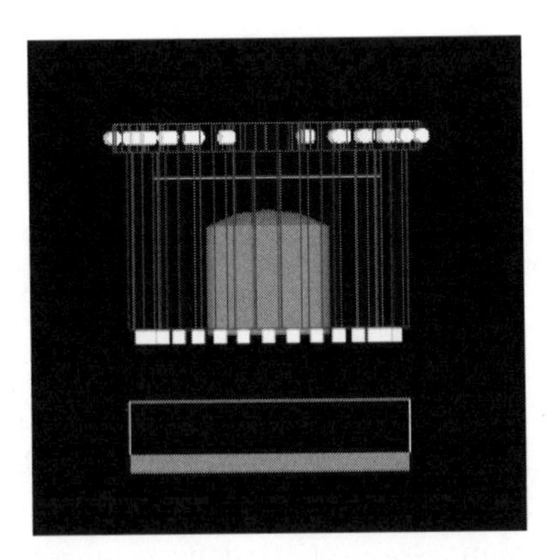

Fig. 4. A screen-shot of some of the major components of the ZEPLIN-III simulation including the full veto array (Pb shielding and lab details are not shown).

4.1. *Tagging Efficiency of the Veto*

The simulation shows that the mean veto tagging efficiency for gamma ray events in the ZEPLIN-III active volume from internal and external background sources is 12%. For neutron events, where events in the xenon may mimic a WIMP signal, the veto is capable of rejecting over 65% of background.

4.2. *Veto Contribution to ZEPLIN-III Background*

The simulation is also used to show the impact of the veto on event rates in ZEPLIN-III itself. Given an energy range of 2-20 keVee (single scatter candidate WIMP event) we expect to see an additional 0.3 gamma-ray events/kg/day/keVee (total) and also an additional 0.1 neutron events/yr (total).

5. Current Status of the ZEPLIN-III Veto

The veto detector has been delivered to the underground facility and is currently being calibrated using multiple energy gamma sources. The data is now being acquired by the Veto DAq system and the whole system is on course to be fully integrated when the upgrade cycle for ZEPLIN-III is completed.

Acknowledgments

The author wishes to acknowledge the support of the Science & Technology Research Council (STFC) for the ZEPLIN-III project and funding provided for the maintenance and operation of the underground Palmer laboratory, hosted by the Boulby Mine. The Boulby Mine is owned and operated by Cleveland Potash Ltd. (CPL), which is a business unit of ICL Fertilizers. ITEP wish to acknowledge joint support of the Russian Federation for the Basic Research Foundation and the Royal Society (grant # 08-02-91851 KO_a).

References

1. D. Yu. Akimov *et al.*, Astroparticle Phys. 27 (2007) 46-60.
2. V. N. Lebendenko *et al.*, arXiv:0812.1150v1 (2008).
3. P. Scovell *et al.*, (in preparation).
4. M. Robinson *et al.*, Nucl. Instrum. Meth. A 511 (2003) 347-353.
5. H. Araujo *et al.*, Nucl. Instrum. Meth. A 545 (2005) 398-411.
6. H. M. Araujo *et al.*, Astroparticle Phys. 26 (2006) 140-153.

PART XIV

Exotics

HYPOTHETICAL DARK MATTER/AXION ROCKETS: DARK MATTER IN TERMS OF SPACE PHYSICS PROPULSION

A. BECKWITH

American Institute of Beam Energy Propulsion
P.O. Box 1907
Madison, Alabama, 35758, USA
650-322-6768
E-mail: abeckwith@uh.edu

Current proposed photon rocket designs include the Nuclear Photonic Rocket and the Antimatter Photonic Rocket (proposed by Eugen Sanger in the 1950s, as reported by Ref. 1). This paper examines the feasibility of improving the thrust of photon-driven ramjet propulsion by using DM rocket propulsion. The open question is: would a heavy WIMP, if converted to photons, upgrade the power (thrust) of a photon rocket drive, to make interstellar travel a feasible proposition?

Keywords: Dark Matter; Photon Rocket; Axions.

PACS: 07.87.+v, 95.85.Ry, 95.35.+d

1. Which DM Candidates are Viable?

Reference 2 provides a ten-point test a new particle has to pass to be considered a viable DM candidate: "(I) Does it match the appropriate relic density? (II) Is it cold? (III) Is it neutral? (IV) Is it consistent with BBN? (V) Does it leave stellar evolution unchanged? (VI) Is it compatible with constraints on self-interactions? (VII) Is it consistent with direct DM searches? (VIII) Is it compatible with gamma-ray constraints? (IX) Is it compatible with other astrophysical bounds? (X) Can it be probed experimentally?" It so happens that WIMPs meet all the above tests. Reference 3 gave a mass value of between 100–300 GeV for WIMPs. A second theory involving undetected particles is that some dark matter is made of hypothetical subatomic particles called "axions." Thus, the difference between the two theories (WIMPs or axions) is that dark matter is either made of a large number of light particles (axions) or a smaller number of heavier particles.

A way to link the two states of DM is to note, as Ref. 4 notes, axions were the whole of DM, that there would be $m_{\text{axion}} \approx 10^{-5} eV$ for a total DM mass $M \approx 10^{12}$ GeV. But if there were WIMPs and axions together, $m_{\text{axion}} \approx 10^{-5}\text{eV}\,|_{\text{no-WIMPs}} \gg m_a|_{\text{WIMPs-exist}}$. For most theoreticians, neutralinos are the preferred SUSY particle of choice for DM. However, in what has startling implications, Refs. 5 and 6 using Mesissner and Nicholai's parameter space arguments obtained WIMP DM masses of between 300 GeV to 400 GeV as an upper range to WIMP masses. What needs to be obtained, is analytical work to fathom an interrelationship between axions and WIMPs as far as DM. Possibly by Bayesian statistical methods for comparing the relative fit between WIMP and axion models, *i.e.*, effective Lagrangian methods as explored by Ref. 7. Axions have been considered as a power source to be scooped up in space because of their estimated thick density in space, and DM candidates for masses considerably above the axion values have been brought up as a way to increase thrust / power for more efficient propulsion. Reference 8 writes that a typical Daedalus's star ship designed for six light years of travel requires 1.7 million metric tons of fuel, which is unrealistic, and Frisbee states that the photon rocket has a travel time of 42 million years to accelerate to one tenth the speed of light.

2. First Principles of an Axion/DM Ramjet

According to Ref. 9 in discussions of the applications of their description of the CAST experiment, axions can be changed by the Primakoff effect into photons, which could theoretically be used as a source of thrust.[9,10] The power density available from axions depends on their mass, the density of axions in space, and the velocity of the vehicle. At 10^{-5} eV, with a velocity of .001c, assuming 200,000 per cc axions in intergalactic space (an axion has a mass 1/400,000,000th of the mass of an electron, so there should be 200,000 per cc in intergalactic space, according to Ref. 11).

$$\text{Power} = 3\text{watts/cm}^2 \times [v/c]/(1 - [v/c]^2). \tag{1}$$

The author used a velocity of .001c, assuming a ramjet is used for inserting axions into a chamber from outer space. At .999c, the power is nearly 1500 watts/cm^2, which appears to be respectable. These calculations assume a density of half a trillion axions per cubic centimeter the vicinity of Earth, more per cc near the galactic center, and only is also assumed that an axion traveling at .1c, hitting a 1 cm-squared region of space, undergoes an 10^{-5} electron volt value to 10^{-3} eV value for the axion mass. This axion mass would then be directly converted into energy, and that there are roughly up to half a trillion axions per cubic centimeter. So how can one obtain a power value of 1500 or so watts per square centimeter at nearly .1c? What is being looked for is how to have a far greater energy power equivalent to 1500 Watts per square centimeter for far slower travel. The reason being that the greater the power output at lower speed of light values, the faster a space craft would be able to accelerate to that speed. Our Dark Matter candidates (WIMPs), instead of being 10^{-3} eV are, instead, 100 to 400 GeV. So a more efficient way is required to reach a power ratio of 1500 or so watts per square centimeter for a rocket. The problem of momentum kick is as follows. As implied by Ref. 12, "Every axion produces a momentum kick of

$$\Delta p = mc \times \gamma \cdot (1 - \beta) \tag{2}$$

where m is the axion rest mass." Special relativity suggests that a low-power-output space craft would take quite a long time to accelerate to almost the speed of light. A viable DM rocket would allow a rocket to have far more power, permitting a more rapid acceleration to at least ten percent of the speed of light in a reasonable time period.

3. And Why This is All Important? Facing Some Serious Problems in Contemporary Cosmology

At the EXTREMA of any function: $dx/dt = 0$. But for Big-Bang, $dS/dt = \infty$ at $S = 0$. And then why would one care about entropy in the first place. It so happens as reported by Ref. 6 as adapted from Refs. 13 and 14. The fact is that the non-SUSY Lagrangian offered by Ref. 5 may not only give the correct mass value for a useful interstellar propulsion candidate, but also tie in with entropy production formalism which avoids $dS/dt = \infty$ at $S = 0$ seen in present cosmology models. It is then appropriate to consider practical applications of Reference 14's quantum 'infinite' statistics. Reference 14 outlines how to get $S \approx N$. Begin with a partition function

$$Z_N \sim \left(\frac{1}{N!}\right) \cdot \left(\frac{V}{\lambda^3}\right)^N. \tag{3}$$

This, according to Ng, leads to an entropy of the limiting value of

$$S \approx N \cdot (\log[V/N\lambda^3] + 5/2). \tag{4}$$

But $V \approx R_H^3 \approx \lambda^3$, so unless N in Eq. (4) above is about 1, S (entropy) would be < 0, which is a contradiction. Now this is where Ng introduces removing the $N!$ term in Eq. (3) above, *i.e.*, inside the Log expression we remove the expression of N in Eq. (4) above. This is a way to obtain what Ng refers to as Quantum Boltzmann statistics, so then we obtain for sufficiently large N, where N is a numerica DM density referred to as $< n >$.

$$S \approx N. \tag{5}$$

4. Conclusion

One can state that near-light speeds, the available axion power would be about 3 watts/cm^2 times $\beta \times \gamma^2$, where $\beta = (v/c)$ is the velocity relative to light, and $\gamma^2 = 1/[1 - \beta^2]$ is the square of the relativistic mass-increase factor. At a velocity of 99.9% c, the available power from axions would be 1500 watts/cm^2, enough power for a modest energy-efficient space drive (the faster it goes, the more such power becomes available). In principle, a photon rocket may be improved upon, using DM/axion destruction via intense E & B fields. In IDM 2008, a mass range for DM up to about 400 GeV, per particle was predicted, adding credibility to a counting algorithm based on Eq. (5).[6] *I.e.*, entropy would then be added due to DM particles which would be produced within the CMBR region of space. *I.e.*, the wave length of DM would be well within the region of space before 380 thousand years after the Big Bang. What is needed, to make a linkage between axions and WIMP DM more understood would be application of Bayesian statistics, as Ref. 7 wrote about. An experimental program of DM applications to space flight may enable investigations into this issue, allow for understanding the genesis of entropy in pre CMBR space, and lead to engineering studies in order to make interstellar travel a possibility.

References

1. A. V. Gulevich, E. A. Ivanov, O. F. Kukharchuk, V. Y. Poupko and A. V. Zrodnikov, *STAIF* **552**, 957 (2001).
2. M. Taoso, G. Bertone and A. Masiero, *Journal of Cosmology and Astrophysics* **JCAP03(2008)022**(March 2008).
3. H. Muramaya, Physics beyond the standard model and dark matter, in *Lectures at Les Houches Summer School, Session 86, Particle Physics and Cosmology: the Fabric of Spacetime*, 31 July-25 August 2006.
4. S. Weinberg, *Cosmology* (Oxford University Press, New York, 2008).
5. K. A. Meissner and H. Nicolai(September 2008), arXiv:0803.2814v3 [hep-th].
6. A. W. Beckwith(24-27 February 2009), arXiv:0810.1493 [physics.gen-ph].
7. F. Feroz, B. C. Allanach, M. Hobson, S. S. AbdusSalam, R. Trotta and A. M. Weber(July 2008), arXiv:0807.4512v1 [hep-ph].

8. R. H. Frisbee, Limits of intestellar flight technology, in *Frontiers Propulsion Science*, eds. M. Millis and E. Davis, Progress in Atronautics and Aeronautics, Vol. 227 (AIAA, 2009) pp. 31–126.
9. J. I. Collar, D. Miller, J. Rasmussen and J. Vieira (2006).
10. R. Bernabei, P. Belli, R. Cerulli, F. Montecchia, F. Nozzoli, A. Incicchitti, D. Prosperi, C. J. Dai, H. L. He, H. H. Kuang, J. M. Ma and S. Scopel, *Physics Letters B* **515**, 6(August 2001).
11. B. Lakic, Search for solar axions with the cast experiment, in *Proceedings of IDM 2008*, (Stockholm, Sweden, 2008).
12. P. Sikivie, *Physical Review Letters* **51**, 1415 (1983).
13. Y. J. Ng, Quantum foam and dark energy, in *International workshop on the Dark Side of the Universe*, 2008. http://ctp.bue.edu.eg/workshops/Talks/Monday/QuntumFoamAndDarkEnergy.pdf.
14. Y. J. Ng, *Entropy* **10**, 441 (2008).
15. A. W. Beckwith(January 2009), arXiv:0809.1454v3 [physics.gen-ph].
16. K. A. Meissner. personal communication, Bad Honnef meeting on "Quantum gravity, new directions and perspectives".
17. P. Giromini, F. Happacher, M. J. Kim, M. Kruse, K. Pitts, F. Ptohos and S. Torre(October 2008), arXiv:0810.5730v1 [hep-ph].

LIST OF PARTICIPANTS, DARK2009

Ahluwalia Dharam Vir
Department of Physics and Astronomy
Rutherford Building
Univ. of Canterbury, Private Bag 4800
Christchurch 8020, NEW ZEALAND
Tel: +64 3 364 2563 (Internal 6563)
Email: dharamvir.ahluwalia@canterbury.ac.nz
http://www2.phys.canterbury.ac.nz/editorial/

Aoki Mayumi
Particle Theory and Cosmology Group
Department of Physics, Tohoku
University, 6-3, aza-Aoba, Aramaki
Aoba-ku, Sendai 980-8578
JAPAN
Tel: +81-22-795-7741
Email: mayumi@tuhep.phys.tohoku.ac.jp

Balázs Csaba
School of Physics, Monash Univ.
Melbourne Victoria 3800
AUSTRALIA
Tel: + 61 3 9902 0328
Fax: + 61 3 9905 3637
Email: Csaba.Balazs@sci.monas.edu.au
http://physics.monash.edu.au/staff/balazs.html
Skype: Csaba_Balazs

Bell Nicole
School of Physics, Univ. of Melbourne
Victoria 3010, AUSTRALIA
Tel: + 61 3 8344 3112
Fax: + 61 3 9347 4783
Email: n.bell@unimelb.edu.au
http://www.ph.unimelb.edu.au/ nfb

Belli Pierluigi
INFN - Sezione Roma "Tor Vergata"
Via della Ricerca Scientifica, 1
I-00133 Roma, ITALY
Tel: +39-0672594543
Fax: +39-0672594542
Email: pierluigi.belli@roma2.infn.it
http://people.roma2.infn.it/dama/
http://people.roma2.infn.it/belli/

Beckwith Andrew
American Institute of Beam Energy Propulsion
P.O. Box 1907
Madison, Alabama, 35758, USA
Tel: 650-322-6768
Email: abeckwith@uh.edu

Barberio Elisabetta
School of Physics (David Caro Building)
The University of Melbourne
VIC 3010, AUSTRALIA
Tel: +61 3 8344 7072
Fax: +61 3 9347 4783
Email: barberio@unimelb.edu.au

Blake Chris
Swinburne University of Technology
PO Box 218, Hawthorn, Victoria 3122
AUSTRALIA
Tel: +61 3 9214 8624
Fax: +61 3 9214 8797
Email: cblake@astro.swin.edu.au

Buote David A.
Department of Physics and Astronomy
4129 Frederick Reines Hall
University of California, Irvine
Irvine, CA 92697-4575, USA
Tel: +1 949 824 6280
Fax: +1 949 824 2174
Email: buote@uci.edu

Buckley Matthew
California Institute of Technology
MC 452-48 1200, E. California Blvd., Pasadena
CA 91125, USA
Tel: 626-379-4337
Fax: 626-568-8473
Email: buckley@theory.caltech.edu
http://www.theory.caltech.edu/people/index.html

Cho Y.M.
Center for Theoretical Physics and School of Physics
College of Natural Sciences, Seoul National University
Seoul 151-742, KOREA and
School of Basic Sciences, Ulsan National Institute
of Science and Technology, Ulsan 689-805
KOREA
Email: ymcho@yongmin.snu.ac.kr

Danninger Matthias
Stockholm University, AlbaNova University Center
Dept. of Physics, 10691 Stockholm, SWEDEN
Email: danning@physto.se

Davis Tamara
School of Physics, University of Queensland
QLD 4072, AUSTRALIA
Tel: +61 7 3346 7961
Email: tamarad@physics.uq.edu.au
http://dark.dark-cosmology.dk/~tamarad/

Eckart Andreas
1. Physikalisches Institut
Zülpicher Str. 77, Köln, 50937
GERMANY
Tel: +49 221/470-3546
Fax: +49 221/470-5162
Email: eckart@ph1.uni-koeln.de

Famaey Benoit
Institut d'Astronomie et d'Astrophysique
Universite Libre de Bruxelles C.P. 226
Boulevard du Triomphe B-1050 Bruxelles, BELGIUM
Tel: +32 2 6502833
Fax: +32 2 6504226
Email: bfamaey@astro.ulb.ac.be
http://www.astro.ulb.ac.be/~bfamaey/

Gebauer Iris
Universität Karlsruhe (TH)
IEKP, Postfach 6980
D-76128, Karlsruhe, GERMANY
Tel: +49-(0)721-608-3454
Fax: +49-(0)721-608-7930
Email: gebauer@ekp.uni-karlsruhe.de

Gentile Gianfranco
Institut d'Astronomie et d'Astrophysique
Université Libre de Bruxelles - CP226
Boulevard du Triomphe
B-1050 Brussels, BELGIUM
Tel: +32-(0)2-6503570
Fax: +32-(0)2-6504226
Email: gianfranco.gentile@ugent.be

Gillard Adam B.
Physics Department, University of Canterbury
Private Bag 4800, Christchurch 8140
NEW ZEALAND
Tel: +64 3 364 2987 ext 4665
Tel: +64 021 103 1351

Fax: +64 3 364 2469
Email: abg22@student.canterbury.ac.nz
or adam.gillard@yahoo.com

Goldman Terry
T-2, MS-B283
Los Alamos National Laboratory, POB 1663
Los Alamos, NM 87545 USA
Tel: 1 505-667-3244
Fax: 1 505-667-1931
Email: tgoldman@lanl.gov

Grande Bardanca Javier
Institut für Theoretische Physik (ITP)
Philosophenweg 16, D-69120, Heidelberg, GERMANY
Tel: +49-6221-54 9446
Email: javiergrand@gmail.com or
j.grande@thphys.uni-heidelberg.de
On leave from:
Departament d'Estructura i Constituents de la Matéria (E.C.M.)
Facultat de Fisica, Universitat de Barcelona
(U.B.) Diagonal, 647, 08028-Barcelona, SPAIN
Tel: 0034 9340 39288
Email: jgrande@ecm.ub.es
Email: javiergrand@gmail.com

Jacholkowska Agnieszka
LPNHE - Laboratoire de Physique Nuclaire et de Hautes Energies
IN2P3 - CNRS - Universits Paris VI et Paris VII
4 place Jussieu
Tour 43 - Rez de chausse
75252 Paris Cedex 05, FRANCE
Email: Agnieszka.Jacholkowska@cern.ch
Tel: +33 (0)1 44 27 41 20
Fax: +33 (0)1 44 27 46 38

Jeltema Tesla
UCO/Lick Observatories
1156 High St. Santa Cruz
CA 95060, USA
Email: tesla@ucolick.org

Howard Ecaterina Marion
6 Casuarina Close, Umina Beach
NSW 2257, AUSTRALIA
Email: khoward@ics.mq.edu.au or
katie@azimuthadv.com
Tel: +61 408 006 295
www.mq.edu.au
Skype ID: katie.howard2009

Kerr Roy Patrick
Department of Physics and Astronomy
Rutherford Building
Univ. of Canterbury, Private Bag 4800
Christchurch 8020, NEW ZEALAND
Email: roypkerr@gmail.com
Email: kerr@icranet.org

Klapdor-Kleingrothaus
Hans Volker
Stahlbergweg 12
74931 Lobbach, GERMANY
Tel: (49) 6226-41088
Email: prof.klapdor-kleingrothaus@hotmail.de
Home-page: http://www.klapdor-k.de
Skype: genius19412

Krivosheina Irina Vladimirovna
Radiophysical Research Institute (NIRFI)
ul. Bolshaja-Pecherskaja 25
603005, Nishnij-Novgorod
RUSSIA
Email: irinaKV57@mail.ru
Home-page: http://www.klapdor-k.de

Kumar Jason
Department of Physics, University of Hawaii
2505 Correa Rd., Honolulu
Hawaii 96822, USA
Email: kumar.jason@gmail.com

Lee Cheng-Yang
Department of Physics and Astronomy
Univ. of Canterbury, Private Bag 4800
Christchurch 8020, NEW ZEALAND
Tel: +64 3 364 2987-7620
Email: cyl45@student.canterbury.ac.nz

Leubner Manfred P.
Institute for Astro- & Particle Phys.
University of Innsbruck
Technikerstr. 25, A-6020
Innsbruck, AUSTRIA
Tel: +43-(0)512-507-6054
Fax: +43-(0)512-507-2923
Email: manfred.leubner@uibk.ac.at
http://homepage.uibk.ac.at/~c706102/

Mankoč Borštnik Norma Susana
Department of Physics, FMF
University of Ljubljana
Jadranska 19, 1000 Ljubljana
SLOVENIA
Email: norma.mankoc@fmf.uni-lj.si
http://www.fmf.uni-lj.si

Marfatia Danny
University of Kansas, Department of Physics
and Astronomy, 1082 Malott, 1251 Wescoe Hall Dr.
Lawrence, KS 66045, USA
Email: marfatia@ku.edu
Tel: 785 864 4591

Martin Benjamin M.S.
Mathematics & Statistics Department
University of Canterbury, Private Bag 4800
Christchurch, 8140, NEW ZEALAND
Tel: +64 3 364 2987 ext 7687
Fax: +64 3 364 2587
Email: B.Martin@math.canterbury.ac.nz
www.math.canterbury.ac.nz/~b.martin/

Motz Holger
Erlangen Centre for Astroparticle Physics
Erwin Rommel Strasse 1, 91058 Erlangen, GERMANY
Email: holger.motz@physik.uni-erlangen.de
Tel: +49 9131 8527076

Moultaka Gilbert
Laboratoire de Physique Théorique et Astroparticules
UMR5207–CNRS, Université Montpellier II
Place E. Bataillon, F–34095 Montpellier Cedex 5, FRANCE
Email: Gilbert.Moultaka@lpta.univ-montp2.fr
www.univ-montp2.fr

Moulin Emmanuel
CEA Saclay, DSM/IRFU/SPP
91191 Gif-sur-Yvette Cedex
FRANCE
Tel: +33 169082960
Fax: +33 169 086 428
Email: emmanuel.moulin@cea.fr

Nakayama Kazunori
Institute for Cosmic Ray Research
University of Tokyo, Kashiwa
Chiba 277-8582, JAPAN
Email: nakayama@icrr.u-tokyo.ac.jp
Tel: +81-4-7136-3166
Fax: +81-4-7136-3165
http://www.icrr.u-tokyo.ac.jp/th/th.html

Neupane Ishwaree
Dep. of Phys. & Astronomy, Univ. of Canterbury
Private Bag 4800, Christchurch 8041
NEW ZEALAND
Email: ishwaree.neupane@canterbury.ac.nz

Philpott Lydia
Theoretical Physics Blackett Laboratory
Imperial College London, London, SW7 2AZ
UNITED KINGDOM
Email: l.philpott06@imperial.ac.uk

Profumo Stefano
Santa Cruz Institute for Particle Physics
ISB-325, University of California, Santa Cruz
CA 95064, USA
Email: profumo@scipp.ucsc.edu
Tel: 1-831-459-3039
Fax: Fax: 1-831-459-5777
http://scipp.ucsc.edu/~profumo/

Puy Denis
University of Montpellier II, Astrophysics group
GRAAL/CNRS UMR 5024, CC72
Montpellier 34090, FRANCE
Tel: +33 46714 3901 or
Cellular: +33 64567 4358
Fax: +33 46714 4535
Email: Denis.Puy@graal.univ-montp2.fr

Rapetti David
Kavli Insitute for Particle Astrophysics and Cosmolgy
and SLAC National Accelerator Laboratory
2575 Sand Hill Road, Menlo Park, CA 94025, USA
or Stanford University Physics Department
452 Lomita Mall Stanford, CA 94305-4085, USA
Email: drapetti@slac.stanford.edu
Tel: 650-736-7884
Fax: 650-725-4096
http://www.stanford.edu/~drapetti/fgas module/

Rizzo Alfio
Vrije Universiteit Brussel, Faculteit Wetenschappen
Dienst ELEM, Pleinlaan, 2
B-1050 Brussels, BELGIUM
Tel: + 32(0)26293220
Fax: + 32(0)26293816
Email: alfio.rizzo@vub.ac.be
http://w3.iihe.ac.be/~rizzo

Roszkowski Leszek
Astro-Particle Theory and Cosmology Group Dept.
of Physics and Astronomy University of Sheffield
Sheffield S3 7RH, ENGLAND
Tel: +44(0)114-222-35-80 or
Tel: +44(0)7799-08-79-38
Fax: +44(0)114-222-35-55
Email: l.roszkowski@shef.ac.uk
http://particle-theory.group.shef.ac.uk/

Saab Tarek
Department of Physics
University of Florida
Gainesville, FL 32611, USA
Tel: 352 328 9568
Fax: 352 328 3591
Email: tsaab@phys.ufl.edu
http://www.phys.ufl.edu/~tsaab/
http://cdms.berkeley.edu/cdms_collab.html

Santoso Yudi
IPPP, Ogden Centre for Fundamental Physics
Department of Physics, University of Durham
Science Laboratories, South Rd
Durham DH1 3LE, UNITED KINGDOM
Email: yudi.santoso@durham.ac.uk
Tel: +44-(0)191-334-3770
Fax: +44-(0)191-334-3658

Schritt Dimitri
Department of Physics & Astronomy
Rutherford Building
Univ. of Canterbury, Private Bag 4800
Christchurch 8020, NEW ZEALAND
Tel: +64 3 364 2563
Email: dsc35@student.canterbury.ac.nz

Scott Patrick
Oskar Klein Centre for Cosmoparticle Physics (OKC) &
Department of Physics, Stockholm University

AlbaNova University Centre
SE-106 91, Stockholm, SWEDEN
Tel: +46 8 5537 8734
Fax: +46 8 5537 8601
Email: pat@physto.se
http://www.fysik.su.se/~pat

Scovell Paul
School of Physics, The University of Edinburgh
Mayfield Road, Edinburgh, EH9 3JZ
UNITED KINGDOM
Tel: +44 (0) 131 650 5284
Fax: +44 (0) 131 650 7003
Email: p.scovell@ed.ac.uk
www.ph.ed.ac.uk/nuclear/darkmatter/

Seo Seon-Hee
Stockholm University, AlbaNova University Center
Dept. of Physics, 10691 Stockholm, SWEDEN
Tel: 46-8-553-78674
Fax: 46-8-553-8601
Email: seo@physto.se
http://www.icecube.wisc.edu/

Shan Chung-Lin
BK21 Frontier Physics Research Division
School of Physics and Astronomy
Seoul National University
San 56-1, Shillim-dong, Gwanak-gu
Seoul 151-747, REPUBLIC OF KOREA
Tel: +82-2-876-2801(O)
Fax: 82-2-875-4719
Email: cshan@hep1.snu.ac.kr
http://dmrc.snu.ac.kr/~cshan/

Smale Peter
Department of Physics and Astronomy
University of Canterbury, Private Bag 4800
Christchurch 8140, NEW ZEALAND
Tel: 03 364 2987 (Ext. 7260)
Email: psm22@student.canterbury.ac.nz

Sobouti Yousef
Institute for Advanced Studies in
Basic Sciences (IASBS), P.O. Box 45195-1159
Zanjan 45195, IRAN
Tel: (+98) 241 415 2259
Fax: (+98) 241 421 4949
Email: sobouti@iasbs.ac.ir

Stoehr Felix S.
ESO, Karl-Schwarzschild-Strasse 2
D-85748 Garching bei Mnchen, GERMANY
Email: fstoehr@eso.org
Tel: +49 89 3200 6283
Fax: +49 89 3200 6703
www.stecf.org/ fstoehr/felix.asc

Tytgat Michel H.G.
Service de physique thorique (PHYSTH) CP225
Universit Libre de Bruxelles Bld du Triomphe
1050 Brussels, BELGIUM
Tel: + 32(0)2 650 5570
Fax: +32(0)2 650 5951
Email: mtytgat@ulb.ac.be

Visser Matt
School of Mathematics, Statistics
and Operations Research Victoria University
of Wellington,PO Box 600, Wellington 6140
NEW ZEALAND
Email: matt.visser@msor.vuw.ac.nz
Tel: +64-4-463-5115
Fax: +64-4-463-5045
http://msor.victoria.ac.nz/

Wiltshire David
Dep. of Phys. & Astron., Rutherford Building
Univ. of Canterbury, Private Bag 4800
Christchurch 8140, NEW ZEALAND
Tel: +64 3 364 2563
Email: david.wiltshire@canterbury.ac.nz
http://www2.phys.canterbury.ac.nz/~dlw24/

AUTHORS INDEX

Ackerman, L., 277
Allen, S. W., 426
Alsing, P. M., 180
Aoki, M., 170

Balázs, C., 87
Barberio, E., 39
Beckwith, A., 557
Belli, P., 3
Bernabei, R., 3
Bolmont, J., 413
Brighenti, F., 264
Bruneton, J.-P., 335
Buckley, M. R., 277
Buote, D. A., 264

Cappella, F., 3
Carroll, S. M., 277
Carter, D., 87
Cattoën, C., 287
Cerulli, R., 3
Cho, Y. M., 97

d'Angelo, A., 3
Dai, C. J., 3
Drees, M., 521

Ebeling, H., 426
Eckart, A., 303
Edsjö, J., 320

Fabian, A. C., 426
Fairbairn, M., 320
Famaey, B., 335

García-Marín, M., 303
Gebauer, I., 218
Gebhardt, K., 264
Gentile, G., 440
Gillard, A. B., 451
Goldman, T., 180
Grande, J., 380

He, H. L., 3
Howard, E. M., 328
Humphrey, P. J., 264

Incicchitti, A., 3
Ishiwata, K., 209

Jacholkowska, A., 413
Jeltema, T. E., 255

König, S., 303
Kamionkowski, M., 277
Kanemura, S., 170
Klapdor-Kleingrothaus, H. V., 137
Krivosheina, I. V., 137
Kuang, H. H., 3
Kumar, J., 18
Kunneriath, D., 303

Leubner, M. P., 194

Ma, X. H., 3
Mankoč Borštnik, N. S., 365
Mantz, A., 426
Martin, B. M. S., 451
Mathews, W. G., 264

Matsumoto, S., 209
Mckellar, B. H. J., 180
Montecchia, F., 3
Moroi, T., 209
Morris, R. G., 426
Motz, H., 504
Moulin, E., 459
Moultaka, G., 66
Mužić, K., 303

Nakayama, K., 233
Neupane, I. P., 116
Nozzoli, F., 3

Pelinson, A., 380
Profumo, S., 243, 255
Prosperi, D., 3
Puy, D., 350

Rapetti, D., 426
Rizzo, A., 494
Roszkowski, L., 51

Saab, T., 535
Santoso, Y., 77
Schmidt, R., 426
Scott, P., 320
Scovell, P. R., 546
Seo, S. H., 482
Seto, O., 170
Shan, C.-L., 521
Sheng, X. D., 3
Sobouti, Y., 356
Solà, J., 380
Stephenson, Jr., G. J., 180
Stoehr, F. S., 344
Straubmeier, C., 303

Tytgat, M. H. G., 28

Visser, M., 287
Vivier, M., 471

Wiltshire, D. L., 397
Witzel, G., 303

Ye, Z. P., 3

Zamaninasab, M., 303